A new view of BRITAIN

John Chaffey

A MEMBER OF THE HODDER HEADLINE GROUP

To
Ruth, David and Helen

The research for this book has involved travelling throughout Britain for many months, from the silent pools of the Flow country in Caithness to the granite cliffs of West Penwith in Cornwall, from the summits of the Cairngorms to the lowest of the Somerset Levels. I would like to thank all of those many people who have given me the benefit of their knowledge and expertise, and who have given up their valuable time to talk to me. Although it would be impossible to mention everyone by name, my special thanks go to Vivien Crump, Sarah Bailey, Joanne Osborn and Julia Morris at Hodders for all their help and advice, to Lindsey Horton for showing me so many of the housing problems of the inner city in Leeds, to Sue Warn for her many constructive suggestions in the early planning of the book, and for providing the material she made so readily available to me, to Mike Marshall-Hollingsworth of the NFU for a very valuable discussion on the current state of the farming industry, to Clyde Medlicott, Estates Manager in the Rhondda Valleys for the many hours he has spent talking to me about industrial regeneration, and to John Breeds, Warden of the Braunton Burrows Nature Reserve, for giving up so much time to show me his work on the Reserve. Finally, I would like to thank my wife, Ruth, for her unfailing support and encouragement throughout the writing of the book.

British Library Cataloguing in Publication Data

Chaffey, John
   New View of Britain
   I. Title
   914.1

ISBN 0 340 55532 7

First published 1994
Impression number   10 9 8 7 6 5 4 3 2 1
Year   1998 1997 1996 1995 1994

Copyright © 1994

All rights reserved. No part of this publication may be reproduced or transmitted in any form or by any means, electronic or mechanical, including photocopy, recording, or any information storage and retrieval system, without permission in writing from the publisher or under licence from the Copyright Licensing Agency Limited. Further details of such licences (for reprographic reproduction) may be obtained from the Copyright Licensing Agency Limited, of 90 Tottenham Court Road, London W1P 9HE.

Typeset by Serif Tree, Kidlington, Oxfordshire.
Printed in Great Britain for Hodder & Stoughton Educational, a division of Hodder Headline Plc, Mill Road, Dunton Green, Sevenoaks, Kent TN13 2YA by Thomson Litho Limited, Scotland.

ACKNOWLEDGEMENTS

The author and publishers would like to thank the following for permission to reproduce copyright materials in this book. Every effort has been made to trace and acknowledge all copyright holders but if any have been overlooked the publishers will be pleased to make the necessary arrangements.

Allan Brockbank for Hartsop Valley photo p. 166; Barrow Borough Council, pp. 60–62; Tony Beale, Liverpool Polytechnic Business School, p. 11; Bolton & Chalkley, *Journal of Rural Studies*, pp 47–48; Bournemouth Borough Council, p. 46, 83; *Bournemouth Evening Echo*, p. 223; I.R. Bowler, Agriculture Under the Common Agricultural Policy, University of Leicester, Dept. of Geography, p. 113; Brecon Beacons National Park Authority, pp. 276–278; British Coal Corporation, pp. 86–87; Cambridge City Council, p. 73; Chester City Council, p. 270; Colwyn Borough Council, p. 231; Cornwall County Council, pp. 149–151; Countryside Commission, pp. 128–130, 133–134, 163–164, 167, 169, 177, 280–281, 284, 288; *Daily Telegraph*, p. 71; Department of Agriculture, p. 175; Department of Employment, p. 75, 152; Department of Employment, *Employment Gazette*, pp. 67, 77; Department of Transport, p. 39; Devon County Council, pp. 142, 144; Dorset County Council, Planning Department, pp. 83, 137, 158; Eliassen and Saltbones, p. 103; *English Nature*, pp. 178, 235–236, 243, 248–249; *Environment Now*, p. 255; *Evening Echo*, Bournemouth, p. 84; Exmoor National Park, p. 170; Fife Regional Council, p. 78; Forestry Commission, pp. 175–176, *The Geographical Association*, pp. 19–20, 44, 55, 72, 79–80, 116, 254; *Geographical Magazine*, pp. 35, 90; George Philip Ltd., p. 271; Glenrothes Development Corporation, p. 77; A. Goudie, p. 251; *The Guardian*, pp. 16, 38, 85, 87, 89, 99, 192, 213, 242, 283; M.J. Healey & B W Ilbery, Location and Change, Oxford University Press, p. 64; Hillman & Whalley, Policy Studies Institute, p. 146; Highland Regional Council, pp. 180, 194; HMSO, pp. 55–56, 89, 95–96, 100, 131; HMSO, Digest of Environmental Protection and Water, pp. 95, 96,100, Crown Copyright Meteorological Chart reproduced with the permission of the controller of HMSO; Institute of Terrestial Ecology, p. 232; IGS, p. 185; *The Independent*, pp. 3, 5, 15, 36, 41, 53, 59, 63, 81, 93, 95, 99, 100, 102, 108–110, 143, 154, 201, 212, 234, 245, 259; Johnston and Gardiner, Changing Geography of UK, Routledge, pp. 90, 94, 104, 107, 147, 156, 199–200; Joint Nature Conservation Committee, pp. 131, 136, 229, 240–241, 249; Leeds City Council Planning Department, pp. 24–25; Leeds Development Corporation, pp. 26–27; Liverpool City Council, pp. 32–34; *Liverpool City Journal*, p. 11; MAFF, pp. 111–112, 117; Merseyside Passengers Transport Executive, p. 43; Motherwell District Council, p. 58; National Rivers Authority, pp. 121–123, 125, 253; NOMIS pp. 81, 83, 148; *The Observer*, pp. 49, 59, 138, 141; O'Dell, p. 85; Oxford City Council, pp. 41–42; Ordnance survey for front cover (1:25 000 Purbeck Coast), back cover (1:50 000 Scotland), p. 12 reproduced with the permission of the controller of HMSO © Crown copyright; *The Planner*, p. 229; Portsmouth City Council, p. 237; Portsmouth 2000, Portsmouth City Council, p. 13; C. Pye-Smith & R. North, Working the Land, Temple Smith, p. 126; *Regional Trends* HMSO, p. 65; *Rhondda News*, p. 70; Robinson, Conflict and Change in the Countryside, Belhaven, p. 162; Rural Focus, p. 54; RSPB, p. 236; J. Sainsbury, pp. 16–18; Scottish Natural Heritage, pp. 161, 207, Snowdonia National Park Authority, pp. 203–205; Stanley and Banks, Batsford, p. 214; *The Sunday Correspondent*, p. 98; Teeside Development Corporation, p. 67; Telford Development Corporation, Commission for the New Towns, p. 54; Thames Water Authority, p. 124; © *The Times Newspapers Ltd 1991*, pp. 40, 59, 65, 94, 97; ULEAC, p. 250; Warren Spring Laboratory, Department of Trade and Industry, pp. 101, 103.

The publishers would also like to thank the following for giving permission to reproduce copyright photographs in this book;

p. 2, Meadowhall Centre Limited; p. 4, Steve Conlan; p. 9, top left and top right, Marketing and Tourism Department, City of Portsmouth; p. 13, Contract Services, Portsmouth City Business Group; p. 15, Lakeside Shopping Centre, Thurrock; p. 27, Leeds Development Corporation; p. 38, J Allan Cash Photolibrary; p. 42, Greater Manchester Passenger Transport Executive; p. 44, Merseytravel; p. 49 middle, Engineering Surveys Ltd, Maidenhead; p. 49 bottom, Geonex; p. 66 bottom, Press Association; p. 67 bottom left, Sealand Aerial Photography; p. 89, Topham Picture Source; p. 92, Environmental Picture Library; p. 97 bottom, J Allan Cash Photolibrary; p. 118 bottom, Forest Life Photographic Library; p. 132, Ecoscene; pp. 134/5, 157, Countryside Commission; p. 162, Robert Harding Picture Library; p. 175 bottom right, Scottish Highland Photo Library; p. 179 top left and middle left, English Nature, Peterborough; p. 180, Swan Photographic Agency; p. 193, Foster Yeoman Ltd; p. 195, J Allan Cash Photolibrary, p. 205, Celtic Picture Library; pp. 211, 214; 230 top right and bottom right, Celtic Picture Library; pp. 235 bottom right, 237 top left, Sealand Aerial Photography; p. 238, University of Cambridge Committee for Aerial Photography; pp. 241 bottom, 250 top, Aerofilms; p. 253 left, Ecoscene; p. 258, J Allan Cash Photolibrary; p. 266, Bath Tourism Marketing Photo Library; p. 272, Ironbridge Gorge Museum; p. 273 middle left, Science Museum; p. 274, Ironbridge Gorge Museum; p. 275, J Allan Cash Photolibrary.

All other photographs supplied by the author.

CONTENTS

INTRODUCTION

URBAN BRITAIN

Two Contrasting Images of Urban Britain   2
The Quality of Life in British Towns   5
Contemporary City and Town Centres   9
The Changing Nature of Retailing   15
The Inner City   21
The Outer Estates   36
Managing Urban Transport Systems   38
Pressures and Conflict at the Urban Fringe   45
Counterurbanisation   50

INDUSTRIAL BRITAIN

Some views of Industrial Britain   52
De-industrialisation   55
Reviving Old Industrial Areas   64
The Advent of High-Technology Industry   72
The Service Industries in Britain   79
Britain's Energy Supplies   85
Industrial Pollution   94

RURAL BRITAIN

A Perspective of Rural Britain   106
Contemporary Issues in British Farming   111
Rural Landscapes   127
Rural Settlements and Services   139
Industry and Rural Development   147
Recreation in Rural Britain   154

UPLAND BRITAIN

Life in Upland Britain   159
The Prospect for Farming in the Uplands   162
Forestry in the Hills   173
Mineral Extraction in the Uplands   184
Water Resources in Upland Britain   195
Recreation in the Uplands   202

COASTAL BRITAIN

The Changing Coastline   210
The Management of Coastal Processes and Landforms   214
Coastal Flooding   229
Managing Estuaries   235
Coastal Ecosystems   243
Pollution around the Coast   252
Recreation at the Coast   258

PROTECTED BRITAIN

Safeguarding Britain's Heritage   262
Conservation in Towns   266
Britain's Industrial Heritage   272
Protecting Britain's Landscapes   275
Protecting Britains Coastline   288

Glossary   291

Index   296

INTRODUCTION

This book aims to present, through the study of several aspects of Britain, some of the major issues and problems that are of current concern. The infinite variety of British landscapes has long been a stimulus to geographers, and it is within this context that a whole range of case studies from various parts of Britain are examined.

The text is organised in six sections. Five major regional themes have been selected to illustrate some of the contemporary questions that dominate the geography of Britain. Some of the themes would appear as natural choices, such as urban Britain, industrial Britain and rural Britain, the others less so. Upland Britain is chosen since a significant part of Britain may rightly be defined as upland. Coastal Britain recognises the importance of our long and varied coastline, and the part that it plays in our national life. The final section, protected Britain, acts as a retrospective review of all of the previous sections. It considers the need to protect and conserve all that is best in our urban and industrial heritage, together with the desire to safeguard our finest landscapes.

Each section begins with an overview, embracing one or more minor case studies that set the scene for the ensuing examination of issues and problems. These are illustrated by examples taken from all parts of Britain, in order to give appropriate regional dimension and flavour. Some issues, such as recreation and pollution run as themes across the sections: other links, such as those between urban and industrial Britain, and rural and upland Britain are obvious and inevitable. Detailed case studies are complemented by shorter ones which illustrate different aspects of the various themes. Everywhere the study material is as up-to-date as possible, with widespread use of newspaper articles, and contemporary reports from local and national bodies. Quotations are widely used throughout the text, to give a clear indication of the attitudes and values that shape spatial decisions.

Each case study incorporates a wide variety of resources for the reader to use. Emphasis is on an enquiry based approach throughout, with the student involved in active learning. A wide variety of exercises is included. Some are short, often involving the interpretation of one resource: others are more extended, and require the handling of a more complex set of data, and may involve the student in decision-making, role play and simulation.

If one theme dominates this text, it is the interaction between people and their environment. There is a progression from the sections on urban and industrial Britain, which are people-dominated, to those on upland and coastal Britain, where the natural environment still prevails, although it is increasingly under threat. In Britain, in the late twentieth century, there is a greater awareness of the environment, and the damage that can be done to it. The need to improve the quality of our built environments, and to maintain the quality of the natural environment is a continuing focus for the early decades of the next century.

URBAN BRITAIN

TWO CONTRASTING IMAGES OF URBAN BRITAIN

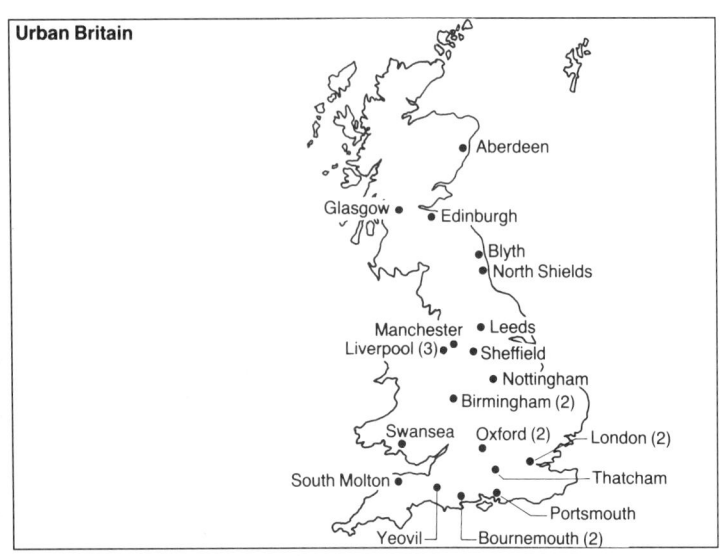

Urban Britain presents a bewildering series of contrasts, even to the unpractised observer. Within individual towns and cities, a journey from the city centre to the outskirts will reveal quite significant variations in the urban fabric, and in its inhabitants. Change is occurring almost everywhere within the urban area, although it is proceeding at a much faster pace in some areas than others.

Two case studies are chosen to illustrate these contrasts: one from the city fringe in Sheffield, the development of the regional shopping centre, Meadowhall and a study of a run-down council estate in North Shields, Tyneside.

In one way or another, urban growth obsesses most of us in Britain almost every day. Most of us are urbanites – 80 per cent of us officially so, more of us on any realistic definition – and we all experience some of the side effects of urbanisation, good and bad.

(Peter Hall, The Containment of Urban England)

MEADOWHALL: REGIONAL SHOPPING CENTRE IN THE DON VALLEY TO THE NORTH-EAST OF SHEFFIELD

Meadowhall is one of a new type of shopping complex developed on a derelict site in the Don Valley (photo Fig. 1.1.1). Similar centres have been developed at Brent Cross in north London (near the junction of the M1 and the M25), the Metro centre in Gateshead adjacent to the A1(M), and the relatively new Lakeside Centre at Thurrock adjacent to the M25. The Meadowhall centre was developed on the site of an old steel works, that finally closed in 1983. It lies within an Enterprise Zone (see p. 65) where there are special advantages that are available to developers. Meadowhall offers a shopping area of 116 thousand square metres, with 223 different stores all in covered malls.

Within Meadowhall there are five different shopping zones, all with traditional names designed to attract the shopper. Integrated with the shopping complex is the Oasis Centre, where shoppers can

Fig. 1.1.1 Meadowhall Shopping Centre, Sheffield. It is located within an Enterprise Zone, and is built on the site of an old steel works, which closed in 1983

eat, and be entertained on the 'Vidiwall'. The centre has been designed with the car-borne shopper in mind, with 12 000 car parking spaces, but there is also space for 300 coaches, and the bus station can handle 120 buses every hour. Two new railway stations serve the centre and in 1994 the new supertram network will link Meadowhall to the city centre of Sheffield. The centre has considerable sales potential, being sited in an area where existing retail provision is particularly low. Meadowhall attracts 53 per cent of its visitors from the top three social groups and 49 per cent are in the 25 to 44 age range. The extract from *The Independent*, 22 February 1992, notes the effect that Meadowhall is having on shops elsewhere in Sheffield (Fig. 1.1.2).

THE MEADOWELL ESTATE; NORTH SHIELDS

For much of the 1980s attention was focussed on the social conditions and the level of deprivation in the inner city areas. In the early 1990s it seems that the attention may well be diverted to conditions on the sprawling outer estates, and the unrest that its likely to occur there.

The centre cannot hold? Sheffield's, as exemplified by 'the hole in the road' in Castle Square, has been hit by the out-of-town Meadowhall development, resulting in many shop closures

There'll always be a High Street

If city centres are to survive, they must become as attractive to shop in as out-of-town complexes. Some of them are trying, as **Nicholas Roe** reports

Sheffield, lunchtime. Bus stops trail ropes of cold-looking people; streets are clogged with cars. Some shop windows wear bright commercial smiles, others – far too many – are blank and empty, like broken teeth in a new set of dentures. Bang goes the smart, welcoming image upon which so much of city's prosperity depends.

Here is an irony. Sheffield is suffering, not just from the recession, but from the effects of a vast shopping centre perched just outside town: Meadowhall, 1.25 million square feet [116 thousand square metres] of Promised Land flowing with low-cost milk and honey. Shoppers love it. And why not? Who would not like to shop in comfort in a bright, modern, purpose-built centre and avoid city-centre parking problems?

But there is a price for this convenience, and it is paid by the city centre. Sheffield's stores are closing down, retail incomes dwindling, city jobs vanishing, the high-street environment is declining. All of which is ironical because elsewhere in the country a move to counter the threat of out-of-centre developments is making progress, spurred by precisely the sort of dismal evidence that Sheffield provides. This counter-action is improving many of our most neglected high streets, and helping to re-craft urban landscapes that have grown devastatingly bland.

In other words, the competition that out-of-town shopping malls represent may be ruining some cities, but in the long run it may save others from their own complacency. The bottom line is that, if they are to compete, city centres have to be made as attractive to shop in as any out-of-town facility.

The expansive Eighties left us with five regional shopping centres – at Sheffield, Thurrock, Newcastle, Brent and Dudley – plus about 250 smaller retail parks. More giants are proposed even now, particularly around the M25, and experts predict that a dozen may eventually spring into glass-atrium life.

Sheffield's boarded-up windows demonstrate the long-suspected truth that, unless towns prepare well in advance against this onslaught, they will suffer.

Fig. 1.1.2 Independent 22 February 1992

The following quotes are taken from an article in *The Independent Magazine* of 5 January 1991. They proved to be particularly prophetic, with major unrest breaking out on the estate in the second week of September 1991.

The Meadowell, known also as the Ridges, Smith's Park: people call the estate all sorts of names. They call it other things, too, like the Bronx, a slum, a bottomless pit to throw money into, and the best place to live in, with the finest community spirit...

The Meadowell is one of North Tyneside's largest council estates, a mile from North Shields and a few hundred yards from the great river. Built in the Thirties to enable the council to clear Bankside, an insanitary warren of shacks and cottages clinging to the steep cliff above North Shield's fish quay. It got off to a bad start... Before the Bankside tenants were allowed into their new homes there was compulsory delousing for all: hardly an uplifting welcome to a new life...

Many people from Bankside brought prolonged experience of unemployment with them when they came to the new estate, but over recent decades things have got much worse. The Meadowell was planned to function rather like a mining village a concentrated dormitory development for the workplaces just down the road – in this case the North sea fishing trade in North Shields, and the shipbuilding and dockyards along the Tyne at Wallsend. Jobs in both places have been disappearing inexorably over the years, with nothing to replace them to the extent that these days 96 per cent of the people on the Meadowell are receiving some sort of state benefit, be it income support, old age pensions, housing benefit or the dole. Virtually no one leaves the estate to go to work, so virtually no one leaves the estate.

'Pull the place down': armed with prejudice you could certainly find plenty of evidence on the Meadowell that some of its residents are doing the job perfectly well themselves. Dozens of houses lie empty, because no one elsewhere in the borough wants to move there. As soon as another falls vacant it is stripped even before the council has time to board it up...

But the key word on the lips of all the community workers on the Meadowell is enabling; enabling its people to achieve a control, an impact on their lives. There are people on this estate, I promise you, says Denise Riach at the Meadowell Rights Centre, who only eat five days in the week. If you are spending all your time bringing up bairns, thinking about how you are going to fill your belly, you haven't got time to think about community awareness...

Dennis March and others spent months doing up the Meadowell Action Group House, a burnt out property donated by the council, to open it as a community centre. For a few months it hosted mother-and-toddler mornings, bingo sessions for pensioners. Then one day a gang of youths stoned an ambulance, called there to help an elderly woman, and the place has been boarded up ever since.

An indoor ski-slope and the other attractions of the Royal Quays (a prestigious new leisure and retail development on the river front of the Tyne) seem a million miles away, not a few hundred yards, from the long queue of pensioners shuffling into the youth centre for their fortnightly hand-out of European Community (EC) butter.

Fig. 1.1.3
Pensioners queuing for E.E.C. butter hand-outs Meadowell Estate, North Shields

1   Discuss the main issues raised by the case studies in turn.

2   How do they reflect contrasting aspects of life in urban Britain?

3   Suggest possible causes of these contrasts.

THE QUALITY OF LIFE IN BRITISH TOWNS

The introductory case studies to this section on urban Britain suggest that there is likely to be a considerable difference in the quality of life between different localities in British towns. Differences will obviously exist between towns too.

1 Study the extract from *The Independent on Sunday* 26 August 1990 (Fig. 1.2.1). This article reviews people's opinions on the quality of life in the two towns rated best and worst in a survey carried out by *Gallup*.

2 Extract from the article as many references to the quality of life as possible, both in Nottingham and Blyth.

3 Try to rank them in some order of importance as they appear in the article.

Fig. 1.2.1
Independent 26 August 1990

Life is not so bad, and that's the bottom line

William Leith finds little to choose between the best and worst towns in Britain

NOTTINGHAM is by no means a horrible place. The floor of the station's main hall is shiny and spotless; there are no beggars and rent-boys. A mile away, the city centre is calm and dignified; even the loitering skateboard gang have nice haircuts and wear Lacoste polo shirts.

The lack of pavement vomit and fast-food debris in the night-club quarter seems almost unnerving.

But the most desirable place in Britain? Just look at its miles of tidy suburbs, and you'll see it for what it is – a bland, smug place that might have been purpose-built for people with weak hearts.

It is the only large city never to have produced a famous pop group. Yet last week, in a Gallup survey for *Moneywise* magazine, Nottingham was voted the most desirable place to live.

The people of Nottingham can't stop talking about it. "It is a great place to live," Phil, a well-dressed engineer, said. "You want for nothing, things are not expensive, there's plenty of work and it is a nice size too – quite big, but not too big."

Helen, a young housewife recently moved from London, is bowled over too: "There's no need to worry about bringing up children here. It's so much safer."

Notice what these people are saying. Nottingham is good because of what it is not; it is not – unlike, say, London (56th in the survey), Liverpool (41st), or Manchester (48th) – full of busy people chasing wealth or poor people jumping out from behind every bush to rob you.

"I'll tell you what Nottingham is," a policeman in the city centre said. "It's a woman's town." He's right – there is something feminine about the city; its small, hygienic centre, the gentle pastel hues of its brick.

People in the surrounding towns hate the place. "People in Sheffield still call us scabs," Woody, a painter and decorator, said, "They hate us as if we came from the soft South, just because our miners wouldn't go on strike."

There is relatively little difference between rich and poor in Nottingham; the way to tell the middle-class area from the council estate is that the council houses all have satellite dishes. A Londoner walking around staid, respectable Mapperley – neat 1930s houses with bow fronts – would find it hard to tell from rock-bottom St Ann's. Mapperley is full of lawn sprinklers and turtle-waxed Mazdas; in St Ann's you hear the growl of souped-up old Ford Capris.

Duncan, an English lecturer who lives in the ultra-staid area by the castle called The Park, said he would love it but for the fact that he was prevented by a by-law from hanging out his washing.

At the other end of the popularity list, Blyth in Northumberland – Gallup's least desirable place – is real enough. It's a man's place, with man-sized industry gone to seed, a place in the shadow of redundant pit-heads and rotting wharves. Most of the men are unemployed. They slump about the town despondently in worn acrylic shirts, arms dark with tattoos, tabloids jammed into pockets.

Blyth made its money from mining in the nineteenth century and shipbuilding in the early twentieth. Now there is nothing. It is a 20-mile drive north of Newcastle along one of the most polluted beaches in Europe. It comes upon you suddenly, at bungalow-level; you see something common only to very poor areas: people sitting about outside tiny houses. The houses are cramped, claustrophobic, not everybody can be comfortably indoors at once.

But Blyth people don't let it get them down. They are used to living in a rotten place; having people tell them so is nothing new. "It's the council," Yvonne said. "They're terrible. If there's anything good in Blyth, they'll ruin it. They're just knocking down the parish church to make way for a shopping centre. What do we want with a shopping centre?"

Fig. 1.2.2a
Central Blyth, Northumberland: note the metal blinds on all the shops windows, fitted to protect against 'ram raiders'

Fig. 1.2.2b
Central Nottingham: the Council House

QUALITY OF LIFE IN BRITAIN

In this section it is proposed to examine the ranking in quality of life of the 38 largest towns in Britain. A team of researchers at the University of Glasgow investigated the quality of life in British cities by conducting a national questionnaire survey. In this survey correspondents were asked to rate the importance of 20 aspects of the quality of life. The rankings achieved for the different elements in the quality of life were as follows:

Fig. 1.2.3

1. Violent crime
2. Non-violent crime
3. Health facilities
4. Pollution
5. Cost of living
6. Shopping facilities
7. Racial harmony
8. Scenic quality
9. Cost of owner-occupied housing
10. Education provision
11. Employment prospects
12. Wage levels
13. Unemployment levels
14. Climate
15. Sports facilities
16. Travel to work time
17. Leisure facilities
18. Quality of council housing
19. Access to council housing
20. Cost of private rented housing

The rankings given by different age groups reveal some interesting differences.

Fig. 1.2.4

Perception of quality of life by age: more than 65 years

1. Violent crime
2. Non-violent crime
3. Health services
4. Pollution
5. Shopping facilities
6. Cost of living
7. Climate
8. Scenic quality
9. Private house costs
10. Quality of council housing

Fig. 1.2.5

Perception of quality of life by age: the young (18–24)

1. Employment opportunities
2. Cost of living
3. Wage levels
4. Violent crime
5. Non-violent crime
6. Health services
7. Shopping facilities
8. Education provision
9. Travel to work times
10. Sports facilities

1 Comment on the choice of criteria for assessing the quality of life in British cities. How do they compare with your findings from the article on Nottingham and Blyth?

2 What factors appear uppermost in people's minds when ranking the criteria?

3 Why are the top ten criteria so different for the two different age groups shown?

QUALITY OF LIFE RANKINGS

Once the twenty criteria had been determined, the researchers sought to match them to relevant environmental, social and economic criteria for the 38 cities. They were thus able to reach a ranking of the cities that combined both a behavioural and an objective element. Few other studies have combined the two elements in this way. The results of the ranking analysis are shown in the list (Fig. 1.2.6).

Fig. 1.2.6

1 Edinburgh
2 Aberdeen
3 Plymouth
4 Cardiff
5 Hamilton-Motherwell
6 Bradford
7 Reading
8 Stoke-on-Trent
9 Middlesborough
10 Sheffield
11 Oxford
12 Leicester
13 Brighton
14 Portsmouth
15 Southampton
16 Southend
17 Hull
18 Aldershot–Farnborough
19 Bristol
20 Derby
21 Norwich
22 Birkenhead–Wallasey
23 Blackpool
24 Luton
25 Glasgow
26 Bournemouth
27 Leeds
28 Sunderland
29 Bolton
30 Manchester
31 Liverpool
32 Nottingham
33 Newcastle
34 London
35 Wolverhampton
36 Coventry
37 Walsall
38 Birmingham

There may be a number of surprises in the final list! The most useful way in which to analyse these results is to plot the cities on a map of Britain, according to their ranking. Plot the first ten cities in one symbol or colour, and repeat this for the other groups of ten, ten and eight. In this way any regional distribution should become clear.

1 Comment on the distribution shown on your completed map.

2 Are there any factors that you can identify that might be responsible for the distribution? Consider the top ten cities first, then the bottom ten, and then the group in between.

IS QUALITY OF LIFE IN ANY WAY RELATED TO ECONOMIC PERFORMANCE?

There is probably no easy answer to this question, largely because of the range of different factors involved. In the first list above some of the surprises may be such a high ranking for cities such as Hamilton-Motherwell, or Bradford, and the relatively modest rankings of such centres as Norwich and Bournemouth. In order to see if there is a relationship between the quality of life rankings and those for economic performance it is necessary to use another survey carried out by researchers at the University of Newcastle. This survey used measures of economic performance at the end of the 1980s, together with measures of change in economic performance in the mid-1980s – the 'static' and 'dynamic' variables. It included all of the Local Labour market areas of Britain (broadly the same functional areas as those used in the Glasgow study). The 38 centres used in the Glasgow study have been withdrawn from the larger list in the Newcastle study and then ranked relative to each other. The rankings for economic performance are as follows:

Fig. 1.2.7

1. Reading
2. Bournemouth
3. Oxford
4. Aldershot-Farnborough
5. London
6. Southend
7. Brighton
8. Norwich
9. Southampton
10. Bristol
11. Luton
12. Portsmouth
13. Cardiff
14. Blackpool
15. Leeds
16. Leicester
17. Edinburgh
18. Plymouth
19. Derby
20. Stoke-on-Trent
21. Manchester
22. Nottingham
23. Coventry
24. Bradford
25. Walsall
26. Hull
27. Bolton
28. Newcastle
29. Aberdeen
30. Birmingham
31. Wolverhampton
32. Middlesbrough
33. Sheffield
34. Glasgow
35. Birkenhead
36. Hamilton-Motherwell
37. Liverpool
38. Sunderland

It is now necessary to repeat the task of mapping the ranking of economic performance of the different towns and cities on an outline of Britain.

1   Map the cities in groups of ten again.

2   Is there a significantly different spatial distribution when it comes to the ranking of towns on economic performance?

3   How might you seek to explain it?

4   Carry out a correlation test between the two sets of rankings.

5   Comment on the result that you obtain, and then write a reasoned explanation of the level of correlation that you have discovered.

CONTEMPORARY CITY AND TOWN CENTRES

(a)

(b)

(c)

(d)

Fig. 1.3.1
City and town centre: (a) Pedestrianised shopping: Portsmouth (b) Enclosed shopping: The Cascades, Portsmouth (c) New Hyatt Hotel, Birmingham (d) Headquarters Building: National and Provincial Building Society, Bradford

City and town centres in Britain have undergone a number of important changes since World War Two. Traditionally they have been the location of the major retailing, business and commerce and entertainment interests. They still maintain these roles in the nineties, but three important influences have been responsible for change.

Firstly, as the nature of retailing, and business and commerce have changed, so central areas have had to respond to these changes. Supermarkets and one-stop shopping have meant refurbishment or rebuilding of retail premises. People now expect to shop in a covered, well-serviced and air-conditioned environment and this has seen the arrival of the shopping mall in the High Street. One estimate is that 88 per cent of all the towns with over 100 000 population had developed major shopping schemes, each with over 46 000 square metres of selling space in the 1965–81 period. There is no reason to believe that the pace has slowed in any way since then: indeed many of the early schemes are fast approaching their first renewal! Business and office technology has also meant that older office buildings have had to undergo substantial refits. Tenants may require raised floors, suspended ceilings or movable partitions to enable new services to be installed and to allow the maximum flexibility that high technology demands.

Alternatively they are replaced by new ones designed to house information and business systems, but often requiring less space because of the continuing development of data storage systems.

Secondly, changes in technology and transport have meant that city fringe or out-of-town locations are appearing as alternative locations for the traditional functions of the city centre. Superstores, retail warehouses, retail parks and regional shopping centres have created a fundamental challenge to the high street retailer. Some retain their city centre premises, but also rent sites on the city fringe; others close their central location and operate one unit on the edge of the urban area. Similarly the needs of businesses may be better served in a new office complex or business park that is again located in more spacious surroundings on the city edge, with better access to trunk routes and airports.

Thirdly, with the vast increase in private transport, most town and city authorities have had to face major problems of traffic management and car-parking provision in the central areas. Several themes emerge here. Pedestrianisation of shopping areas is common to most centres of cities and large towns now. Multi-storey or underground car-parking is now essential provision in the centre. Traffic calming or traffic restriction measures have had to be introduced. Inner relief roads have been built to improve the flow of traffic to and around the centre. Many of our largest urban centres are now considering mass transit systems as a method of moving the enormous number of commuters and shoppers that use the city centre daily.

Three studies follow:

- Liverpool: a major regional shopping centre
- Portsmouth: a sub-regional centre
- Yeovil: a free-standing country town.

LIVERPOOL: FUNCTIONAL ZONING IN THE CITY CENTRE

The City Centre of Liverpool is one of the most important in Britain outside of London. It is a major regional shopping centre (the latest available figures show it to have one of the highest retail turnovers of any shopping centre in England and Wales) serving a population of four million in an area extending from Preston southwards to North Wales and Cheshire. It is also one of the country's largest office centres, with some 1.2 million square metres of office accommodation and containing such national headquarters as Littlewoods and Royal Insurance. Within the City Centre there is a range of tourist attractions, including its maritime heritage, its links with the rock world and the Beatles and an architectural and urban heritage that draw one to two million visitors each year. Finally, it provides nearly 40 per cent of all employment within the Liverpool area.

The map below (Fig. 1.3.3) shows the land uses displayed on Liverpool City Council's Strategy Review.

Geographers would wish for a more closely defined view of the City Centre than that shown in the Strategy Review.

After some class discussion, decide how you could delimit the functional City Centre more closely.

1   Make your own tracing or copy of the map. Mark a tentative boundary of the City Centre on your map.

2   What problems did you find emerged in deciding where to put the boundary?

3   How would you attempt to verify your results in a field survey?

The central shopping area

The main artery of the central shopping area extends along Lord Street and Church Street, with Whitechapel making an intersection between the two. Further shopping exists in the streets adjoining the main axis. The highest order shopping is found in Church Street, which is regarded as the prime location, with greater demands for space than that currently available. Lord Street suffered at one time from being separated from Church Street by the busy Whitechapel road, which was a major bus route into the City Centre. In the western section Matthew Street has developed into something of a specialist shopping centre, which has no doubt been enhanced by the opening of the famous Cavern Walks complex, with its shop units and office accommodation. The Old Post Office is the site of major shopping refurbishment, and there has been some interesting small-scale redevelopment in the William Street/Tarleton Street area. Two major shopping centres have been built at the eastern end of the central shopping axis, the recently refurbished St John's Centre and the newer Clayton Square Centre (see photo Fig. 1.3.2). Most of the main shopping streets have been pedestrianised, including Lord Street, Church Street, Parker Street and part of Whitechapel and most of the adjacent streets to the west. In the streets which have not been pedestrianised there has tended to be a decline in the shopping quality, and their best hopes for prosperity in the future lie in developing alternative service and entertainment functions.

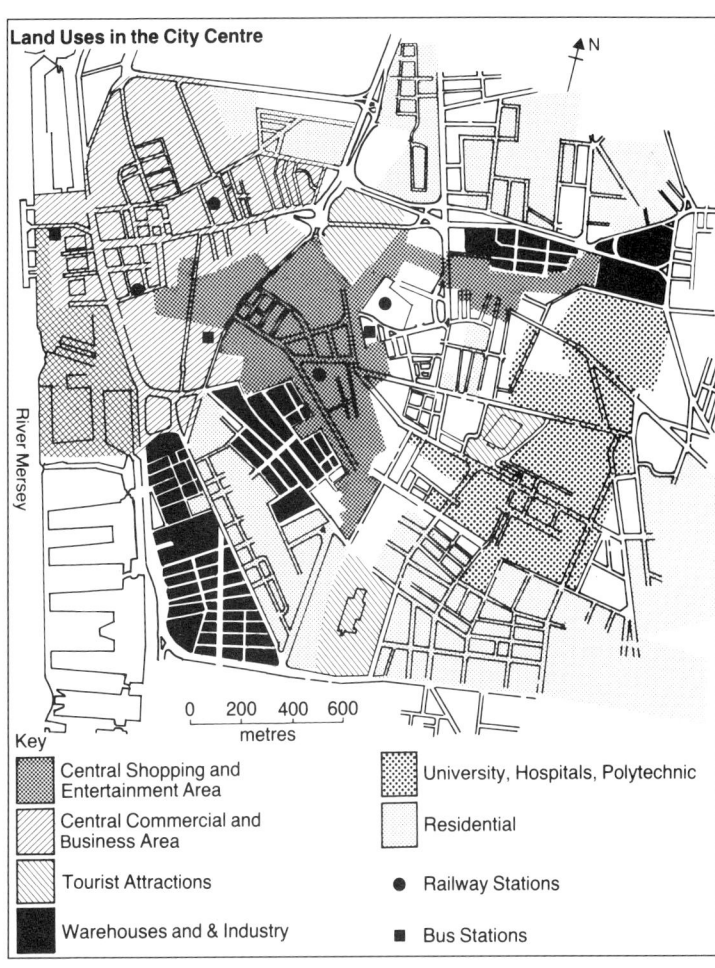

Fig. 1.3.2
Liverpool City Centre: St. John's Shopping Centre

Fig. 1.3.3
Strategy Review Nov 1987, Liverpool City Journal

Figure 1.3.4 City Centre Shoppers: Attitudes towards Liverpool City Centre

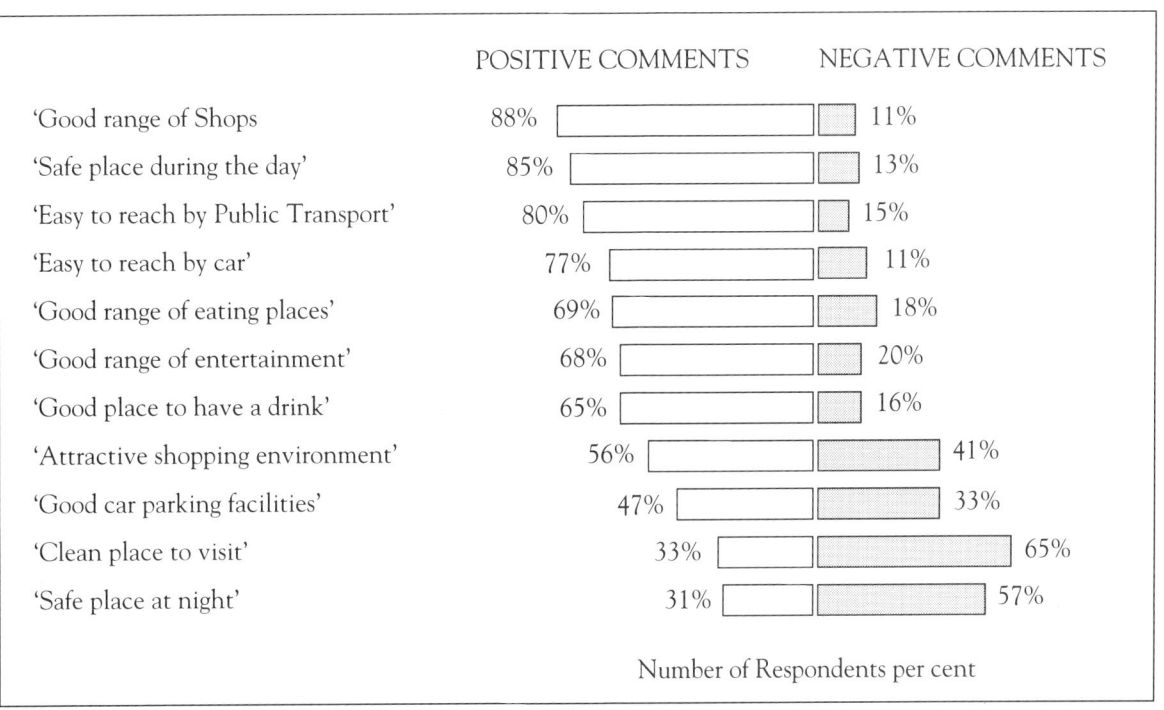

	POSITIVE COMMENTS	NEGATIVE COMMENTS
'Good range of Shops'	88%	11%
'Safe place during the day'	85%	13%
'Easy to reach by Public Transport'	80%	15%
'Easy to reach by car'	77%	11%
'Good range of eating places'	69%	18%
'Good range of entertainment'	68%	20%
'Good place to have a drink'	65%	16%
'Attractive shopping environment'	56%	41%
'Good car parking facilities'	47%	33%
'Clean place to visit'	33%	65%
'Safe place at night'	31%	57%

Number of Respondents per cent

PORTSMOUTH: PROMOTING AND IMPROVING THE COMMERCIAL ROAD SHOPPING CORE

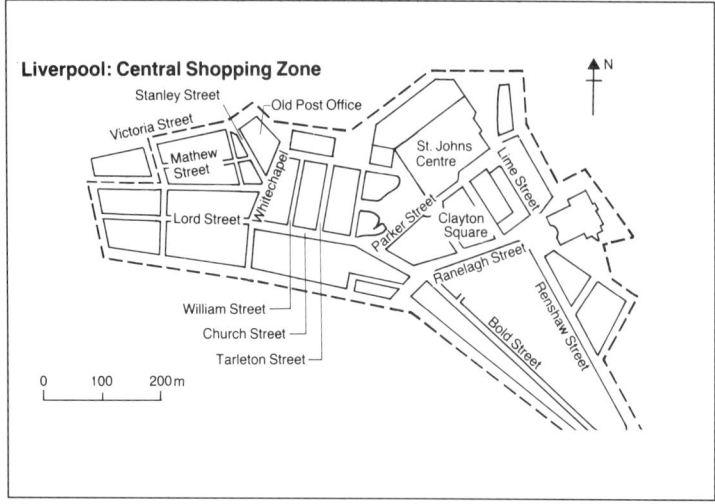

Fig. 1.3.5

Using the information in the account above, and the outline map (Fig. 1.3.5) attempt to map the main functional features of Liverpool's central shopping area.

1   How far does it display the contemporary features of a modern shopping centre as described in the introduction?

2   What problems may it face in the future, and how might it seek to solve them? The City Centre Shoppers' attitude survey (Fig. 1.3.4) will help in answering these questions.

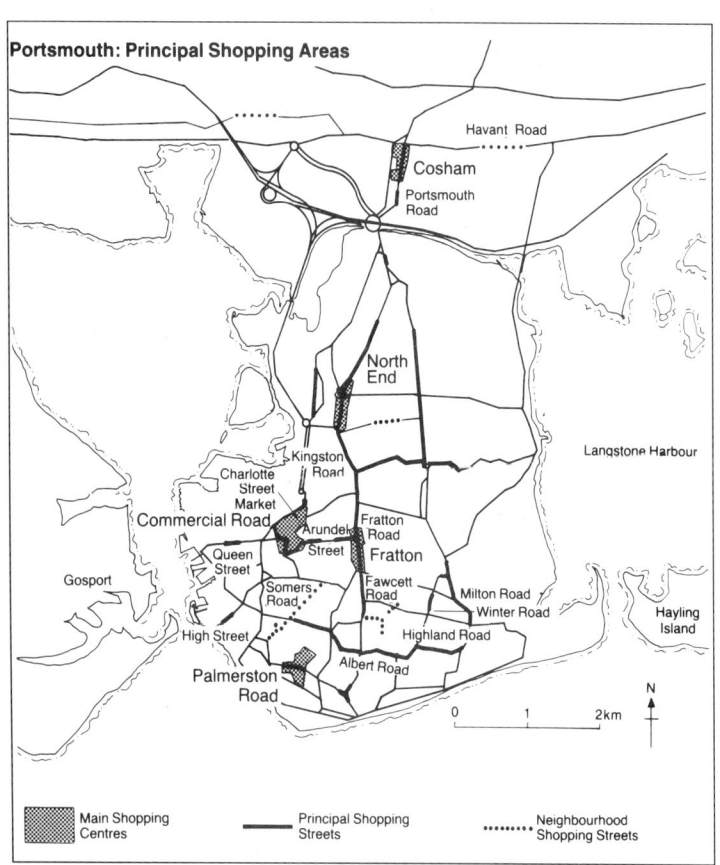

Portsmouth is the major shopping centre for south-east Hampshire.

Commercial Road is the principal shopping centre of the five recognised on Portsea Island and on the adjoining mainland at Cosham (Fig. 1.3.6). It attracts some 17 million shoppers a year, with a 1991 turnover of £170 million. It provides employment for some 4500 people. Its continuing promotion by the City Council is part of a general campaign by the city, which also includes publicising the maritime heritage image, and stressing the spectacular growth as a Cross Channel Ferry port.

Portsmouth City Council is very conscious of the threat posed to its city centre shopping areas by the growth of out-of-town shopping, and improvements in shopping facilities in neighbouring towns and cities (Fig. 1.3.7). Three main priorities were mentioned in the Central Retail and Commercial Association's report to the City Council. They were:

• additional car parking

• improvements to the image of Portsmouth e.g. refurbishment of the Tricorn Centre (a concrete structure combining shopping and a multi-storey car park described by some as the ugliest building in Britain!)

• up-grade and cover in the precinct.

Study the aerial photograph (Fig. 1.3.8) which identifies 22 different initiatives for the up-grading of the Commercial Road shopping area. Excluding the proposals for improvements to the Tricorn and the up-grading of the precinct, select five other improvements from the photograph that you think should be given a high priority. Explain why you think such priorities should be given. (The Cascades Centre and Sainsbury's superstore have now been completed, so these should also be excluded from your list.)

Fig. 1.3.6

CONTEMPORARY CITY AND TOWN CENTRES

Commercial Road, Portsmouth: its Regional Setting and Competitors

Fig. 1.3.7

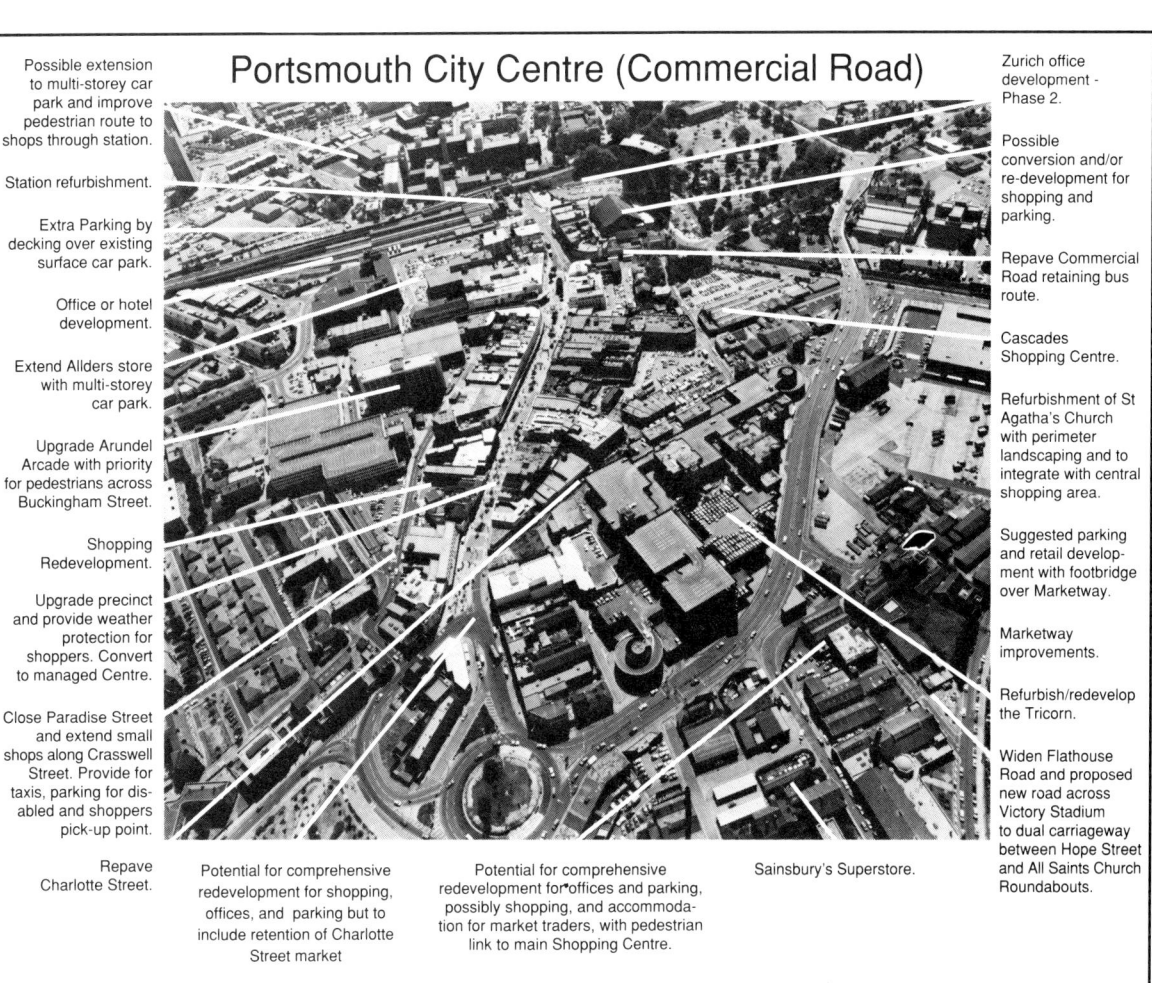

Fig. 1.3.8 Portsmouth City Centre (Commercial Road)

Possible extension to multi-storey car park and improve pedestrian route to shops through station.

Station refurbishment.

Extra Parking by decking over existing surface car park.

Office or hotel development.

Extend Allders store with multi-storey car park.

Upgrade Arundel Arcade with priority for pedestrians across Buckingham Street.

Shopping Redevelopment.

Upgrade precinct and provide weather protection for shoppers. Convert to managed Centre.

Close Paradise Street and extend small shops along Crasswell Street. Provide for taxis, parking for disabled and shoppers pick-up point.

Repave Charlotte Street.

Potential for comprehensive redevelopment for shopping, offices, and parking but to include retention of Charlotte Street market

Potential for comprehensive redevelopment for offices and parking, possibly shopping, and accommodation for market traders, with pedestrian link to main Shopping Centre.

Sainsbury's Superstore.

Zurich office development - Phase 2.

Possible conversion and/or re-development for shopping and parking.

Repave Commercial Road retaining bus route.

Cascades Shopping Centre.

Refurbishment of St Agatha's Church with perimeter landscaping and to integrate with central shopping area.

Suggested parking and retail development with footbridge over Marketway.

Marketway improvements.

Refurbish/redevelop the Tricorn.

Widen Flathouse Road and proposed new road across Victory Stadium to dual carriageway between Hope Street and All Saints Church Roundabouts.

CHANGE IN THE CENTRE OF A COUNTRY TOWN: YEOVIL

The map (Fig. 1.3.9) shows the position of Yeovil in South-west England. It functions as the main service and shopping centre for a considerable area of south-east Somerset and north-west Dorset. After Taunton, Yeovil is the largest shopping centre in Somerset and serves an area which has seen a considerable growth in population in the second half of the twentieth century. South Somerset district saw an increase in population of 33.6 per cent between 1961 and 1991. Yeovil's catchment area for convenience goods is 120 000 people, and for comparison goods it is 165 000. It thus occupies an intermediate position in the shopping hierarchy.

Centres such as Sherborne and Crewkerne offer a small range of shops and generally lower order goods, and the larger, more distant centres, such as Bournemouth, Bristol and Exeter offer a much wider range of higher order comparison goods (see Fig. 1.3.9).

In 1969 the Town Plan booklet identified the main problem in the centre: 'Traffic has become the dominant feature of the centre of the town and people are prevented from moving freely. Through traffic, local traffic, delivery vehicles and customers' cars all compete for road space in the centre of the town.' The 1969 Town Plan showed how matters might be improved, with the establishment of an inner ring road system from which a series of distributor roads led to a pedestrianised town centre. In the plan the possibility of shopping expansion to the north of the centre is noted.

In the 1969 plan it was not felt necessary to call for any major expansion of shopping. Changes in retailing and greater affluence within the catchment area showed this to have been too pessimistic. The site for shopping expansion identified in the 1969 plan was later described as 'fairly unsightly with untidy car parks and the backs of shops,' but it did have the two churches (St John the Baptist and the Methodist Church) at the west and east ends of Vicarage Street respectively. These two features were seen as being able to act as two focal points at either end of the new pedestrian mall. So the nucleus of an idea that was to see fruition as the Quedam Shopping Centre was born.

The Quedam Centre (see Fig. 1.3.10) completed in 1985, added 10 225 square metres to the Yeovil centre total of 42 000. It houses three main stores and 34 smaller shop units. In a small country town particular efforts have to be made to ensure that the centre blends in with the existing urban environment. The photo shows the use of local stone and brick in the buildings, and the neat landscaping of the sloping site. The opening of the Quedam has meant that existing retailers in Yeovil's central area have, in many cases, taken the opportunity to move into the new centre, allowing other retailers to move into the central area of Yeovil. Such has been the success of the Quedam that on one occasion before Christmas 1991, police actually had to stop would-be shoppers on the roads into Yeovil because the centre car parks were full to capacity!

Consider the advantages and disadvantages of the development of shopping centres such as the Quedam in free-standing country towns such as Yeovil.

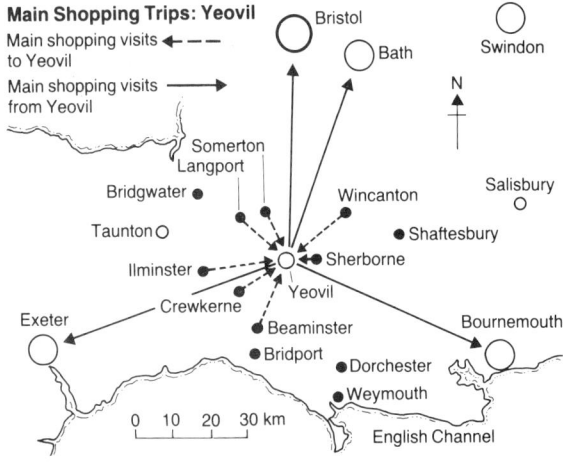

Fig. 1.3.9

Fig. 1.3.10
The Quedam Shopping Centre, Yeovil

THE CHANGING NATURE OF RETAILING

After studying a picture of the city centre in the late twentieth century, it is necessary to examine some of the changes in retailing in the second half of the twentieth century and the forces responsible for these changes.

Fig. 1.4.1
The Independent 18 November 1991

Shoppers are increasingly forsaking high streets and traditional town centres, for vast, out-of-town shopping centres built on sites that are legacies of Britain's industrial past.

There shoppers decant in enormous parking lots and head indoors, where the atmosphere is warm, bright, safe and synthetic – complete with patrolling security guards and splashing fountains.

There are obvious advantages over traditional town centres: no crawling through heavy traffic, no hunting for space in gloomy multi-storey car parks, no vandalism or graffiti. And, as almost everyone reaches the malls by private transport, no urban poor to unsettle the happy shopper....

Out-of-town shopping began in the late 1960s and has transformed Britain's towns. It started in industrial warehouses and old mill buildings, taken over by food stores or carpet emporiums. Soon it grew into clusters of what some estate agents call 'crinkly tin sheds', springing up near new ring roads, selling furniture, 'white goods', and do-it-yourself products. Out-of-town shopping is usually in town but on the fringe or in the suburbs ... Brent Cross in north London, the first of the huge, new shopping centres, with dozens of stores under one roof, occupies the site of an old dog track. It opened in the mid 1970s.

It is part of the post war decentralising of Britain. People are willing, and able, to travel ever further to work, to play and to shop. Once they moved mainly between suburb and centre; now they are as likely to journey around the periphery of town or to other towns. The trend has created its own momentum. It is the result of rising car ownership, growing car mileages and spreading trunk roads and motorways.... Even if no more are built, the move out-of-town will continue. Supermarket chains will continue to open millions of feet of superstores each year, according to estate agents Hillier Parker.

Fig. 1.4.2
Lakeside Shopping Centre, Thurrock, a vast enclosed shopping centre in London's commuter belt

David Stathers, director of estates for Boots, said 'where town centres have good road access and parking, a pleasant and well-managed environment, they can compete very effectively.' Character, history, culture and variety should give town centres a natural advantage but many have lost these advantages in their attempt to attract shoppers in cars. Inner ring roads, drab multi-storey car parks and indifferent concrete shopping centres have despoiled them.

The Independent 18 November 1991.

Study the extracts from the article in *The Independent*.

1   What are the main stages of change through which retailing in Britain has passed?

2   What forces have been responsible for these changes?

3   Why have town and city centres now become so disadvantaged?

4   What measures can town and city centres take to reverse the trend?

J. SAINSBURY: THE GROWTH OF A SUPERMARKET CHAIN

Fig. 1.4.3
The Guardian, 19 March 1991

Today, Sainsbury's opened its three hundreth supermarket. It will add another 20 stores in the rest of the financial year which began yesterday. Recently, its arch-rival Tesco, raised £572 million so that it could launch a similar number of what it prefers to call 'superstores'. And these two are merely the leaders of a food industry which seems constantly to be short of space.

How long can this go on? Even in an era when local authorities fall over themselves to grant planning permission, how many more stores can Britain's crowded cities and Britons' wallets support?

Given the length of time it has taken J. Sainsbury to complete its latest century, the question might seem academic. London's Grocer reached its first hundred in 1907, after 38 years in business. It rang up the second 100 in 1931 with London's Westbourne Grove shop. Sixty years seems an awful long time to wait for number 300, in Crawley near Gatwick airport. The answer is, of course, that the last 60 years have seen something of a change in retailing.

Westbourne Grove (which was closed in 1962) sold about 400 lines in about 280 square metres of space. Shoppers at Crawley and similar new stores have 15 000 product lines and 3800 square metres to tramp around....

The increase in store size explains why the number of stores has edged up so slowly. In some years they have closed more stores than they have opened. This might seem a little extravagant when a new store can cost up to £30 million, depending on the cost of the land. Building the shell costs around £5 million, and another £5 million goes on fitting it out — especially with the increasing number of refrigerated units.

A cheap plot of land in a relatively poor location can be obtained for as little as £5 million, but the perfect site could cost as much as £20 million, bringing the total, (even without interest costs) to anything between £15 million and £30 million....

Big stores attract more shoppers, who spend more on fancier products which produce higher margins. Crawley has 150 000 people less than a 15 minute drive away.

Larger stores are also getting cheaper to run. Getting the products to the store is more efficient – and is getting more so, with the cost per case falling by two per cent per annum. Computer systems are as crucial to these distribution improvements, as they are to pushing up productivity inside the store (even though Crawley will employ more than 500 people)....

The Guardian March 1991

Fig. 1.4.4
J Sainsbury. Store Development and Social Indicators

Date	Total Number* of stores	Number of* new stores	Number of* super-markets over 372 M²	Average sales area new stores (M²)	Approx. number* products sold	Households owning cars (per cent)	Households owning refrigerators (per cent)	Households owning microwaves (per cent)
1900	48	–	–	93	130	–	–	–
1950	244	1	–	28	550	7	2	–
1960	256	5	7	397	2500	30	40	–
1970	225	19	99	1070	3800	50	61	–
1980	231	9	197	1379	7000	58	92	3
1990	291	22	289	3001	14 000	73	95	42

* Figures refer to Sainsbury's financial year end

Study the extract from *The Guardian*, the statistics presented, the map showing the distribution of Sainsbury's stores, and the list of store openings for 1991–2 and planned openings for 1992–3. (Figs 1.4.3 to 1.4.5a and b.)

1 What appear to be the main trends in superstore development in the late twentieth century?

2 What are the main factors responsible for such trends?

3 Plot the distribution of the new stores opening in 1991–2 and 1992–3 on a blank map of Britain and compare it with the map showing the existing distribution of stores. Comment on the spatial distributions revealed on the two maps.

4 Compare the distributions with a population density map in a good atlas. Comment on any relationships, or lack of relationship that you can discover.

THE CHANGING NATURE OF RETAILING

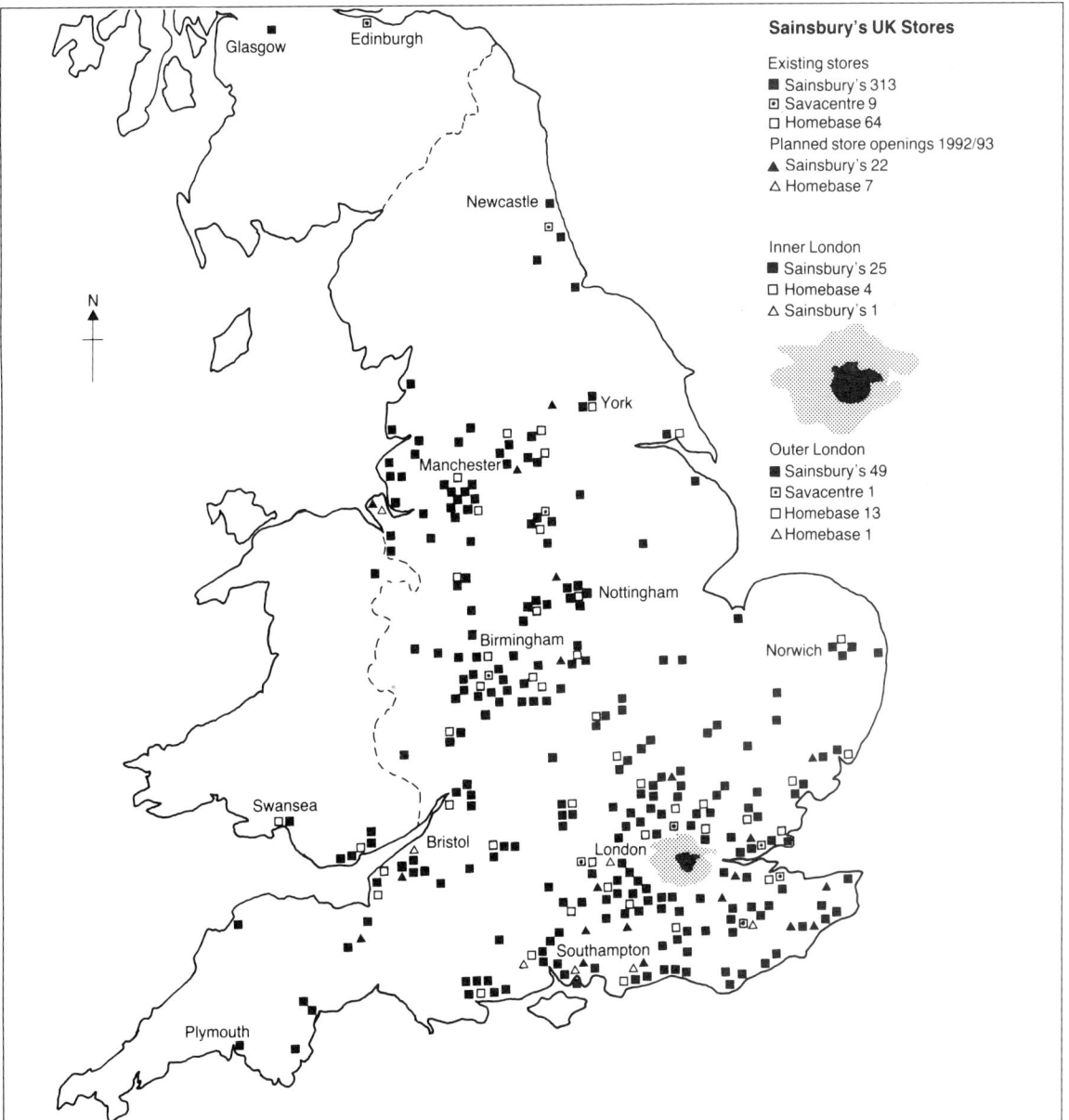

Fig. 1.4.5a

J SAINSBURY: NEW STORE OPENINGS

STORES OPENED 1991–2

SAINSBURY'S

Crawley	Wrexham
Thetford	York
Barnstaple	Bracknell
Shrewsbury	Kidlington, Oxford
Rustington	Lincoln
Chelmsford	Torquay
Maidstone	Liverpool
Macclesfield	Glasgow
Alperton	Preston
Haywards Heath	Stafford
Hedge End, Southampton	

New Sales Area .... 732 700 sq ft
[68 140 sq m]

HOMEBASE

Crawley	Oxford
Chelmsford	Blackheath

New Sales Area .... 130 100 sq ft
[12 100 sq m]

Fig. 1.4.5b

PLANNED STORE OPENINGS 1992–3

SAINSBURY'S

Dulwich	Ashford
Godalming	Leicester
Staines	Alton
Portsmouth	Ripley
Sevenoaks	Ipswich
Bristol	Taunton
Camberley	Whitstable
Folkestone	Huddersfield
Northfleet	Hove
Wirral	Stevenage
Basildon	Harrogate

New Sales Area .... 715 100 sq ft
[66 500 sq m]

HOMEBASE

Tunbridge Wells	Hove
Southampton	Wirral
Richmond	Bristol
Portsmouth	

New Sales Area .... 222 500 sq ft
[20 690 sq m]

A NEW VIEW OF BRITAIN

SUPERSTORE DEVELOPMENT IN THE BOURNEMOUTH AREA

The map and data (Fig. 1.4.7) show the changing distribution of Sainsbury's stores in the Bournemouth area in the twentieth century.

1 Use diagrammatic methods to plot this data, on a tracing or a copy of the map, to show the changing distribution in the most effective way.

2 What are the main trends in superstore location in a large conurbation, such as Bournemouth – Poole – Christchurch? In the light of the previous work in this section, how might they be explained?

3 What would appear to be the main considerations in choosing the latest site at Ferndown?

Fig. 1.4.6
New Sainsbury's Superstore: Talbot Heath, Poole, Dorset

Fig. 1.4.7

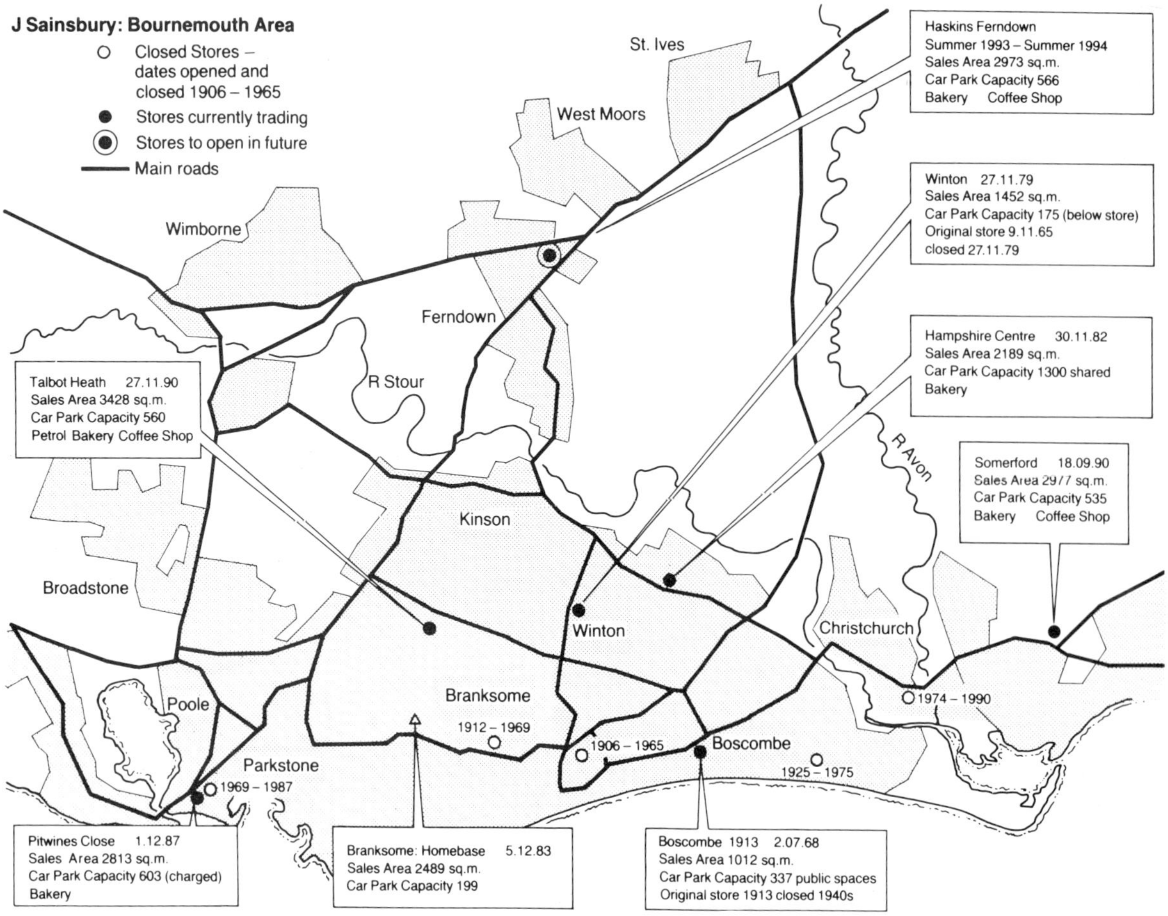

J Sainsbury: Bournemouth Area
○ Closed Stores – dates opened and closed 1906 – 1965
● Stores currently trading
⊙ Stores to open in future
— Main roads

Haskins Ferndown
Summer 1993 – Summer 1994
Sales Area 2973 sq.m.
Car Park Capacity 566
Bakery   Coffee Shop

Winton   27.11.79
Sales Area 1452 sq.m.
Car Park Capacity 175 (below store)
Original store 9.11.65
closed 27.11.79

Hampshire Centre   30.11.82
Sales Area 2189 sq.m.
Car Park Capacity 1300 shared
Bakery

Somerford   18.09.90
Sales Area 2977 sq.m.
Car Park Capacity 535
Bakery   Coffee Shop

Talbot Heath   27.11.90
Sales Area 3428 sq.m.
Car Park Capacity 560
Petrol Bakery Coffee Shop

Pitwines Close   1.12.87
Sales Area 2813 sq.m.
Car Park Capacity 603 (charged)
Bakery

Branksome: Homebase   5.12.83
Sales Area 2489 sq.m.
Car Park Capacity 199

Boscombe 1913   2.07.68
Sales Area 1012 sq.m.
Car Park Capacity 337 public spaces
Original store 1913 closed 1940s

Parkstone 1969 – 1987
Branksome 1912 – 1969
1906 – 1965
Boscombe 1925 – 1975
Christchurch 1974 – 1990

RETAIL CHANGE IN THE GREATER SWANSEA AREA

Fig. 1.4.8a
The Quadrant Shopping Centre, Swansea: one of the city centre's enclosed shopping malls

Fig. 1.4.8b
Retail warehouses: Fforest Fach, Swansea: these warehouses are typical of those located on out-of-town sites, and include DIY stores (shown) as well as furniture and carpet warehouses, grocery superstores and hypermarkets

After examining the growth of a supermarket chain, and its impact on one area in the South of England it is now necessary to look at a total view of retail change in a much larger region.

Swansea dominated its region in 1978, with a retail floor space of 139 350 square metres, five times as large as each of the four district centres of Llanelli, Moriston, Neath and Aberafan (Port Talbot). This represented a pattern consistent with urban growth in the region, and, as yet, untouched by pressures for major retail change. Since the mid 1970s these pressures have made their presence felt in this region as in every other one. Out-of-town superstores, retail warehouses and retail parks have all made their appearance, although these have been countered to a certain extent by the advent of central shopping developments in Swansea and Port Talbot in the 1970s and the 1980s. Many of the developments have been on derelict 'brownfield sites', which have had environmental as well as commercial implications.

Study the three maps (Figure 1.4.9a – c) which show:

- The shopping hierarchy in the Greater Swansea region in 1978

Fig. 1.4.9a
Geography, Volume 74 Number 3

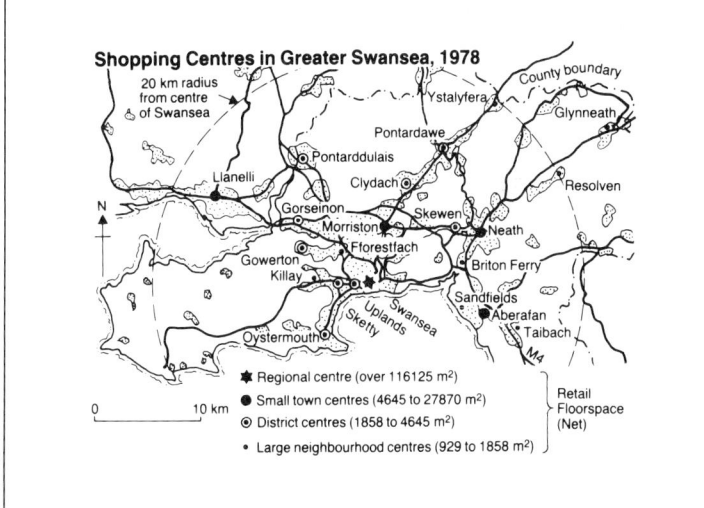

- The distribution of superstores in Greater Swansea in 1988

Fig. 1.4.9b
Geography, Volume 74 Number 3

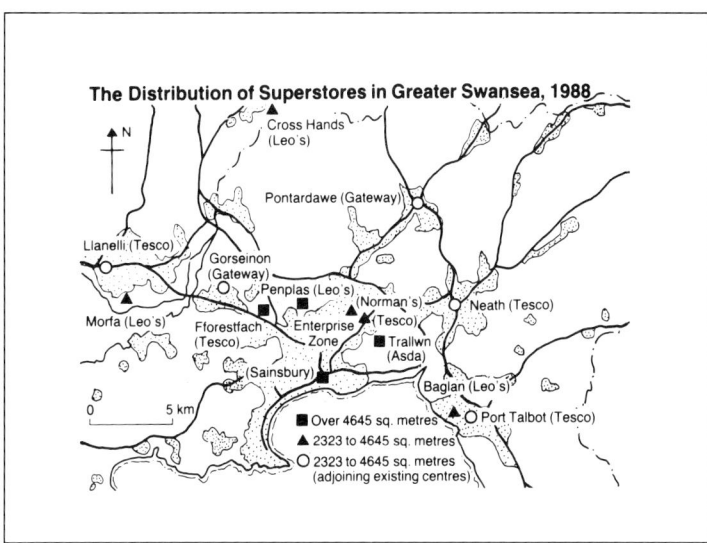

- The distribution of retail warehouses and retail parks in Greater Swansea in 1988.

IS THERE ANY HOPE FOR THE NEIGHBOURHOOD CENTRE?

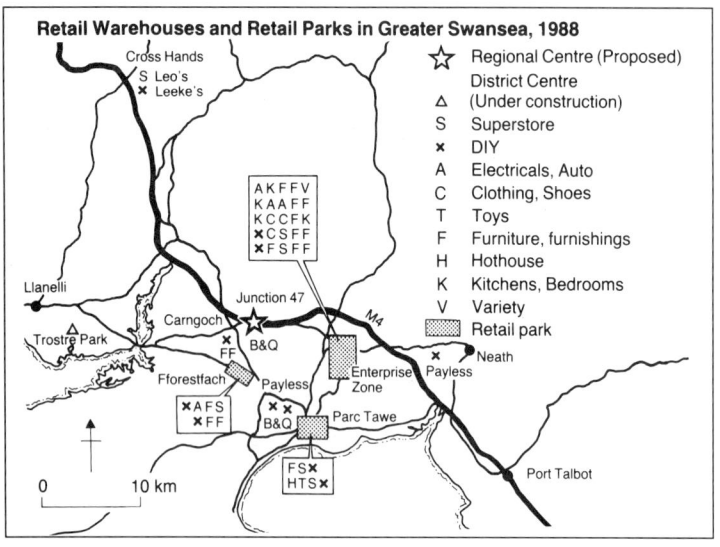

Fig. 1.4.9c
Geography, Volume 74 Number 3

Change in retailing means that some of the smaller centres are bound to suffer. More often than not these centres are likely to be in inner city locations. Here they still fulfil an important function, serving those that are unable to use the more sophisticated facilities elsewhere. However a variety of pressures has threatened these inner city neighbourhood centres. In the early 1980s expenditure by Birmingham residents fell by 12 per cent in real terms in inner city centres. In the Lozells Road Centre expenditure fell by nearly a third, and at Ladypool Road it fell by 70 per cent. By way of contrast other centres showed general increases in expenditure of up to 50 per cent, and in some outer centres expenditure doubled. A spiral of decline became evident, the result of lack of demand, the overprovision of out-moded shop premises, poor shopping environments, high vacancy rates and a lack of new investment.

West Midland County Council and Birmingham City Council launched an initiative in the early and mid 1980s to bring about improvements in shop premises and shopping environments in the inner city areas.

The types of improvements to the premises are shown in the diagram (Fig. 1.4.10).

1 Comment on the three different distributions shown on the maps and attempt to explain them.

2 What would be the main advantages in locating a major regional centre at Junction 47, and what would be the major objections to such a proposal?

Fig. 1.4.10

THE INNER CITY

Fig. 1.5.1
Examples of old and renewed housing in inner city Leeds

To the casual observer the change that takes place in urban character beyond the edge of the city centre is clear enough. Along the main roads the well-stocked and attractively presented window displays of the national multiples are replaced by small independent stores, offering a range of convenience goods and small specialist shops that cannot afford the high rents of the city centre. Symptomatic of the area, many shop premises display 'For Sale' notices, or are simply boarded up against vandals. Those that venture down the side streets often enter an environment that owes its origins to the quickening pulse of the city growth in the late nineteenth century. Narrow, ill-repaired streets are fringed by poorly maintained dwellings, whose walls often bear the graffiti born of the frustration of the local inhabitants. Interspersed with the dwellings are the remains of early industry, small dingy workshops, scrap metal yards and even abandoned warehouses, empty-windowed and with roofs open to the sky. Disused railways, and underused canals thread through the area, and open spaces are usually littered with burnt-out wrecks of cars, and used as informal rubbish tips.

Yet this image, still true of parts of many of our cities, does not reflect the real attempts that have been made at local and national level to improve this environment. Elsewhere one might equally find neatly refurbished and newly painted facades to houses, well-paved roads, lined with newly planted trees, leading to attractive squares and courtyards, with well-designed parking spaces and thriving community centres and playgrounds.

But the people who live in these inner city areas are still amongst the most deprived in the country. Many are trapped in a vicious circle of poverty, doomed often to spend the rest of their lives in such unpromising surroundings. Many are unemployed, often long-term, and suffer ill health, the result of poor provision of basic services in their overcrowded homes, and inadequate diet. High levels of crime and vandalism stem from such poor living conditions. The old, the low paid, those on welfare support and those that belong to ethnic minorities have to bear the daily burden of living in the inner city.

Attempts to improve physical and social conditions in the inner city have a long history. Slum clearance began in the early part of the twentieth century, and in the post war period successive governments have introduced inner city initiatives.

- Labour Governments in the 1960s and 1970s introduced the Urban Programme, and the idea of partnerships between national and local government to operate on a joint basis in the inner city.

- Conservative Governments in the 1980s took up the challenge with a wide range of measures and initiatives. City action teams (groups of civil servants specifically seconded to deal with inner city problems) and Task Forces (aimed at large-scale projects, such as garden festivals) were the main response to the series of riots that occurred in 1981.

- The two most significant of the initiatives introduced by the Conservative Governments were the Enterprise Zones and the Urban Development Corporations.

- The Enterprise Zones aimed at encouraging the growth of industry in inner city zones, took on a much wider spatial significance with examples being created in locations far removed from the typical inner city.

- The Urban Development Corporations sought to catalyse development in inner city areas, with powers to acquire land, and create the conditions necessarily favourable for significant injections of private capital.

- Mrs Thatcher's famous election night pledge in 1987 to 'do something about the inner cities' spawned the much heralded Action for Cities in 1988, which hardly mentioned local Government, but contained a string of proposals from government departments.

- In the post-Thatcher era Michael Heseltine launched his City Challenge in 1991, in which a number of competing local authorities had to bid for a share of £82 million to be spent on schemes within their area. In autumn 1992 new commitments under the Urban Programme ceased and the City Challenge Scheme was wound up.

LEEDS: RENEWAL IN THE INNER CITY

Fig. 1.5.2
Inner city Leeds: Woodhouse

WOODHOUSE (see Fig. 1.5.2)

'Environmental conditions are universally poor in Woodhouse ... and the surrounding housing areas suffer from neglected roads and footpaths, collapsing walls and toilet blocks, derelict sites, unmade back streets, rubbish and general dumping. The Wharfedales have totally unmade streets on a steeply sloping site.'

Fig. 1.5.3
Inner City Leeds: Chapeltown

CHAPELTOWN (see Fig. 1.5.3)

'... severe neglect of walls, gardens, open spaces, footpaths and road surfaces, which will continue to get worse unless improvements are undertaken. The densely built-up area from Shepherd Lane northwards is going downhill fast in environmental terms. Bankside Street suffers from unregulated car repair businesses creating major eyesores and traffic problems.'

Fig. 1.5.4
Inner city Leeds: Beeston

BEESTON (see Fig. 1.5.4)

'The area is very densely built up with no open space within it, and drab environmental conditions prevail throughout the whole area. The Longroyds streets are probably the least bad with the Beverleys, the Garnetts and the Harlechs amongst the worst. There are a number of derelict buildings in the area.(For example Burton House and old school kitchens on Burton Road.)'

'Leeds has so much going for it; a buoyant economy, a skilled and settled work force, an unprecedented demand for office accommodation, **a varied and attractive range of housing**, and its undisputed position as the best serviced motorway city in the United Kingdom, with the M1, M62, M621 and the proposed A1/M1 link all within minutes of the city centre.'

The first three quotations come from a planning report on environmental conditions within three of the ten Urban Renewal areas established by Leeds City Council in 1985. The fourth quotation is from the introduction to the Leeds Development Corporation's Strategic Plan in 1988.

What are likely to be the main causes of the environmental decay referred to in the planning report on conditions in the inner city areas within Leeds?

The extract from the Strategic Plan of the Leeds Development Corporation expresses a far more optimistic view of the city. Explain the apparent anomaly shown in the bold type and suggest reasons for the very positive approach. Why does it differ so much in style and content from the other extracts?

In common with all of the other industrial cities of Britain Leeds is faced with major social, economic and environmental problems in its inner city areas. These areas are very much a legacy of the Industrial Revolution, which led to an unprecedented expansion of a wide range of factory employment, particularly in the fields of engineering, textiles and clothing. Most of this manufacturing was concentrated in the valley of the Aire and its tributaries, together with the canals, railways and the main roads which serviced it. This growth of industry was accompanied by the need for, and the provision of housing for the growing population. Leeds became notorious for the widespread building of back-to-back housing over much of its inner area.

Inner city renewal and renovation is clearly concerned with more than the provision of decent, adequate housing. It embraces not only the physical upgrading of dwellings, but also the improvement of the street and neighbourhood environment, and the creation of new and lasting employment opportunities in what is often an area in which factories and service provision are fast disappearing. This can establish a more balanced and purposeful social order.

In this case study we shall be looking at two aspects of renewal: the Urban Renewal programme of Leeds City Council, with its emphasis on improvements in housing and the environment, and secondly the work of the Leeds Development Corporation, which sets itself an altogether wider brief covering environmental and residential developments and the provision of employment, leisure and recreational facilities.

Housing renewal

There are basically two main methods that are available to the urban planner and decision-maker when renewal of housing is being considered.

Fig. 1.5.5
(a) Canalside redevelopment central Leeds (b) Renewed housing Woodhouse

Clearance and comprehensive redevelopment is one option, and the alternative is to follow an intensive programme of improvement to existing housing stock. The two alternatives are clearly not mutually exclusive on a city-wide basis.

Housing renewal in Leeds dates back to the 1930s, and does seem to have passed through a distinct series of phases:

1   Large scale slum clearance in the 1930s.

2   Clearance and improvement operating in tandem in the 1950 to 1960s (at a time when other authorities were far more concerned with extensive redevelopment).

3   A reduced clearance programme: intensified improvement programme targeted on areas with the greatest need in the 1970 to 1980s.

In Leeds, area-based policies of housing improvement and renovation have been followed since the 1960s. Leeds initiated the Improvement Grant area that gave rise to two important national programmes:

1   The General Improvement Area (GIA) established by the Housing Act of 1969. This area approach was designed to boost areas of fundamentally sound housing, through the use of improvement grants, together with the introduction of a series of measures that sought to improve the street environment, such as pedestrianisation, landscaping, the provision of play-space and off-street parking.

2   The Housing Action Area (HAA) initiated by the Housing Act of 1974. These areas were targeted because they had some of the very worst housing conditions and also, inevitably, were areas of multiple deprivation and social stress.

In 1985 Leeds City Council set up ten Urban Renewal Areas (URAs) in the inner city. These were much larger than the GIAs and the HAAs (some of which were subsumed within the URAs). These URAs provided a framework for targeting improvement grants, but also resources from Housing Associations and the Government funded Urban Programme.

The map (Fig. 1.5.6) shows the distribution of the URAs in the inner city of Leeds.

It is now proposed to look briefly at representative wards in each of the URAs, which were described in the opening quotations in the section on Urban Renewal: Woodhouse, Chapeltown and Beeston.

The three profiles of the City Wards (Fig. 1.5.7) in which the three URAs are located are shown on the three tables (University – Woodhouse; Chapel Allerton – Chapeltown; Beeston). In the ward, statistics are shown by means of a comparison with the mean figures for Leeds City.

1   Comment on the effectiveness of this method of displaying socio-economic profiles of the three City Wards.

2   Which of the three wards do you consider to be the most representative of the inner city? Give reasons to justify your decision.

3   Which of the three wards appear to have the best socio-economic conditions, and could be described as the least representative of the inner city? Justify your reasoning.

4   Which criteria would be most useful to the City Housing Department in deciding that the ward should be declared a URA.

Fig. 1.5.6

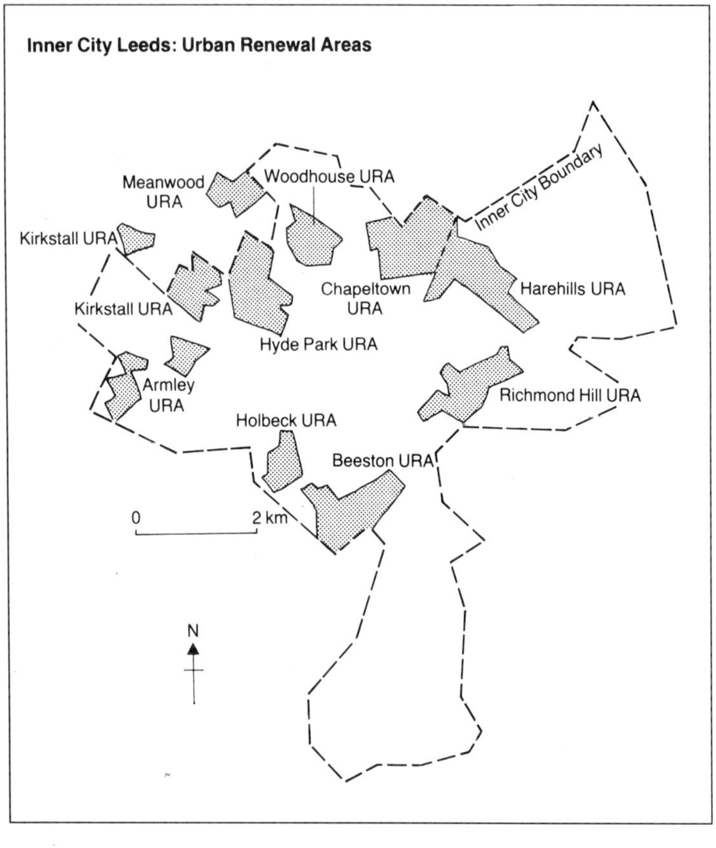

Inner City Leeds: Urban Renewal Areas

Fig. 1.5.7
Leeds City Council: Planning Department

BEESTON (BEESTON)

VARIABLES	LESS THAN 0.25 × CITY MEANS	0.25–0.5	0.5–0.75	0.75–1	CITY MEAN	1–1.5	1.5–2	2–3	MORE THAN 3 × CITY MEANS
Resident Population (on right of census)					445.00	●			
Children under 5					5.78	●			
Children aged 5–15				●	16.37				
Pensioners					18.01	●			
Residents aged 75+					5.82	●			
Ethnic Origin					4.04	●			
Lone parent households				●	2.44				
Overcrowded household spaces				●	3.34				
Shared dwellings				●	3.67				
Owner-occupied households					53.45	●			
Council-rented households				●	37.39				
Private-rented households					8.88	●			
Unemployment rate total					10.31	●			
Unemployment rate 16–24 year olds				●	16.08				
Unemployment rate 45–64 year olds					9.24	●			
Households without car					47.97	●			

UNIVERSITY (WOODHOUSE)

VARIABLES	LESS THAN 0.25 × CITY MEANS	0.25–0.5	0.5–0.75	0.75–1	CITY MEAN	1–1.5	1.5–2	2–3	MORE THAN 3 × CITY MEANS
Resident Population			●		445				
Children under 5					5.78	●			
Children aged 5–15				●	16.37				
Pensioners					18.01	●			
Residents aged 75+					5.82	●			
Ethnic Origin					4.04			●	
Lone parent households				●	2.44		●		
Overcrowded household spaces					3.34		●		
Shared dwellings					3.67				●
Owner-occupied households		●			53.45				
Council-rented households					37.39		●		
Private-rented households					8.88			●	
Unemployment Rate Total					10.31	●			
Unemployment Rate 16–24 year olds					16.08	●			
Unemployment Rate 45–64 year olds					9.24	●			
Households without car					47.97	●			

CHAPEL ALLERTON (CHAPELTOWN)

VARIABLES	LESS THAN 0.25 × CITY MEANS	0.25–0.5	0.5–0.75	0.75–1	CITY MEAN	1–1.5	1.5–2	2–3	MORE THAN 3 × CITY MEANS
Resident Population				●	445.00				
Children under 5					5.78	●			
Children aged 5–15					16.37	●			
Pensioners				●	18.01				
Residents aged 75+				●	5.82				
Ethnic Origin					4.04				●
Lone parent households					2.44		●		
Overcrowded household spaces					3.34			●	
Shared dwellings					3.67				●
Owner-occupied households				●	53.45				
Council-rented households					37.39	●			
Private-rented households					8.88	●			
Unemployment rate Total					10.31		●		
Unemployment rate 16–24 year olds					16.08		●		
Unemployment rate 45–64 year olds					9.24		●		
Households without car					47.97	●			

N.B. Values along top represent ranges e.g. "2–3" represents range two to three times city mean etc.
All values % except resident population (= No. of people resident on night in each census area)   Position of dot indicates status of City Ward

Fig. 1.5.8
Homes built by Housing Association, Woodhouse, Leeds. Housing Associations are now responisble for a major share of house-building in Britain.

Under new legislation (1991), HAAs and GIAs cease to exist. URAs remain the prime locations for change in the inner city in Leeds. Since April 1991 each new URA has to undergo a Neighbourhood Renewal Assessment (a socio-economic study) before designation. Another important new initiative is the co-operation that has developed between Leeds City Council and Housing Associations (currently the most important provider of new homes in the city (see Fig. 1.5.8)). Leeds Partnership Homes is the joint co-operative body, with the City Council providing the land, and the Housing Association building the dwellings.

Leeds Development Corporation: a new approach

Leeds Development Corporation is one of the latest series of Urban Development Corporations set up by the Government to encourage and stimulate regeneration in selected inner city areas. In Leeds the Development Corporation is responsible for some 536 hectares of underused land in two areas, the Kirkstall Valley to the west of the City Centre, and South Central Leeds, immediately to the south of the City Centre. The locations of the two areas are shown on the map (Fig. 1.5.9).

The Corporation was set up in June 1988 and is designed to have an operational life span of five years. It has been allocated a budget of £15 million, with a possible similar amount from the Government-funded City Grant, specifically designed to foster private sector development. Amongst its many aims are included the facilitating of new industrial growth and development, the improvement of transport links, particularly to the motorway system, the carrying out of a wide range of environmental measures, the promotion of retail, leisure and recreational facilities and the provision of a variety of residential accommodation.

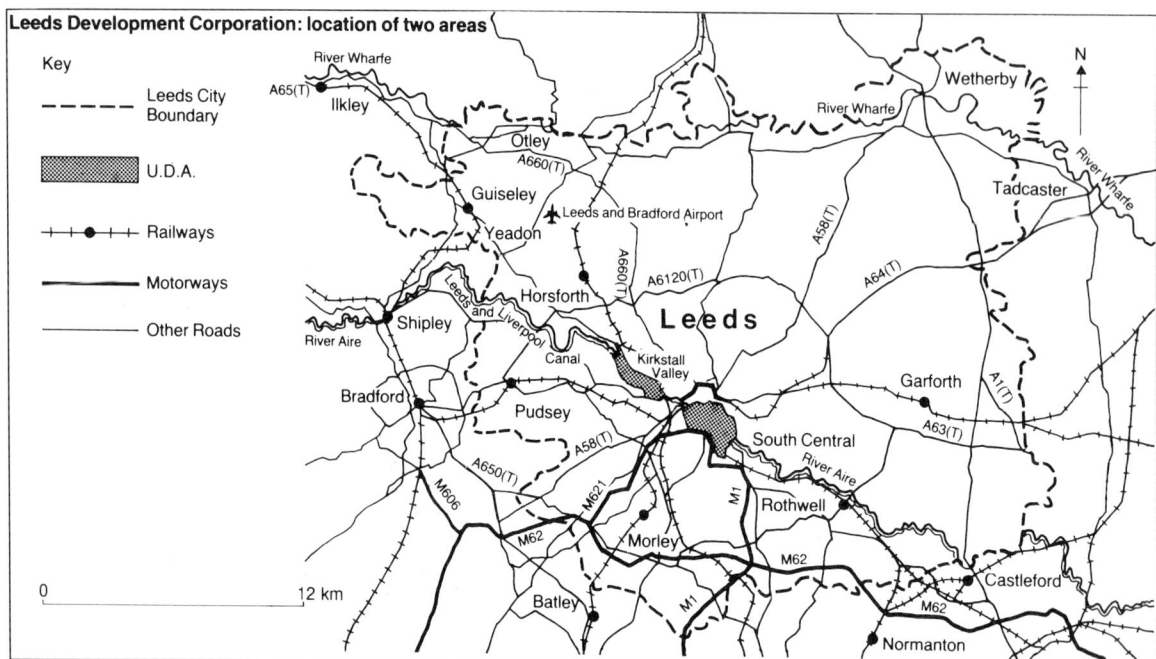

Fig. 1.5.9

Fig. 1.5.10
Kirkstall Valley, Leeds

THE KIRKSTALL VALLEY

The area of the Kirkstall Valley that is part of the Urban Development Area (UDA) is shown on the map (Fig. 1.5.9) and the aerial photograph (Fig. 1.5.10).

1 Using the photograph analyse the existing land use of the part of the valley that is shown.

2 Draw a sketch of the area shown in the aerial photograph, and indicate on it the main types of land use that you can distinguish (both within the area of the UDA and surrounding it). Within the area of the UDA pay particular attention to derelict areas and those that appear to be underused (prime targets for development).

The proposed plans for the zone of the Kirkstall valley shown in the photograph include:

- an integrated retail development on district centre scale
- major leisure facilities
- a business park
- provision of a number of residential units
- increased public accessibility within the area
- provision of environmentally attractive open space, with particular reference to the waterside areas along the River Aire and the canal.

Fig. 1.5.11
Industry in the Leeds Development Corporation Zone: Tetley's Brewery

SOUTH CENTRAL LEEDS

This zone of the UDA lies between the two industrial suburbs of Holbeck to the south-west, and Hunslet to the south-east and the City Centre to the north.

The River Aire flows across the area in an arc from north-west to south-east, and the main motorway links into the city terminate along its south-western boundary.

In many ways this is an inner city area different from those that we examined in the previous section where the emphasis was very much on residential land-use. South Central Leeds established a strong industrial function in the nineteenth century, and this tradition is still very important today (see Fig. 1.5.11). Renewal and regeneration in this zone presents a different set of problems and opportunities.

It is often said that the creation of Urban Development Corporations has led to conflict between the aims of the corporation and those of the locally elected Council and its appointed officials. The following have been amongst some of the criticisms of the UDCs:

1 The Corporations have been imposed by central Government and are given very considerable power over land and its development.

2 Their policies, reflecting Government strategies are often at variance with those of locally elected councils, leading to the charge that the UDCs are undemocratic.

3 One of the principal aims of the UDCs is to channel funds from the private sector into the inner city. Private sector capital may encourage development that is not in the interest of the local community in the inner city.

4 Although one of the principal aims of the UDCs was the creation of employment in the inner city, the jobs that have been created have often gone to outsiders who have the skills and expertise that is lacking in the local labour force.

5 Housing development, largely the responsibility of private developers has often not met the demands of the local community, which finds itself unable to compete for properties beyond its financial reach.

6 Transport funding has been concerned mainly with road construction and not with the provision of better public transport.

THE INNER CITY

1 What arguments can be put forward to counter each of the above criticisms?

2 How might conflicts between local Government bodies and UDCs be reduced?

'In February 1988 I wrote to all city tenants saying that I saw this partnership between public and private sectors as offering a golden opportunity to achieve real and lasting benefit for local people.'

Councillor Stan Austin (Chairman of Housing Management Committee, Birmingham City Council)

HEARTLANDS: A STRATEGY FOR THE REGENERATION OF EAST BIRMINGHAM

'Heartlands is centred on the East Birmingham District of Nechells (see Fig. 1.5.12). Formerly a thriving industrial area, it bears all the scars of inner city decline. Thousands of jobs disappeared in the recession of the early 1980s. The population of 13 000 lives mainly in council housing, and male unemployment stands at almost 30 per cent. The old skills simply do not match the new jobs.'

Peter Heatherington, The Guardian 2 February 1990.

DEPRIVATION STATISTICS FOR NECHELLS

CENSUS ENUMERATION DISTRICTS WITH SIGNIFICANT DEPRIVATION

	NECHELLS		BIRMINGHAM	
	No.	%	No.	%
In worst 2.5% in England and Wales	23	46.9	324	16
In worst 2.5–5% in England and Wales	10	20.4	163	8.1
In worst 5–10% in England and Wales	12	24.5	186	9.2

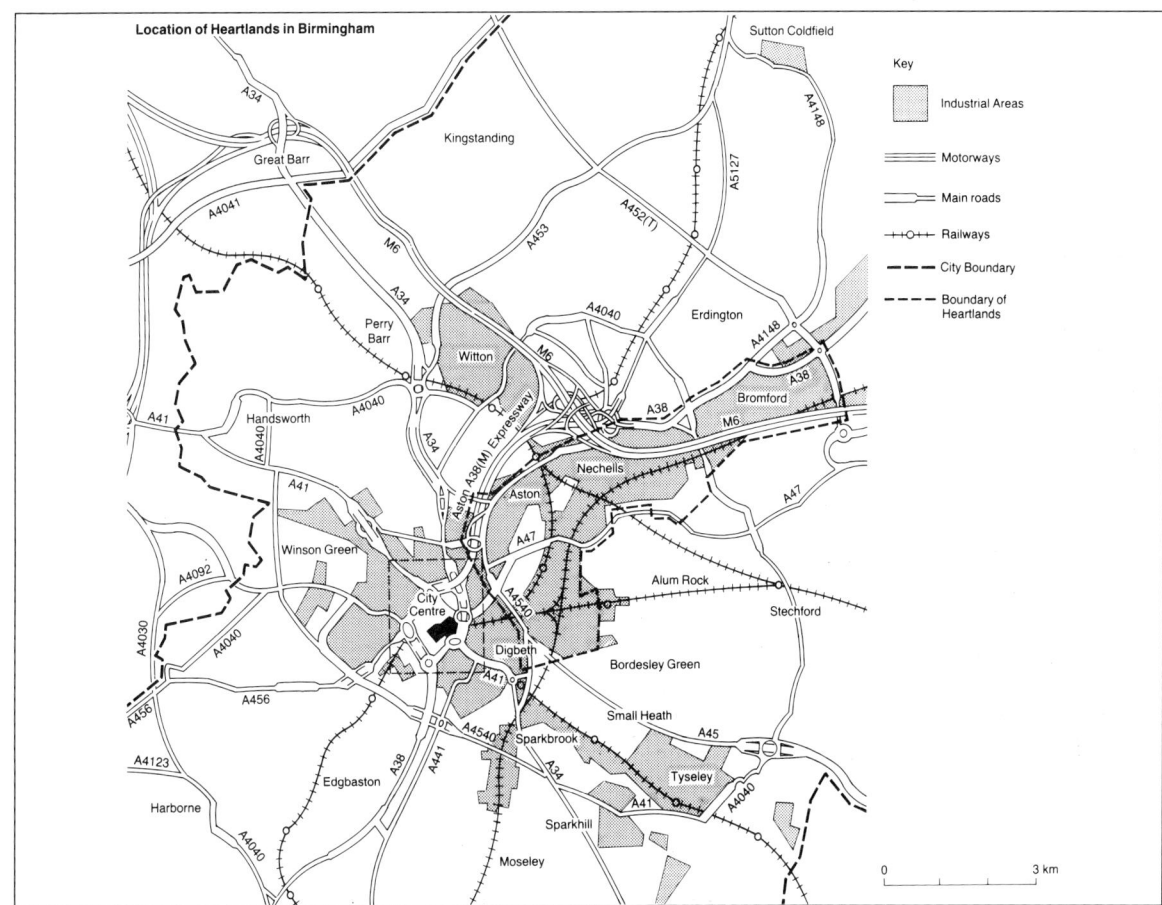

Fig. 1.5.12

'The environment of Nechells is poor:

- It gives people who live and work in the area, and those who travel to and through it a poor image of Nechells. It affects the quality of life.

- Heavy industry, derelict and vacant sites, scrap yards, and major roads and railways are highly visible and close to residential areas.

- The planning and designing of homes is now regarded as poor. People now complain that within housing estates too much open space is seen as impersonal and nobody's responsibility. Litter dumping, vandalism and other unsocial behaviour close to people's homes are the result. Maintenance is difficult, and often poor.

- There is a lack of usable open space. In the past it has been unimaginatively landscaped and is inhuman in scale.'

Nechells Development Framework

Nechells lies in the centre of the Heartlands redevelopment area in East Birmingham (see Fig. 1.5.12). The above extracts and basic statistics give some idea of the problems that have to be faced in the area, and the quote from Councillor Austin hints at the main thrust behind the new strategy for the regeneration of the area.

Use the four photographs (Fig. 1.5.13), all taken within the Heartlands area to discuss some of the major environmental problems that will have to be overcome in the regeneration of the area.

Birmingham Heartlands was first redeveloped through Heartlands Limited, a private company which operates in partnership with the Chamber of Commerce and Industry, and Birmingham City Council. The shareholders in Heartlands Limited include important private companies such as Tarmac and Wimpey, together with Birmingham City Council and Birmingham Chamber of Commerce.

Fig. 1.5.13
Environmental problems in the Heartlands area, Birmingham: (a) Derelict housing (b) Derelict factory (c) Workshops with difficult access (d) Scrapmetal merchants

(a)

(b)

(c)

(d)

Birmingham Heartlands: Main Land Use Proposals

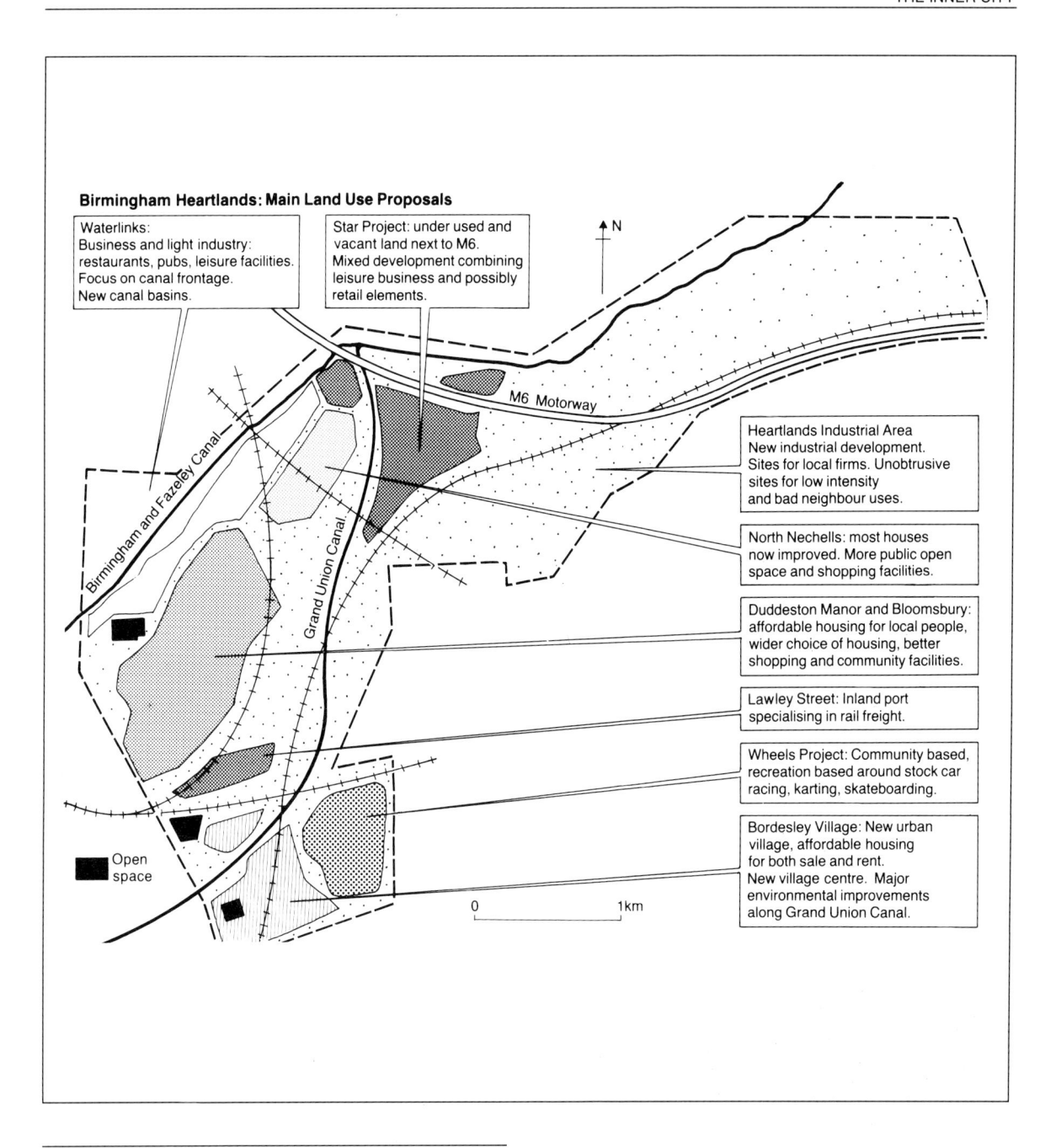

Waterlinks: Business and light industry: restaurants, pubs, leisure facilities. Focus on canal frontage. New canal basins.

Star Project: under used and vacant land next to M6. Mixed development combining leisure business and possibly retail elements.

Heartlands Industrial Area New industrial development. Sites for local firms. Unobtrusive sites for low intensity and bad neighbour uses.

North Nechells: most houses now improved. More public open space and shopping facilities.

Duddeston Manor and Bloomsbury: affordable housing for local people, wider choice of housing, better shopping and community facilities.

Lawley Street: Inland port specialising in rail freight.

Wheels Project: Community based, recreation based around stock car racing, karting, skateboarding.

Bordesley Village: New urban village, affordable housing for both sale and rent. New village centre. Major environmental improvements along Grand Union Canal.

Fig. 1.5.14

In March 1992 Heartlands Limited was replaced by Birmingham Heartlands Development Corporation which receives Government funding.

Heartlands covers an area of 2500 acres (1212 ha). Nearly 13 000 people live there, mostly in council properties that were built in the 1950s and 1960s. Unemployment is high (32 per cent) compared with Britain as a whole (9.3 per cent) in February 1989. It is dominated by manufacturing industry, with such firms as Freight Rover/Daf, Rover Group, S.P. tyres, Hoskyns Computers, Dunlop, Jaguar, Metro-Cammel, British Telecom, Delta and Lucas (with an obviously heavy reliance on the traditional transport of the West Midlands).

1. The map (Fig. 1.5.14) shows the main features of the Heartlands area. Given the task of promoting the area to potential developers, what would you consider to be the most prominent advantages that would need to be stressed? What obvious disadvantages would you seek to minimise?

A NEW VIEW OF BRITAIN

Fig. 1.5.15
New housing and office developments in Birmingham Heartlands

2   Discuss the ways in which the four photographs (Fig. 1.5.15) could be used in promotional literature aimed at attracting new business to Heartlands.

HEALTH IN THE INNER CITY IN LIVERPOOL

'The despoliation of our cities concerns me not just as a Londoner, but as a doctor because it generates a great deal of illness, depression and family disruption. It is compounded by the reversal over the same decade of the century-long trend towards greater equality of income and the prising open of the already wide gap between the rich and the poor. The consequences of that economic process are presented in human terms … in our surgeries every day. Patients existence which is exhausting them and making them ill…' David Widgery in 'Some Cities'.

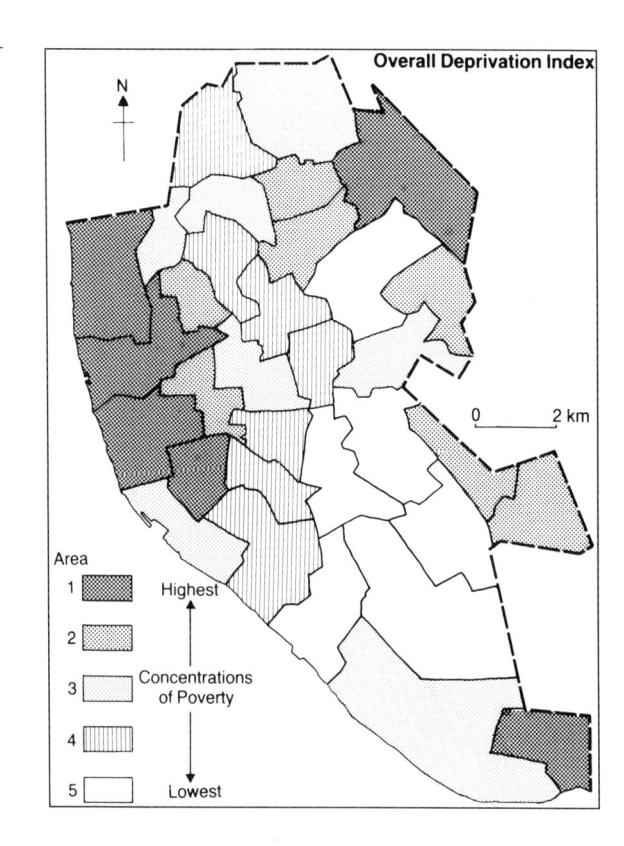

Fig. 1.5.16

THE INNER CITY

The map of Liverpool (Fig. 1.5.16) shows an overall deprivation index for all the wards in Liverpool. The index is the combined score of four indicators which are good measures of poverty or deprivation:

- the overall unemployment rate
- the level of car ownership as an indicator of income
- the proportion of non-owner-occupiers as an indicator of relative wealth
- the proportion of overcrowded households as a measure of housing problems.

A second map (Fig. 1.5.17) shows the distribution of social areas in the city, according to income and housing.

1. Identify the inner city areas on the map (use the map of the ward boundaries (Fig. 1.5.18) to identify which wards are the ones that suffer from the highest degree of deprivation).

2. Now examine the pie chart (Fig. 1.5.19a) showing the main causes of death in Liverpool in 1984. Identify the main causes of death

Main Causes of Death in Liverpool in 1984

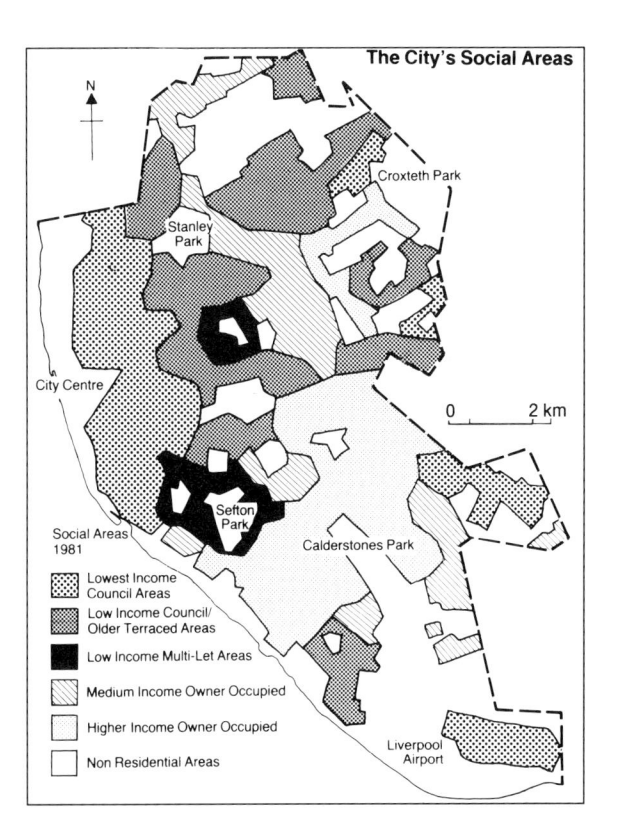

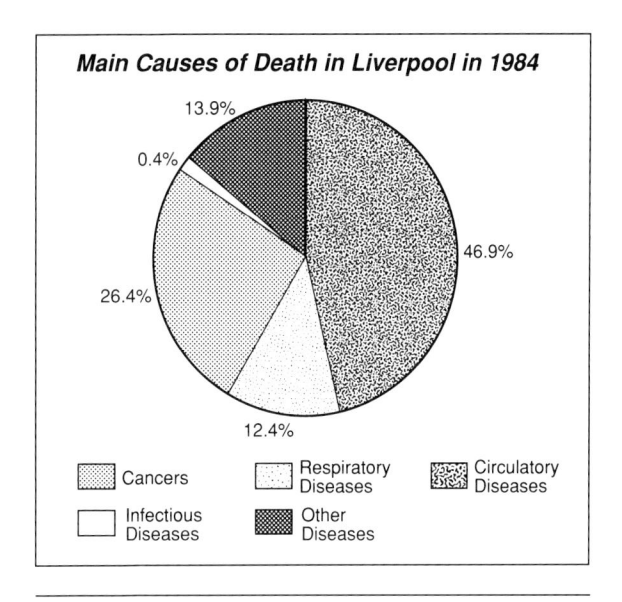

- Cancers: 46.9%
- Respiratory Diseases: 12.4%
- Circulatory Diseases: 26.4%
- Infectious Diseases: 0.4%
- Other Diseases: 13.9%

Fig. 1.5.19a

from the chart, examine their distribution in Liverpool as shown on the maps (Fig. 1.5.19b) and study the scattergraph (Fig. 1.5.19c).

3. What broad conclusions can be drawn between the distribution of deprivation, social areas and the total deaths?

The City's Social Areas

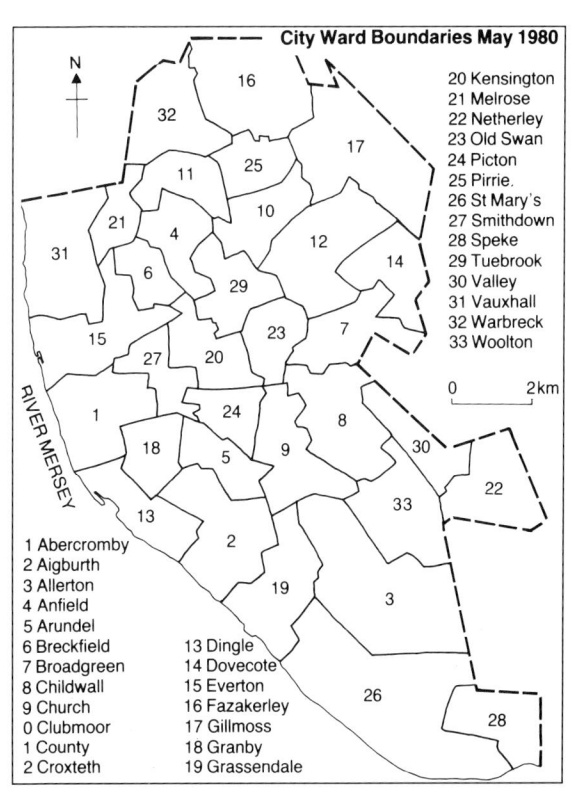

Social Areas 1981:
- Lowest Income Council Areas
- Low Income Council/Older Terraced Areas
- Low Income Multi-Let Areas
- Medium Income Owner Occupied
- Higher Income Owner Occupied
- Non Residential Areas

Fig. 1.5.17

City Ward Boundaries May 1980

1 Abercromby
2 Aigburth
3 Allerton
4 Anfield
5 Arundel
6 Breckfield
7 Broadgreen
8 Childwall
9 Church
0 Clubmoor
1 County
2 Croxteth
13 Dingle
14 Dovecote
15 Everton
16 Fazakerley
17 Gillmoss
18 Granby
19 Grassendale
20 Kensington
21 Melrose
22 Netherley
23 Old Swan
24 Picton
25 Pirrie
26 St Mary's
27 Smithdown
28 Speke
29 Tuebrook
30 Valley
31 Vauxhall
32 Warbreck
33 Woolton

Fig. 1.5.18

Fig. 1.5.19b

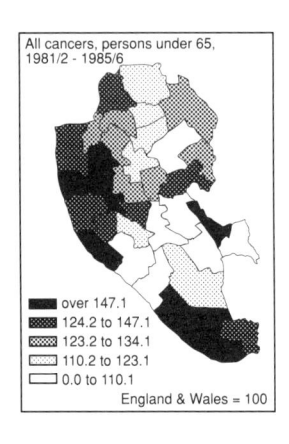

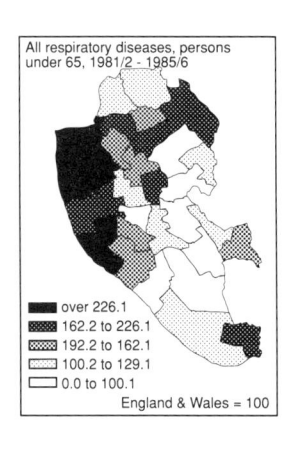

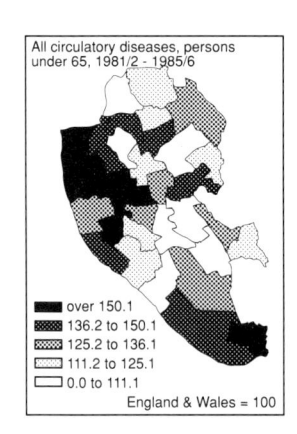

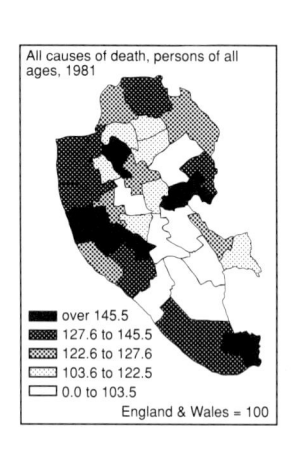

All cancers, persons under 65, 1981/2 - 1985/6
- over 147.1
- 124.2 to 147.1
- 123.2 to 134.1
- 110.2 to 123.1
- 0.0 to 110.1

England & Wales = 100

All respiratory diseases, persons under 65, 1981/2 - 1985/6
- over 226.1
- 162.2 to 226.1
- 192.2 to 162.1
- 100.2 to 129.1
- 0.0 to 100.1

England & Wales = 100

All circulatory diseases, persons under 65, 1981/2 - 1985/6
- over 150.1
- 136.2 to 150.1
- 125.2 to 136.1
- 111.2 to 125.1
- 0.0 to 111.1

England & Wales = 100

All causes of death, persons of all ages, 1981
- over 145.5
- 127.6 to 145.5
- 122.6 to 127.6
- 103.6 to 122.5
- 0.0 to 103.5

England & Wales = 100

Fig. 1.5.19c

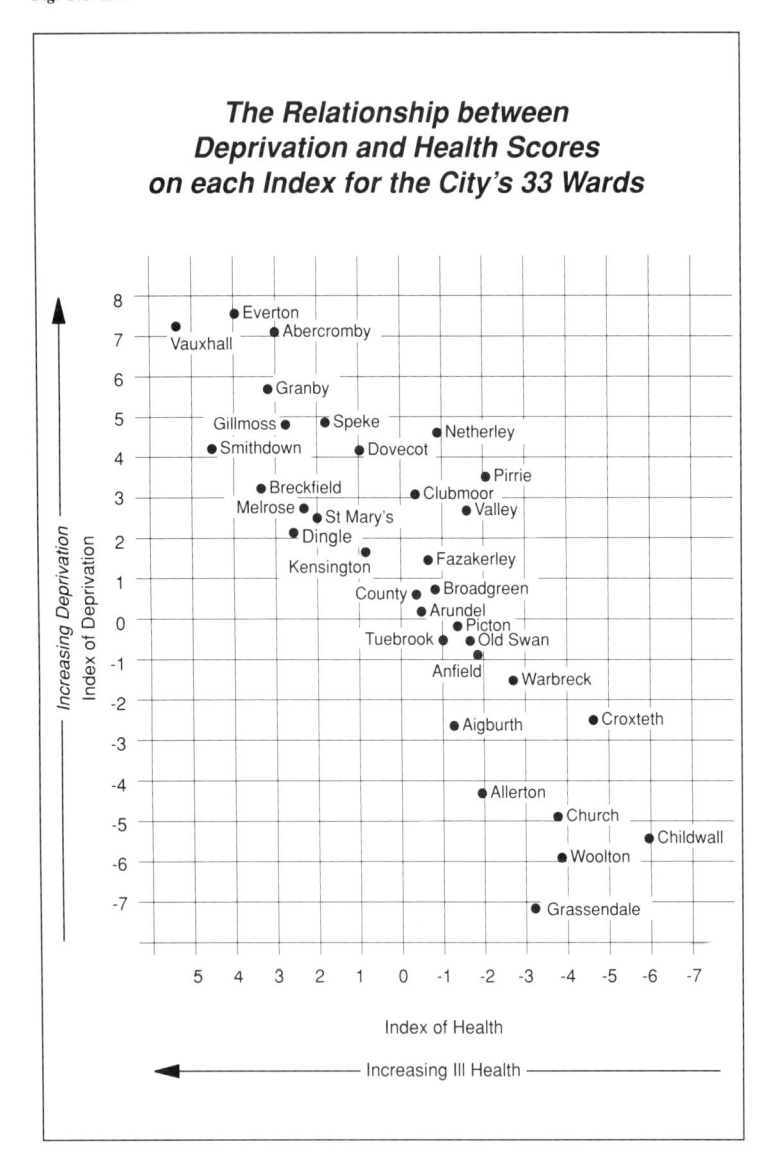

The Relationship between Deprivation and Health Scores on each Index for the City's 33 Wards

IS THE PROBLEM OF DEPRIVATION CONFINED TO THE INNER CITY? A VIEW FROM SCOTLAND

It will have not escaped notice that the worst areas for deprivation in Liverpool are not confined to the inner city wards. Fig. 1.5.20 shows that two wards, Speke and Gillmoss which suffer from high levels of deprivation are on the edge of the city. Both of these areas are mainly occupied by large council estates.

Sprawling local authority estates are commonly found on the outskirts of Britain's cities. Research has been done on deprivation in Scotland's cities to see if the problem of deprivation, and associated difficulties is also common on the outer estates.

Three maps (Fig. 1.5.20) show levels of multiple deprivation (derived from 59 different variables) in the three largest Scottish Cities). Multiple deprivation is shown in census enumeration districts, and is mapped according to whether each district is in the worst 0–1 percent, 1–5 per cent, 5–10 percent and 10–20 percent.

1. Study each of the three maps. Make tracings of the outline of the boundary of each of the city areas. Then circle those areas that appear to have the greatest concentration of deprivation.

2. Comment on your results. Do they seem to confirm the idea that deprivation is now, in Scotland, as much a problem on the big outer city estates as it is in the inner city areas?

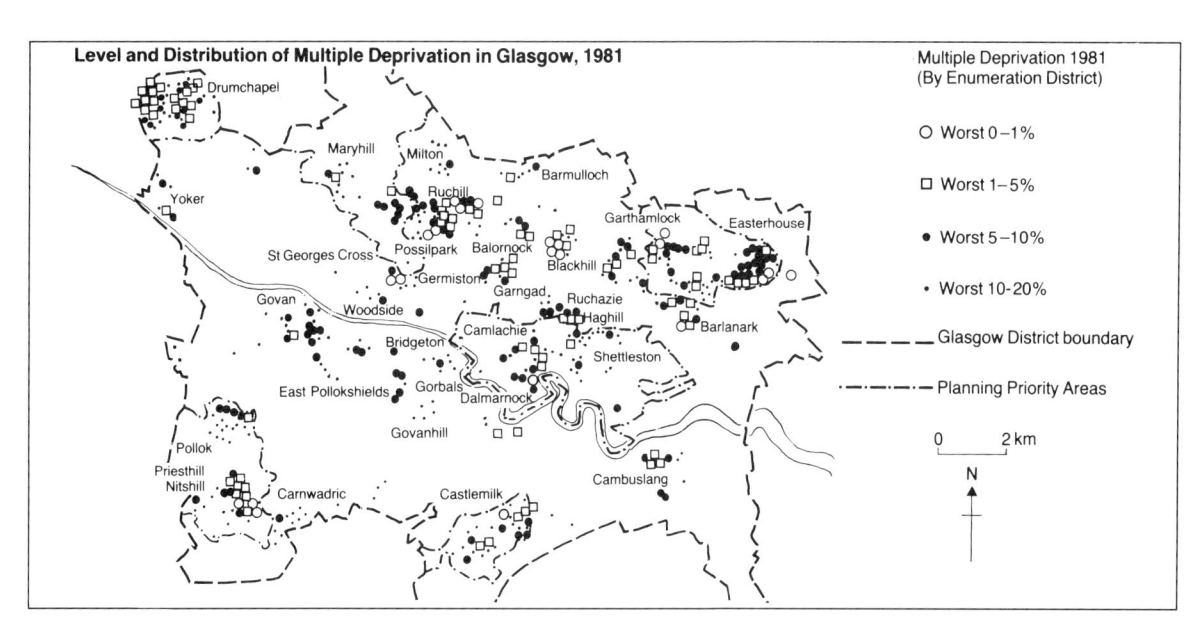

Fig. 1.5.20
Scottish Geographical Magazine volume 105 November 2 1989

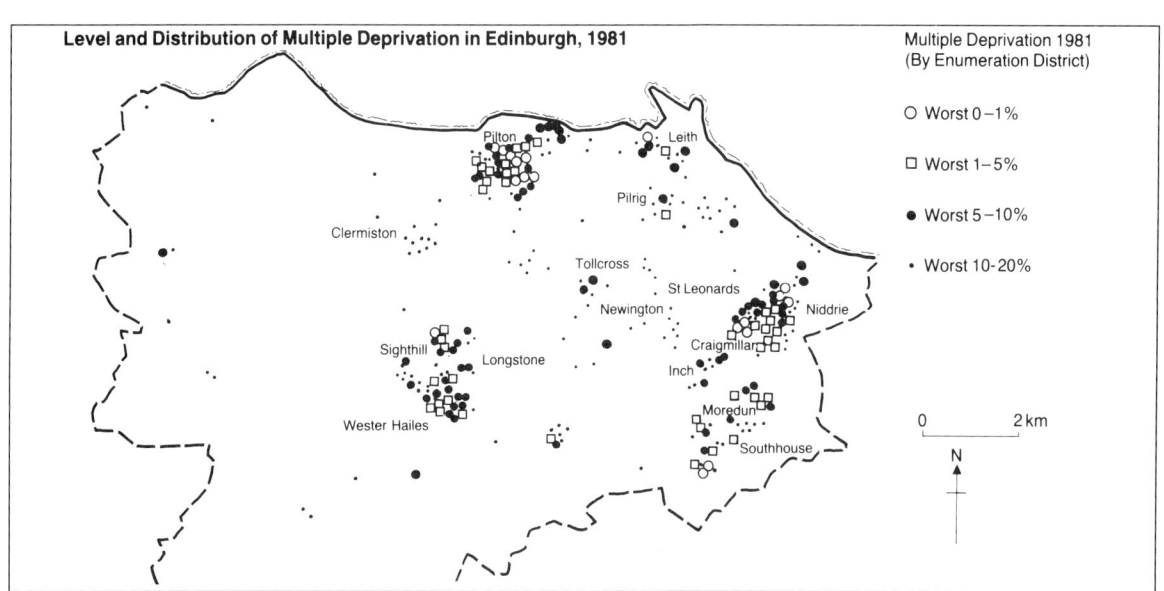

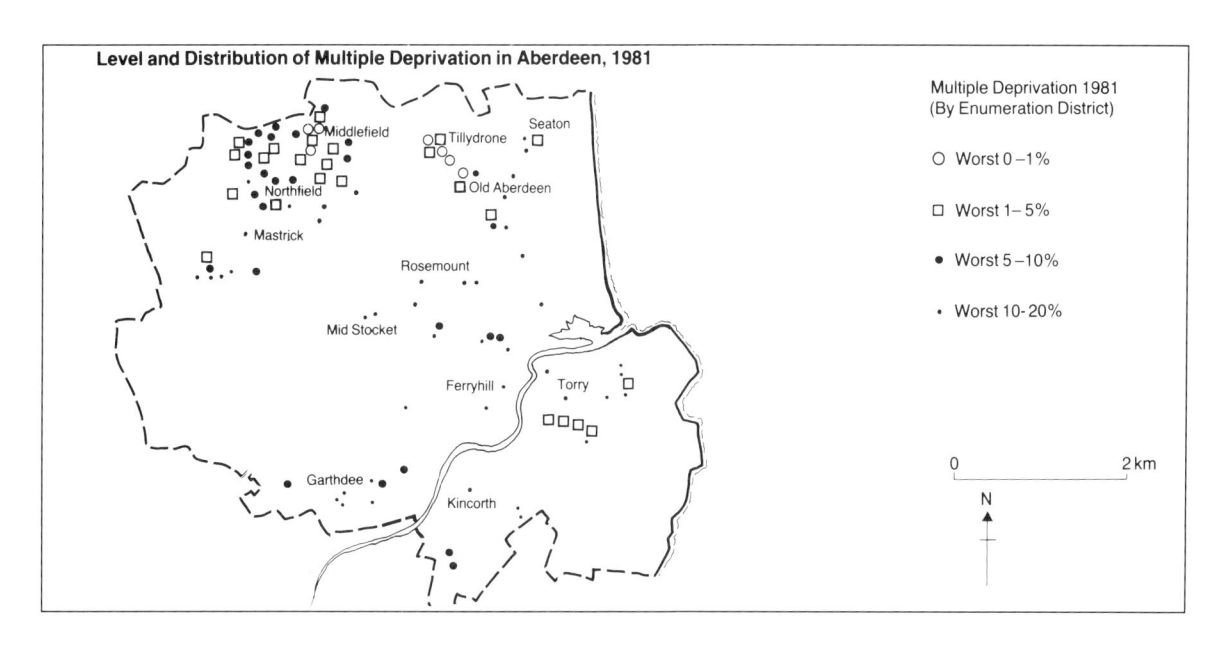

THE OUTER ESTATES

In the previous section on the inner city, the exercise on multiple deprivation in Scottish cities suggested that, although deprivation in the inner city was far from being eradicated, it was also a major problem on the huge outer council estates in our major cities. Meadowell, in the introductory section is a case in point. Many of these estates date from the post-war period, when there was an urgent need to rehouse people from the inner areas, either as a result of slum clearance, or wartime damage. Professor Duncan McLennan, wrote,

'The poorest home owners do not live in the heart of the city. At best the conventional wisdom of core decline which has influenced policy perceptions in the 1980s is not proven', Paying for Britain's Housing, a 1990 report.

In 1981 riots were widespread in Britain's inner cities – in Toxteth, Liverpool; in St. Paul's Bristol; in Chapeltown, Leeds and in a number of other cities. Ten years later similar riots were to occur, but in a different location – in the outer estates in Newcastle, Oxford and Cardiff.

BLACKBIRD LEYS ESTATE: OXFORD

Blackbird Leys estate lies some six miles from the centre of Oxford (see Fig. 1.6.1). It was built on a greenfield site in the late 1950s to accommodate families from slum clearance in inner Oxford and workers in the car works at Cowley (Fig. 1.6.2). Similar to other outer estates it appears to be well-planned but lacks any real centre, apart from a small shopping parade and a community centre. There are two pubs, a sports centre and a youth club, two primary schools and a middle school. The population grew to 11 000 in the late 1960s but has now declined to about 9000.

In the first week of September 1991, Blackbird Leys achieved national prominence through the publicity of 'hotting' – the racing of stolen high performance cars along its streets. This led to some violent incidents and a high police profile in the estate.

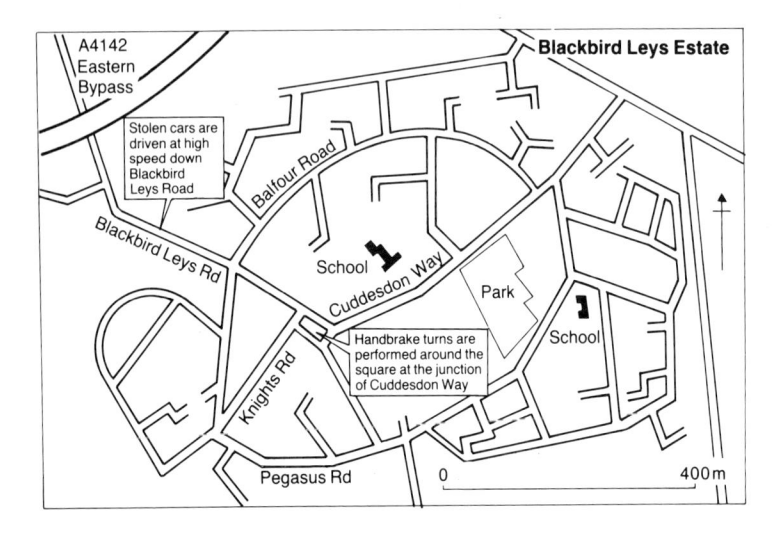

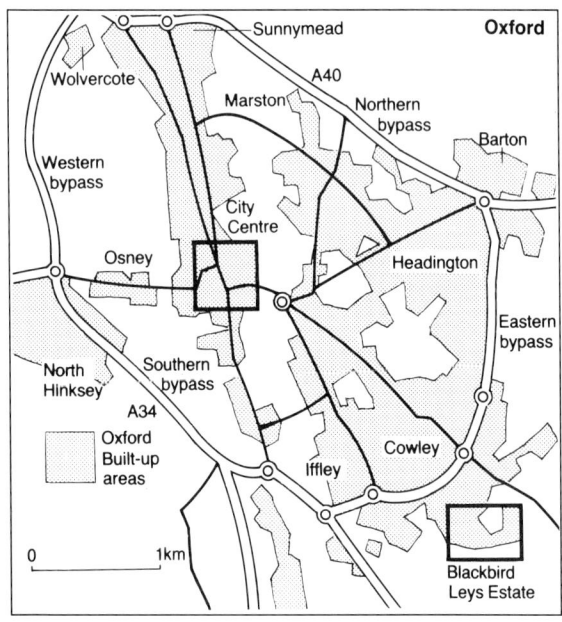

Fig. 1.6.1
Independent 3 September 1991

THE OUTER ESTATES

Fig. 1.6.2
(a) Shopping Centre: Blackbird Leys Housing Estate, Oxford (b) Blackbird Leys; note ramps in road to prevent racing of stolen cars

Opinions of Blackbird Leys:

Rev. James Ramsay: 'The background problems of Blackbird Leys need to be taken into account; boredom, little money and overcrowding. Levels of violence have increased markedly following recent job losses at Rover. There has been no investment for retraining in the community. The pressure on family life is tremendous. The numbers of couples splitting up has risen considerably. The estate is outside the Oxford ring road and there are only two access roads. This leads to the area being blocked off, which creates a feeling of being separated from the prosperity of Oxford.'

Val Smith, city councillor: 'I feel entirely safe in Blackbird Leys. Many people who were removed here after inner-city slum clearance were given the option of returning to the centre of Oxford when redevelopment there was completed. Very few decided to.'

Elderly resident: 'You know, despite everything that is going on, I don't think I could have finer neighbours. I would trust them with the key of the house and they always call and ask if I need anything from the shops. I intend to die here peacefully.'

55-year old woman: 'I'm a bundle of nerves. It's not safe to come out at night. I look out of my bedroom window and see them skidding the cars and I saw the riot from the same window. There were petrol bombs and shouts and broken glass.'

Peter Snow, Oxford journalist: 'In Oxford terms at least, Blackbird Leys is the pits ... a quarter of its residents are terrified to walk the streets at night, one in five of its boys 10 to 16 have been apprehended by the police.'

Spokesman for the 'hotters': 'The youth club is naff. There's no harm in what we do. We don't kill anybody. We're not on drugs. We're not criminals. The police should keep out of it. It's none of their business'.

Unemployed Blackbird Leys dweller: 'It's not about hotting any more. It is about who controls Blackbird Leys. It's our territory. We live here. They just can't come with their riot gear. They're not wanted.'

Blackbird Leys resident: 'Well, the kids have got nothing to do, they've got to have some fun, haven't they? I don't see why the police can't leave them alone. All those people are insured anyway. I know they are losing their cars but it can't really hurt them. The kids aren't doing any harm.'

Superintendent Ralph Perry, Oxford Police: 'We have received countless complaints about the hotting from residents on Blackbird Leys. This has got to end and our tactics are working.'

(Sources: various)

The above statements illustrate a range of values concerning the crime, violence and unrest on Blackbird Leys housing estate.

1   Analyse the range of values held in the statements, and try to classify them into groups.

2   Are there any social and physical differences between conditions in the inner city and those on the outer estates?

3   How do they reflect conditions in urban society in Britain in general?

MANAGING URBAN TRANSPORT SYSTEMS

Fig. 1.7.1 Commuters

The calm before the storm.... The first commuters board the 07.57 from Orpington in Kent to London's Charing Cross, at the start of their gruelling daily journey into work on one of Britain's most overcrowded lines.... It might be all right if no-one gets on after Orpington. Though seats are full, there is sufficient leg room and just enough space to negotiate a newspaper. But at the first stop, Petts Wood, the deluge begins. A dark line of passengers, two or three deep greets the train. It squeaks to a halt, the doors are open and the human horde climbs in.... At the third stop, Elstead Wood, the real crush is on. People worm their way into any space that is available, wedged against doors, windows other bodies....

Edward Pilkington, The Guardian 24 November 1989.

The spectre of 'gridlock' looms large... Gridlock is a traffic jam that crosses back on itself to block off all possible escape routes. As one 'grid' of roads is blocked, so more junctions are closed off and a city is slowly brought to a standstill by the volume of its own traffic. The situation in London is so critical that just one vehicle snarl-up at a strategic road junction could bring the city to its knees in certain conditions.... Local conditions in provincial city centres are little better.

Patrick Donovan, The Guardian, 4 January 1991.

The above quotations from newspaper articles are typical condemnations of the increasing chaos that affects Britain's urban transport systems. Neither private nor public transport seems capable of solving these problems at the moment. Car ownership has reached its highest level, yet in a city like London, only 14 per cent of all commuters drive to work. The remainder rely on a public transport system that increasingly fails to carry its passengers in reasonable conditions, or deliver them on time. In British cities parking a car has become a major problem for both shoppers and commuters, despite important building programmes of multi-storey and underground facilities. Suburban streets are equally crowded with vehicles parked because of inadequate garaging facilities.

Finding a solution to these problems has exercised town planners and traffic consultants throughout the post war-period. There seem to be four main lines of approach to the difficulties.

1   Introducing new road schemes that restrict access to certain areas, or speed the flow of traffic.

2   Introducing traffic management schemes that deter the private motorist from entering the central areas of cities, e.g. operating road pricing schemes, or park-and-ride schemes.

3   Integrating and streamlining present public transport systems so that they become more efficient.

4   Introducing new mass transit systems to provide a new level of low-cost public transport from the suburbs to the city centre, such as the supertram in Sheffield and the Metrolink in Manchester.

With forecasts of road transport increasing by between 27 per cent and 47 per cent by the year 2000, and by as much as 142 per cent by 2025, a number of new schemes have been put forward. The map (Fig. 1.7.2) shows the main trunk road plan for the London area.

Fig. 1.7.2

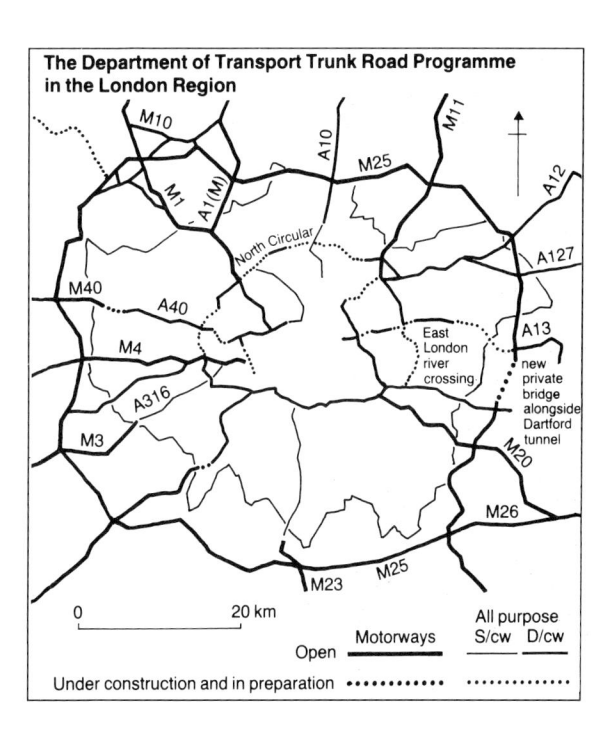

NEW ROAD SCHEMES FOR LONDON

London is grinding to a halt. Chronic congestion has cut average traffic speeds to 11 mph.... Car numbers in London will rise from 2.2 million to 2.8 million in the next decade. Motor traffic will grow by 1 or 2 per cent a year. The 'gridlock'... will become routine. The Independent on Sunday 1 May 1990

London has lived with its chronic traffic problems for a long time. In the 1960s and 1970s new ring and radial routes were advocated as solutions. Alternative emphasis was on the need to promote public transport in central London and concentrate on road construction in the outer suburban areas. With the abolition of the (GLC) Greater London Council responsibility for the trunk roads in London has passed to the Department of Transport. Traffic flows increased by seven per cent from 1983 to 1987 and peak traffic speeds had been reduced from 20.7 km per hour in 1972 to 17.6 km per hour in 1990.

During the period from 1984–90 four major assessment studies were carried out for the Department by traffic consultants. These proposed far-reaching changes in the network that would have resulted in the demolition of 6000 homes (and costing £3.5 billion) thus bringing a commercial blight on significant swathes of the capital. In 1990 the four assessment study area proposals were abandoned.

Some of the major road schemes outside of the proposals, such as the Hackney–M11 link and the East London River Crossing were still to go ahead (see Fig. 1.7.4). However, the Government decided to abandon the plans for the East London River Crossing in July 1993, and the environmental damage in Oxleas Woods was averted.

One survivor from the schemes of the 1980s is the Red Routes Plan, which was inaugurated on 7 January 1991 (see Fig. 1.7.5).

THE TIMES MONDAY JANUARY 7 1991

Red double-deckers, not red routes

John Adams believes London's new clearways, introduced today, will add to the jams and delay confronting the car

Since the formal abandonment of London's motorway programme almost two decades ago, tens of millions of pounds have been invested in an incoherent search for a solution to the city's transport problems. The four and a half miles of "red route" inaugurated today are the pathetic culmination of all this effort.

The red route is the most recent mutant thrown up by the process of policy evolution. It is a stunted version of the nearly extinct species known as the urban motorway, and can be distinguished from its cousin, the urban clearway, mainly by its marginal markings, which are red instead of yellow. Like its predecessors, its most notable characteristic is that it provides more road space for cars.

So far it runs only from Hampstead Garden Suburb to Islington, via Archway Road. Soon it is to be extended eastward and eventually will cover 300 miles of road that the department refers to as London''s Priority Route Network.

The department claims that this network will benefit not only motorists, but pedestrians, cyclists and bus users. It may help a few, but for most people who live, work and shop along the red routes, life will get worse as more traffic is inevitably attracted to the extra road space. Since almost all the vehicles using the red routes will begin and end their journeys on the other 8000 miles of road in London that are not red, most of these other roads will also become more congested, noisy, dangerous and polluted.

The impossibility of meeting demand in urban areas by building more roads has been accepted by virtually all transport planners for at least 20 years. Clearways and red routes are worse than pointless. They will make existing traffic problems worse, and postpone the day when the underlying problem can no longer be ignored.

Most of London was built up around a highly efficient rail system that was supplemented by a bus service for shorter journeys. For those using this system the centre represented a peak of mutual accessiblity. Policies fostering dependence on the car have turned it into a sink of congestion.

Even if it were possible to rebuild London completely to accommodate the car, the experience of southern California suggests that the result would be a social and environmental disaster. London's transport problems can be solved only by improving the public transport system that served it so well in the past. A recent report by the (Conservative-dominated) London Boroughs Association demonstrates convincingly that the cheapest, quickest, and most effective way of doing this is by taking road space from cars and giving it to buses. An extended, rigorously enforced network of bus lanes, with priority at junctions, would simultaneously improve the bus service, relieve overloaded rail services, and reduce car traffic.

For such a policy to succeed without draining the life out of the inner city, one further step is essential. If the welcome being extended to motorists in suburban London and beyond is not withdrawn, people and jobs will abandon the inner city for low density suburbs. The East London river crossing and the upgrading of the North Circular Road, proposals to double the capacity of the M25, the greatly expanded road building programme in non-urban areas, and the continued granting of planning permission for car-dependent commercial and residential developments are all undermining attempts to improve London by restraining the car.

What has changed dramatically since 1973 is not so much the argument about what to do with Archway Road but the political weight of the lobby for improved public transport and restraint of the car. Many more people now doubt the wisdom of policies that would increase further the country's dependence on the car. A growing minority advocate *reducing* this dependence.

The red routes could turn out to be the final mutation before the transport department's car-dominated policies for London become extinct.

The author is Reader in Geography at University College London.

Fig. 1.7.3
Times 7 January 1991

1 Read the article from *The Times* 7 January 1991, (Fig. 1.7.3).

2 What are the main criticisms of the red routes?

3 What recommendations does the author make about the improvement of transport systems in London?

MANAGING URBAN TRANSPORT SYSTEMS

Fig. 1.7.4

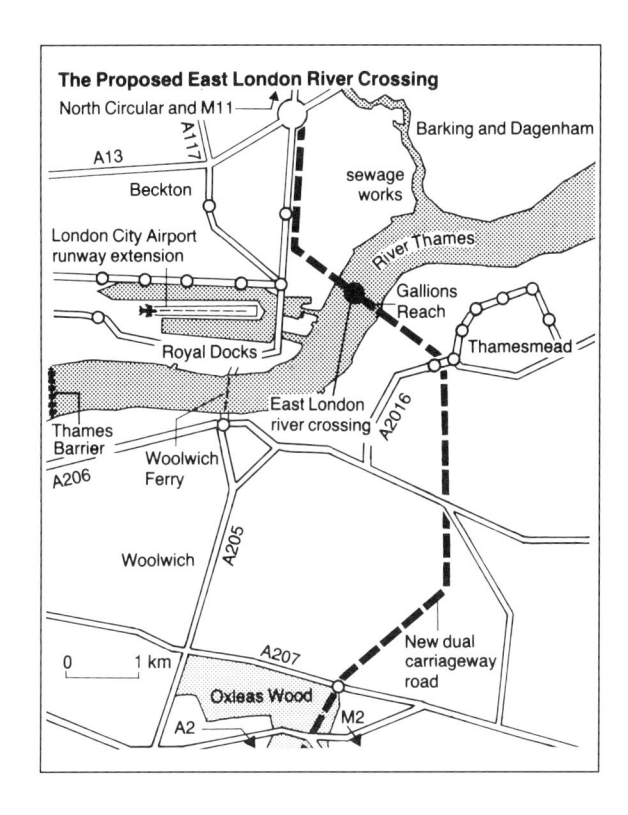

The Proposed East London River Crossing

PARK AND RIDE IN OXFORD

Like many provincial cities, Oxford suffers from severe traffic congestion. It lies at the intersection of two major trunk routes – the A40 London to South Wales road, and the A34 Southampton to Birmingham road. Successive by-passes have been constructed to reduce the flow of traffic through the city, culminating in the building of the ring road in 1967. In 1972 plans for a massive inner relief system, costing £250 million were put forward. High financial costs and the probable severe damage to the urban environment in Oxford led to the abandonment of plans for these new roads, and the creation instead of 'The Balanced Transport Policy' for the city. Since 1973 the following measures have been introduced as part of the Balanced Transport Policy:

- on-street parking controls
- restrictions on movement e.g. pedestrianisation
- cycle routes
- traffic management e.g. traffic calming measures
- support for public transport e.g. financial support for Park-and-Ride
- Tight planning controls on provision of parking spaces in the city centre.

Fig. 1.7.5
Red route: North London Highgate Hill: these routes were introduced to prevent on-street parking on some major through routes in London in January 1991

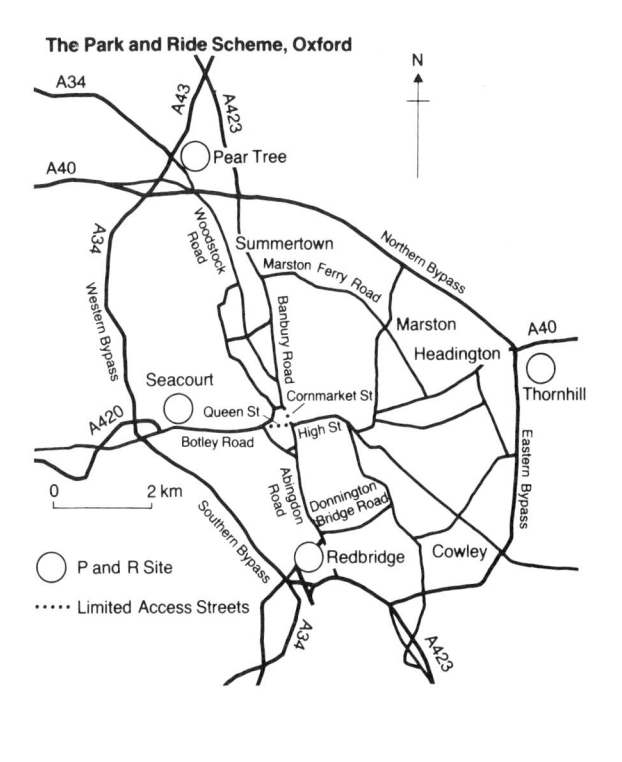

The Park and Ride Scheme, Oxford

Fig. 1.7.6

41

Fig. 1.7.7
Park and Ride in the City of Oxford: Thornhill Car Park on the eastern outskirts of the city. Commuters and shoppers park their cars here, and take the bus into the centre of the city where car-parking is severely restricted.

PARK AND RIDE

THE CAPACITY OF THE PARK AND RIDE CAR PARKS

Pear Tree	820
Redbridge	1250
Seacourt	550
Thornhill	485
Total Park and Ride	= 3105

Fig. 1.7.8b

The Park and Ride scheme is shown on the map (Fig. 1.7.6). It began in 1974, with the construction of the Redbridge park at the southern end of the city. Seacourt followed in 1974, Pear Tree in 1976 and Thornhill in 1985 (see Fig. 1.7.7). East-west, and north-south routes are now possible between the four different parks. The services (either limited stop express or hail-and-ride stopping service) are contracted out to different bus companies, after appropriate tendering.

1 How successful is the Park and Ride scheme in Oxford?

Using Figs. 1.7.8a and b, comment on the level of success of the scheme.

2 What criteria would you use for selecting new sites for car parking at the edge of the city?

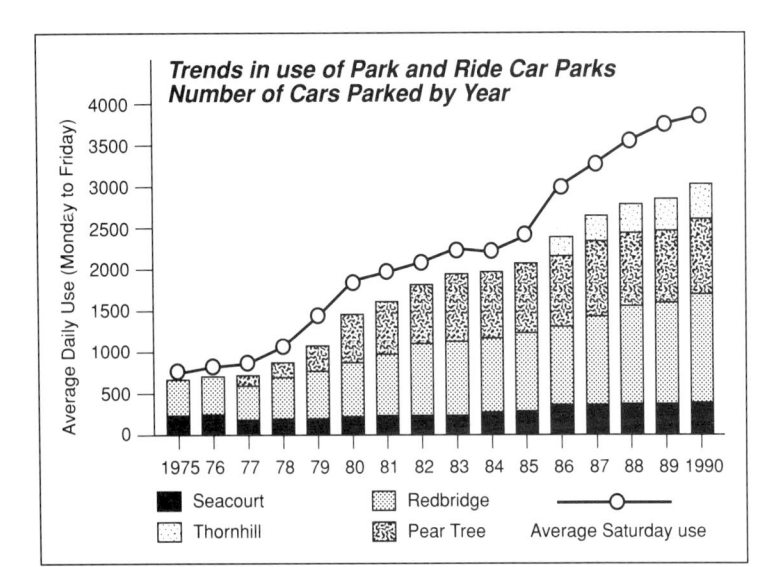

Fig. 1.7.8a

INTEGRATED PUBLIC TRANSPORT: MERSEYTRAVEL

Passenger Transport Authorities were set up under the 1968 Transport Act, charged with running and maintaining an integrated and efficient system of all public road and rail transport within their area. Merseytravel, set up under the Merseyside Passenger Transport Authority, is responsible for public transport throughout Greater Merseyside. It embraces four elements of travel in the region: the Mersey tunnels, the Mersey ferries, Merseyrail and the Merseytravel bus services. It operates in a difficult physical environment, serving the Wirral peninsula as well as the remainder of metropolitan Merseyside to the east.

Fig. 1.7.9
Merseyrail commuter train

In some ways Merseyrail (Fig. 1.7.9) is the linchpin of the system. One-third of Merseyside's population lives within a kilometre of a Merseyrail station, and the sytem is intensively used for access to the city centre. The network consists of three closely knit lines (see Fig. 1.7.10):

Merseyrail Facts and Figures

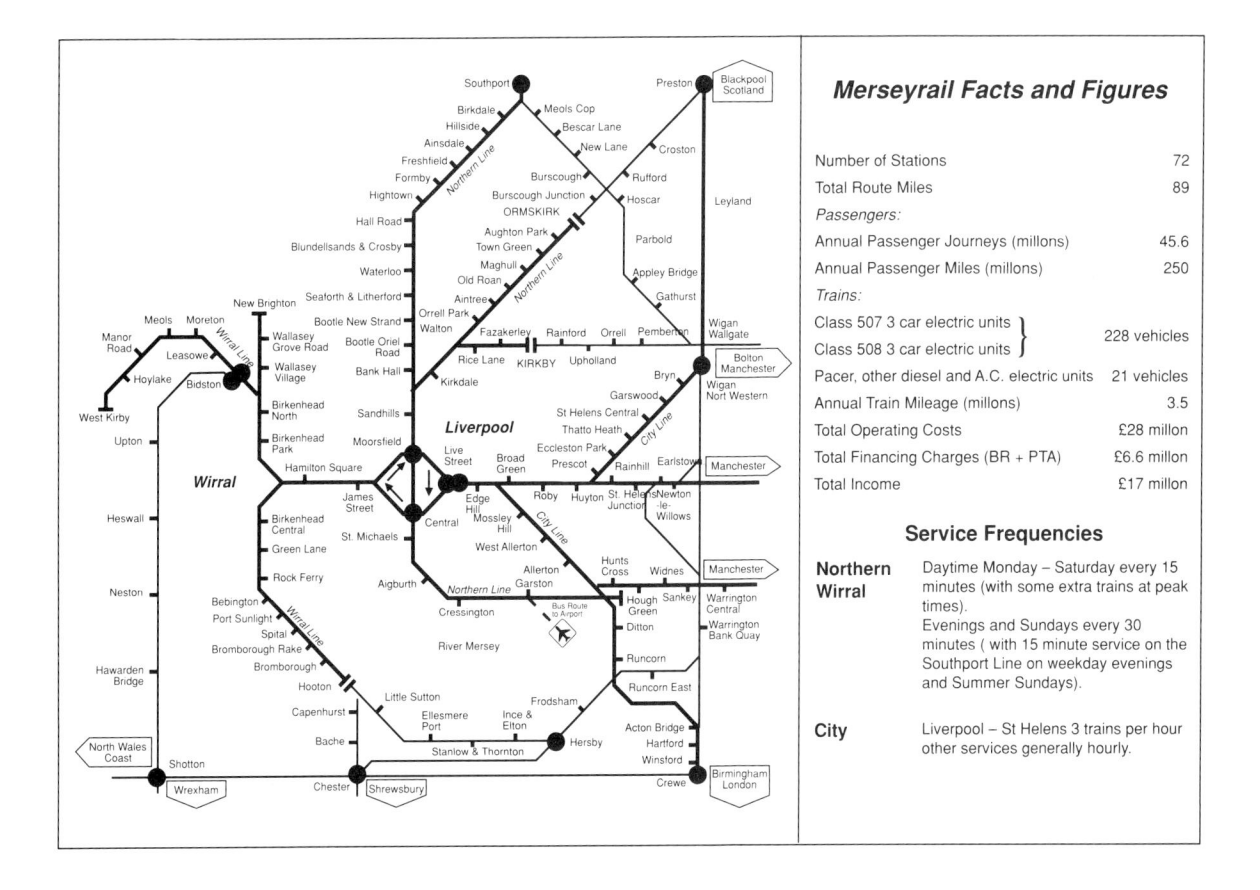

Number of Stations	72
Total Route Miles	89
Passengers:	
Annual Passenger Journeys (millons)	45.6
Annual Passenger Miles (millons)	250
Trains:	
Class 507 3 car electric units } Class 508 3 car electric units }	228 vehicles
Pacer, other diesel and A.C. electric units	21 vehicles
Annual Train Mileage (millons)	3.5
Total Operating Costs	£28 millon
Total Financing Charges (BR + PTA)	£6.6 millon
Total Income	£17 millon

Service Frequencies

Northern Wirral — Daytime Monday – Saturday every 15 minutes (with some extra trains at peak times). Evenings and Sundays every 30 minutes (with 15 minute service on the Southport Line on weekday evenings and Summer Sundays).

City — Liverpool – St Helens 3 trains per hour other services generally hourly.

- the electrified Wirral line that links Liverpool City Centre with New Brighton, West Kirby and Hooton

- the Northern line, wholly electric to Kirby and Ormskirk and between Southport and Hunts Cross

- the City line to Wigan and Crewe.

Current initiatives include, the opening of new stations such as Whiston, improvements to existing stations, 'Railside Revival' a scheme to promote an improved railside environment and a new centralised communication and control system for the Wirral and Northern lines.

On Merseyside 34 different bus operators undertake to provide bus services. Commercial bus services make up the bulk of the system, but a supported bus service makes up deficiencies in the commercial service. Specialised door-to-door buses serve the disabled. Two ferry routes cross the Mersey, from Liverpool to Birkenhead, and from Liverpool to Wallasey, though both operate at a deficit. Increased orientation towards the leisure/tourism market could improve their financial position. Both Mersey tunnels are operated by Merseytravel. Traffic seems to increase by about a million vehicles a year, which has now resulted in the tunnels operating at a profit.

Fig. 1.7.10

MASS TRANSIT SYSTEMS: THE MANCHESTER METROLINK

Increasingly, city authorities are looking to surface mass transit systems to solve their public transport dilemmas. Trams disappeared from city streets in Britain many decades ago, but their successful reappearance, albeit in more modern form, on the European scene, has convinced many transport committees in Britain of the need for this form of urban transport in the future. In cities such as Duisburg and Dusseldorf in Germany, and Grenoble in France, light mass transit systems have eased traffic congestion, and provided a more effective public transport system. A number of British cities including Manchester, Sheffield, Birmingham and Bristol have plans for these systems, and several others such as Leeds and Portsmouth are seriously considering them as an option (Fig. 1.7.11).

Fig. 1.7.11
Geography, Volume 77 (part 1) Number 334, Light Mass Transit Developments

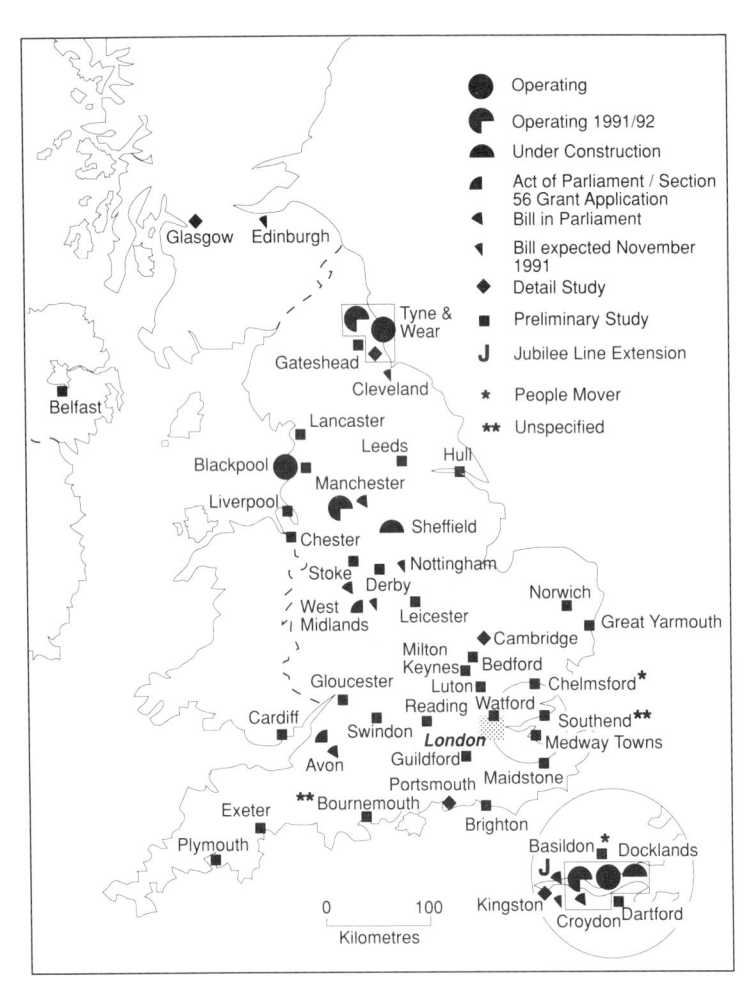

Fig. 1.7.12
Metrolink tram, Manchester. A new mass transit system, that will eventually link most of suburban Manchester to the city centre

The initial line will thus link the two suburban areas of Altrincham and Bury to the city centre, and as the map shows there are possible extensions planned to link other suburban and outer centres to the centre of Manchester.

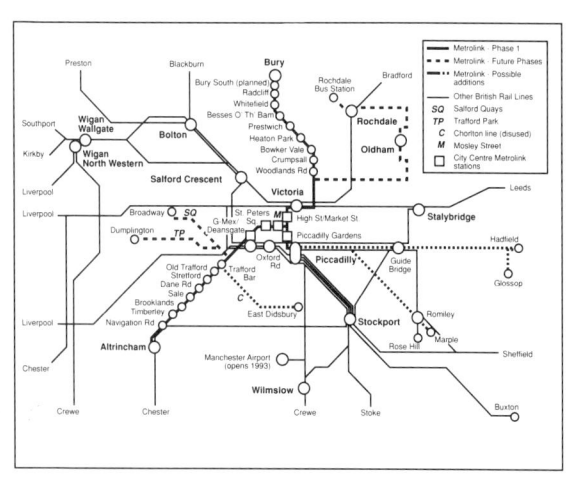

Phase 1 of the Manchester Metrolink is due to open in phases between February and June 1992 (see Fig. 1.7.12). It will convert 25 km of British Rail track to the north and south of the city centre, and link these sections with a new 2.5 km street level track running through the centre of the city (Fig. 1.7.13). The new Metrolink will run between Altrincham in the south and Bury in the north serving 18 stations on the British Rail network, and six new street level stations in the city centre. Interchange facilities will be provided with British Rail at Manchester's Victoria and Piccadilly stations. The Metrolink will operate at a frequency of five minute intervals during the peak period, and ten to fifteen minute intervals during less busy periods. Major engineering works have been necessary in the refurbishment of the Cornbrook viaduct, the building of a new viaduct at the G-Mex exhibition centre and some disruption has been inevitable in the city centre during the laying of the new tracks.

Fig. 1.7.13

Before embarking on a major new urban transport scheme such as the Metrolink, a wide range of preparatory studies have to be carried out in order to produce a consultative document.

1   Discuss the range and scale of the studies that would be necessary.

2   Which main points would have to be made in a draft introduction to the consultative document explaining why the Metrolink is necessary for Manchester?

PRESSURES AND CONFLICT AT THE URBAN FRINGE

The spread of urban growth into the countryside has been a constant theme for much of the present century. In the inter-war period so-called ribbon development along major transport arteries consumed significant areas of the countryside. The introduction of controls in the form of the designation of Green Belts around the major conurbations in the post-war period brought some relief to rural areas on the urban fringe. It did, however, transfer the problem to the areas beyond the Green Belt, which themselves came under pressure from the property developers.

The need for containment of urban growth in Britain was recognised as long ago as the beginning of the century by such people as Ebenezer Howard and Raymond Unwin. The Green Belt around London was officially designated and approved in the first development plans for the London area in 1947. Four main functions of Green Belts have been recognised:

- to check the spread of further urban development
- to prevent neighbouring towns from coalescing
- to preserve the special character of towns
- to assist in urban regeneration.

Green belts may thus be defined as 'areas of open and low density land use surrounding existing major settlements where urban extension is to be strictly controlled.' Green belts now cover 1.82 million hectares, having doubled in size since 1980, and now occupy some 12 per cent of England and Wales (see Fig. 1.8.1). There is no doubt that Green Belts have done much to contain urban sprawl, but one recent report suggests that they may have outlived their usefulness. The problems created in the more distant countryside have been outlined. It is suggested that some of the land in the Green Belt is now 'brown belt' – up to 20 per cent of London's Green Belt is damaged and of little environmental value. Up to 100 000 manufacturing jobs may have been lost in the Metropolitan Green Belt as a result of the restrictions on development. Some authorities, like Tayside Regional council, have actually abandoned the Green Belt around Dundee in favour of more general countryside measures. Pressures in the urban fringe are examined in three case studies:

one location lies on the eastern outskirts of Bournemouth on the urban side of the Green Belt, another lies within the Glasgow Green Belt, and a third lies in the zone beyond the Green Belt in Berkshire.

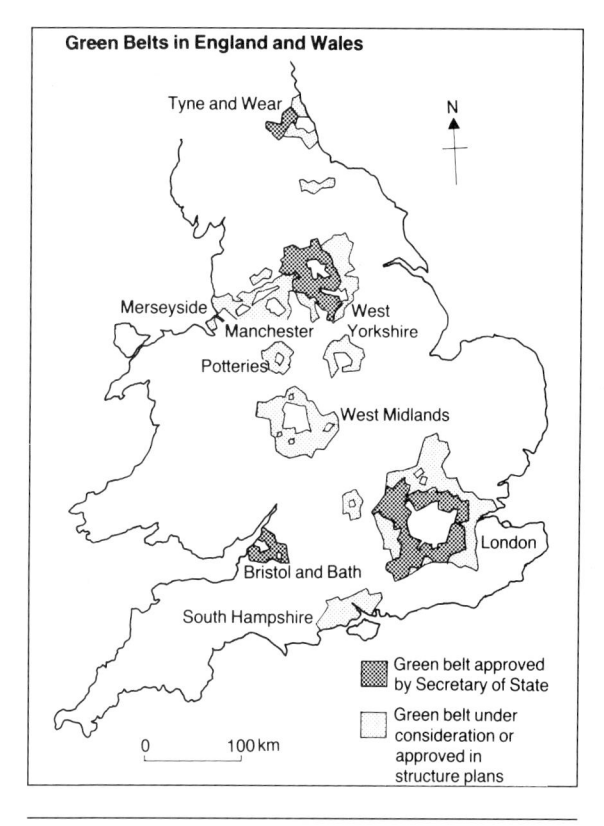

Fig. 1.8.1

DEVELOPMENT PRESSURES ON THE EASTERN FRINGE OF BOURNEMOUTH

The Bournemouth-Poole conurbation is one of the most rapidly expanding areas along the south coast of Britain, and is set to pass the half million mark early in the next century. Green Belt protects land around the conurbation. Within the belt to the west there is a variety of land uses, including the

A NEW VIEW OF BRITAIN

Fig. 1.8.2

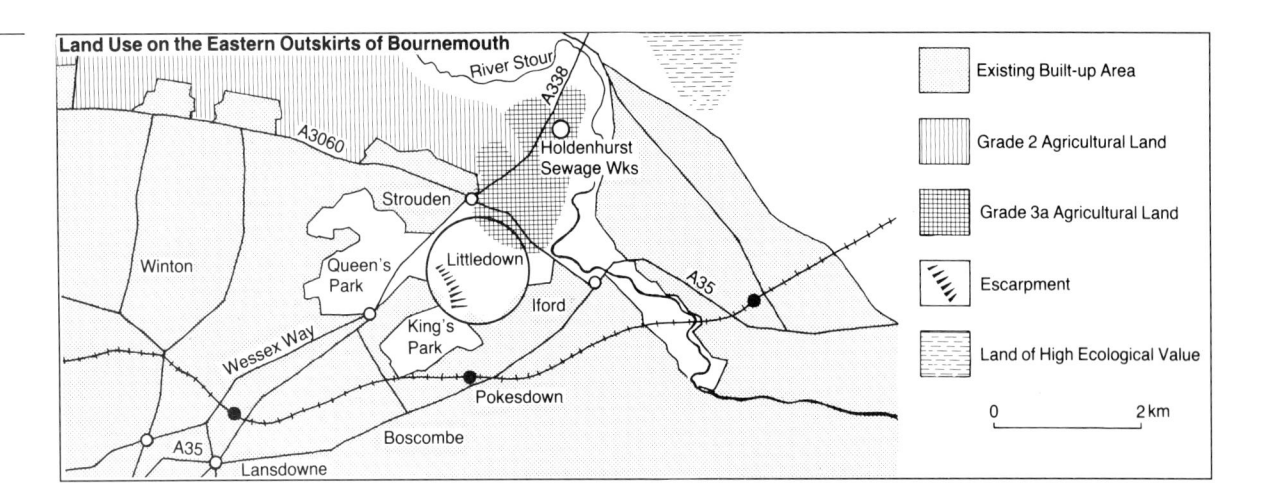

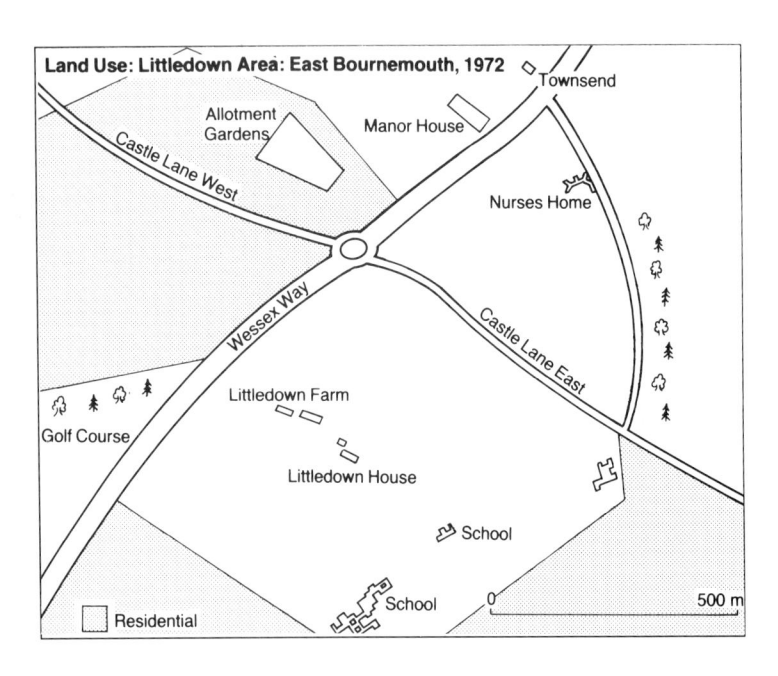

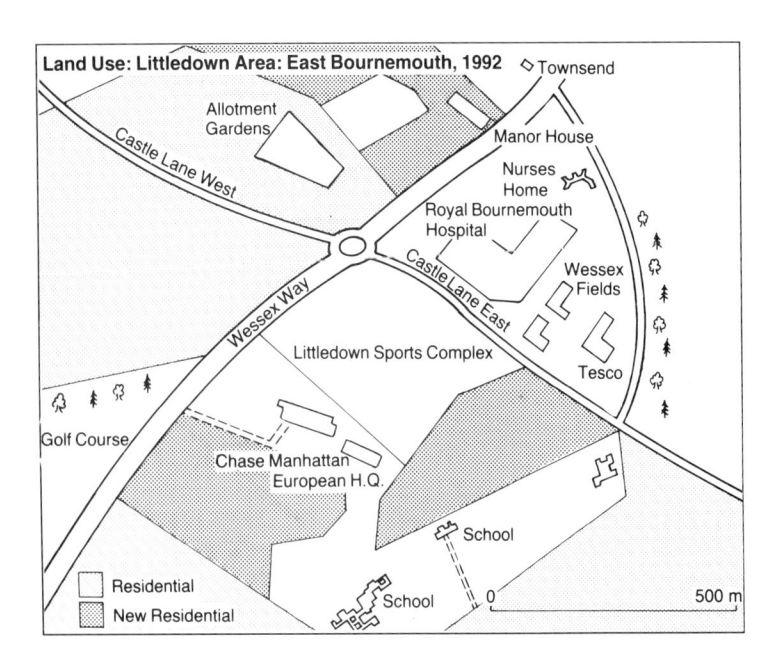

Land Use on the Eastern Outskirts of Bournemouth

Legend:
- Existing Built-up Area
- Grade 2 Agricultural Land
- Grade 3a Agricultural Land
- Escarpment
- Land of High Ecological Value

Land Use: Littledown Area: East Bournemouth, 1972

Land Use: Littledown Area: East Bournemouth, 1992

ecologically valuable and sensitive remnants of the Dorset heathland, whilst to the east the belt extends towards the New Forest. The map (Fig. 1.8.2) shows the variety of land use on the eastern edge of the urban fringe of the Bournemouth area. Of particular interest is the changing land use of the former Littledown Estate, indicated on the map by the circle. It lies just inside the inner edge of the Green Belt to the east of Bournemouth.

The larger scale maps (Figs. 1.8.3 and 1.8.4) show the way in which land use has changed in this area between the early 1970s and the early 1990s (Fig. 1.8.5). The Littledown Estate was originally farmed as one unit which embraced land to the north and south of Castle Lane. The land to the north of Castle Lane was sold off to be used for housing, the building of a new district hospital and an office campus; much of this work has now been completed, together with the building of a new Tesco superstore (which opened in December 1991). The land to the south went through a series of development proposals, but the final resolution of the land use is seen on the 1992 map. Two large areas have been devoted to housing, another to the Littledown Centre (a major sports complex), and most significantly, the remainder of the land is now occupied by the European Headquarters of the Chase Manhattan Bank (which has created 1500 new jobs in the Bournemouth area).

Fig. 1.8.3 & 4

PRESSURES AND CONFLICT AT THE URBAN FRINGE

(a)

(b)

(c)

(d)

Fig. 1.8.5
The Littledown area on the eastern fringe of Bournemouth in the 1990s: different types of land use
(a) amenity land (b) Wessex Fields: new office accommodation (c) the new Royal Bournemouth Hospital (d) new Tesco superstore under construction (opened December 1991)

1 What would have been the main pressures influencing Bournemouth Borough Council's decision to develop this area and change its land use from farming and amenity, to housing and service industries?

2 Which local bodies would object to these proposals, and what would their arguments be?

CONFLICTS WITHIN THE GLASGOW GREEN BELT

The maps (Figs. 1.8.6 and 1.8.7) show the extent of the Green Belt around Glasgow, and the location of new private developments in Strathkelvin district. Although the need for a Green Belt around Glasgow was first recognised in 1946, final implementation was not until the 1980s. Local authorities in Strathclyde have maintained a very strict policy of resistance to the development of greenfield sites for residential purposes. In 1975 84 per cent of planning permission for private housing was on greenfield sites. By 1985 60 per cent of private housing

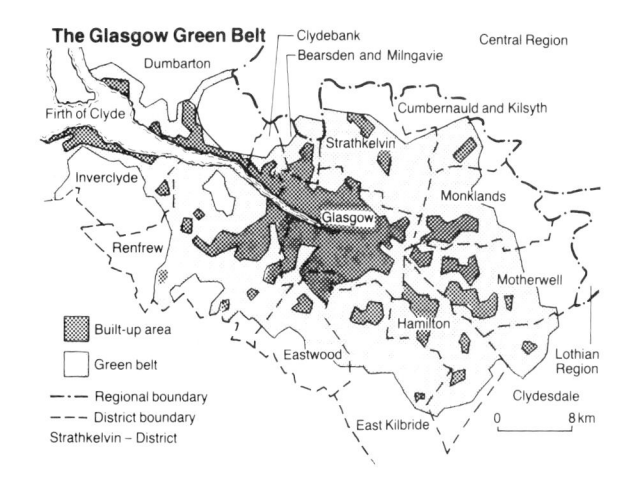

Fig. 1.8.6
M Pacione Journal of Rural Studies 1990/91

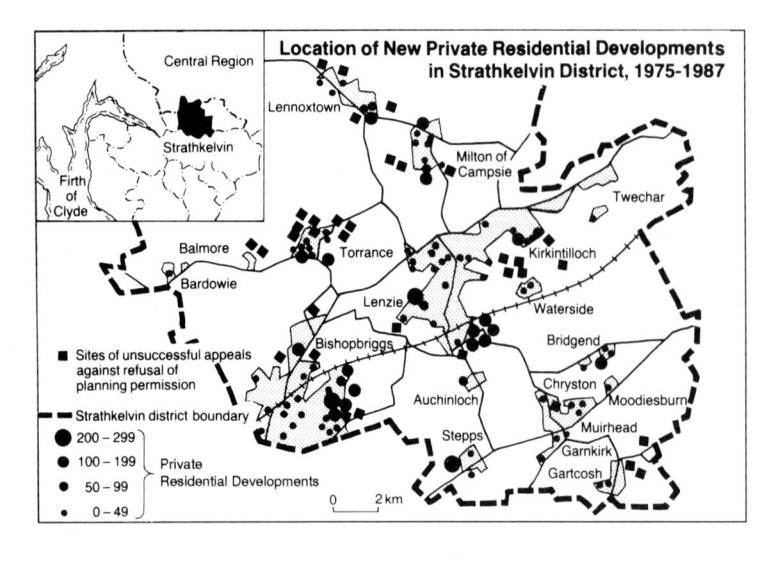

Fig. 1.8.7
M Pacione Journal of Rural Studies 1990/91

completions were on brownfield sites, and 90 per cent of planning permissions were on brownfield sites. Much of the development within the Strathclyde region in the last two decades has been in the Strathkelvin district.

1   Calculate the percentage change in the populations of Strathclyde region and Strathkelvin district.

2   Now study the map (Fig. 1.8.7). Comment on the pattern of new residential population and planning permission refusals that the map shows. Can it be related to the population changes shown in the table (Fig. 1.8.8) and the percentage changes that you have calculated?

POPULATION CHANGE IN STRATHCLYDE AND STRATHKELVIN 1961–1991

	POPULATION			
	1961	1971	1981	1991
Strathclyde	2 583 975	2 575 421	2 404 532	2 218 219
Strathkelvin	58 756	77 429	87 161	83 616

Fig. 1.8.8

One particular instance of conflict over Green Belt land was on the outskirts of the village of Torrance. In 1986 Strathkelvin District Council stated that it was the intention of the district council to bring to an end the growth of the village that had doubled the population in ten years. The Green Belt boundary had been drawn tightly around the edge of the built up area and no further extension to the village was proposed, be it in the form of a housing estate or even an individually constructed dwelling.

The map (Fig. 1.8.9) shows the location of the Tower Farm site in the village of Torrance. Originally the site was zoned for residential development and public open space in 1971. Developers purchased the land in 1973 and submitted plans in 1976 (which were refused). In 1980 detailed proposals for the construction of 28 houses on 2.4 hectares at the eastern end of the site were submitted. Both applications were called in. A public enquiry resulted in the refusal of both proposals. A major stumbling block was the need for additional sewerage infrastructure. The developers then offered to provide the sewerage, though the local authority would be expected to take responsibility for it eventually. In addition the developers offered sites within the proposed development for sheltered housing, a playing field and for recreational use. In the subsequent appeal the developers' case was built on the following arguments:

- a shortage of sites for good quality housing in Strathkelvin

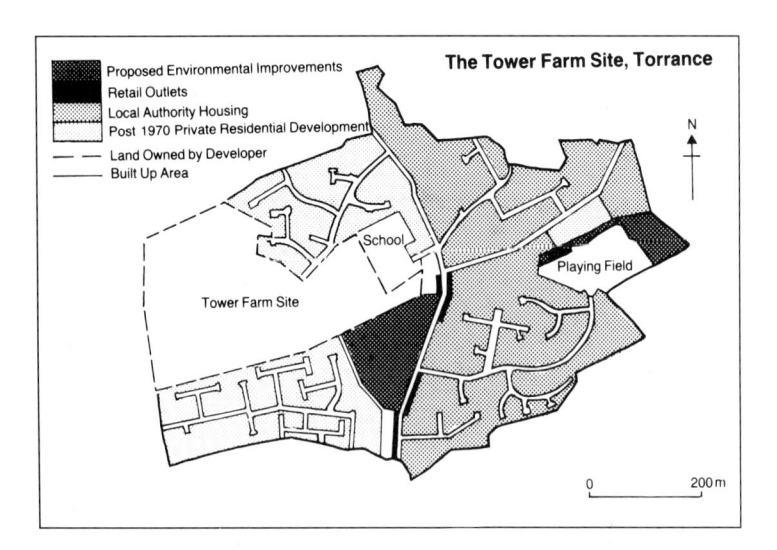

Fig. 1.8.9
M. Pacione Journal of Rural Studies 1990/91

PRESSURES AND CONFLICT AT THE URBAN FRINGE

- Tower Farm site was purchased on the basis of the 1971 zoning
- the proposed development was the logical rounding off of the village
- the major infrastructure problem could be solved with the construction of sewerage at the developers' expense
- part of the site could be used for sheltered housing and recreational purposes.

The appeal was subsequently lost in 1988.

1 Take each of the points made by the developers and reply to them from a planner's point of view.

2 List any additional points that would be made in support of refusing the appeal.

CHANGING LAND USE IN THATCHAM

Thatcham is a settlement in the western part of Berkshire, in the zone beyond the Green Belt shown on the map (Fig. 1.8.10). It lies in the middle of the 'sandwich' in Berkshire. In the eastern end of the county, the London Green Belt restricts development, in the west of the county there is an Area of Outstanding Natural Beauty ... 'the centre is disappearing under concrete'.

Study the diagram (Fig. 1.8.11) and the two photographs (Fig. 1.8.12).

1 Comment on the effectiveness of the technique used for showing the growth of population (households) in the central part of Berkshire.

2 Describe the land use in Thatcham in 1963 and 1986 as shown in the two photographs. What percentage of land was occupied by housing in 1963 and in 1986? What have been the other main land use changes?

What do these three case studies tell us about the different attitudes to housing and industrial development within Green Belts and in areas outside?

Fig. 1.8.12
Aerial photograph: Thatcham 1963 and 1986

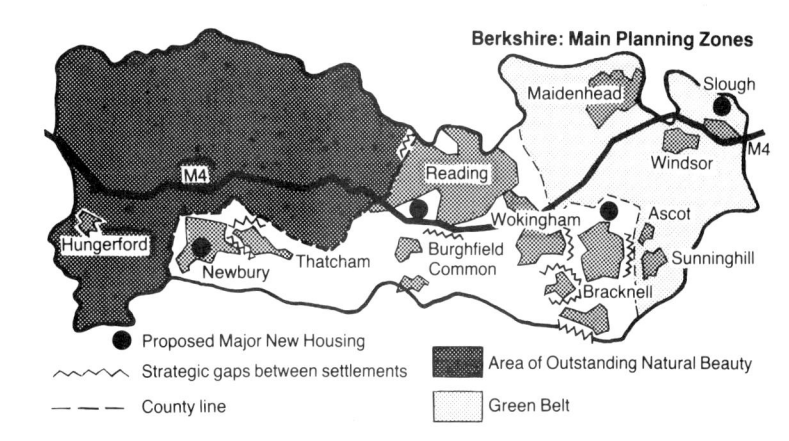

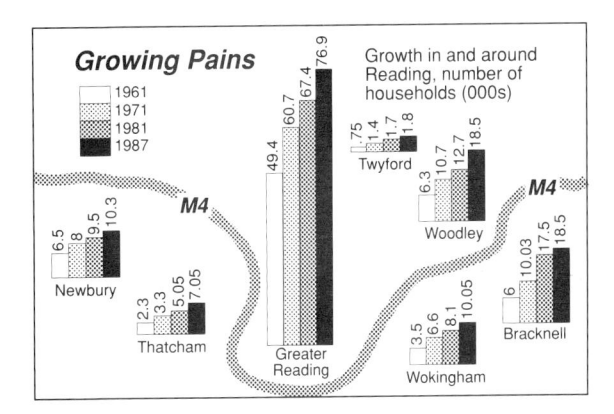

Fig. 1.8.10

Fig. 1.8.11
Observer 11 June 1989

COUNTERURBANISATION

In Britain the period since the 1950s has shown that there is no longer a tendency towards population concentration in the major conurbations, cities and towns. There has been a regular and progressive intensification of population movement away from these major centres. This movement of population into the more rural and less heavily populated areas is known as counterurbanisation. This is not just a reflection of people moving from the cities when they reach retirement age, since it involves the whole of the age range of the population, and thus must mirror the changing pattern of job opportunities. Decentralisation in industry, the increased use of private transport, and improvements in the trunk road system mean that there are much greater opportunities to live in the more rural areas, away from the large city concentrations. People are thus able to escape the increasingly congested and polluted cities, and enjoy the benefits of living in a cleaner and less frenetic environment. The movement of population to these reception areas will clearly have a number of important social, economic and environmental consequences.

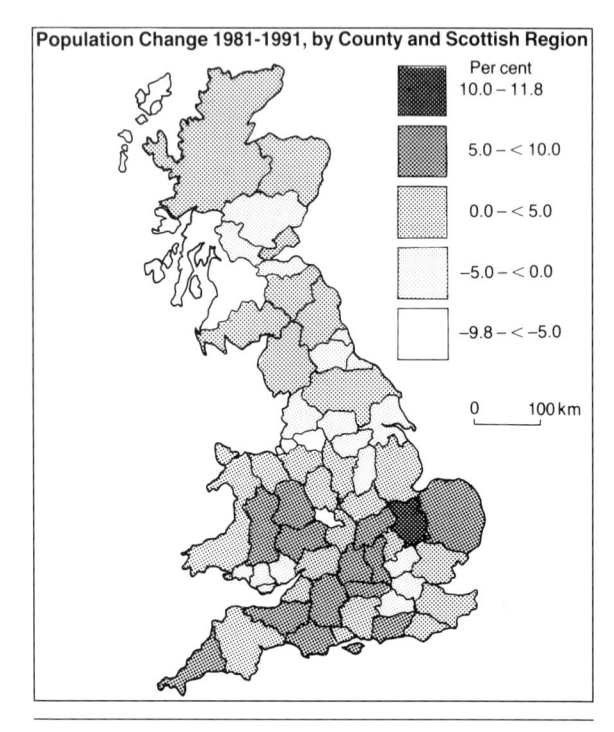

Fig. 1.9.1
Teaching Geography, January 1992

Study the map (Fig. 1.9.1) which shows population change in Great Britain for the period 1981–91.

1 What does the map tell us about the areas of greatest loss of population (more than 5%)?

2 Where are the areas of modest loss of population (under 5%)?

3 Where are the areas of greatest gain of population (10% and over)?

4 Where are the areas of modest gain of population (0–10%)?

5 Use your answers to assess the evidence for a continuing trend of counterurbanisation in the 1980s.

NEWCOMERS TO NORTH DEVON

The population of North Devon District is one of those areas shown in Fig. 1.9.1 that shows a modest growth of population in the period 1981–91. The table (Fig. 1.9.2) shows population change in South Molton District 1951–91.

Fig. 1.9.2
HMSO Census 1951, 1961, 1971, 1981 and 1991

POPULATION CHANGE IN SOUTH MOLTON RURAL DISTRICT 1951–1991

	1951–61	1961–71	1971–81	1981–91
Population change (no.)	−1041	−93	+1790	+1753
Population change (%)	−9	−1	+16	+13
No. of parishes with increasing population	2	9	24	23
No. of parishes with decreasing population	27	20	5	6

Preliminary figures for the 1991 census show that the larger settlements in the area, such as South Molton, have continued to grow, although this has not been matched in the smaller and more remote ones.

South Molton itself has maintained healthy growth into the early 1990s with an increase of 10.3 per cent in its population in the 1981–91 period. There has been an interesting growth of manufacturing industry in the district, much of it in South Molton itself. This has been largely the result of the Assisted Area status that it had until the mid 1980s. Similarly the service sector has grown, although not at the same rate as manufacturing. Most of the newcomer households were generally youthful or middle aged, only eight per cent were of retirement age. Eighty-nine per cent of the newcomers were owner-occupiers. Twenty-eight per cent of the newcomers were in the managerial group, many running their own businesses.

The map (Fig. 1.9.3) shows the newcomer households' place of former residence, and the table (Fig. 1.9.4) shows an analysis of reasons for moving to North Devon.

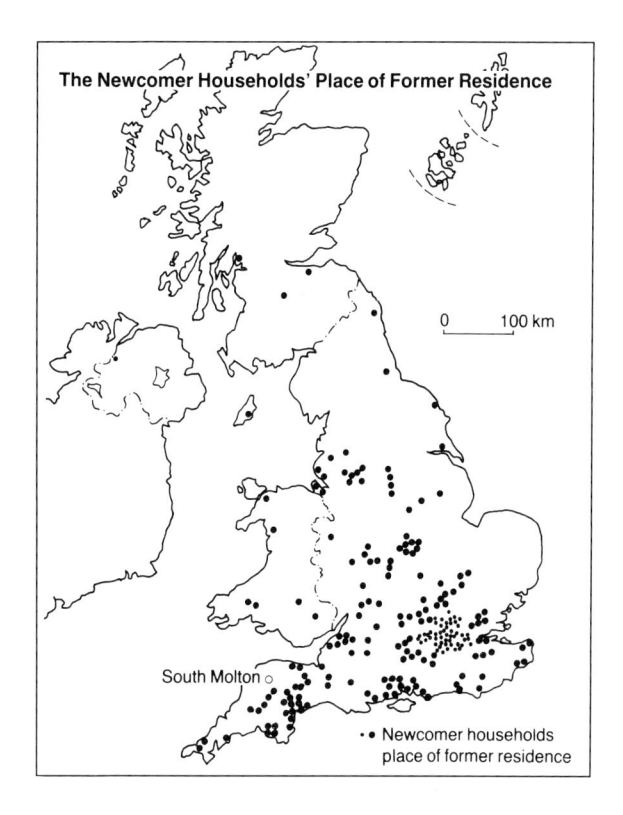

Fig. 1.9.3

Use the map (Fig. 1.9.3) and the table (Fig. 1.9.4) to write a short analysis of the spatial nature of the move and the reasons for moving. How might the two be related?

Fig. 1.9.4
Bolton and Chalkley, 1990

	REASONS FOR MOVING		
ECONOMIC REASONS	FOR LEAVING	FOR CHOOSING	FOR CHOOSING
	PER CENT	NORTH DEVON PER CENT	SOUTH MOLTON PER CENT
Move required by existing employer	10	10	0
Voluntary job/career related	12	30	29
Unemployment	10	2	0
Housing/property	5	15	54
Economic reasons Total	37	57	83
NON-ECONOMIC REASONS			
Lifestyle change	23	1	0
Retirement	14	0	0
Family/health	11	18	8
Social and physical environment	10	24	8
Non economic Total	58	43	16
Other reasons	4	0	0
Total	99	100	99

INDUSTRIAL BRITAIN

SOME VIEWS OF INDUSTRIAL BRITAIN

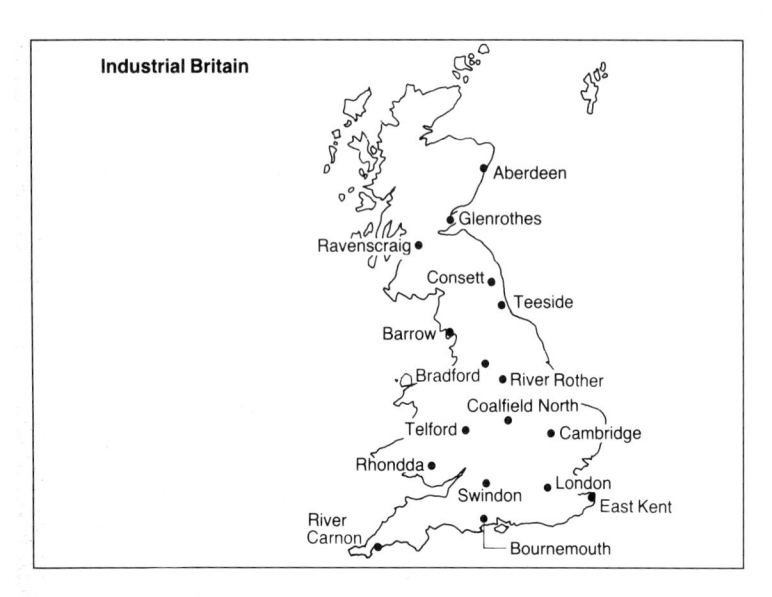

Industrial Britain

[Map showing locations: Aberdeen, Glenrothes, Ravenscraig, Consett, Teeside, Barrow, Bradford, River Rother, Coalfield North, Telford, Cambridge, Rhondda, Swindon, London, East Kent, River Carnon, Bournemouth]

Since the mid-1970s, the United Kingdom, in common with other industrialised European countries, has been experiencing one of the most traumatic periods of industrial change in its twentieth century history. (David Keeble in the preface to W.F. Lever, Industrial Change in the United Kingdom)

In the early 1990s the patterns of Industrial Britain are the result of a number of trends for change, some long-standing, some more recent.

- De-industrialisation has seen the absolute decline of manufacturing industry, with consequent job losses and stricken communities.

- Industrial revival, brought about by both private and public capital, has made an uneven impact, with some areas still suffering from the social and economic effects of large scale closures.

- Manufacturing industry now employs far fewer workers than the growing service industries (a trend referred to as tertiarisation).

- The changing nature and requirements of new industries have seen rural areas gaining new jobs more rapidly than urban ones.

- High technology has spawned its own manufacturing locations, and has affected working practices throughout industry.

- Concern for the environment is now a major factor in decision-making at all levels, although some industries have been much slower to adapt to new and more demanding standards than others.

Three introductory case studies are used to illustrate some of the faces of urban Britain in the early 1990s.

- Consett is an old steel centre, which suffered massive job losses in 1981 when British Steel closed their plant in the town (Fig. 2.1.1 and Figs. 2.1.2(a) and (b)).

- Telford, Shropshire is one of Britain's new towns, originally designated in 1963, and expanded in 1968. The planning and construction of the town was controlled by the Telford Development Corporation, which was wound up in 1991 (Fig. 2.1.3).

- Teleworking is likely to have an increasingly important effect on work patterns in the late 1990s and the early twentieth century (Fig. 2.1.4).

For each of the case studies, read the resource material, and where appropriate, study the diagrams.

1 Consett: Study Fig. 2.1.1; identify which of the trends or processes affecting manufacturing industry has been important at Consett. How successful have recent initiatives been?

2 Telford: What do the graphs (Fig. 2.1.3) tell us about the growth of industry, employment and population during the period of Telford's growth as a New Town? How does it reflect the trends indicated in the introductory paragraph?

Consett strives to keep disaster at bay

RECESSION WATCH By David Bowen

A VICIOUS wind was whipping snow along the streets of Consett last week, but inside the Derwentside Industrial Development Agency (DIDA) there was comfort to be found.

"There is a high degree of uncertainty about the first quarter of 1991," said Eddie Hutchinson, chief executive, "but quite a number of businesses are holding up exceedingly well. There have not been any major failures and I do not think there will be any."

It is 10 years since the steelworks that dominated this hill-top County Durham town were closed down. Since then, Consett has become something of a cliche as film crew after film crew has come first to measure the effect of the closure, which pushed unemployment up to 27 per cent, then to track the town's revival.

Through the combined action of local authority, government and the private sector, 190 companies have set up in or moved to Consett, cutting unemployment from 9 000 in 1982 to 3 400 now, and creating a diversified and modern manufacturing base.

The most famous arrival is Derwent Valley Foods, maker of "Phileas Fogg" snacks in Medomsley Road, Consett, as the advertisement has it. But its neighbours (actually on Number One Industrial Estate) include makers of supercomputers, nappies and plastic milk bottles.

The North-east has been more resilient to recession than southern England.

The principal reasons for this are the region's low costs and the fact that it missed the extravagant spending boom of the late 1980s. But Eddie Hutchinson believes that Consett's carefully reconstructed economy will give it two specific advantages.

Almost half the 5100 jobs created with DIDA help in the past decade have come from start-ups. That should protect Consett from the "branch factory syndrome" that used to affect it so badly.

When there was a downturn in the past, peripheral factories always suffered first. In 1980, British Steel closed its Consett works, Ever Ready slashed 1 000 jobs locally, and the ballbearing maker RHP closed its plant. This time, the syndrome is more likely to work the other way round.

The supercomputer maker Integrated Micro Products, for example, has slimmed down its Silicon Valley factory and concentrated all research and production in Consett.

Second, Mr Hutchinson says, many businesses have strong balance sheets based on long-term capital rather than bank lending. Much of this has been public money, from British Steel, British Coal and the European Steel and Coal Community, but most new companies have also been supported by venture capital.

"We have seen examples of venture capitalists doing everything to keep good businesses going when they wobble," Mr Hutchinson says. There are companies, though, that have found it difficult to renegotiate with their bank and have even had to switch from one venture capitalist to another.

This year, DIDA is unlikely to bring more than 30 projects to maturity, against the 35 to 40 it has been completing up to now. But he is confident that his long-term plan, to bring unemployment down to the national level or below by 1995, will not be derailed.

Fig. 2.1.1
Independent on Sunday 13 January 1991

Fig. 2.1.2
(a) Industrial Estate, Consett, Co. Durham (b) new Safeway superstore, Consett. Both of these locations are close to the old British Steel Plant at Consett, and illustrate the bringing of new jobs to the town

A NEW VIEW OF BRITAIN

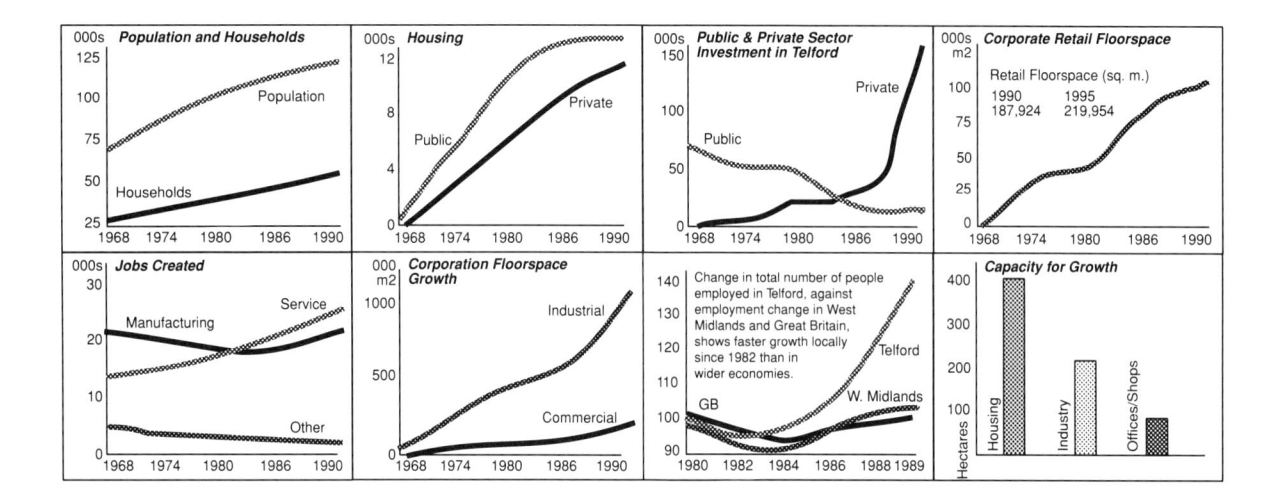

3   Teleworking: Study Fig. 2.1.4: how will teleworking be likely to affect the future pattern of industry in Britain? What impact is it likely to have on rural communities?

Fig. 2.1.3
Telford, Shropshire: Elements of Growth, copyright: reproduced by permission of the Commission for the New Towns

Fig. 2.1.4
Rural Focus, Volume 4 Issue 3, Autumn 1990

Teleworking in the Countryside

Mary downloads the last of her day's data-process assignment: the updating of a mailing list for a London-based professional institution. She signs off, closes down her PC, and leaves the house with a baby in the buggy to walk to the village shop.

On the way she passes the home of her friends, John and Anne, who have not yet finished their day's work. She can see the glow of the monitor screen through the window. They are into book-keeping and accountancy and maintain spread-sheet data on contract for several local businesses.

There'll be a chance to catch up with them during the evening at the pub where the talk is as likely to be about software, new electronic mailing methods and networks, as about village preoccupations.

Her husband Jim will make up a foursome for the evening. He's on his way back from a two-day advanced computer-skills enhancement course at the agricultural college. He has been on training release from his managerial job in the Teleworking Business Centre which is housed in converted farm buildings on the edge of the village. Part of his function at the centre is to run training courses in word processing and other Information Technology skills for villagers.

The village could be anywhere in rural England, and the time not so far into the future. This decade, in fact, when a sizeable proportion of the rural population could be earning its living through teleworking to customers in London and other British urban centres, and perhaps places like New York and Frankfurt, or by supplying Information Technology services to local businesses and community groups.

But home-based teleworking is not enough to ensure that rural England gains maximum benefit from the IT possibilities.

Consequently, a great deal of interest has focused upon an idea pioneered by the Swedes: the concept of the "telestuga', or, as it translates into English, telecottage. By which is meant a "workshop" equipped with computers, electronic mailing facilities, fax, photocopiers, desk-top publishing, and database access to provide an IT centre serving a village, or group of villages. It provides community services for local organisations, support for small businesses, and freelancing individuals, including "distance" teleworkers, and acts as a training centre.

The idea is catching on fast and could lead to rural England becoming one of the main national locations for Information Technology activities.

One centre that is up and running is the Moorlands Telecottage, opened on December 14 in the village of Warslow, near Leek, in the Staffordshire Moorlands Rural Development Area, and housed in a former middle-school library of what is now the Manifold Primary School. The funding was from the Rural Development commission, the Staffordshire County Council, the Staffordshire Educational Computer Centre and the Leek College of Further Education and School of Art.

DE-INDUSTRIALISATION

'I've heard of only two guys who've found other jobs – there's just no work around here any more. Everything seems to be closing or shedding labour.' Ravenscraig worker.

The dictionary definition of de-industrialisation is a dismal one: the cumulative weakening of the contribution of manufacturing industry to a country's economy. This is not just a matter of factory closures, industrial bankruptcies and heavy job losses, but it also concerns the future of whole communities, their morale, their well-being and their health. The signs of de-industrialisation in Britain began to appear in the 1960s and the trend since then has been remoreseless.

WHAT ARE THE MAIN INDICATORS OF THIS DECLINE?

Such indicators might include:

- the absolute loss of jobs in manufacturing industries
- the contribution of manufacturing industry to the Gross Domestic Product
- levels of investment in manufacturing industry
- overall levels of unemployment
- the balance of payments.

Study Figs 2.2.1 to 2.2.3 which show indicators of de-industrialisation in Britain.

1 Discuss the relative merits of these indicators as evidence for de-industrialisation.

2 Take each indicator in turn and assess the evidence for de-industrialisation.

3 What other indicators could be used to assess the amount and rate of de-industrialisation?

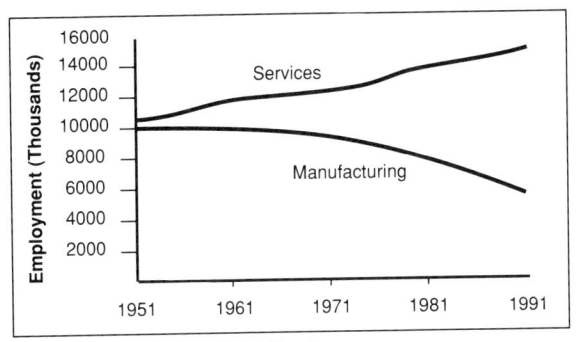

Employment in Manufacturing and Services

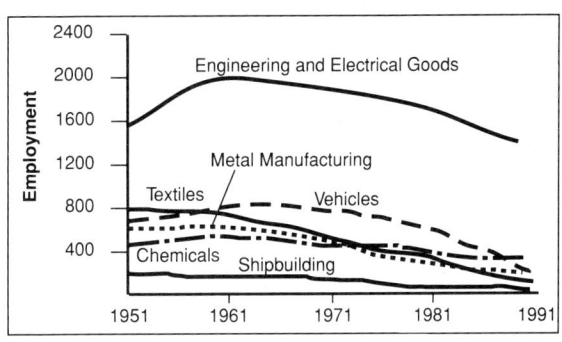

Employment in Selected Manufacturing Industries

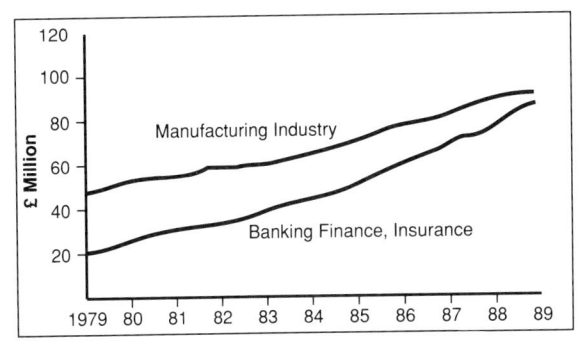

Contribution to GDP of Manufacturing Industry and Banking, Finance and Insurance

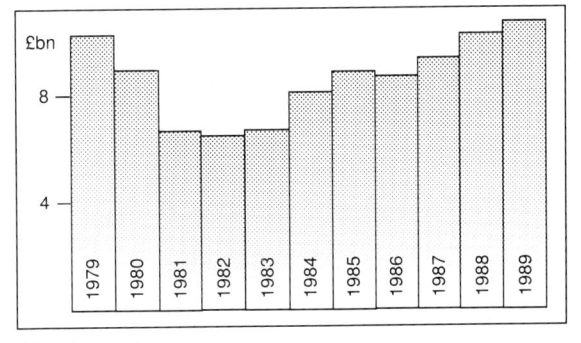

Manufacturing Investment

Fig. 2.2.1a–c

Fig. 2.2.1d
Geography, Volume 75 (Part 4), 1990

Fig. 2.2.2

Percentage of UK Workforce Unemployed 1931-1991

Fig. 2.2.3

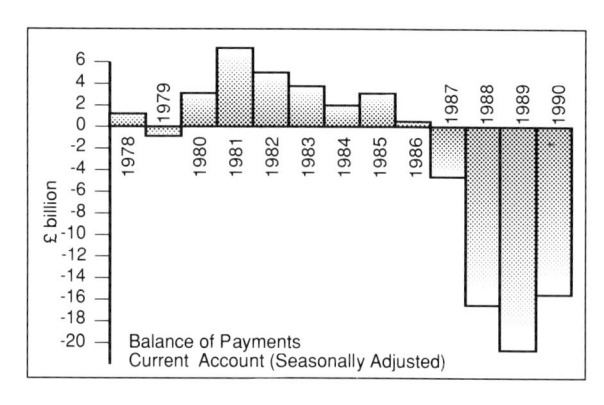

Balance of Payments Current Account (Seasonally Adjusted)

Fig. 2.2.4
Ravenscraig:
The end
approaches

It is now proposed to examine briefly the background to de-industrialisation in Britain. The main signs of an absolute decline in Britain's manufacturing industry began to appear in the 1960s. By this time there was evidence of decline in a range of important industries – iron and steel, engineering and textiles. The loss of jobs in these industries continued unabated into the 1970s, a decade of much harsher business conditions, when the world economy grew more slowly. Levels of demand fell, and, at the same time, there was increasing competition coming from the newly industrialising countries. Industry in Britain could respond in a number of ways. Reorganisation and restructuring of companies increased productivity and efficiency, but was bought at the price of considerable job losses. The role of the mass-production factory was questioned, and more flexible forms of working began to appear, with sub-contracting of supplies and services becoming significant. In the 1980s the trend continued, particularly in the recession of the early 1980s. From 1971 to 1989 2.8 million manufacturing jobs were lost, one million alone disappearing between 1979 and 1981. In the early 1990s there is no clear sign of an end to the process of de-industrialisation, with a second recession resulting in continuing job losses.

The bell of closure tolls early, but inevitably, at Ravenscraig

11000 Ravenscraig jobs at risk

Steel men fear survival deal will kill Ravenscraig

MacGregor lands £2bn steel deal to save Ravenscraig

A test of mettle as Ravenscraig closes

A £500m-plus steel rescue

RAVENSCRAIG: THE END FOR STEEL-MAKING IN SCOTLAND

Ravenscraig (Fig. 2.2.5) and its steel-making complex never seem to be far from the headlines. Scotland's only integrated steelworks has been under periodic threat of closure throughout the 1980s. Despite a promise to keep it open until 1994, British Steel announced in January 1992 that the plant was to close in September 1992, with a loss of a further 1 200 jobs from a plant that at one time employed 13 000.

DE-INDUSTRIALISATION

Ultimately British Steel closed Ravenscraig on 26 June 1992 which effectively brought to an end steel-making in Scotland.

The map (Fig. 2.2.6) shows the location of the five steel making complexes that were in production in the United Kingdom (early 1992), annotated to show the major sources of supply to the industry, and the broad areas of the major markets within the United Kingdom.

1 Using the information on the map attempt to rank the complexes in terms of their long term suitability as viable locations for manufacturing iron and steel and their associated range of products.

2 Why has Ravenscraig proved to be so vulnerable?

The decision to build a major steel plant at Ravenscraig, near Motorwell, was taken by the Macmillan Government in 1962. In many ways this was a political decision, since a simultaneous move was made to build a similar plant in south Wales, on the coast of the Severn Estuary at Llanwern. Thus, both Wales and Scotland were to get new steel plants. Although the decision may have been politically astute, it was very questionable economically. There was not a large enough market to absorb products of both of these plants, and they were thus faced with the threat of unviability from the start. Llanwern stood a better chance of survival because it was located near the large markets for

Fig. 2.2.6

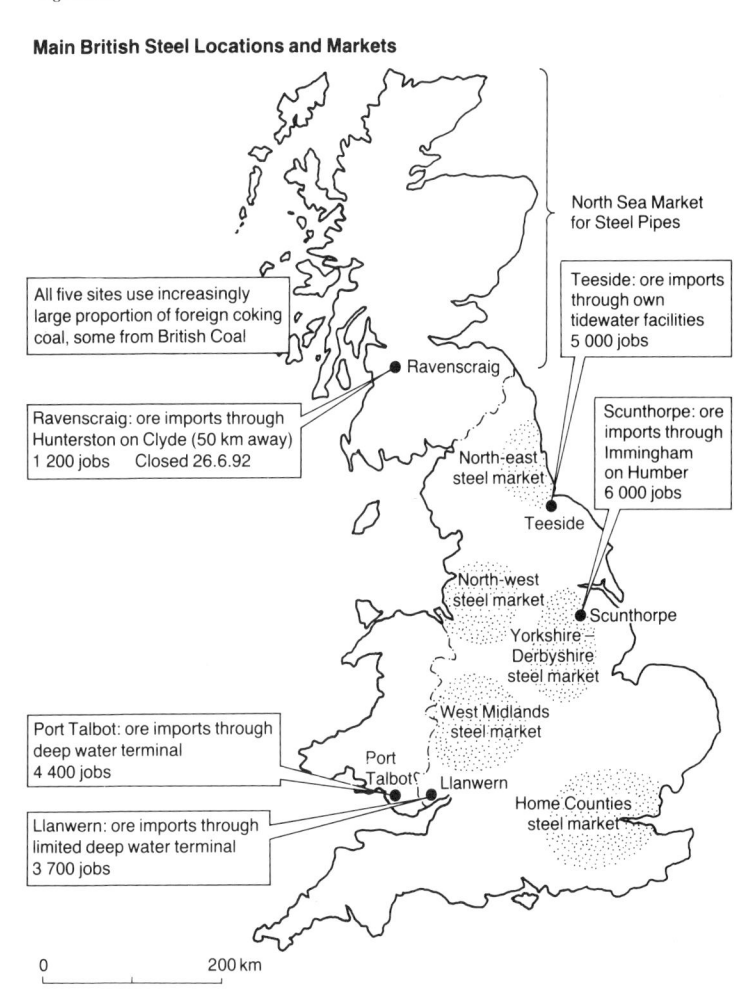

Main British Steel Locations and Markets

- All five sites use increasingly large proportion of foreign coking coal, some from British Coal
- Ravenscraig: ore imports through Hunterston on Clyde (50 km away) 1 200 jobs Closed 26.6.92
- Teeside: ore imports through own tidewater facilities 5 000 jobs
- North Sea Market for Steel Pipes
- Scunthorpe: ore imports through Immingham on Humber 6 000 jobs
- Port Talbot: ore imports through deep water terminal 4 400 jobs
- Llanwern: ore imports through limited deep water terminal 3 700 jobs

Markets shown: North-east steel market, North-west steel market, Yorkshire-Derbyshire steel market, West Midlands steel market, Home Counties steel market

0   200 km

sheet steel in the English Midlands. In order to create a market for the Scottish steel, the Government persuaded two car manufacturers to locate plants in Scotland at Linwood and Bathgate. Both plants survived for a number of years, but finally succumbed to the harsh economic climate of Scotland, thus depriving Ravenscraig of two of its major customers.

There was much concern over the future of Ravenscraig in the early 1980s, with even hints from the Scottish Minister that he might resign if Ravenscraig were to close. In December 1987, prior to the privatisation of British Steel, the company pledged, 'subject to market conditions', to keep open all its main steel – producing plants, including Ravenscraig, for at least the next seven years. However, in May 1990 it announced that the hot stripmill at Ravenscraig would close in 1991, and a further announcement came in June 1990 that the entire Ravenscraig operation would close in 1994, or earlier, depending on market conditions.

Fig. 2.2.5
British Steel, Ravenscraig which finally closed in the summer of 1992

Fig. 2.2.7

EMPLOYMENT IN SCOTTISH STEEL SCOTLAND 1974–1989

PLANT	1974	1982	1986	1989	JOBS LOST 1982–9	JOBS LOST 1974–89	NOTES
Clydebridge	2410	800	120	100	−700	−2310	
Clyde Iron	1191	0	0	0	0	−1191	closed 1978
Hallside	554	0	0	0	0	−554	closed 1980
Gartcosh	1164	637	0	0	−673	−1164	closed 1986
Glengarnock	1257	251	0	0	−251	−1257	closed 1985
Hunterston	0	116	100	100	−16	100	jobs gain
Imperial	567	500	429	400	−100	−167	
Calder	521	150	0	0	−150	−521	closed 1986
Various RDL	2342	400	350	320	−80	−2022	privatised*
Others	2022	200	0	0	−200	−2022	
Scotland Total	**26 765**	**12 169**	**6865**	**6225**	**−5944**	**−20 540**	

*All RDL establishments were privatised; most jobs have now been lost with the exception of Westburn (Trafalgar House) which closed in 1990.

Fig. 2.2.8

EMPLOYMENT IN SCOTTISH STEEL MOTHERWELL DISTRICT 1974–1989

PLANT	1974	1982	1986	1989	JOBS LOST 1982–9	JOBS LOST 1974–89	NOTES
Ravenscraig	5712	4581	3270	3200	−1381	−2512	
Dalzell	2275	820	650	600	−220	−1675	
Lanarkshire	1342	0	0	0	0	−1342	closed 1979
Mossend Engineering	343	0	0	0	0	−343	closed 1980
Mossend App. Train.	285	82	43	0	−82	−285	
Mossend RDL	305	0	0	0	0	−305	closed 1977
Fullwood	266	201	157	150	−51	−116	
Craigneuk/Clydeshaw	1567	1022	270	155	−867	−1412	
Clydesdale	2642	2373	1476	1200	−1173	−1442	
Motherwell District	**14 737**	**9079**	**5866**	**5305**	**−3774**	**−9432**	

The tables (Figs. 2.2.7 and 2.2.8) show the employment in Scottish steel 1974–89, and employment in Scottish Steel in the Motherwell district 1974–89 (Motherwell figures are not included in the first table).

1 Use a range of techniques to prepare these figures for presentation on a television programme.

What points would you wish to highlight?

2 Which key national figures would discuss the demise of steel-making in Scotland on such a programme?

In Scotland total employment in 1989 was 1 967 500, with 20.5 per cent (403 337), employed in manufacturing. The total number of people employed in Motherwell District in 1989 was 52 200, with 18 113 employed in manufacturing industry. When steel-making ceases in Scotland 18 137 jobs will be lost, 10 169 of them in Motherwell.

Compare the level of projected job losses from the proposed final closures in Scotland, at the national scale, with those in Motherwell, at the local scale. What evidence is there of de-industrialisation at the local scale?

Ravenscraig's closure: the announcement

The final decision to close Ravenscraig was made public on 8 January 1992. The following day there was widespread comment in the newspapers, and more in the Sunday newspapers on 12 January. Three extracts are quoted below:

It is odd that a people as canny as the Scots should have continued to use the long-doomed Ravenscraig steel plant as a political virility symbol. Strip steel plants had a certain potency as status objects when the plant was conceived and built in the late fifties and early sixties. But, by the seventies, steel production, however sophisticated, had begun to have a historical look.... Commercially Ravenscraig was always misconceived... Ravenscraig's fate has been clear for the past ten years. It was clinched by the privatisation of British Steel in 1988, and accelerated by the recession. The Scots should have devoted less energy to trying to reverse the tide of history, and more to a pre-emptive diversification of the local economy in the boom years of the 1980s.

The Independent 9 January 1992

The only legitimate complaint that can be laid against British Steel's decision yesterday to close the Ravenscraig works in Scotland is that it has been too long delayed. Ravenscraig, a product of the interventionist years of the Macmillan government should probably never have been built. Any remaining rationale for its existence disappeared when the car plants which it was supposed to supply were closed in the 1980s. Yet it was allowed to stagger on, though subject to a series of salami cuts that has reduced the remaining workforce to 1150 men from some 13 000 in its heyday....

These are harsh economic facts. They do not mean that sympathy is not due to the steelworkers, who, like some 200 000 of their predecessors in the industry, now find that they can no longer use their skills. Fortunately their future prospects need not be grim. Corby and Consett are now both reasonably prosperous towns, despite the steel closures that cost many more jobs than will go in Ravenscraig.

The Times 9 January 1992

The faces of men, invariably chiselled, grimy, suspicious images of manual labour.... We could make albums of them, gleaned from decades: shipyards, car works, pits, engineering works, steel mills... the faces always. This time after the blizzard of redundancy notices in recent years, they are mostly the faces of young men, men in their twenties and thirties, yet cast in the same mould as their elders. There is nothing for them. Unemployment in parts of Motherwell is already touching 25 per cent.... Faces. The men are familiar, part of the iconography. The women and children are seen less often, but it is they who will bear the burden of breadwinning and holding families together. There will be no editorials on such everyday heroism, no reports of the commonplace tragedies, no front page photographs, that tell us that Ravenscraig is an outmoded symbol, a mark of self-esteem made redundant by a changing world. But no-one demanded subsidy, there was no special pleading. In the absence of figures from British Steel – which has held them close to its chest like a card sharp holding a marked deck – the best educated guess is that the plant has been profitable these last few years. Ravenscraig was real and true; others played false.

Ian Bell: The Observer 12 January 1992

Different views of the impending Ravenscraig closure are expressed in the above extracts (although two are making the same point).

1   Analyse the values that underly the views being expressed.

2   Why is there obvious conflict between the value positions?

3   Which do you consider to be the more valid?

What of the future for Ravenscraig?

'We are not the kind of people who go around with our heads bowed low after a setback. We have got the makings of a diverse economy already in place. We are resilient. We've been through this before.'

Motherwell District Council official

'After the Craig there is nothing.'

Worker for a subcontractor at Ravenscraig

'At the moment, even with the closure, a lot of people will be on one big party. It will take a year for all this to have a full impact. By then the redundancy money may have gone and families will have to get used to unemployment benefit. For many that might be impossible. Redundant workers are now on cooking and computing courses where there will be little chance of a

job after finishing. The right kind of training that this area needs is not being properly looked at.'

Joe Eley, *adviser at the Craigneuk Development and Support Office*

The Government's immediate reaction to the announcement by British Steel was the promise of the setting up (with the aid of Community funds) of an Enterprise Zone in Lanarkshire – not an entirely new idea since it was first mooted in 1991. One of the most encouraging developments is the creation of the Mossend freight terminal near Motherwell which will serve as the Scottish freight terminal for the Channel Tunnel. Although only 150 jobs will be created initially, the surrounding 320 hectares will be developed as a new industrial complex with a likely investment of £50 million and an employment potential of 10 000 jobs. There are a number of other recommendations that have been made.

1 Ravenscraig could be sold off, as a going concern to a buyer in the private sector.

2 Use the funds available through British Steel Industry (an organisation set up to provide capital for new industries in areas hit by steel closures) to encourage a range of smaller industries to set up in the Motherwell area, as has happened at Consett and Corby.

3 Encourage the development, with the help of the latest technology, of a super-mini direct-smelting mill (latest state-of-the-art mill) at the coastal location of Hunterston, which is the present deep water ore terminal for Ravenscraig.

Evaluate each of these three options, and then rank them in order of viability. You may have options of your own – include these within your evaluation.

Reclamation of the site will be an expensive undertaking, since much of it is contaminated with the waste materials from 35 years of steel making. Estimates range as high as £200 million and the cleaning-up process is likely to last into the first decade of the next century.

With the final closure of steel-making at Ravenscraig in June 1992, British Steel will concentrate its production on its four remaining sites – Port Talbot and Llanwern in South Wales, Redcar on Teeside, and Scunthorpe. Scunthorpe is probably the most vulnerable: the existing plate mill at Scunthorpe is scheduled to close with the opening of a new £300–400 million plate mill on Teeside. Latest indications (January 1992) are that British Steel will increasingly concentrate its future

Fig. 2.2.9

EMPLOYMENT: BARROW-IN-FURNESS

TABLE 1 EMPLOYMENT SECTORS BARROW TRAVEL TO WORK AREA (TTWA)

	TTWA PER CENT	NUMBER
Agriculture	1.0	385
Manufacturing	49.0	18 865
Service	47.0	18 095
Construction	3.0	1 155
Total	100	38 500

TABLE 2 EMPLOYMENT SECTORS, BARROW BOROUGH

	PER CENT	NUMBER
Agriculture	0.3	
Energy and Water Supply	1.5	
Manufacturing and Construction	56.3	
Distribution and Commerce	21.4	
Other Services	20.5	
Total	100	31 700

TABLE 3 MANUFACTURING EMPLOYMENT

LOCATION	PERCENTAGE
Barrow-in-Furness TTWA	49.0%
Barrow Borough	48.9%
Cumbria	27.9%
Northern Region	26.0%
England and Wales	25.9%

TABLE 4 SERVICE SECTOR EMPLOYMENT

AREA	PERCENTAGE
Barrow-in-Furness TTWA	40%
Barrow Borough	45.9%
Cumbria	61.4%
Northern Region	64.1%
England and Wales	65.1%

MAJOR FURNESS EMPLOYERS (THOSE WITH MORE THAN 50 EMPLOYEES)

Fig. 2.2.10

Vickers Shipbuilding & Engineering Limited (VSEL)	14 000 – 14 500	Shipbuilding
Glaxochem	1750 – 2000	Pharmaceuticals
Scott	1000 – 1250	Tissue Making
Ashley-Rock	751 – 1000	Electrical Fittings
K Shoes	501 – 750	Footwear
BCL Cellophane	351 – 400	Packaging
Sovereign	251 – 300	Building Chemicals
Camille Simon	201 – 500	Detergents
Lister Mutual Yarns	201 – 250	Textiles
Furness Footwear	151 – 100	Footwear
Burlington Slate	101 – 500	Quarried Products
Oxley Developments	101 – 500	Electronics
Furness Footwear	101 – 250	
Telemeter Engineering	101 – 150	Engineering

Service

South Cumbria Health Authority	1000 – 1500	Local Government
Barrow Borough Council	1000 – 1500	Health Care
Cumbria Education Authority	1000 – 1500	
Asda Stores	250 – 500	Retail
College of Further Education	101 – 500	Education
Tesco Stores	101 – 300	Retail
Norweb/North West Gas	100 – 300	
Cumbria County Council	N/Available	Local
Simmonds	101 – 200	Retail
Marks and Spencer	101 – 150	Retail

Retail (Cumulative Totals)

Hotel and Catering	2300
Construction	1300
Transport	1300

investment at Port Talbot, in South Wales, and Teeside. The other centres such as Scunthorpe, and Llanwern will be run as finishing operations like Shotton in North Wales. Once British Steel employed 250 000 workers, in late 1992 the figure will be 47 000.

Fig. 2.2.11
VSEL Shipyard in Barrow-in-Furness

A NEW VIEW OF BRITAIN

BARROW: STEEL, WARSHIPS AND ENGINEERING

'Though ore mining and smelting were active in Furness as early as the twelfth century, it was in 1859 that the first blast furnaces were built on a coastal site to the west of the town, where the present works still stand. Today, with the exhaustion of the Furness fields, ore comes in by sea through Barrow docks. Four blast furnaces produce haematite pig iron, and the company also carries out a wide range of iron casting and general engineering. A nearby steel works produces steel ingots from its seven open-hearth furnaces. The production of iron and steel is, now, however, subordinate to shipbuilding and marine engineering, and 11 000 workers are now employed by the two associated firms of Vickers–Armstrong (shipbuilding and engineering).'

F J Monkhouse *Cumbria*, in *Great Britain: Geographical Essays 1962*

Iron and steel manufacturing has long since ceased in Barrow-in-Furness, Cumbria and the loss of its steel industry is evidence of a measure of de-industrialisation in the area.

1   Study the tables (Figs. 2.2.9 to 2.2.10) which show the main employment statistics for Furness and Barrow-in-Furnesss.

2   Use these figures to present an analysis of the present pattern of employment in the Barrow area, relating it where possible to the regional and national pattern. Where does the pattern suggest some vulnerability to future change?

BARROW-IN-FURNESS JOB CUTBACKS ANNOUNCEMENTS 1991

EMPLOYER	NUMBER
VSEL	1500
VSEL	2500–4500*
VSEL Apprentices	300
Ashley-Rock	50
College of Further Education	13
Commercial Office Equipment	10
Camille Simon	40
Glaxochem (Ulverston)	200**
Empat Removals	6
Furness Travel	2–4
Total at 15 April 1991	4623–6623
Total in 1990	2205
Overall total 1990–91+	6823–8823
Unemployment level March 1991	2832
Unemployment potential	9655 to 11 655

** Volunteers over 55 ...

Fig. 2.2.13

Fig. 2.2.12

BARROW-IN-FURNESS JOB CUTBACKS 1990–1991

JOB CUTBACK ANNOUNCEMENTS 1990

Scott Limited	400
Listers Mutual Yarns	270
VSEL YTS	68
Ashley Rock	108
K Shoes	25
DHSS	23
Prudential	4
Strand	20
Hindpool	10
Griffin Aggregates	16
Barrow Office Equipment	11
VSEL	700
VSEL (to March 1991)	550
Totals	2205

3   Now study the job cutback announcements for 1990–91 (Figs. 2.2.12 and 2.2.13) and read the article, 'Barrow pays high price of defence slump' (Fig. 2.2.14).

4   In what ways do these statistics confirm any ideas that you may have had about vulnerability? Most recent figures for unemployment in Barrow (1989) showed a level of 6.2% (about the national average). Given the job losses for 1990–91 and a 1989 employed labour force of approximately 38 500, what sort of unemployment levels are likely in the next few years from 1989?

5   Apart from the problems associated with Barrows's employment structure, the article indicates that Barrow's location is far from ideal. Using a good atlas, draw a sketch map, suitable for presentation to a visiting delegation from the European Community's Regional Policy Directorate, to show the weaknesses in Barrow's accessibility, together with a series of recommendations on how it might be improved.

With the winning of the order for the fourth Trident submarine on 7 July 1992, hundreds of jobs at VSEL have been safeguarded. The keel was laid on 1 February 1993. Nevertheless manning levels at VSEL's yards have been halved to approximately 7000 in the last few years.

Barrow pays high price of defence slump

THE CRANES rise over the Barrow-in-Furness skyline like watch towers, their chains swinging noisily above terraced houses.

Once it was a comforting sight and sound for people living within a few minutes' walk of the shipyard gates. Now they are not so sure. In one of the seedier parts of the town centre, posters declaring "I can work, I will work" have special significance for the shipyard men.

Barrow is facing massive job losses. Between 5500 and 6000 shipyard workers will have to go on the dole in a few years' time, swelling the town's unemployment rate from 6 per cent to 27 per cent and more. About 60 per cent of the town's male population depend on shipbuilder VSEL for their living.

The company announced on Wednesday that because of the slump in Ministry of Defence orders, it is to cut its workforce from 12 500 over the next four years. Two years ago, 14 300 people were working at the Barrow shipyard on the Trident submarine programme and other long-term production lines.

VSEL says that manpower requirements within the UK warship building yards will fall from about 21 000 last year to only 8000 by 1995.

Everyone in Barrow has known that job cuts were inevitable. Men have been talking about the numbers in the yard, streets, pubs and clubs for some time. But the scale of the job losses has taken the town's breath away.

With a population of about 73 000, Barrow has some glaring problems. The first is the town's location. It is remote, out on a limb on the west Cumbrian coast, with poor communications. This makes it almost impossible for people to travel to other towns for work.

Next, its industrial and economic life almost completely relies on the shipyard, with nearly 85 per cent of the company's employees living in Barrow.

The less obvious problem is that cuts will leave young people who had planned an apprenticeship in the yard no longer able to depend on such employment. There is nowhere else for them to go except away from Cumbria.

Bob McCulloch, the chief executive of Barrow borough council, will pull no punches when he meets Edward Leigh, Trade and Industry Minister, on Monday.

Mr McCulloch said: "The scale of the job loss was a shock, but we are not going to sit here and do nothing about it. There are many things that must be done for the long-term future of the town, and the key will be how much government support and help is given."

The town's leaders say they urgently need grant aid and a special enterprise zone needs to be created. They want assistance to improve road and rail links and the local airport to re-open, with links to Manchester and Northolt in London. They also want the DTI to help them gain access to the European Community's regional assistance funds, so far denied to them. Barrow lost its assisted area status in 1982 and almost no new speculative industrial development has taken place since.

The area needs industrial development if its skill base is not to be lost. Frank Ward, the district secretary of the Amalgamated Engineering Union, who met local MPs yesterday, is angry at the scale of the job losses and worried about the town's future. He said "One of the most upsetting aspects of all this is the long-term drop in manning levels at the yard and the future for young people. Everybody knew there would be cuts, but they have panicked, and it is one hell of a shock."

Fig. 2.2.14
Independent 23 March 1991

REVIVING OLD INDUSTRIAL AREAS

Since de-industrialisation has been such a powerful process over the last three decades, it has become increasingly important to stimulate new industrial development and economic growth in the regions affected. The need for some sort of regional policy, aimed at promoting such improvements, was recognised as long ago as the 1930s. Government policies now form only part of the range of measures that are available in industrial districts suffering from decline and decay. The main forms of assistance are:

- Official regional policy of the government of the day. This has varied considerably since the 1930s. The map (Fig. 2.3.1a) shows the changes since World War Two. Regional policy was most vigorous in the period between 1963 and 1977, with a wide range of incentives on offer, including grants, loans, tax allowances and general and industrial infrastructure provision. To a degree regional policy was dismantled during the 1980s as a result of the Thatcher Government's monetarist and free-market approach. Only Development Areas and Intermediate Areas remain, and one commentator has noted that, since no area has been removed from the list, the policy does not appear to be very effective.

However, in July 1993, the Major Government introduced significant changes (see Fig. 2.3.1b). For the first time areas in the south-east, including two in inner London were given assisted status. This is a reflection of high unemployment levels in such areas as Thanet, and the Isle of Wight, and such towns as Hastings and Harwich. The south-east suffered particularly badly in the recession of the early 1990s, and there was much lobbying by local authorities for assisted status. Another notable change was the removing of some areas from the assisted list. Towns like Telford, Shotton, parts of southern Scotland, and of the Highlands will no longer receive special help. It was judged that they had made sufficient economic progress, in terms of new factories built, new businesses established and new jobs created, to no longer need the special financial help that comes with assisted area status.

Fig. 2.3.1a

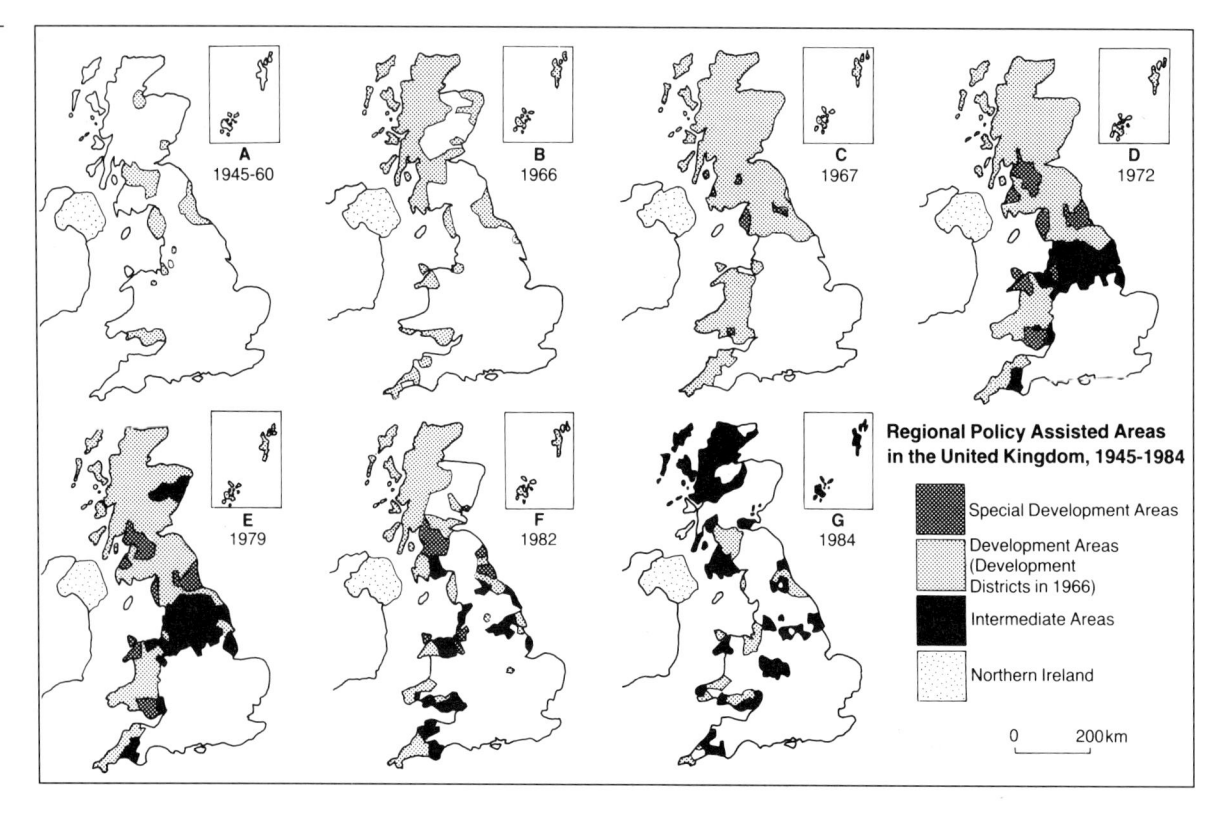

Regional Policy Assisted Areas in the United Kingdom, 1945-1984

- Special Development Areas
- Development Areas (Development Districts in 1966)
- Intermediate Areas
- Northern Ireland

A 1945-60, B 1966, C 1967, D 1972, E 1979, F 1982, G 1984

0 200km

- The European Regional Development Fund: grants from this fund were restricted to areas within member countries where national regional policies were already operating. After the Brussels summit in 1988, national governments, once they had been granted aid, had to contribute up to 50 per cent of the total assistance available. Most of the funds spent in Britain were on public infrastructure projects.

- Industry specific Enterprises: with heavy job losses in coal, steel, and shipbuilding areas, companies were set up to provide assistance in these regions. Thus British Steel (Enterprise) Ltd. (1975), British Coal (Enterprise) Ltd. (1985) and British Shipbuilders (Enterprise) Ltd. (1986) were established. They have proved to be effective in providing new jobs in areas like Consett and Corby.

- Government Development Agencies: both Welsh and Scottish Development Agencies are autonomous bodies, responsible for a wide range of activities to revive industry in the two respective countries. The Welsh Agency has been particularly important in initiating a range of developments in the South Wales Valleys, whilst in Scotland the Agency there has been responsible for programmes within different industries, and was particularly prominent in

Fig. 2.3.1b
The Times, 24 July 1993

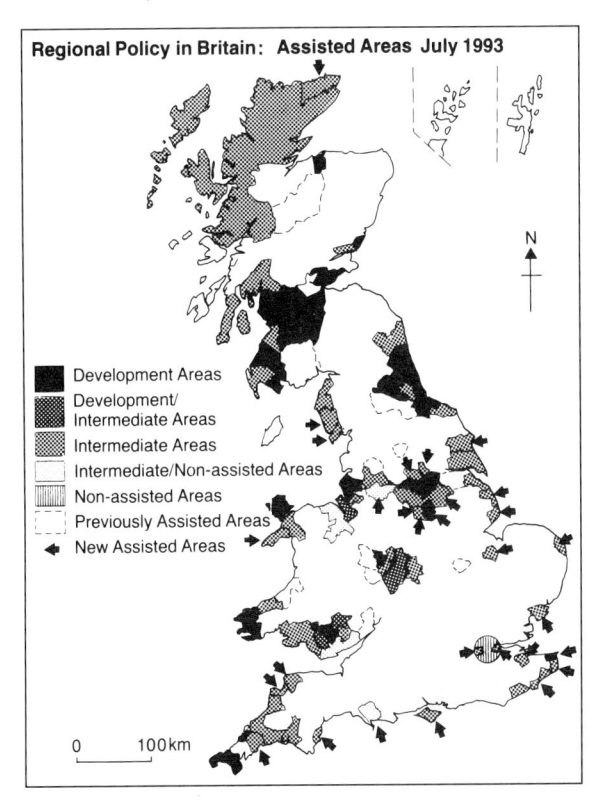

Fig. 2.3.2

the urban renewal work of GEAR (Glasgow Eastern Area Renewal Project).

- Enterprise Zones: (Fig. 2.3.2) these were created under the Finance Act in 1980. These were seen as enclaves of free-market activity within which companies could enjoy a whole range of benefits, including no payment of rates for the first ten years, and tax allowances on capital spending. The distribution of Enterprise Zones is shown on the map, although some of the early ones are about to lose some of their incentives after 10 years.

Fig. 2.3.3

Urban Development Corporations
- Tyne and Wear
- Teeside
- Leeds
- Central Manchester
- Merseyside
- Sheffield
- Heartlands (Birmingham)
- Black Country
- Cardiff Bay
- London Docklands
- Bristol

- Urban Development Corporations (Fig. 2.3.3): these have been dealt with in the section on Urban Britain, but they have obviously been important in a number of districts that have suffered heavily from de-industrialisation. From the point of view of industry they have been instrumental in providing a business environment in which the private sector could prosper.

- Local initiatives: these have been initiated either by local authorities (Enterprise Boards) or local groups of businessmen (Enterprise Agencies). These offer a range of support facilities for new or existing industries. Enterprise Boards have developed and managed industrial estates, and provided loans and training. Enterprise Agencies aim to help small businesses through a range of support services.

THE REVIVAL OF TEESIDE

Fig. 2.3.4 Mrs Thatcher on derelict land. Teeside

'*If only Mrs. Thatcher could see us*' he said. He lives on an estate where 91 per cent of heads of households are unemployed. Colin Armstrong, unemployed steel worker, interviewed by *The Guardian*, 1984.

In many ways Teeside is a classic area of serious de-industrialisation. It was an industrial region dominated by heavy industry, with its iron and steel dating from the 1850s, (later branching out into heavy engineering) and its chemical industry from the 1920s. With help from regional development policies it seemed to have good prospects in the 1960s and early 1970s. However, serious haemorrhage of jobs began in the 1970s and continued through the recession of 1979–81 into the late 1980s.

In some districts, employment in iron and steel fell by over a half in the 1970s. In the 1970s and 1980s, employment in the chemical industry fell by one-third. Overall, manufacturing industry lost nearly half its workforce between 1971 and late 1980s.

Teeside Development Corporation

The main measures taken to promote renewed industrial growth in the Teeside region in the 1980s were firstly the establishment of Enterprise Zones in Hartlepool in 1981, and in Middlesbrough in 1983. Teeside Urban Development Corporation was designated in May 1987, the largest in Britain with responsibility for 49 square kilometres of the region – the largest continuous area of de-industrialised land in Europe. It is useful to view the unemployment figures for Cleveland, (Fig. 2.3.5b), shortly after it was established. For comparison, unemployment trends for the 1978–88 period are shown (see Fig. 2.3.5a).

Fig. 2.3.5a

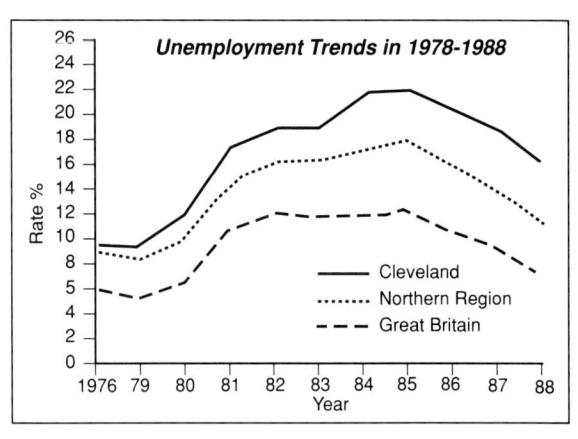

Unemployment Trends in 1978-1988
— Cleveland
······ Northern Region
– – – Great Britain

UNEMPLOYMENT IN CLEVELAND AT THE LAUNCH OF TEESIDE DEVELOPMENT CORPORATION

Fig. 2.3.5b Department of Employment Employment Gazette September 1987

UNEMPLOYMENT INDICATORS: SEPTEMBER, 1987

	UNEMPLOYED CLAIMANTS			UNEMPLOYMENT RATES(%)		
	MALE	FEMALE	TOTAL	MALE	FEMALE	TOTAL
Hartlepool	6319	1968	8287	23.8	11.9	19.2
Langbaurgh	8516	2934	11450	20.2	11.8	17.1
Middlesbrough	10748	3374	14122	26.2	13.2	21.2
Stockton-on-Tees	9554	3690	13244	18.8	11.5	16.0
CLEVELAND	35137	11966	47103	24.5	12.1	19.5
Northern Region	151721	59471	211192	17.7	10.2	14.7
Great Britain	1880838	859373	2740211	11.8	7.8	10.1

The map (Fig. 2.3.7) shows the extent of the area controlled by the Teeside Development Corporation, together with the two Enterprise Zones

Fig. 2.3.6
(a) Chemical Industries, Teeside   (b) Iron and steel plant on Teeside

Fig. 2.3.7

Comment on the relative unemployment pattern for Cleveland Boroughs, Cleveland County, the larger regions and Great Britain in 1987.

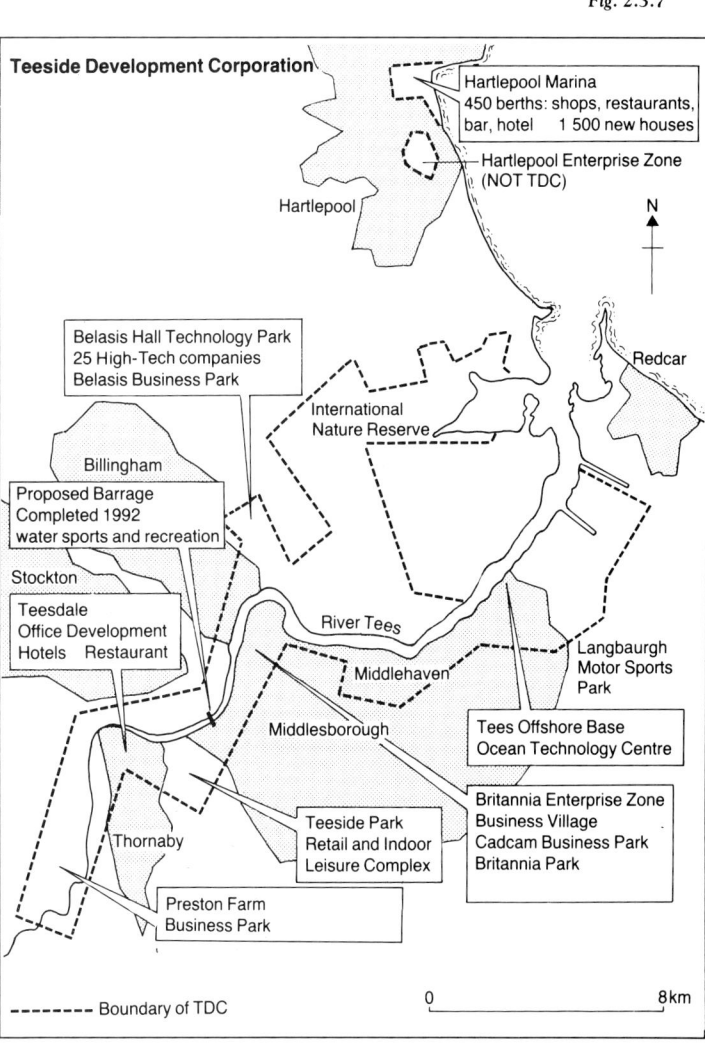

in Hartlepool and Middlesbrough (the Britannia Enterprise Zone). The Development Corporation has identified a number of 'flagship' projects, which are identified on the map. These ten projects can all be developed as discrete units, independent of one another. The Britannia Enterprise Zone lies within the Teeside Development Corporation's area, with the Hartlepool one operating as a separate entity. Within the first three years 6500 square metres of industrial, commercial and retail space have been completed and occupied. A further 139 000 metres are under construction or planned. In three years 90 new companies, creating 8000 jobs, have been attracted to the area.

After two and a half years of operation a report on the impact of the Teeside Development Corporation (TDC) on the local area identified the following areas of concern.

- More consultation between TDC and the local authorities was necessary, in order for local opinions to be voiced, and for local expertise to be fully used.

- More could be done to support local community groups.

- Most of the flagship projects involve complex technical problems: these have to be approached with caution and proper research carried out into the difficulties.

- New developments within the TDC area need to be seen in relation to the whole of the Teeside Region; overcapacity may develop in such areas as retailing and provision of leisure facilities.

- A sensitive and imaginative approach to architectural styles on the flagship sites needs to be employed.

- Some proposals for waste incinerators would lead to unacceptable levels of atmospheric pollution outside of the TDC area.

1 By referring to the list of forms of assistance available to de-industrialised areas at the beginning of this section, assess the extent to which Teeside is well-placed to take advantages of these forms of financial help.

2 Study the flagship proposals on the map of the Teeside Development Corporation. How are they likely to alter the employment structure of the area? How far will they solve the unemployment brought about by de-industrialisation?

3 Urban Development Corporations often cause conflict with local authorities. Suggest how the measures of concern expressed in the report might be resolved.

RENEWAL IN RHONDDA

The Rhondda Valleys in South Wales achieved their maximum output of coal in 1913. From 60 pits 41 000 miners produced 10 million tons of coal. In the late nineteenth and early twentieth century the Rhondda Valleys emerged as a unique industrial landscape in Britain. The twin valleys, deeply cut in the South Wales moorland plateau, were hardly the ideal location for coal mining (Fig. 2.3.8).

Fig. 2.3.8
Coal-mining landscape, Cwm Parc, Rhondda 1967. In common with the rest of Rhondda, Cwm Parc has now lost its mines, and the whole process of landscape reclamation has done much to improve the appearance of this tributary valley of the Rhondda Fawr (see back cover).

Collieries had to be located in the narrow valley bottoms, and often in the gloomy tributary valleys. Room had to be found also for the sidings and railway lines to the coast, for the seemingly endless rows of miners cottages, and for the huge quantities of spoil from the mines. Few areas in Britain were so congested or so polluted. From its heyday in 1913 Rhondda was to see a steady decline in its markets for coal, and progressive closure of its pits was inevitable (Fig. 2.3.9). With the final closure of the Maerdy pit in the Rhondda Fach, coal mining ceased in 1990. De-industrialisation was complete.

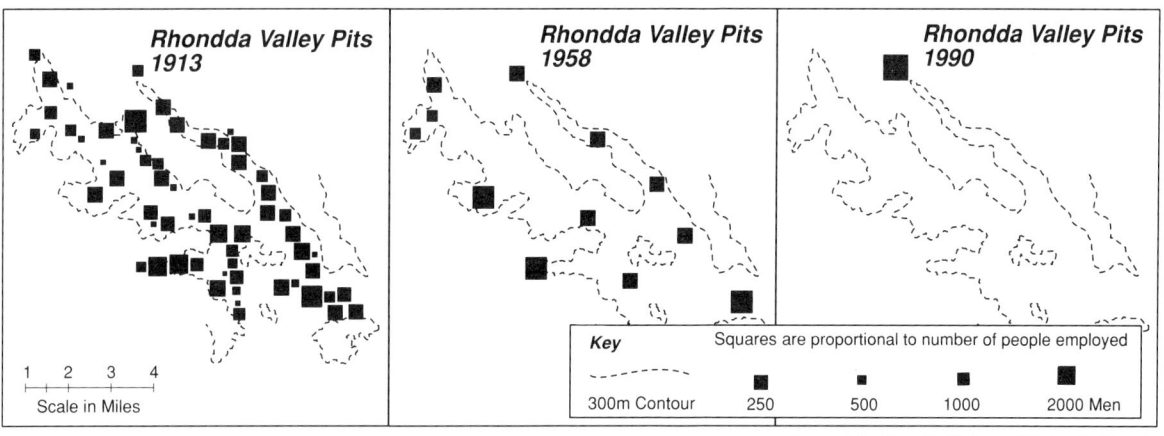

Fig. 2.3.9

Rhondda was in the first group of areas to qualify for government assistance in 1934, and has remained in this position ever since. The renewal of such an area presents enormous problems, both economic and environmental. Throughout the long decline, increasingly large parts of the infrastructure became redundant, particularly the intricate rail network, with its frequent tunnels, bridges and viaducts. Spoil heaps were unsightly hazards, and little was done to improve their appearance. Much of the housing stock was antiquated, suffered from subsidence and lacked basic services. Levels of pollution went unmonitored and the health of the valleys' inhabitants inevitably suffered, the more so because of inadequate and unsuitable diets. It is against this background that successive governments, local councils and businessmen have sought to bring about improvements.

In 1972 it was noted that 'by 1964, after over 25 years of government encouragement, there were still more than twice as many people employed in coal-mining as in manuacturing in the valleys ... over half the manufacturing jobs that had been provided internally were for women, so that in the production sector of the economy nearly 80 per cent of the male employment was in coal-mining.' Other writers noted that the levels of occupancy of factories on industrial estates were very low. Twenty years on, problems still have to be faced. In a Guardian article, Ted Merrette, managing director of AB Electronics, a company that employs 5000, mainly in the valleys, comments. *'We employ mainly female labour because the work is quite delicate, and not really suited to big, burly ex-miners – but we do employ a lot of their wives.'*

Nevertheless the Rhondda valleys have benefited considerably from recent Government programmes, notably the Valleys Initiative of the Welsh Development Agency. Many of the old colliery sites have been reclaimed and landscaped. Since 1976 over a million pounds a year have been spent by the Welsh Development Agency on reclamation in the Rhondda Valleys. By 1988, 21 sites (some 360 hectares) have been reclaimed. In Clydach Vale, the site of the old Cambrian Colliery forms part of a vast reclamation scheme (Fig. 2.3.10). Industry and modern housing now occupy the derelict site, together with amenity areas, and playing fields. The Forestry Commission has also been active in restoring Rhondda to some of its pre-Industrial Revolution attractiveness.

Fig. 2.3.10
Reclaimed land and new industry, Clydach Vale, Rhondda 1992. Much of the valley bottom was formerly occupied by the site of the Cambrian Colliery.

Where are the jobs to come from?

Fig. 2.3.11

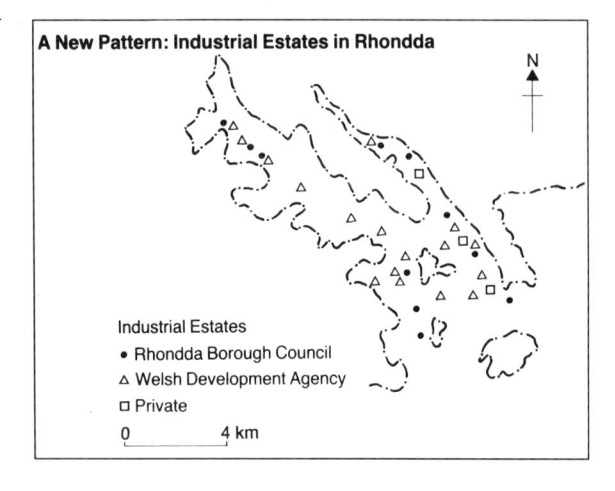

A New Pattern: Industrial Estates in Rhondda

Industrial Estates
- ● Rhondda Borough Council
- △ Welsh Development Agency
- □ Private

0 — 4 km

A wide range of new industrial sites has been provided and the much needed industrial diversification is beginning to appear. Rhondda now has nearly 300 new firms, mostly located on industrial estates (Fig. 2.3.11).

Study the Newspaper extract (*Rhondda News* May 1987) 'In tune with the Japanese' (Fig. 2.3.12)

1   What appear to be the main ingredients of success indicated in the article?

2   A number of other factors may have been important in the success of the company. Suggest what they may have been.

3   In what ways do you consider Rhondda to still be disadvantaged as a location for new industries?

Fig. 2.3.12

Focus on the changing face from a Valley's Industrial Revolution

In tune with Japanese
That's the Valdon way

THE JAPANESE connection now dominates so many specialist, high-tech areas of Welsh industry and commerce – and that is the opportunity spotted and expertly exploited by Valdon, Ltd.

Managing Director Colin Jones says with justified pride: 'From 1980 when we started with a few directors and three or four workers, we have grown to a workforce of 90.

"Turn over? From about £130 000 to about £3 000 000!"

Valdon's success has largely stemmed from an exclusive contract with National Panasonic based in Cardiff.

"We make all the flat surround outers for their television sets," he says. "That involves moulding and spraying – plus manufacturing some interior parts for their sets."

The Valdon expansion goes on and on.

Another branch of the Japanese multi-national, Kyushu Panasonic, has opened up in Newport, making typewriters and printers. Valdon has started producing for them as well, with the demand for its quality products keenly appreciated by a company well known for its demanding standards.

"We have a second company, Valdon Enterprise, which means that we are major suppliers to Hoover of Merthyr," adds Colin. "We make all the work-top lids for their washing machines plus moulding some internal parts."

There is not a shred of doubt over the policies of Colin and co-director Brian Trott.

They are backing Rhondda and they are committed to a long-term, expanding progress.

"We are putting our money on the future," he says.

"We are investing a great deal of money and it will be several months yet before all the renovations and landscaping is complete on the former Conatec factory, with several acres we have bought in Llwynypia."

"These factories cover 52 000 and 30 000 sq. ft. respectively, and we are halfway moving into these enlarged premises from our current base in Treorchy. The overall investment goes into hundreds of thousands of pounds."

"We feel totally justified in the growth we are anticipating."

"The purchase of premises and machinery means we are confident of a prosperous future with a high demand for products into which we put every ounce of our skill, determination and professional pride."

REVIVING OLD INDUSTRIAL AREAS

Corridor to Europe where no one stops
Winner by John Grigsby

TO THOSE who drive through it on the way to the Channel ports, East Kent, is part of the prosperous South East.

Behind the oast houses, the hop fields, the marinas and the places associated with Charles Dickens, however, lies one of the most deprived areas in Britain.

The district of Thanet, for example, which covers the towns of Margate, Ramsgate and Broadstairs, the home town of Sir Edward Heath, has the sixth highest unemployment in Great Britain – at more than 16 per cent of the workforce.

In England, only South Tyneside and Hartlepool have proportionately higher numbers of people out of work.

Over the past few years, East Kent has lost thousands of jobs in farming, the Kent coalfield, hotels and catering, manufacturing and on the Channel ferries.

The opening of the Channel Tunnel, while it should help other parts of the county, is likely to make matters worse in the short term. More jobs will be at risk at the ferry terminals of Dover and Folkestone.

Before yesterday's announcement, international firms were expected to by-pass much of East Kent and choose towns such as Dartford, Ashford and Maidstone for their business parks, technology centres and warehousing.

Sir Alastair Morton, chairman of both Eurotunnel and the East Kent Initiative, which argued for assisted area status for Swale, Thanet, Dover and Shepway, said the area would now be able to compete more effectively with France's Nord Pas-de-Calais region for investment and development associated with the tunnel.

American and British companies employed 12,500 people in grant-aided Nord Pas-de-Calais in 1991 and the total is rising.

An initial survey in different parts of East Kent last year identified £130 million worth of projects which might be eligible if the area qualified for a grant. This would help to create 3,000 local jobs.

Mr Martin Hemingway, general manager of the East Kent Initiative, said that Whitstable and Herne Bay, which were excluded from the assisted area map, would benefit from the spin-off which the new status would bring to the area.

City of wool fears shearing of jobs
Loser by John Grigsby

FEW Victorian cities have done more to change their economy – and their image – over the past decade than Bradford.

The dark city dominated by chimneys, once the wool and textile capital of the world, has become a tourist centre – with the help of almost £70 million in direct Government grants and European aid over the last eight years. Much of the funding depended on the city's assisted area status.

Its Indian and Pakistani restaurant trail attracts gourmets from all over Britain. Its National Museum of Photography, Film and Televison has a world-wide reputation.

Bradford produced artist David Hockney and his work is on show in Sir Titus Salt's model industrial village of Saltaire.

Haworth, home of the Bronte family and the second most visited literary shrine in Britain, is within the council area.

But Bradford, which has a population of 468,000, has also broadened its manufacturing base into high-technology industry, banking and insurance.

Mr Gerry Sutcliffe, the Labour council leader, said yesterday: 'This virtually brings to an end Government support for manufacturing investment in the city.

"This will cost the district millions of pounds and lose us thousands of jobs. The business community in Bradford will be horrified."

The city says it has lost 10,500 jobs since the recession began. Unemployment is now running at more than 12 per cent.

The council estimates that over the past eight years, direct Government grants resulting from assisted area status, having safeguarded 11,080 local jobs and supported £153 million of investment in local firms.

A major worry for the council is that it may no longer be eligible for European aid. Although the Government has stressed that the designation of assisted areas is not linked to European funding, many in local authorities believe the Government aims to be the arbiter over which councils receive EC help.

If they are right, Bradford could lose £30 million in European grants over the next five years.

In July 1993 important changes were made to the areas receiving Government financial assistance (see Fig. 2.3.1b). East Kent (Thanet) became a Development Area, whilst Bradford lost its Intermediate status.

1 Study the two extracts from *The Daily Telegraph* above. How will these adjustments to the assisted areas affect the lives of the people, and the environment in East Kent and Bradford?

Fig. 2.3.13
Daily Telegraph
24 July 1993

THE ADVENT OF HIGH-TECHNOLOGY INDUSTRY

It is the view of some geographers that a new phase of industrial growth (or re-industrialisation) is now taking place. In the European context it involves the emergence of new and relatively small firms, the rapid growth of high-technology industry, the widespread use of new computer-based technology (over 60 per cent of UK factories use microchips to automate production now) and an increase of inward investment by trans-national firms. In reality there is a considerable inter-linking of all of these elements. This section deals more specifically with high-technology industry.

Fig. 2.4.1
D. Keeble, Geography, Volume 75, (Part 4)

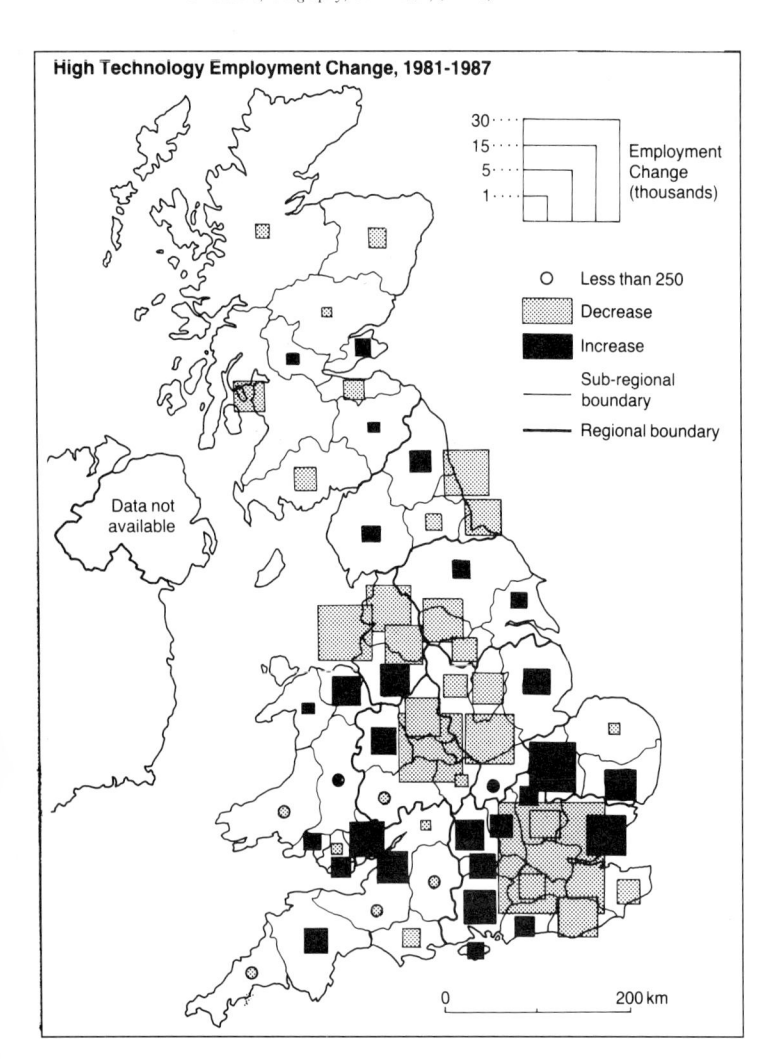

Much discussion has centred on precise definitions of high-technology industry. It is now accepted that industries under this heading will include those that:

- employ advanced scientific and technological practices
- are involved in the manufacturing of scientific and technologically advanced products
- are often involved in advanced research and development.

One industrial geographer, Keeble, recognises four main characteristics of high-technology in Britain at the current time.

- Very rapid technological change in products, with correspondingly short product life-cycles (the amount of time the product retains a profitable share of the market).
- A rapid growth in market demand (Government defence expenditure has been a significant component in this growth, but recent defence cuts could diminish its importance).
- Great reliance on highly qualified staff research scientists, engineers and managers who are involved in research and development.
- Changes in industrial structure, involving the expansion of trans-national companies, and the growth of new, often small, indigenous companies.

Industrial and commercial brochures for most towns and cities now stress the importance of 'hi-tech' industries, and a varied range of incentives are offered in order to attract what is widely seen as a remedy for industrial structures that need modernisation. There does, however appear to be a clear geographical distribution of high-technology industry in Britain. Its traditional base has been in Greater London and South-east England (44 per cent of national employment in 1981).

Study the map (Fig. 2.4.1) which shows high-technology employment change 1981–87.

1 What were the main changes in employment in high-technology during this period?

2 Account for the nature and level of change in:

a South–east England

b Scotland and Wales.

THE CAMBRIDGE PHENOMENON

The Cambridge Phenomenon was first described in a report (1985) by a firm of consultants based in the city. It refers to the marked concentration of high-technology industry in the Cambridge region. In 1989 there were 600 high-tech companies in south Cambridgeshire, employing 20000 workers. There has been a remarkably steady growth in the creation of new firms in the Cambridge region – an average rate of 50 a year. Many of the firms have been founded by local research workers and scientists (in 1984, 75 per cent were locally owned), most of them highly educated and technologically qualified. Much of the work is inevitably biased towards research and development; Cambridge is not so well suited to production since it lacks a large labour pool, does not possess a range of low cost factories and does not enjoy the incentives possessed by towns in Assisted Areas. Much of the employment created was in high level and high income occupations, but there was a considerable multiplier effect, because of the spending power generated being responsible for growth in the retailing, leisure and service industries.

Why did these companies find Cambridge such an attractive location? The table (Fig. 2.4.2) shows the response to a survey of high-tech companies in the mid 1980s (indigenous refers to companies originating in Cambridge, mobile refers to those moving in from outside).

1   Why does Cambridge appear to be such an attractive location for these companies?

2   Are there any significant differences between the responses of the indigenous companies and the mobile ones?

The Cambridge Science Park

Much of the growth of the high-tech industry in the Cambridge area owes its origin to the establishment of the Cambridge Science Park in 1973 (Fig. 2.4.3). The Science park is defined as 'a collection of high-technology industrial companies or research institutes in attractive, well-landscaped surroundings, developed to a very low density, situated near a major scientific university – and enjoying significant opportunities of interchange with that university'. It was established on land owned by Trinity College on the outskirts of the city, in response to Government pleas for much

Fig. 2.4.2

THE REASONS FOR LOCATING IN THE CAMBRIDGE REGION
IMPORTANT FACTORS

FACTORS	HIGH-TECH PER CENT		CONVENTIONAL PER CENT	
	INDIGENOUS	MOBILE	INDIGENOUS	MOBILE
Availability of labour	70	77	41	41
– graduates from the University	35	47	8	15
– scientific/technical staff	65	70	15	19
– managerial staff	10	23	11	19
– other staff	9	23	8	19
Availability of premises	67	80	66	78
Attractive area to live in	63	83	50	59
Good reputation of area	63	83	45	59
Good communications	52	67	46	59
Robust local economy	43	63	41	41
Proximity to similar organisations	42	57	23	22
Need to establish R and D links	38	53	15	26
Need to retain existing staff	32	32	32	22
Proximity to customers	30	30	59	59
Proximity to suppliers	28	17	29	22
Government assistance	7	3	6	–

Fig. 2.4.3
Cambridge Science Park: high technology campus on the northern outskirts of the city

Fig. 2.4.4

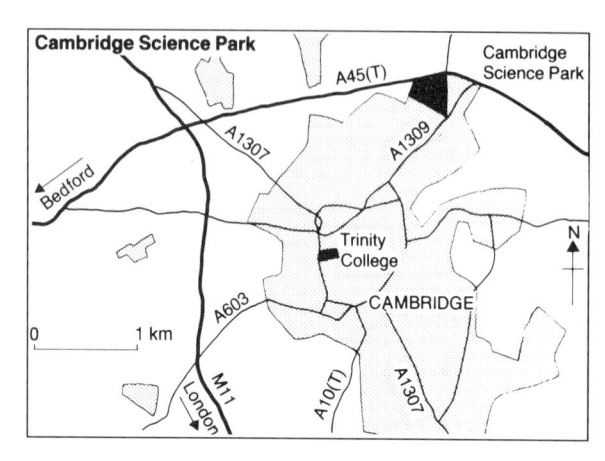

closer links between universities and high-technology industry. Eighty companies, employing some 2800 workers now occupy the park. Among the many fields of research and development carried on in the Science Park are: optical fibre development, miniaturisation of integrated circuits, CADCAM (Computer aided design and computer aided manufacture), and biotechnology. The park's location is shown on the map (Fig. 2.4.4).

1   Suggest some of the advantages of this location on the fringe of Cambridge.

2   What are the effects on the social, economic and natural environment of the growth of high-technology industry in Cambridge?

HIGH-TECH IN THE M4 CORRIDOR; THE CASE OF SWINDON

'It sounds pretty daft, but in Swindon at the moment, they seem to be recruiting a lot of people from outside, especially with skills, and bringing them all in, and unemployed who have lived here virtually all their lives, they are finding it difficult to get work because it's all specialised industry.'

The M4 corridor, stretching from London through Berkshire, Wiltshire and Avon into South Wales (Fig. 2.4.5) is popularly associated with the growth of high-technology industry. Swindon lies roughly half-way between what many consider to be the heartland of high-tech industry (Berkshire) and the

Fig. 2.4.5

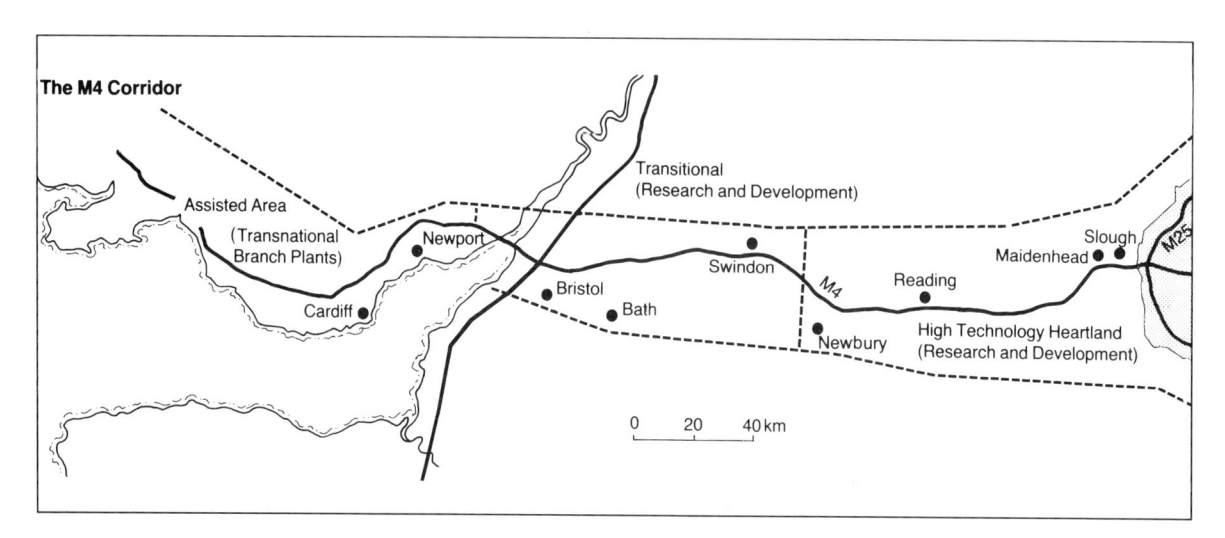

THE ADVENT OF HIGH TECHNOLOGY INDUSTRY

assisted areas of south Wales that can enjoy the benefits of a range of incentives. In many ways Swindon was, with its Great Western Railway works, established in 1848, an outlier of the industrial north sitting uncomfortably in rural Wiltshire. Swindon was very much a one-industry town, with the railway works (see Fig. 2.4.6) at one time accounting for one job in three. In the 1950s a measure of diversification into a range of other industries, particularly into engineering, saw the creation of some 14 000 jobs. Manufacturing peaked in the late 1960s, and since that time there has been a decline, which culminated in the closure of the railway works in 1987.

Study the table (Fig. 2.4.7) which shows the industrial structure of Swindon in 1968, 1984 and 1987.

1 Calculate location quotients for the different sectors of employment in 1984 and 1987. Comment on your results.

2 What have been the significant changes over the twenty year period?

Since 1975, 44 high-tech companies have established themselves in Swindon. Although this is relatively modest growth compared with Cambridge, it nevertheless represents a significant strengthening of the industrial base of the town. Three examples of the type of firm moving into Swindon are:

- Intel, a leading micro-processing company which has its UK sales and European marketing headquarters in Swindon

Fig. 2.4.6
Swindon's Railway works, seen in the left-centre of the photograph

- Logica VTS word-processing systems
- National Semiconductors (microprocessor) centre for European marketing and distribution.

Significantly, both Intel, and National Semiconductors are among 15 American based firms that have moved into Swindon since 1975 (see photos of Great Western railway works and Intel for comparison Figs 2.4.6 and 2.4.8).

There are some interesting differences that exist between hi-tech industry in Swindon, and in Cambridge.

INDUSTRIAL (OCCUPATIONAL) STRUCTURE SWINDON (1968, 1984, 1987) AND UK (1984 AND 1987)

Fig. 2.4.7

OCCUPATIONAL GROUP	SWINDON'S WORK FORCE 1968 (PER CENT)	SWINDON'S WORK FORCE 1984 (PER CENT)	UK WORK FORCE 1984 (PER CENT)	SWINDON'S WORK FORCE 1987 (PER CENT)	UK WORK FORCE 1987 (PER CENT)
Agriculture	3.2	1.9	1.78	0.3	1.4
Manufacturing Industry	47.1	37.2	33.5	26.0	24.0
Construction	6.8	4.7	5.5	4.8	4.7
Distribution	11.1	15.6	12.2	20.0	20.0
Banking/Insurance	1.0	4.7	5.3	11.4	10.9
Other services	30.8	35.9	41.8	37.5	39.0
Total	100.0	100.0	100.0	100.0	100.0

Fig. 2.4.8
Intel, international micro-processor's European headquarters in Swindon

HIGH-TECHNOLOGY IN SCOTLAND: THE CASE OF GLENROTHES

It has been mentioned that high-technology firms have shown a willingness to locate in Wales and Scotland, in Assisted Areas where financial incentives are available. The name Silicon Glen has been given to the region between Glasgow and Edinburgh, which has attracted much of this investment.

The industry is dominated by the trans-nationals, with the USA having a 77 per cent share. Japanese components make up 30 per cent of the input to the industry, although this will fall with new EC regulations. Nearly 11 per cent of all Scottish manufacturing employment is in electronics, and is responsible for about 15 per cent of Scottish manufacturing output.

1   Suggest what some of these differences might be.

2   Which of the two centres might be better equipped to maintain its high-tech growth into the next decade?

It should be noted, in addition to becoming an important centre for high-tech industries, Swindon has seen a broadening of its base in the service industries, with a number of firms, such as Allied Dunbar, and Nationwide Anglia moving major administrative functions into the town.

Glenrothes

Glenrothes, in Fife, a new town established in 1948 (Fig. 2.4.9) has developed an important high-tech sector. Study the employment structure figures for Glenrothes and the United Kingdom (1990), (Fig. 2.4.10).

Fig. 2.4.9
Glenrothes, Fife

THE ADVENT OF HIGH TECHNOLOGY INDUSTRY

GLENROTHES EMPLOYMENT STRUCTURE (TOTAL JOBS AT 31 MARCH 1991)

STANDARD INDUSTRIAL CLASSIFICATION	SIC	JOBS IN GLENROTHES					UNITED KINGDOM (a)
		MALES	FEMALES	ALL	F.T.E.	PER CENT	
All Industries and Services	0–9	10 1000	8075	18 175	16 761.5	100.00	100.00
Production Industries	0–4	5242	2627	7869	7705.0	43.30	26.09
Agriculture, Forestry and Fishing	0	28	2	30	27.5	.17	1.37
Energy and Water Supply	1	448	111	559	555.0	3.08	1.98
Other Mineral and Ore Extraction, etc	2	247	55	302	300.5	1.66	3.23
Metal Goods, Engineering and Vehicles	3	2896	1545	4441	4335.5	24.43	10.31
Other Manufacturing Industries	4	1623	914	2537	2486.5	13.96	9.20
Construction and Services Industries	5–9	4858	5448	10 306	9056.5	56.70	73.91
Construction	5	1035	93	1128	1115.0	6.21	4.76
Distribution, Hotels, Catering and Repairs	6	936	1512	2448	1867.5	13.47	21.25
Transport and Communication	7	439	128	567	521.0	3.12	6.17
Banking, Finance, Insurance, etc.	8	352	370	722	651.0	3.97	12.06
Other services	9	2096	3345	5441	4902.0	29.94	29.67

Note: (a) At September 1990
F.T.E. – Full Time Employment

Fig. 2.4.10
Employment Gazette April 1991

1 Draw Lorenz curves for both the United Kingdom and Glenrothes to show the degree of diversification of their industrial structure.

2 Compare your results for Glenrothes and the United Kingdom. What are the significant points of comparison?

The table (Fig. 2.4.11) shows the number of high-tech firms in Glenrothes in 1991.

3 Analyse the table (Fig. 2.4.11)

a to show the degree of foreign investment and,

b to show the amount of indigenous growth.

4 How secure is the base of high-tech industry in Glenrothes?

It is important to examine the position of Glenrothes within the Fife region as a whole. Study the employment figures for Fife for the period 1981–7 (Fig. 2.4.12).

1 What appear to be the most significant trends in the employment structure of Fife as a whole in the 1980s?

2 How might you attempt to explain these trends?

Fig. 2.4.11
Number of High-Technology Firms in Glenrothes 1991

SIZE (EMPLOYEES)	TOTAL COMPANIES	TOTAL (HT)	ORIGIN (HT)	
500+	2	1		N. America
300–499	3	–		
200–299	5	1		N. America
100–199	13	2	1	N. America
			1	Glenrothes
50–99	24	4	2	UK
			1	Scotland
			1	Glenrothes
25–49	27	5	2	N. America
			2	UK
			1	Glenrothes
Under 25	145	14	2	UK
			5	Scotland
			7	Glenrothes

3 Use the employment figures, a good atlas map of central Scotland, and the three photographs (Fig. 2.4.13) to help you write a short promotional leaflet, encouraging industry to come to Fife. You will need to use the employment figures very skilfully in order to display them to the best advantage. Remember to incorporate references to the photographs in your leaflet.

Fig. 2.4.12

EMPLOYMENT IN FIFE

INDUSTRY	1987	1984	1981	CHANGE 1981–7	CHANGE 1981–87 (PER CENT)
Agriculture, Forestry and Fishing	1436	1823	2275	−839	−36.90
Energy and Water Supply Industries	4941	8513	9679	−4738	−49.00
Extraction of Minerals and Ores other than Fuels	219	1486	2061	−1842	−89.40
Manufacture of Metals and Minerals	1723	435	542	+1181	+217.80
Chemicals	137	127	976	−839	−86.00
Manufacture of Metal Goods	950	780	784	+166	+21.20
Mechanical Engineering	4143	3838	3197	+946	+29.60
Electrical and Electronic Engineering	6924	8061	8939	−2015	−22.50
Vehicle and Transport Equipment	6326	7544	7266	−940	−12.90
Instrument Engineering	199	168	267	−68	−25.50
Food, Drink and Tobacco	2620	3111	4555	−1935	−42.50
Textiles and Clothing	3164	3927	4248	−1084	−25.50
Paper and Printing	3319	3411	3605	−286	−7.90
Other Manufacturing	2882	2929	3543	−661	−18.70
Construction	6256	10120	7931	−1675	−21.10
Wholesale and Retail Distribution	13084	14164	14397	−1313	−9.10
Repair of Consumer Goods and Vehicles	976	1187	977	−1	−0.10
Hotels and Catering	5957	5680	5456	+501	+9.20
Transport and Communication	4082	4528	4463	−381	−8.50
Banking Insurance Finance and Business Services	5815	5441	4513	+1302	+28.90
Public Administration, Medical Services etc	26 360	24 825	24 572	+1788	+7.30
Other Services	9376	7147	7016	+2360	+33.60
Total	110 889	119 245	121 262	10 373	+ 8.60

Fig. 2.4.13
(a) High-Technology Plant at Glenrothes (b) Loch Leven and Bishop Hill near Glenrothes (c) The fishing village of Crail, on the coast to the east of Glenrothes

THE SERVICE INDUSTRIES IN BRITAIN

Since World Two there has been a persistent decline in employment in manufacturing industry, and an equally persistent increase in employment in the service industries. In the 1950s employment in the service industries overtook that in the manufacturing sector. The gap between the two has widened to the extent that in 1991 over 70 per cent of Britain's labour force was employed in the service sector.

The table (Fig. 2.5.1) shows the numbers in employment in the manufacturing and service industries in the late 1980s. Broadly, employment in the service sector may be divided into three groups, the producer services, consumer services and public services. Producer services exist to serve other organisations, for instance a computer consultancy will advise other organisations on the most suitable computer system for their accounting, a bank will provide a range of financial services to manufacturing industry. Retailers provide consumer services, as do hotel chains and leisure organisations. Public services provide for both the producer and the consumer: they include provision for social needs (health, social services and education); infrastructure needs (road building, subsidies to public transport); administration and regulation of private and commercial conduct (monitoring of environmental impacts) and the provision of public security.

Study the table (Fig. 2.5.2) showing UK service employment in 1987.

1   Comment on the patterns shown by these statistics.

Fig. 2.5.1

UK EMPLOYMENT IN THE MANUFACTURING AND SERVICE INDUSTRIES 1982–1989 (000s EMPLOYED)

	1982	1983	1984	1985	1986	1987	1988	1989
Manufacturing	5897	5525	5409	5362	5227	5152	5215	5129
Service Industries	13406	13500	13836	14108	14298	14594	15168	15323
All Industries and Services	21059	21067	21238	21423	21387	21584	22226	22232

Geography, Volume 75, (part 4), October 1990

Fig. 2.5.2

UK EMPLOYMENT IN THE SERVICE INDUSTRIES 1987

TYPE OF SERVICE	EMPLOYMENT ('000)	PERCENTAGE OF TOTAL	CHANGE 1981–7 (PER CENT)	PART-TIME (PER CENT)	FEMALE (PER CENT)
Producer					
private business	2280	15.5	+29	14	50
private distribution	1511	10.3	+1	11	29
Consumer					
private	5046	34.3	+12	45	62
Public					
services	5043	34.2	+6	40	64
distribution	819	5.6	−10	5	20

Geography, Volume 75 (part 4), October 1990

REGIONAL COMPONENTS OF SERVICE EMPLOYMENT CHANGE, 1981–1987

REGIONAL DEVIATION FROM NATIONAL EMPLOYMENT CHANGE, FOR EACH SECTOR, AS PERCENTAGE OF 1981 EMPLOYMENT

	PRODUCER		CONSUMER				PUBLIC	
	B/FS	WH.	T/C	RET.	H/C	MIS	P/A	SOC.
National Change	+35	+5	–7	+4	+12	+34	+8	+10
Regional Deviations								
Greater London	–3	–11	–8	–1	–8	–16	–6	–12
Rest of SE.	+17	+7	+7	+1	+17	+3	–4	–2
South West	+10	0	+4	+1	–5	+10	+4	–7
East Anglia	+11	+8	+21	+6	+10	+5	–10	+12
East Midlands	–3	+7	+2	+1	+10	+5	+12	+6
West Midlands	–4	+6	+8	0	–7	+13	–10	–4
North West	–14	–7	–7	–7	–6	–17	–4	–6
Yorkshire/Humberside	–16	–2	–3	–3	+3	+7	0	–3
North	–14	–14	–6	–11	–12	+8	+4	+4
Scotland	–13	–9	–3	–9	–13	–22	–4	–7
Wales	–15	–11	–10	–1	–3	+3	–10	–1

Producer Services
B/FS: Business and Financial services.
Wh: Wholesale and materials-handling services.
T/C: Transport and Communications

Consumer Services
Ret: Retailing and consumer services.
H/C: Hotels and Catering services.
Mis: Miscellaneous (social organisations, recreation and personal) services.

Public Services
P/A: Public Administration and defence.
Soc: Social (educational, health) services.

Fig. 2.5.3
*Geography,
Volume 75 (part 4), October 1990*

Study the table (Fig. 2.5.3) showing regional components of service employment change 1981–7.

1 Discuss the main patterns shown in this table.

2 Use a suitable correlation technique to see if there is any relationship between selected components in the producer, consumer and public sector.

3 Comment on any relationship that you discover.

THE CHANGING PATTERN OF SERVICE INDUSTRIES IN CENTRAL LONDON

Consider a situation where more than a million people travelled to and from work on a small island that was a few square kilometres in area. It would be a spectacular achievement that would require significant incentives to persuade them to do so: an impressive infrastructure to allow them to do so: and a flexible system of development on the constrained island site that could cope with the problem of physically housing an employment concentration of that magnitude. The parallel between this example and that of central London is not exact, but it is close.

(M.E. Frost and N.A. Spence, *Geography Journal* Vol 157 part 2 July 1991)

Central London represents the biggest concentration of employment in Britain, with about 5.5 per cent of its labour force working in a few kilometres of densely packed urban area. Here lies one of the world centres of finance, business and commerce, and thus the biggest concentration of service employment in Britain. Although there was an overall loss in jobs in Central London in the 1970s there was a modest growth in the 1980s, although this conceals some important changes within the groups employed.

Study the table (Fig. 2.5.4) which shows changes in employment within Central London between 1981 and 1987 for a number of selected employment categories.

1 Comment on the growth and decline in the various sectors shown.

2 What factors would have led to the decline in manufacturing industry?

THE SERVICE INDUSTRIES IN BRITAIN

EMPLOYMENT CHANGES IN CENTRAL LONDON 1981–87

CHANGE IN EMPLOYMENT BY SELECTED INDUSTRY DIVISIONS

	\multicolumn{5}{c}{DIVISION}
	2–4	6	7	8	9
Central London					
1981 (total)	127 000	194 000	135 000	343 000	307 800
1981–1984 (absolute)	–5000	+6300	–14500	+31200	+16100
(per cent)	–4.0	+3.3	–10.7	+9.1	+5.2
1984–1987 (absolute)	–43 000	–3100	–17200	+73400	–12600
(per cent)	–33.1	–1.5	–14.3	+19.6	–3.9
1987 (total)	81600	197 200	103 300	447 700	311 300
1987 (share per cent)	6.9	16.7	8.8	38.0	26.4

Key 2–4 Manufacturing Industry
6 Distribution, Catering and Repairs
7 Transport and Communications
8 Banking, Insurance and Finance
9 Other Sources

Fig. 2.5.4

The buoyancy of banking, insurance and finance sector can be related to a number of factors:

- the growth of producer services in finance and business

- the restructuring of financial markets organisation

- the increased need for international banking, accountancy and legal services, in which Central London has special expertise.

There may, however, be certain constraints on the continued growth of this sector in Central London – the difficulties of:

- providing a supply of a highly qualified and technologically skilled labour force

- providing and extending suitable office space (Central London costs per unit area are some of the highest in the world)

- devising and introducing improvements in public transport that can cope with an increase of up to 150 000 new commuters in Central London in the next decade.

The diagrams (Fig. 2.5.5) show the take-up of office space in the City of London in the 1980s and early 1990s, and the projected demand and availability in the early 1990s.

In the light of the constraints mentioned above, discuss the main implications of these diagrams.

Fig. 2.5.5
The Independent on Sunday

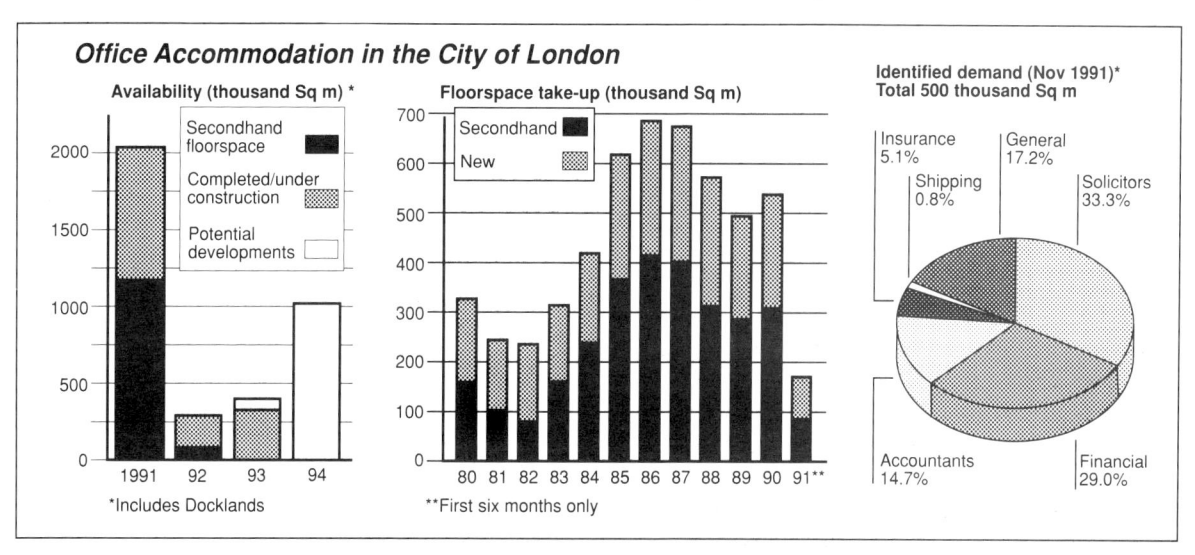

Office Accommodation in the City of London

Availability (thousand Sq m) *
 — Secondhand floorspace
 — Completed/under construction
 — Potential developments
(1991, 92, 93, 94)
*Includes Docklands

Floorspace take-up (thousand Sq m)
 — Secondhand
 — New
(80 81 82 83 84 85 86 87 88 89 90 91**)
**First six months only

Identified demand (Nov 1991)*
Total 500 thousand Sq m
- Insurance 5.1%
- Shipping 0.8%
- General 17.2%
- Solicitors 33.3%
- Financial 29.0%
- Accountants 14.7%

DECENTRALISATION AND SERVICE GROWTH ON THE DORSET COAST: THE CASE OF BOURNEMOUTH

Bournemouth has a long history of servicing leisure and recreation since its creation at the beginning of the nineteenth century. It has added an important new dimension to its service sectors in the 1980s and 1990s, and now ranks as one of the leading business and financial centres on the south coast. Its traditional tourist function provides permanent work for 12 000 people and extra seasonal work brings the summer total to 20 000. Spending by the two million tourists who visit Bournemouth each year is estimated at £250 million and supports a whole range of additional employment in the retail, leisure and entertainment sectors. The growing conference industry has generated extra employment in the service sector, and serves to even out the seasonal nature of tourism. In the last twenty years, however, it is the rapid expansion of office based employment that has created additional employment in the service sector (Fig. 2.5.6). Five companies have their national headquarters in the town, and Chase Manhattan has established its European headquarters on the eastern outskirts.

The table (Fig. 2.5.8) shows the main employers in the Bournemouth area and the table 2.5.7 shows office rents for a number of centres in the South of England.

Fig. 2.5.6
Abbey Life Headquarters, Bournemouth: it occupies a central position in Bournemouth's new office quarter

THE SERVICE INDUSTRIES IN BRITAIN

Fig. 2.5.7 (below)
Office Rents: Southern England

LOCATION	OFFICE RENTAL 1991 ($£/M^2$)
City of London	646
London Victoria	538
Brighton	215
Bristol	205
Southampton	183
Winchester	188
Exeter	156
Bournemouth	140

Study the employment structure figures for the Bournemouth TTWA (travel-to-work area) shown in Fig. 2.5.9.

1 Calculate the location quotients for the various groups for 1989.

2 Use a graphical method to show the percentage changes that have occurred in the period 1980–9.

3 Comment on the changing nature of the employment structure in the Bournemouth area, both in relative and absolute terms.

Fig. 2.5.8

MAJOR EMPLOYERS IN BOURNEMOUTH AND POOLE 1991

ORGANISATION	DATE MOVED TO AREA	NUMBER OF EMPLOYEES
Manufacturing Sector		
Davy McKee Corporation (P)	1974	550
Devilbiss Co. (B)	*	330
Dolphin Packaging (P, N)	*	450
Powell Duffryn (Hamworthy Engineering) (P)	*	2000
Metal Box (P)	*	550
Merck (P)	*	1100
Revlon Group (B)	*	1050
Siemens Plessey (P)	*	2000
Sunseeker International (P, N)	*	300
W L Miller & Son (P, N)	*	950
Services Sector		
Abbey Life (B, N)	1975	1300
American Express (P)	1981	160
Barclays Bank (P)	1975	1500
Chase Manhattan (B)	1985	1100
Frizzell Insurance (B, P, N)	1970	1750
Gresham Group (B, N)	1975	260
General Accident (B)	*	100
Health First (B, N)	1985	210
Link House Publications (P)	1975	290
Lloyds Bowmaker (B, N)	*	600
McCarthy and Stone (B, N)	*	360
Municipal General Insurance (B)	1989	200
RNLI (P, N)	1974	310
Teachers Assurance (B)	1971	200

Notes: *= Pre 1970   N = National Headquarters B = Bournemouth P = Poole

Fig. 2.5.9 (below)

EMPLOYMENT STRUCTURE FOR BOURNEMOUTH TTWA* 1980 AND 1989

1980 STANDARD INDEX CLASSIFICATION DIVISION	PERCENTAGE OF TOTAL			
	1980	1989	DORSET 1989	UK 1989
0 Agriculture, Forestry and Fishing	0.8	0.4	0.2	1.4
1 Energy/Water Supply Industries	0.9	1.0	1.3	2.0
2 Extraction/Manufacture: Mins/Metals	3.0	1.8	2.1	3.3
3 Metal Goods/Vehicle Industries etc.	11.2	10.1	11.7	10.4
4 Other Manufacturing Industries	5.1	4.3	6.1	9.4
5 Construction	4.5	3.8	3.8	4.8
6 Distribution, Hotels/Catering: repairs	32.2	29.8	26.2	21.0
7 Transport/Communication	6.2	6.0	5.2	6.0
8 Banking, Finance, Insurance, Leasing etc	11.3	18.0	13.7	11.8
9 Other Services	24.7	24.7	28.0	28.0

*TTWA = Travel to work area

A NEW VIEW OF BRITAIN

The Chase Manhattan transfer from London

Fig. 2.5.10
New European and African Headquarters of the Chase Manhattan Bank on the eastern outskirts of Bournemouth at Littledown.

Some Comments:

European Systems Chief Executive, Vince Grant, *'What I did was to say, why don't we go where people would go of their own accord – where people might like to go, and dream of going once they pack up their work. I was pushing the thought process to the south coast, and by coming to Bournemouth we found that place'.*

P.J. Challen, Borough Planning Officer, *'The relocation of the operations side of Chase Manhattan Bank to Bournemouth together with the opening of the Bournemouth International Centre promises to reactivate interest in the town as a major centre on the south coast, providing the employment, shopping and recreation needs for a large area.'*

Discuss the main local social, economic and environmental implications of the move of Chase Manhattan to Bournemouth.

Fig. 2.5.11

Manhattan transfer

Special report by Martin Webster

THE Chasside operation of Chase Manhattan Bank has grown on former grazing land in Littledown in just six years to become one of the most important centres the banking giant has in Europe.

Every day of the week, every day of the year, US$75 billion of business is transacted in the green glasshouse sitting off the Spur Road as you drive into Bournemouth.

Like other recent developments in the area – the Littledown Centre, Bournemouth General Hospital, and now the Wessex Fields office development – Chase has transformed the landscape.

And it has transformed the working lives of people it used to employ in London and transformed the lives of people who, six years ago, may not even have known what Chase Manhattan was...

As one of the major banks in the world, Chase's decision to come to Bournemouth to set up its operations and administration sections was hailed by all locally as good news.

Local jobs – 1000 by the year 1996 Chase promised. Local housing – with 500 jobs being relocated from London, all would need somewhere to live. Local traders – 1500 people by the year 1996, working in Bournemouth and with money to spend in local shops.

But Chase didn't stick to their word on the jobs front – they bettered it.

The 1500 figure by 1996 will, in fact, be reached by the end of this year. It's been a phenomenal growth, beyond even the wildest dreams of top Chase directors, and Bournemouth is now looked upon by the bank's chiefs in New York as a role model for other operations to emulate.

And all this from an office which was originally set up as the back room branch for London. It has turned the tables on London, with Bournemouth by the end of this year employing more people than their City colleagues.

The company has already received planning permission to build a third phase on the Littledown site, but says there are no plans yet to do so. It is a precaution just now in case business outgrows space again, as it did with phase one.

The growth of Chase has been one of the factors in helping to elevate Bournemouth into being the fifth most important financial centre in the country.

Evening Echo, Bournemouth, 22 October 1990

BRITAIN'S ENERGY SUPPLIES

The mix of fuels used in the production of Britain's energy supplies has shown important variations over the last fifty years. The different proportions of the fuels used reflect a whole range of economic, technological and political factors.

The table (Fig. 2.6.1) shows the changing percentage contribution of the fuels used at ten year intervals from 1937–87.

1 Comment on the varying proportions of fuels used over the 50 year period.

2 What explanations might there be for these variations?

The table (Fig. 2.6.2) shows similar figures for the current period and forecasts into the next century.

1 What do you consider to be the main trends forecast in the next century?

2 What events could lead to a substantial alteration in these forecasts?

THE FUTURE FOR COAL

The proportion of coal in the energy mix has remained between 32 and 37 per cent since the early 1980s, and the figures for the early years of the next century suggest a very slight increase in 2020. The future for the British coal industry is, however, a far from certain one. The industry faces privatisation in the 1990s, and this could lead to an even greater slimming down of the industry than has already taken place. The main market for British coal, the power stations, is unlikely to be sustained at the present level as the generating companies, National Power and Powergen, look to new technologies and alternative sources of cheaper fuel.

The table (Fig. 2.6.3) shows figures for Britain's coal production, sales and productivity for the period 1947–92.

1 Produce a series of composite line graphs to show trends in a) production, b) sales and c) productivity.

2 Comment on the main trends that are apparent from your graphs.

Fig. 2.6.1
Energy Mix: UK 1937–1987

YEAR	TOTAL (MTCE)	COAL (PER CENT)	OIL (PER CENT)	NATURAL GAS (PER CENT)	OTHERS (PER CENT)
1937	202	74	26	–	–
1952*	232	90	10	–	–
1957	247	85	15	–	–
1967	276	59	39	1	1
1977	285	38	41	19	2
1987	308	33	37	24	6

(*1947 not shown as immediately post war and not representative; others include hydroelectric power (HEP) and nuclear. MTCE – million tonnes of coal equivalent.)

YEAR	TOTAL (MTCE)	COAL (PER CENT)	OIL (PER CENT)	NATURAL GAS (PER CENT)	NUCLEAR (PER CENT)	HEP (PER CENT)	RENEWABLES (PER CENT)
1985	320	32	36	24	7	1	–
1990	344	33	34	23	7	2	1
2005	438	30	33	27	7	2	1
2020	517	34	32	28	4	1	1

Fig. 2.6.2
Energy Mix: UK 1985–2020, Guardian 17 November 1989

UK COAL PRODUCTION SALES AND PRODUCTIVITY 1947–1992

	1947	1950	1955	1960	1965–6	1970–1	1975–6	1979–80	1982–3	1985–6	1988–9	1991–2
Production												
Deep-mined (million tonnes)	188	206	211	187	177	136	115	109	105	88	85	71
Open cast (million tonnes)	12	14	14	10	9	9	11	14	16	16	19	17
Total	200	220	225	197	186	145	126	123	121	104	104	88
Power Station Sales												
Million tonnes	28	34	44	53	70	75	76	89	81	86	81	81
Percentage of output	14	15	20	27	38	52	60	72	70	83	78	92
Number of Collieries	958	901	850	698	483	292	241	219	191	133	86	50
Productivity												
Manpower (000's)	718	689	695	589	456	286	248	232	208	139	80	44
Output per man/shift (tonnes)	1.1	1.2	1.3	1.4	1.8	2.2	2.3	2.3	2.4	2.7	4.1	5.3

British Coal Annual Report and Accounts 1991–2

Fig. 2.6.3

3 Which sets of statistics would you use to convince a sceptical friend that there was a future for the British Coal mining industry?

Fig. 2.6.4
New colliery in the Selby coalfield set in arable landscape of North Yorkshire

Colliery closures

It will be seen from the table that the rate of colliery closure has increased considerably in the 1980s. In early 1992 there were only 53 pits left, employing 49 300 miners. In the early 1970s the Plan for Coal insisted that the best chances for coal lay in the Midlands and Yorkshire where geological conditions were best. The more peripheral coalfields in Scotland, south Wales and the North-east of England were to be starved of investment even though they had some of the best coking coals and anthracite. Investment in the 1980s has been concentrated in the fields (mainly Nottinghamshire and Yorkshire) that were producing coal for the main market, the electricity generating industry. This investment has not met with much success and increasing numbers of colliery closures are occurring in this heartland.

In mid-September 1991, a report from Rothschilds, the merchant bankers, proposed the most drastic set of reductions for the early 1990s. The coal industry would be reduced to 14 pits employing 11 000 miners. Coal mining would be wiped out in South Wales and the North-west, and only one pit would remain in the North-east and in Scotland.

No great corporation in Britain has been improving productivity as impressively as British Coal (BC). But the faster the improvement (12 per cent last year), the more the ground slides under its feet as Britain's energy industry is increasingly opened up to market forces. The crunch day is 31 March 1993, when existing contracts with the electricity authorities, which take about 80 per cent of BC's output, expire. Then they will be able to buy coal from the cheapest sources in the world while pushing ahead with combined cycle gas-fired stations. Yesterday Powergen announced a £40 million terminal in Liverpool – for completion by late 1993 – capable of handling up to 4.9 million tonnes of imported coal, to add to the 16.7 million tonnes imported last year (against exports of 2.1 million tonnes). Employment, once over a million when King Coal ruled the world, is down to 49 300 in 53 pits. If the report commissioned from Rothschild is true, the number may dwindle to under a dozen: and history will remember Arthur Scargill as an optimist.

Guardian 24 January 1992

1 What are the main implications of this extract for

a coalmining communities throughout Britain?

b electricity generation?

2 Are there any other scenarios that could be envisaged for the future of coal mining in Britain?

In October 1992, miners' fears, as prompted by the Rothschild report, became a reality. The Government announced a massive closure programme for the coal-mining industry: 31 out of 50 collieries were to close by the end of March 1993, with a consequent loss of 30 000 jobs from a present total of 53 000. Closures would occur in all existing coalfields (except Scotland), but both Yorkshire and Nottinghamshire would be particularly seriously hit. Four of the pits would be 'mothballed', in order to secure their reserves if there were better times ahead. The main reasons given for this considerable cutback in mining capacity was that there was simply no market for the coal that was being produced from the threatened pits. Very considerable quantities of coal were being stockpiled both at the pitheads and at the coal-burning power stations that are the main market for the coal. Furthermore no contract had been signed with the privatised power companies, National Power and Powergen, for supplies of coal beyond March 1993.

The reaction to the announcement was quite unprecedented. Fierce anger was expressed in the coalfields, but this was also echoed in a sense of outrage, and support for the miners throughout the country. Within a week the closure programme had been modified by the Government, largely as a result of public opinion and pressure from its own backbenchers in the House of Commons. Ten pits were to remain on the closure list, subject to the statutory consultation period. A reprieve was given to the remaining 21, pending the result of a wide ranging Government review. This review would consider not only the future of the threatened pits,

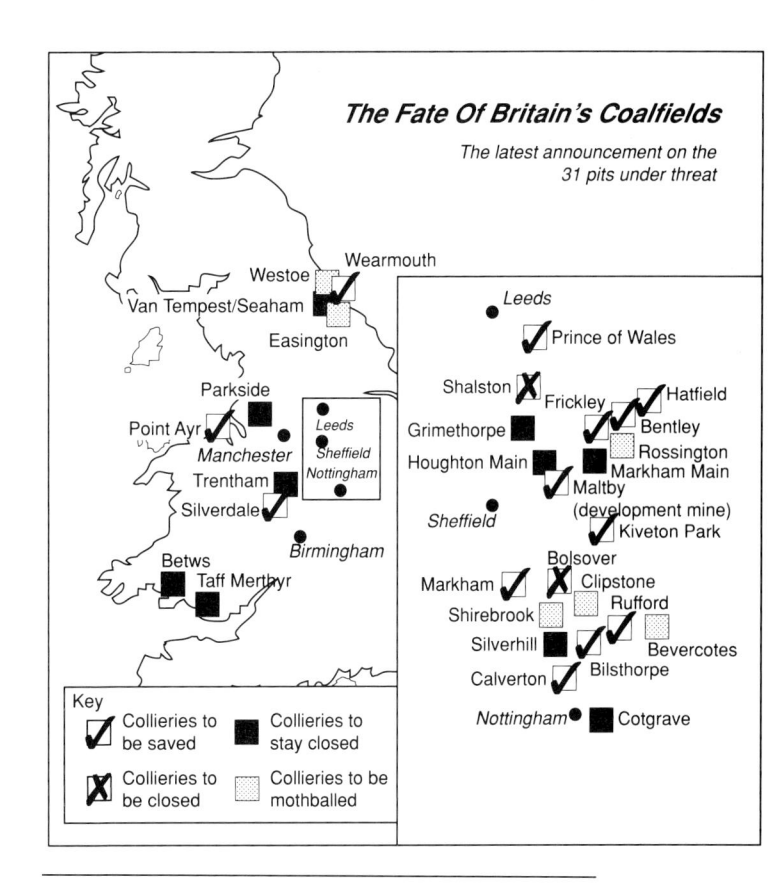

Fig. 2.6.5
Guardian 26 March 1993

Fig. 2.6.6
Deep-mined open-cast coal production 1986–1992

YEAR	DEEP-MINED	OPENCAST	TOTAL
	(MILLION TONNES)		
1986	90.3	14.2	104.5
1987	85.9	15.8	101.7
1988	83.7	17.9	101.6
1989	79.6	18.6	98.2
1990	71.5	17.8	89.3
1991	72.3	17.0	89.3
1992	71.0	16.7	87.7

but also the broader, and relevant, issues of the fuel market within the privatised electricity generating industry, the position of nuclear power (which was not privatised because of the high cost of decommissioning the power stations) and the consent procedure for licensing gas-fired power-stations.

In late March 1993, the Government announced its final decision on colliery closures. The details as summarised in *The Guardian* 26 March 1993, are shown below.

- Pits: 12 to close, 12 saved, 6 'mothballed' (care and maintenance basis), 1 (Maltby) to be a development mine.
- Another 5 800 miners to join the 8 000 who have left since October 1992.
- £13 billion, five year contracts with the generators.
- French electricity link to stay open, nuclear power and dash for gas to continue.
- Government subsidy for extra coal sales.
- Privatisation at earliest possible opportunity.
- Former British Coal pits to go private.
- Working practices must change. More hours underground.
- £200 million aid package for communities.

The map (Fig. 2.6.5) shows the final decision on the 31 pits.

Open cast coal mining

The table (Fig. 2.6.6) shows the production of deep-mined and opencast coal for the five years 1986–90.

There are two different kinds of operation in the mining of opencast coal. British Coal Executive operates one set of sites; the licensed sites are run by private contractors. About 80 per cent of the production goes to power stations. Opencast sites are usually cheap and easy to run, and employ 7500 directly and another 7500 indirectly. Whereas deep mines in 1989 only yielded profits which represented three per cent on capital invested, opencast operations profits represent 158 per cent return on capital investment. To this extent British coal will rely heavily on opencast operations in reaching its financial targets.

Increasing open cast production appears to be an attractive option for the future, but the mining does pose severe environmental problems, both in the mining stage, and in restoration. The average site is some 200 hectares, and has a life of five years.

Fig. 2.6.7
New open cast pit: Coalfield North, Leicestershire

Coalfield North (Fig. 2.6.7) in north-west Leicestershire occupies 182 hectares, of which 146 had been used for agricultural land. The site had proven reserves of eight million tonnes, and extraction rates in the late 1980s were one million tonnes a year. Although the site was due to cease operation in late 1991, new proposals to develop an adjacent site were announced in 1989. FOIL (Fight Opencast in Leicestershire) was formed as a local protest body. Amongst the arguments put forward by FOIL are:

- shallow mining creates only a relatively small number of jobs, compared to deep mining
- its very presence discourages any local investment
- house prices have fallen as a result of the proposal
- increases in noise and dust levels
- although restoration of the landscape is always a part of the contract for mining, 400 hundred year old landscapes cannot be re-created in the 50 year restoration period.

What are the main arguments that British Coal could deploy for the opening of this new site?

BRITAIN'S HYDROCARBON RESOURCES: THE FUTURE

Petroleum has been produced from the North Sea fields since 1975, with peak production of 2.5 million barrels a day being achieved in 1985.

BRITAIN'S ENERGY SUPPLIES

Fig. 2.6.8
Piper Alpha Disaster

Original forecasts of a steady decline from that figure are now being revised in the early 1990s. In the mid-to-late 1980s there was something of a dismal outlook, with falling oil prices being followed by the Piper Alpha disaster in 1988 in which 167 workers died (Fig. 2.6.8) and considerable production lost. In 1990 production was only 70 per cent of its 1985 levels (Fig. 2.6.9).

OIL PRODUCTION (MILLION TONNES) 1977–92			
1977	37.5	1985	128.2
1978	53.3	1986	128.6
1979	77.9	1987	125.5
1980	80.5	1988	114.4
1981	89.5	1989	91.6
1982	103.4	1990	91.6
1983	114.9	1991	91.6
1984	125.9	1992	96.6*
			*estimate

Fig. 2.6.9

Now, however, there are confident predictions of a second peak being achieved in the late 1990s. 1990 was a record year for drilling and discoveries almost equalled total production for 1990 itself. About fifty oilfields are currently in production and some of these will be abandoned in the 1990s as production levels fall.

Optimistic forecasts, however, suggest that new fields may come on stream at a rate of four to ten a year for the next 20 years. Another estimate is that up to 300 new fields could be developed in the next 30 years. More conservative estimates put the figure of new oilfields at an additional 60 by 2015. It is likely that many of the new fields to be developed will be smaller than existing ones, and new technology will be important in their development. The new Lyell field, 300 miles north-east of Aberdeen will be developed with a submarine manifold, an installation that lies on the sea bed, and sea bed wells. From the manifold the oil will flow to a platform on the existing Ninian field and after processing, will be pumped ashore. Up to 40 per cent of new developments are likely to use submarine manifolds. Saltire, another field that is scheduled to begin production in 1992, will use an existing platform to process and deliver the oil to the mainland. In the wake of the Piper Alpha disaster, the Cullen report has resulted in an upheaval of safety regimes throughout the North Sea. This could hit production in the short term and shorten the lives of fields in the long term.

Study the diagram (Fig. 2.6.10), which appeared in *The Guardian* in the mid-1980s.

Fig. 2.6.10
Guardian

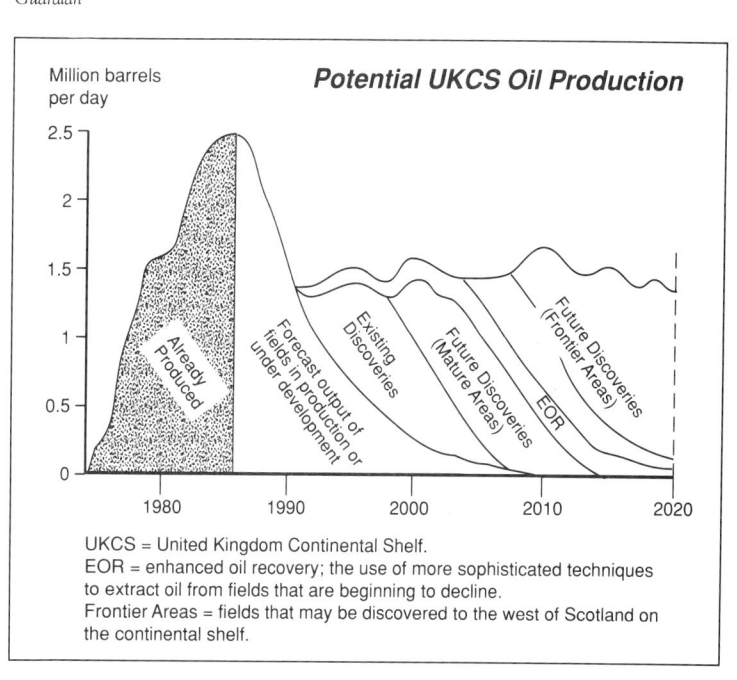

Potential UKCS Oil Production

UKCS = United Kingdom Continental Shelf.
EOR = enhanced oil recovery; the use of more sophisticated techniques to extract oil from fields that are beginning to decline.
Frontier Areas = fields that may be discovered to the west of Scotland on the continental shelf.

In the light of the last two paragraphs suggest ways in which the diagram might be modified to take in recent developments.

The falling oil prices of 1986 and the encouraging results from drilling in the early 1990s have opposite effects on the prosperity of the onshore areas that service operations in the North Sea.

Professor Alex Kemp, Aberdeen University:
'Aberdeen is dangerously dependent on oil. We missed out on the new technology industry which came to central Scotland because of the oil boom. I think we are in for a hard time over the next year or two. People are apprehensive and those involved in drilling are feeling very down.'

'Aberdeen lost 25 per cent of its oil-related jobs between 1985 and 1987. House prices fell six per cent in the Aberdeen area, while the value of residential property elsewhere in Britain soared by 33 per cent. Now the price of oil has recovered, so has Aberdeen. The 10 000 jobs lost in 1986 have been recovered and house prices have increased 20 per cent in the past two years.

1 Discuss the full range of effects that slump or optimism in the North Sea can have on towns like Aberdeen.

2 What are the long term effects likely to be, when oil production begins to fall dramatically?

Fig. 2.6.11

Natural Gas Distribution System

A new boom for natural gas?

Natural gas was first discovered in the North Sea in 1964 and was first brought ashore in 1967. Gas is now delivered at four terminals along the North Sea coast at Bacton, Theddlethorpe, Easington and St Fergus in Scotland. The South Morecambe field in the Irish Sea sends its gas to the terminal at Barrow, and into the gas grid (Fig. 2.6.11). Although the fields in the southern North Sea are now past their peak production, the shortfall has been eliminated by supplies from the Frigg field and 'associated' gas from fields such as Brent.

New gas discoveries in the renewed round of drilling in the North Sea are as encouraging as the oil finds, with 1990 figures showing 15 per cent more cubic metres than UK consumption for that year. Drilling in the Irish Sea, with its important new base at Barrow is also likely to yield new finds like the one off Rhyl. This could lead to a production of seven million cubic metres a day, with a processing plant at Point of Ayr.

In the late 1980s natural gas regularly supplied 25 per cent of Britain's energy requirements. Forecasts for the early years of the next century suggest that this pattern will be maintained. New commercial deals will allow Britain to take increasing amounts of natural gas from Norwegian fields in the 1990s since there is likely to be a serious shortfall by the mid-1990s. One significant development is the combined cycle gas-fired power station (CCGP), which uses exhaust gases from the primary turbine to generate

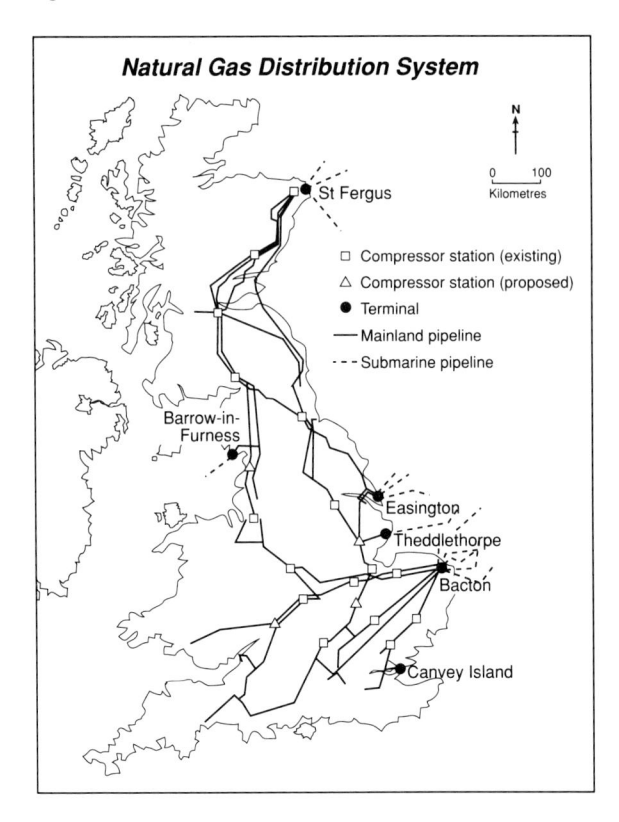

Fig. 2.6.12

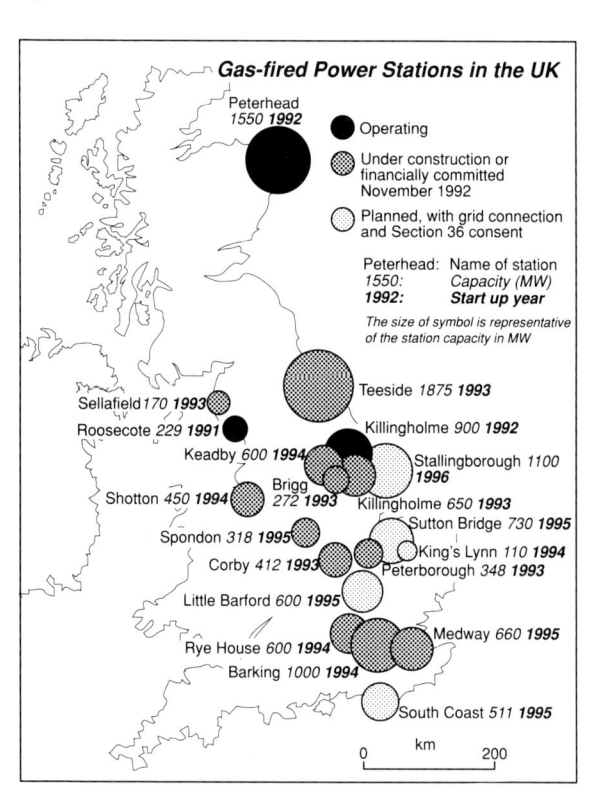

Gas-fired Power Stations in the UK

steam for a second – a much more efficient use of the fuel. Powergen are building a new station at Killingholme, south Humberside, but the main 'dash for gas' is being led by independent companies that are building CCGP stations that will sell power to the electricity boards (Fig. 2.6.12). Lakeland Power, with its stations at Roosecote, near Barrow-in-Furness is the first of this new group of power stations to come on stream. Environmentally these new stations have advantages over coal. They produce virtually no sulphur dioxide, about a fifth of the nitrogen oxides that coal stations produce and only 60 per cent of the carbon dioxide produced from the equivalent heat output from coal. However it is argued by some that the development of the new stations is a profligate waste of resources (British gas reserves could be exhausted in 25 years), since this extra capacity is not needed and will further handicap the coal industry. Many debates took place during the colliery closure crisis of October 1992 concerning the relative cost of electricity generated from coal and that generated from gas. Few experts agree on which is the cheaper commodity.

A FUTURE FOR NUCLEAR POWER?

'The Government has for some time recognised that our nuclear power is more costly than power from fossil-fuelled power stations.'

John Wakeham, Secretary of State for Energy, House of Commons, 9 November 1989.

This statement came in the run-up to the privatisation of the electricity industry, after years of confident prediction that nuclear power would be responsible for increasingly large amounts of the electricity generated in this country. Throughout the late 1970s and the 1980s it was asserted that by the turn of the century up to 22 per cent of our primary energy demand would be met from nuclear sources. None of the nuclear power stations were to be included in the privatisation of the electricity industry and a separate Government run company, Nuclear Electric was set up to run the nuclear programme.

At the time of the decision, the nuclear power industry operated the following power stations (Fig. 2.6.13):

Fig. 2.6.13

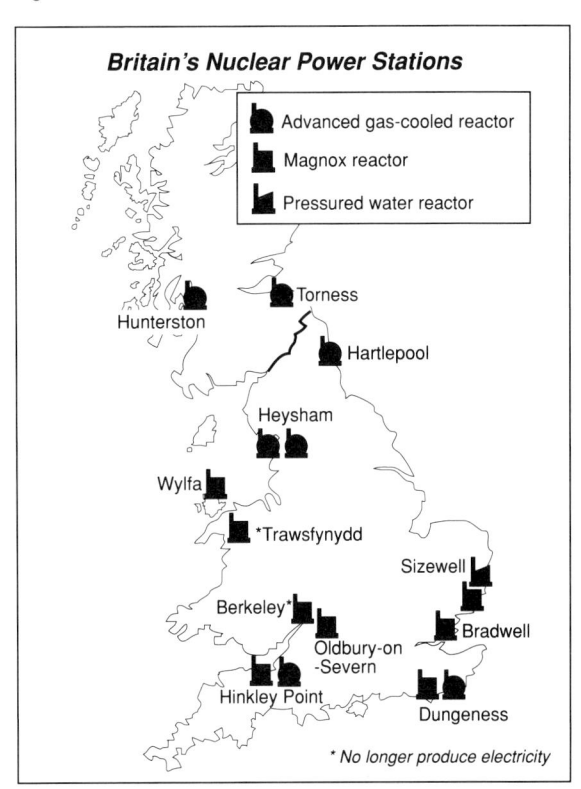

Britain's Nuclear Power Stations

- Advanced gas-cooled reactor
- Magnox reactor
- Pressured water reactor

Torness, Hunterston, Hartlepool, Heysham, Wylfa, *Trawsfynydd, Berkeley*, Sizewell, Oldbury-on-Severn, Hinkley Point, Bradwell, Dungeness

* No longer produce electricity

- eight ageing Magnox power stations (two have now ceased producing electricity: Berkeley and Trawsfynydd)

- seven Advanced Gas-cooled Reactor stations (AGR).

Sizewell B (Fig. 2.6.14), the first of a planned new programme of pressurised water reactors (PWR) was half finished, late in construction and far over its original budget cost. The programme of both Magnox and AGR stations had long faced a whole series of technical, economic and operational problems described in *The Sunday Times* of 12 November 1989 as 'Britain's longest running industrial disaster'. In the future, the industry would also have to face massive expenses, estimated in 1989 at between £8 billion and £15 billion for decommissioning and dismantling these stations and disposing of waste.

Some two years on from the decision not to include nuclear power stations in the privatisation programme the mood has changed somewhat. The prospect of increased emissions of carbon dioxide from fossil fuel burning power stations has now received much more publicity. James Hann, chairman of Scottish Nuclear Ltd summed up a changing perspective, *'More and more people, who still want their energy as cheaply as possible provided the lights don't go out, now worry about the impact the energy industries are having on our environment.*

Fig. 2.6.14
Sizewell B

Coupled with images of forests and lakes dying from acid rain, unacceptable air pollution, vast oil slicks and fires raging through Kuwait's oilfields there is the ultimate terrible vision of our planet withering beneath the sun's rays, trapped in a giant greenhouse.... All of these images now jostle with the crumbling radioactive concrete of Chernobyl for a share of the world's environment conscience.'

Whether this reflects widely held opinions is less clear, but there are signs that the nuclear industry in Britain may show a resurgence in the mid-1990s. A government review of the nuclear industry is due in 1994, and there is some optimism that new PWRs will be ordered. Planning permission exists for a PWR at Hinkley Point in Somerset. There are possibilities for others at Sizewell (in addition to Sizewell B) and at Sellafield and Chapelcross in Scotland. UK Nirex, the nuclear waste management company has chosen an undergound site near Sellafield for the disposal of all low and intermediate level waste although questions arise about its suitability. However, doubts will still remain: a report released in early 1992 shows a significant correlation between leukaemia deaths and workers in the British nuclear industry.

1   What do you consider to be the strongest arguments for the renewed building of nuclear power stations in the mid-1990s?

2   What assurances would need to be given before a new programme of construction began?

ALTERNATIVE SOURCES OF ENERGY

Estimates of the proportion of our electricity supplies that could be generated from alternative sources tend to proliferate: some are indicated below:

- Combined onshore and offshore wind energy resources could provide well over 60 per cent of the UK's energy requirements.

- The Severn Barrage could provide 75 per cent of the electricity consumption in England and Wales.

- Britain could generate half of its electricity from natural sources within 20 years.

- Britain could produce ten per cent of its electricity from wind, waves and the sun by the turn of the century.

A policy initiative by the Government in 1988 promised to 'stimulate the development and application of renewable energy technologies wherever they had prospects of economic viability and environmental acceptability'. The Environment White paper of 1990 promised to work towards producing 1000 megawatts (MW) of electricity from renewable sources – two per cent of Britain's needs, by the end of the century. It seems however, that applications from companies to develop electricity from renewables far outstripped the programme for 1991 (150–200MW).

Wind energy

The option of energy from wind power seems to have attracted much publicity recently. There does seem a reasonable chance that schemes could soon be in operation at a number of sites in Britain.

The two maps (Figs. 2.6.15 and 2.6.16) show the potential for generating electricity from wind in Europe and the relationship between areas of high wind speed and high landscape value in Britain.

Study the two maps and then:

1   Comment on the relative potential for wind-generated power in Britain compared to the rest of Europe.

2   In what ways would you modify your conclusions in 1 after studying the second map?

Fig. 2.6.15
Independent, January 1992

In practice wind power would become both expensive and awkward if it were to exceed supplying ten per cent of the nation's electricity. Providing ten per cent of Britain's electricity requirements would require more than 1000 wind farms. It is likely that fewer than half of the 58 projects the Government has agreed to subsidise will actually be built. In Scotland there is no subsidy for wind farms, yet it has the greatest potential. The most effective farms would be those built in offshore positions, but this is unlikely to occur until well into the twentieth century. By mid-1993 several sites in Cornwall and Wales were operational.

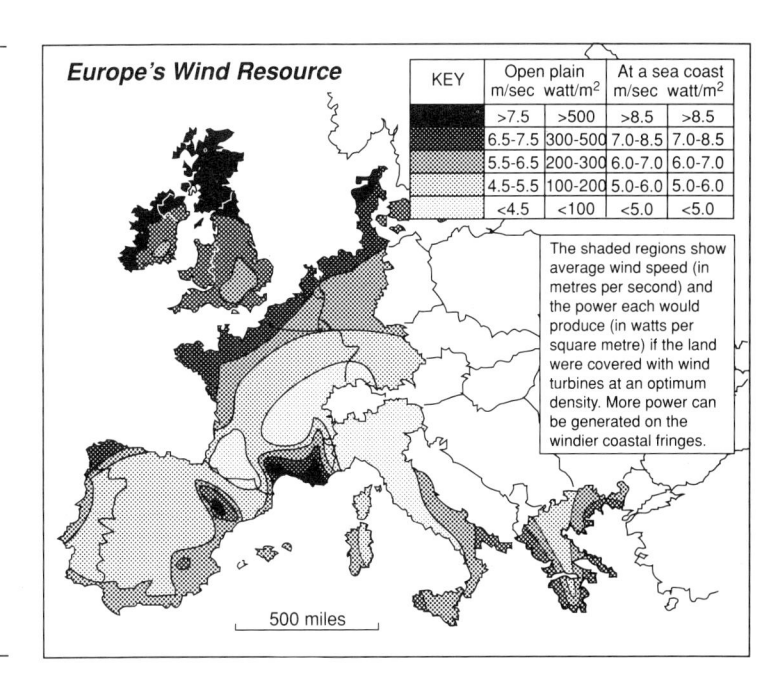

Europe's Wind Resource

KEY	Open plain		At a sea coast	
	m/sec	watt/m²	m/sec	watt/m²
	>7.5	>500	>8.5	>8.5
	6.5-7.5	300-500	7.0-8.5	7.0-8.5
	5.5-6.5	200-300	6.0-7.0	6.0-7.0
	4.5-5.5	100-200	5.0-6.0	5.0-6.0
	<4.5	<100	<5.0	<5.0

The shaded regions show average wind speed (in metres per second) and the power each would produce (in watts per square metre) if the land were covered with wind turbines at an optimum density. More power can be generated on the windier coastal fringes.

500 miles

Fig. 2.6.16
Independent, January 1992

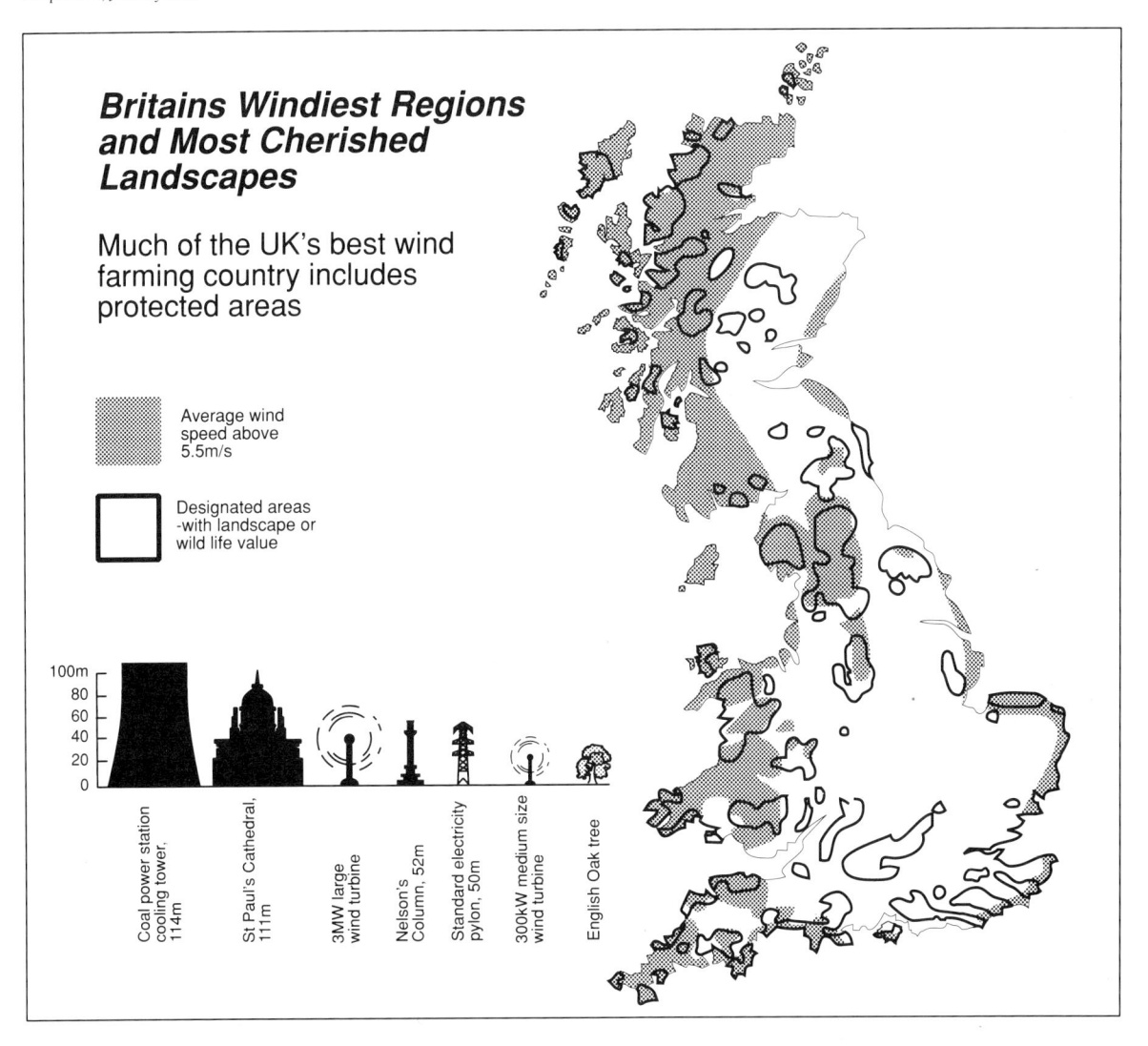

Britains Windiest Regions and Most Cherished Landscapes

Much of the UK's best wind farming country includes protected areas

- Average wind speed above 5.5m/s
- Designated areas -with landscape or wild life value

Coal power station cooling tower, 114m
St Paul's Cathedral, 111m
3MW large wind turbine
Nelson's Column, 52m
Standard electricity pylon, 50m
300kW medium size wind turbine
English Oak tree

INDUSTRIAL POLLUTION

The environmental problems that will be prominent in the 1990s are just the same as those at the beginning of the 1980s. Climatic change, acid rain, tropical deforestation, toxic waste management, ozone depletion, the loss of species, and all the rest will still be with us as we go into the next millenium and beyond. All that we have done in this last decade is to recognise that they require political solutions. But that is simply to arrive at the start gate. We have yet to get going.

Tom Burke: The Times 21 December 1989

PERCEPTION OF POLLUTION OF THE ENVIRONMENT

Fig. 2.7.1

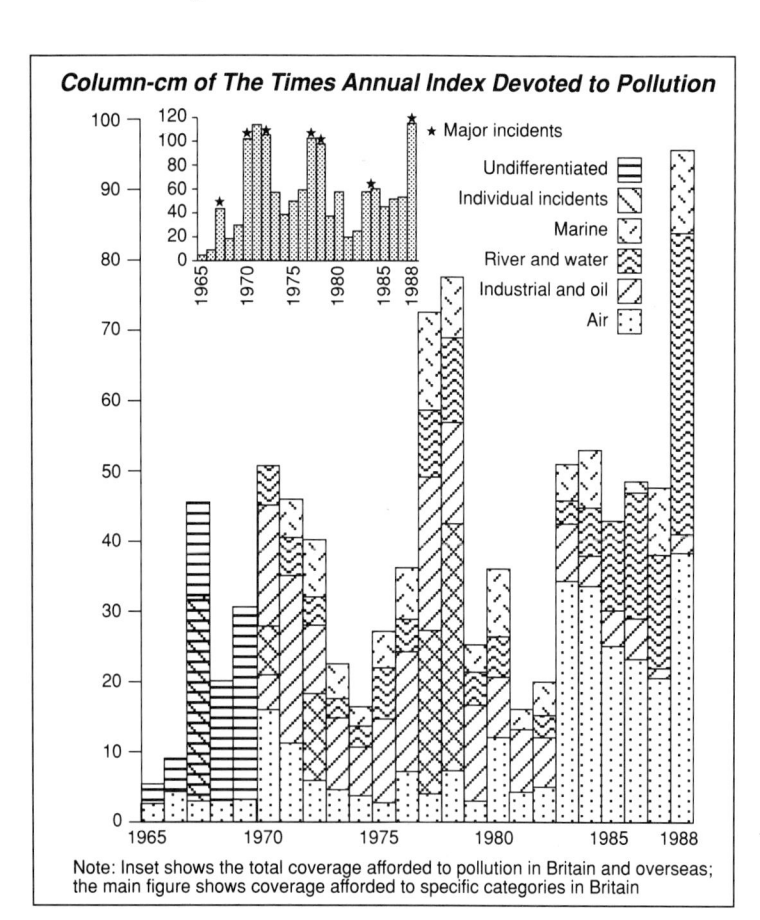

Column-cm of The Times Annual Index Devoted to Pollution

★ Major incidents
- Undifferentiated
- Individual incidents
- Marine
- River and water
- Industrial and oil
- Air

Note: Inset shows the total coverage afforded to pollution in Britain and overseas; the main figure shows coverage afforded to specific categories in Britain

Pollution of the environment has increasingly attracted the attention of the media and the public in general. Figure 2.7.1 shows the number of column centimetres devoted to pollution of the environment in *The Times* 1965–88 and Figure 2.7.2 records a survey of people's concerns over the environment and what they would be prepared to do to help.

1 Study figure 2.7.1. Analyse the trends that are apparent over the years between 1965 and 1988. Are there any problems with the classification used in this diagram?

2 Study figure 2.7.2. What factors would have to be taken into consideration when analysing these responses?

Classify the responses in the left hand column so that they may be compared with the column -cm in *The Times*. How do people's concerns compare with media reporting of pollution of the environment as shown in *The Times* survey?

WATER QUALITY IN RIVERS

Conflicting views exist on the degree of pollution of Britain's rivers. Successive issues of *The Sunday Times* in February and March 1989 illustrate this point. On the 26 February the papers front page lead commented, *'Britain's rivers are being polluted at a faster rate than at any time since national records began.'* A week later, a Government minister, writing in the same paper, noted, *'Let us be clear that we start from a position of respectability. Ninety per cent of our rivers are in a good or fair condition'*.

The diagrams (Fig. 2.7.3) show water quality condition for the rivers of England and Wales for the mid 1980s.

1 Comment on the pattern of water quality as shown in the diagrams? What are the significant regional variations?

2 Would it be possible to reconcile the two statements above after a study of these diagrams?

INDUSTRIAL POLLUTION

Focus: The Environment

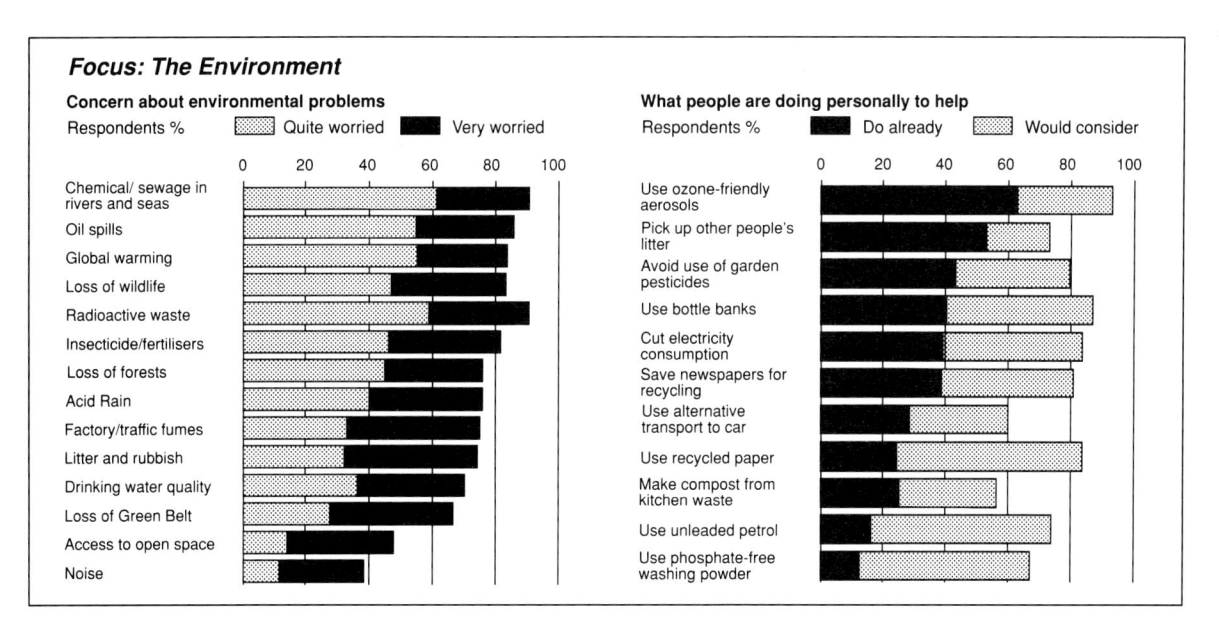

Concern about environmental problems
Respondents % — Quite worried / Very worried

- Chemical/sewage in rivers and seas
- Oil spills
- Global warming
- Loss of wildlife
- Radioactive waste
- Insecticide/fertilisers
- Loss of forests
- Acid Rain
- Factory/traffic fumes
- Litter and rubbish
- Drinking water quality
- Loss of Green Belt
- Access to open space
- Noise

What people are doing personally to help
Respondents % — Do already / Would consider

- Use ozone-friendly aerosols
- Pick up other people's litter
- Avoid use of garden pesticides
- Use bottle banks
- Cut electricity consumption
- Save newspapers for recycling
- Use alternative transport to car
- Use recycled paper
- Make compost from kitchen waste
- Use unleaded petrol
- Use phosphate-free washing powder

Fig. 2.7.2
Independent on Sunday

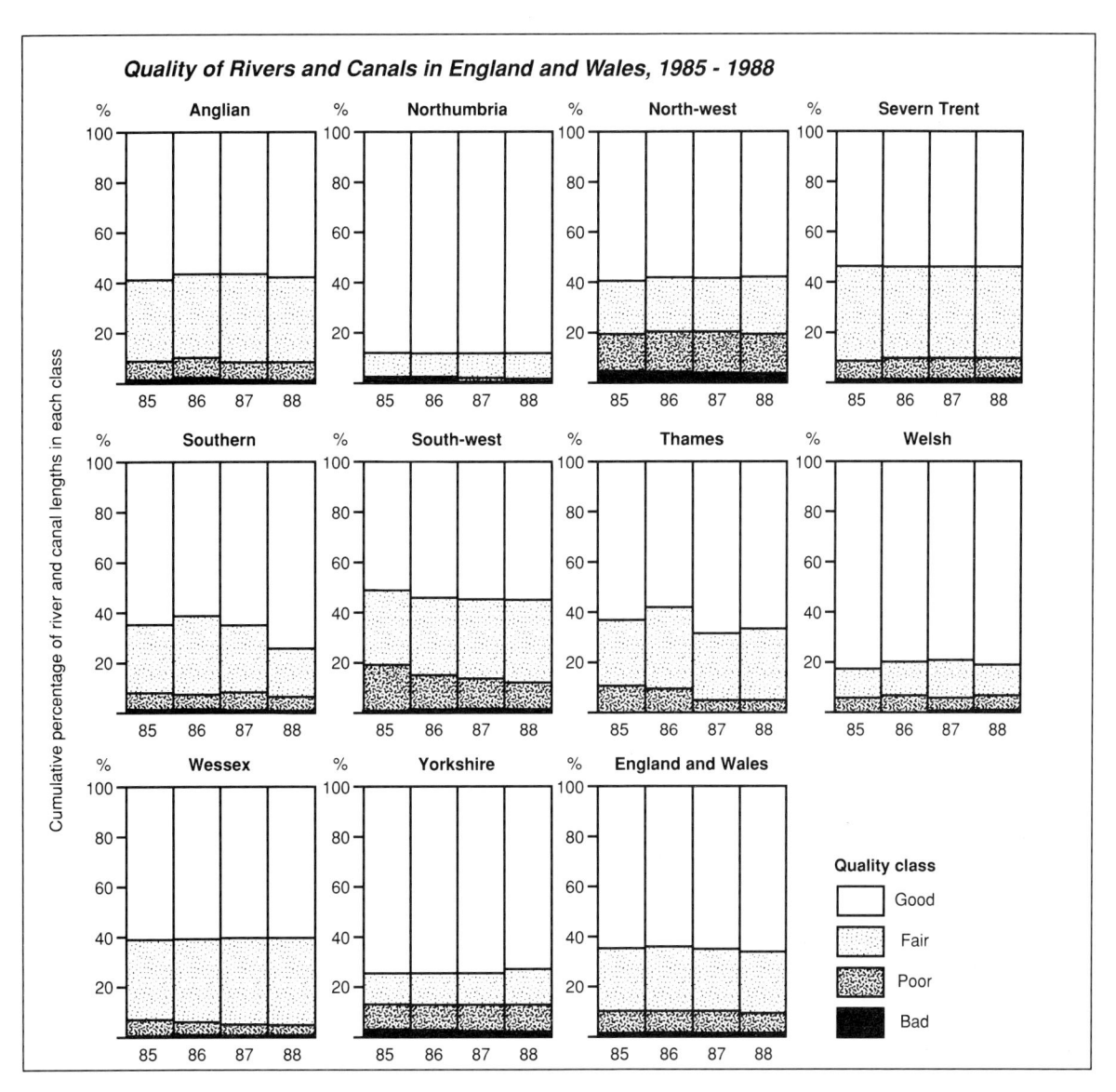

Quality of Rivers and Canals in England and Wales, 1985 - 1988

Regions shown: Anglian, Northumbria, North-west, Severn Trent, Southern, South-west, Thames, Welsh, Wessex, Yorkshire, England and Wales (years 85, 86, 87, 88)

Cumulative percentage of river and canal lengths in each class

Quality class:
- Good
- Fair
- Poor
- Bad

Fig. 2.7.3

The statistics (Fig. 2.7.4) show the number of pollution incidents recorded on rivers in England and Wales in the period 1981–8.

1 Draw a series of line graphs to show the number of incidents in each river authority's area during this period.

2 How does this help to assess the comments in *The Sunday Times*?

The other group of statistics (Fig. 2.7.5) focuses on industrial pollution in the rivers of the water authorities.

1 Calculate from the statistics, the percentage of pollution incidents that are caused by industry.

2 For each water authority, calculate a pollution index by dividing the total number of industrial pollution incidents by the length of rivers in the authority's area. Comment on the regional variations shown in your results.

Fig. 2.7.4

WATER POLLUTION INCIDENTS AND PROSECUTIONS IN ENGLAND AND WALES

WATER REGION	1981	1982	1983	1984	1985	1986	1987	1988	LENGTH OF RIVERS (KM)
Anglian	1095	1077	1288	1544	1707	1468	1605	1446	4453
Northumbria	509	544	613	654	722	729	671	7956	2785
North-West	1350	1288	1385	2241	2202	2480	2965	3365	5900
Severn Trent	2401	2681	3354	4372	4524	4497	4435	5292	6167
Southern	1300	1327	1400	1574	1668	1725	1795	1742	2161
South-West	1143	1227	1639	1685	1796	2220	2251	2760	2601
Thames	1810	2120	2345	2486	2695	2890	2969	3925	2418
Welsh	–	–	–	1418	1681	1619	2489	2707	4781
Wessex	844	790	966	1125	993	1332	1339	1920	2466
Yorkshire	1136	1020	1165	1536	2006	2444	2738	2974	6034
England and Wales	12 600	13 100	15 400	18 635	19 994	21 404	23 257	26 926	39 776

Fig. 2.7.5

WATER POLLUTION INCIDENTS REPORTED IN 1988: BY CAUSE

| WATER REGION | INDUSTRIAL | | | FARM | SEWAGE | | SEWERAGE | OTHER | TOTAL |
	OIL	CHEMICAL	OTHER		OWN	OTHER			
Anglian	478	98	71	203	←— 373 —→			223	1446
Northumbrian	135	←— 66 —→		80	←— 273 —→			141	695
North-West	508	←— 338 —→		612	100	115	399	495	2567
Severn Trent	1300	567	541	582	73	73	626	1530	5292
Southern	459	←— 182 —→		120	98	114	133	409	1515
South-West	254	←— 341 —→		840	←— 488 —→			463	2386
Thames	1256	←— 323 —→		188	←— 610 —→			997	3374
Welsh	197	79	274	582	122	105	249	577	2185
Wessex	435	←— 160 —→		392	←— 168 —→			506	1661
Yorkshire	403	←— 620 —→		353	←— 459 —→			698	2533
England and Wales	5425	←— 3660 —→		3952	←— 4578 —→			6039	23 654

Two case studies: the Rother and the Carnon

Sources of Pollutants in the River Rother

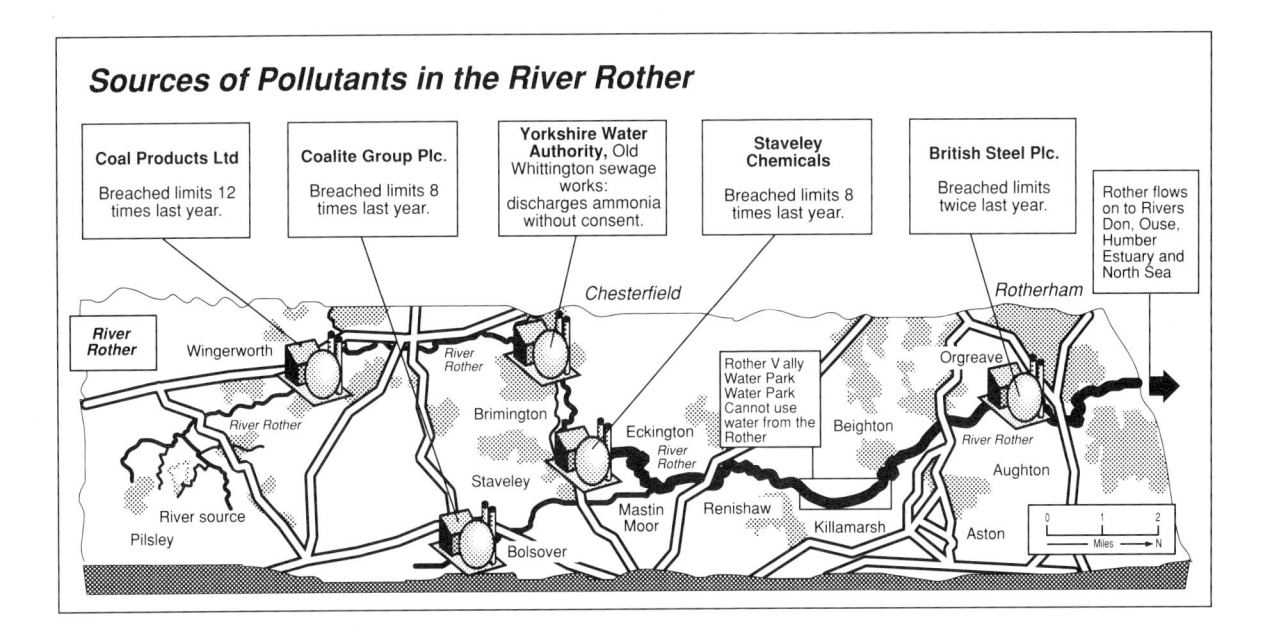

- **Coal Products Ltd** — Breached limits 12 times last year.
- **Coalite Group Plc.** — Breached limits 8 times last year.
- **Yorkshire Water Authority**, Old Whittington sewage works: discharges ammonia without consent.
- **Staveley Chemicals** — Breached limits 8 times last year.
- **British Steel Plc.** — Breached limits twice last year.
- Rother flows on to Rivers Don, Ouse, Humber Estuary and North Sea.

Locations shown: River Rother, Wingerworth, Pilsley (River source), Brimington, Chesterfield, Staveley, Bolsover, Mastin Moor, Eckington, Renishaw, Killamarsh, Beighton, Rother Valley Water Park (Cannot use water from the Rother), Orgreave, Aughton, Aston, Rotherham.

Fig. 2.7.6
Sunday Times 26 February 1989

The map (Fig. 2.7.6) shows the course of part of the River Rother in Yorkshire and the main sources of the pollutants. In 1989 the Rother was one of the most polluted rivers in Britain (Fig. 2.7.7). Over much of its length it was classed as grade four or poor. In 1991 it was still referred to as Britain's dirtiest river, with 33 km of dead water. It still receives industrial pollutants from the four major chemical companies. The Rother flows through the centre of the Rother Valley Water Park, yet its water is too polluted to be used in the central, focal lake in the park. It has to be supplied from a nearby, unpolluted brook. The National Rivers authority (NRA) now has a firm commitment to bring coarse fish back into the river by 1994. Yorkshire Water are committed to building an improved sewage treatment works and all the industries have agreed expensive pollution reduction targets.

The Carnon

Flowing through a major coalfield and industrial area, the Rother is polluted from a number of sources. In Cornwall the River Carnon flows only twelve kilometres from its source in the tin mining country near Redruth into Carrick Roads, a large estuary on which Falmouth is situated (Fig. 2.7.8) Effluent and waste from the old tin mines are likely to have caused the pollution that has affected the river for a number of years. Serious pollution in the river was the result of a chain of events in late 1991 and January 1992:

- One of Cornwall's two working tin mines, Wheal Jane closed in 1991, hard hit by the slump in tin prices. Department of Trade and Industry loans were not to be renewed.

- The owners, Carnon Consolidated, obtained planning permission to turn the area into a leisure park, which allowed it to avoid going into liquidation and to continue to run South Crofty, Cornwall's other working tin mine.

Fig. 2.7.7 Pollution in the River Rother

- Carnon then sold the pumps that drained Wheal Jane, and water levels began to rise. In November 1991 the water reached the surface and began to flow into the Carnon river. NRA money was used to pump the water into a surface lagoon, filled with toxic waste from 15 years of tin mining.

- Carnon ceased pumping on 4 January 1992 because the lagoon was becoming dangerously full.

- NRA threatened to stop paying for the pumps unless Carnon started them again.

- Carnon restarted the pumps, but probably too late.

- 13 January the build up of water in the mine burst through the Nangiles adit and began to flood into the River Carnon with a highly toxic content of heavy metals.

1   The Government's Policy is 'The Polluter pays'. Who should pay in this case?

a   Carnon Consolidated.

b   The previous owners of Wheal Jane Rio, Tinto Zinc and Consolidated Goldfields.

c   Joint responsibility should be accepted.

Fig. 2.7.9 Sunday Correspondent 27 May 1990

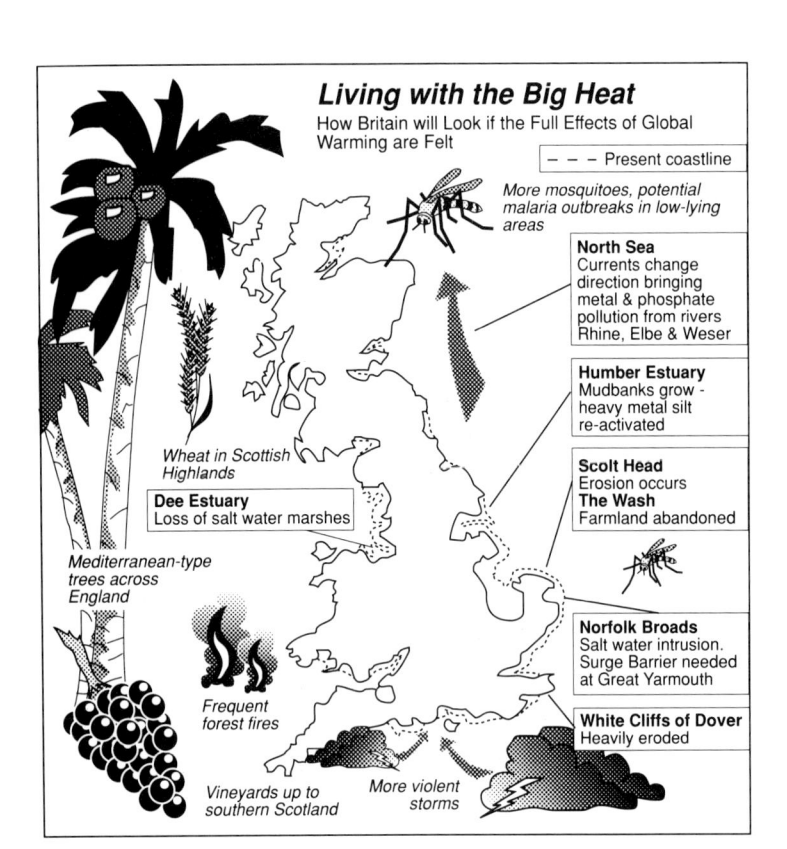

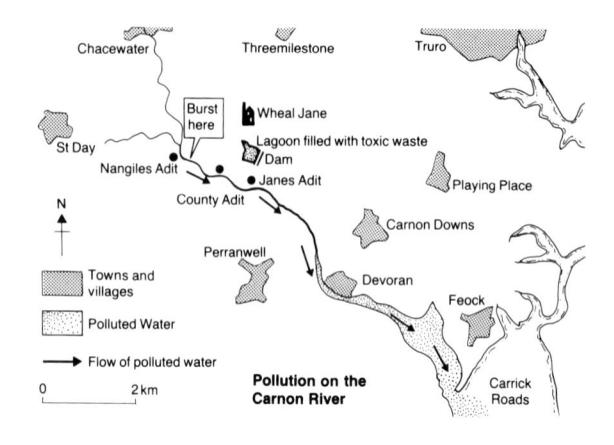

Fig. 2.7.8

ATMOSPHERIC POLLUTION

Concern for the increasing levels of atmospheric pollution is confirmed in the graphs shown at the beginning of this section. Global warming ranked third and acid rain ranked eighth out of the fourteen concerns expressed. Global warming may appear high on the list because it has achieved a high government and media profile. Acid rain certainly received much attention in the early eighties and although it is still very much a pollution issue, global warming has overtaken it in terms of media time and space.

The greenhouse effect is a natural phenomenon, caused by the gases in the earth's atmosphere. The atmosphere allows the passage of short wave radiation from the sun, which then heats up the earth's surface. Long wave radiation from the earth's surface is absorbed by gases in the atmosphere. Some of this energy is then radiated back to earth, thus warming its surface. The most important of the greenhouse gases is carbon dioxide, others include water vapour, methane, nitrous oxide and tropospheric ozone. These gases are responsible for maintaining a balance between incoming and outgoing radiation. They keep the earth's temperature more or less constant and much higher than it would be without them. In the last two hundred years increased emissions of carbon dioxide from the earth and the introduction of gases such as the chlorofluorocarbons have upset this balance. These emissions of gases have resulted in an increase in the earth's temperature of some 0.5°C to 0.7°C on average in the last 150 years. Even without reduced emissions of these gases temperatures are set to rise between 1.5°C and 4.5°C by 2050. Some of the consequences of this increase in temperature for Britain are shown in figure 2.7.9.

INDUSTRIAL POLLUTION

1 Comment on the changes shown in the diagram.

2 What additional implications of global warming are not shown in the diagram?

Fig. 2.7.10
Independent 17 April 1990

Greenhouse Gases and Rising Sea Levels

Increase in greenhouse gases measured in CO2 equivalent, parts per thousand by volume

Rising temperatures and sea levels
- Temperature left scale
- Sea level right scale

Contribution of greenhouse gases to climate change: 1980–1990

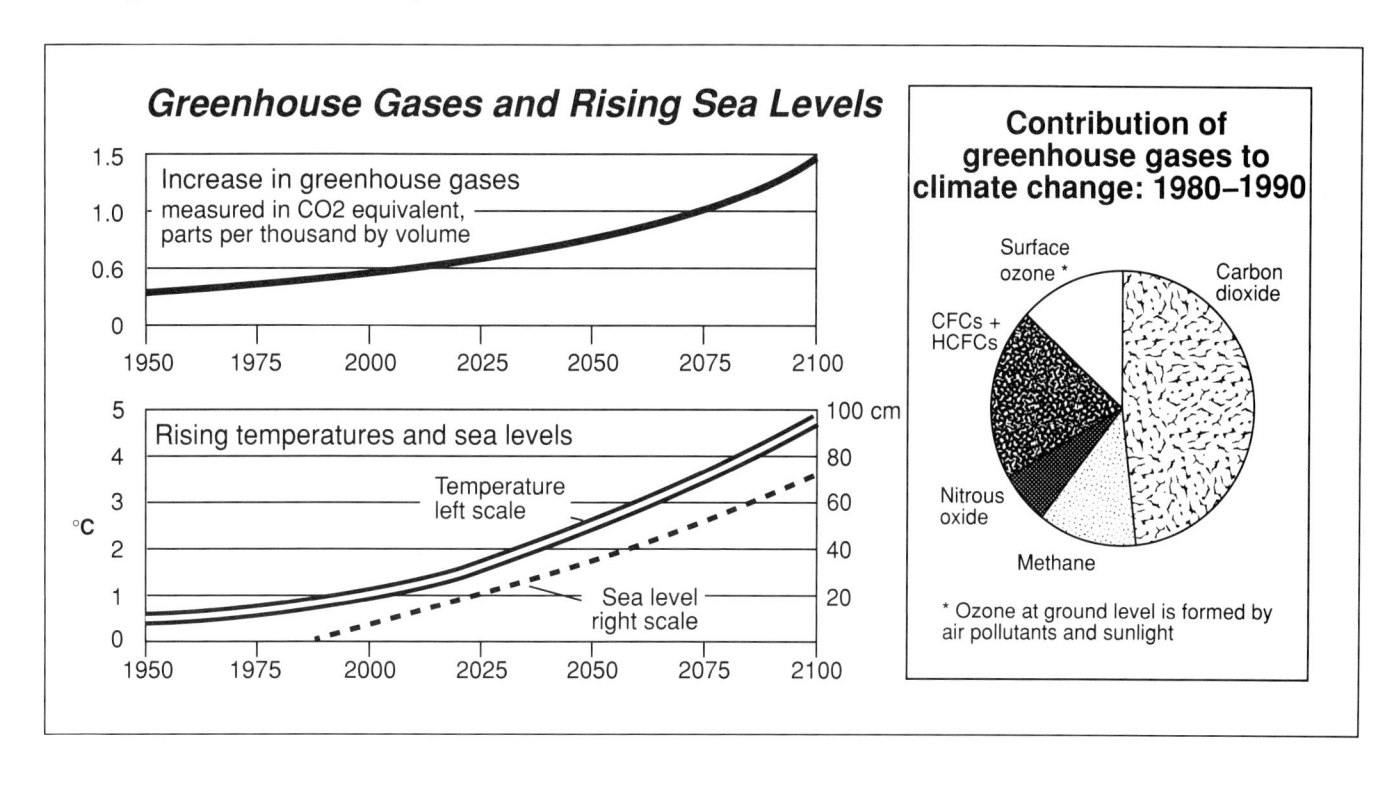

- Surface ozone *
- Carbon dioxide
- CFCs + HCFCs
- Nitrous oxide
- Methane

* Ozone at ground level is formed by air pollutants and sunlight

Figure 2.7.10 shows the greenhouse gases contribution to climatic change. The overall global increases in temperature and rise in sea level resulting from these increases are also indicated. Figure 2.7.11 shows the source of carbon dioxide emissions in Britain, with projections of likely future emissions in the next century.

1 Comment on the pattern of emissions in Britain.

2 What are the global implications of the other two graphs?

It is obvious that some fairly urgent action is necessary in order to reduce the emission of greenhouse gases. Britain produces about three per cent of the world's emissions, with one per cent of the world's population. In order to reduce our emissions to the world average (Latin American countries), emissions would need to be cut by two thirds. If the world is to achieve a 50 per cent reduction in emissions, British levels should be cut to about one sixth of the present level – about an 84 per cent reduction.

Carbon Dioxide Emissions in UK 1960 - 2020

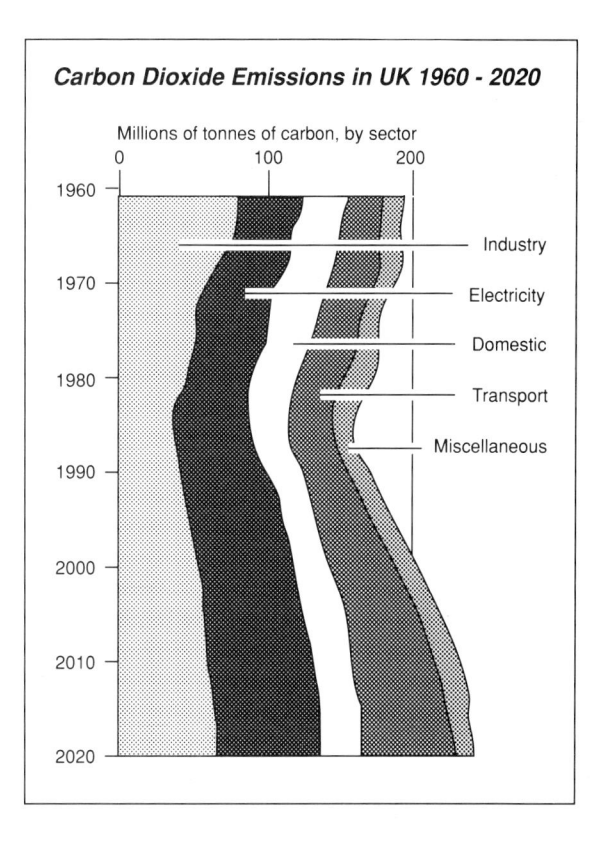

Millions of tonnes of carbon, by sector

- Industry
- Electricity
- Domestic
- Transport
- Miscellaneous

Fig. 2.7.11
Guardian
1 June 1990

Fig. 2.7.12
Climate Action Network UK from The Independent 29 October 1990

CARBON DIOXIDE TARGETS

COUNTRY	TARGET DETAILS	CONTRIBUTION OF WORLD CO_2 EMISSIONS (PER CENT) CURRENT ESTIMATES
No CONTROLS		
US	Not in favour of emission controls despite a vague commitment by President Bush to stabilise at unspecified levels in February 1990	22.0
USSR	Not in favour of emission controls at present	18.4
STABILISERS		
Japan	Stabilise at 1990 levels by 2000	4.4
UK	Stabilise at 1990 levels by 2005	2.8
Canada	Stabilise at 1990 levels by 2000 'as first step'	2.0
Italy	Stabilise at 1990 levels by 2000. Parliamentary resolution for 20% cut by 2005	1.8
Belgium	Stabilise at 1988 levels by 2000	0.5
Austria	Support stabilisation at 1990 levels by 2000. 20% cuts proposed by environment minister	0.3
Finland	Stabilise at 1990 levels by 2000, at least	0.26
Sweden	Stabilise at 1988 levels by 2000	0.22
Norway	Stabilise at 1990 levels by 2000	0.22
Switzerland	Has suported stabilisation at 1990 levels by 2000	0.2
Ireland	Support stabilisation at current levels by 2000	0.14
REDUCERS		
Germany	25% reduction on current levels by 2005. Agreed by cabinet but not yet ratified by parliament	3.2
Australia	20% reduction by 2005	1.1
Netherlands	Stabilise by 1995, 3% to 5% reduction by 2000	0.65
Denmark	20% reduction by 2000, up to 50% by 2030	0.3
New Zealand	20% reduction by 2000	0.1

All EC member states except UK agreed overall Community stabilisation at current levels by 2000 at an informal meeting of Environment Council Ministers on Sept 23 1990, under a formula which would allow Spain, Greece, and Portugal initial CO_2 increases.

The table (Fig. 2.7.12) shows agreed carbon dioxide target details for the major industrialised countries.

1 Devise a way of presenting these statistics more effectively.

2 Comment on the relative merits of the British commitment.

3 Refer back to the diagram (Fig. 2.7.11) which shows the sources of carbon dioxide emission in Britain. Suggest a range of proposals that would lead to the stabilisation in current emissions by 2005.

4 Many say that this commitment is not enough. What additional measures could be introduced to reduce present levels of emission? You may like to consider in detail Sir Walter Marshall's comment in *The Independent* 3 Janaury 1990. '... the solution to the problem of greenhouse warming will need a massive investment in nuclear power'.

Chlorofluorocarbons and methane are two other important contributors to the increased greenhouse effect. ICI is responsible for about ten per cent of the world production of chlorofluorocarbons (coolants for fridges, air conditioners, foam blowing agents, solvents and aerosols) and Britain is responsible for about six per cent of the world's consumption. The main sources of methane in Britain are shown in the diagram (Fig. 2.7.13).

1 Suggest ways in which the output and consumption of chlorofluorocarbons could be reduced.

2 Comment on the sources of methane in Britain. How could its production be reduced?

Fig. 2.7.13

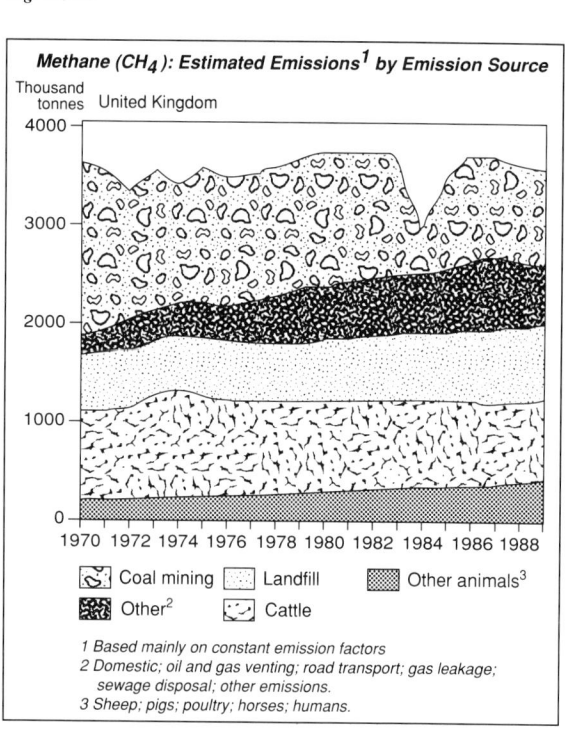

Methane (CH_4): Estimated Emissions[1] by Emission Source
Thousand tonnes — United Kingdom

Legend: Coal mining; Landfill; Other animals[3]; Other[2]; Cattle

1 Based mainly on constant emission factors
2 Domestic; oil and gas venting; road transport; gas leakage; sewage disposal; other emissions.
3 Sheep; pigs; poultry; horses; humans.

ATMOSPHERIC POLLUTION AND ACID RAIN

The main cause of acid rain is the conversion within the atmosphere of sulphur dioxide and nitrogen oxides emitted from coal-fired power stations and vehicle exhausts. These oxides are transformed into sulphuric acid and nitric acid. Acid deposition can occur in both dry and wet forms. Dry deposition occurs relatively close to the sources of emission and involves aerosols, gases and fine particles. Wet deposition is associated with precipitation and is found much further downwind. The main features are shown in the diagram (Fig. 2.7.14).

Damage caused by acid rain is widespread: it affects natural systems such as forests, lakes and rivers, is harmful to crops and disfigures buildings and other structures. Acidification of lakes and rivers can have serious effects on fish, with losses in both total populations and species diversity. Forests and trees suffer from the acidification of soils by rainfall. Liberation of toxic metals in the soil damages root systems and affects trees' intake of water which can cause dieback and death.

The maps (Fig. 2.7.15) show the acidity of rainfall in Britain and the wet deposited acidity in Britain in 1989 (the total acid load).

Fig. 2.7.14

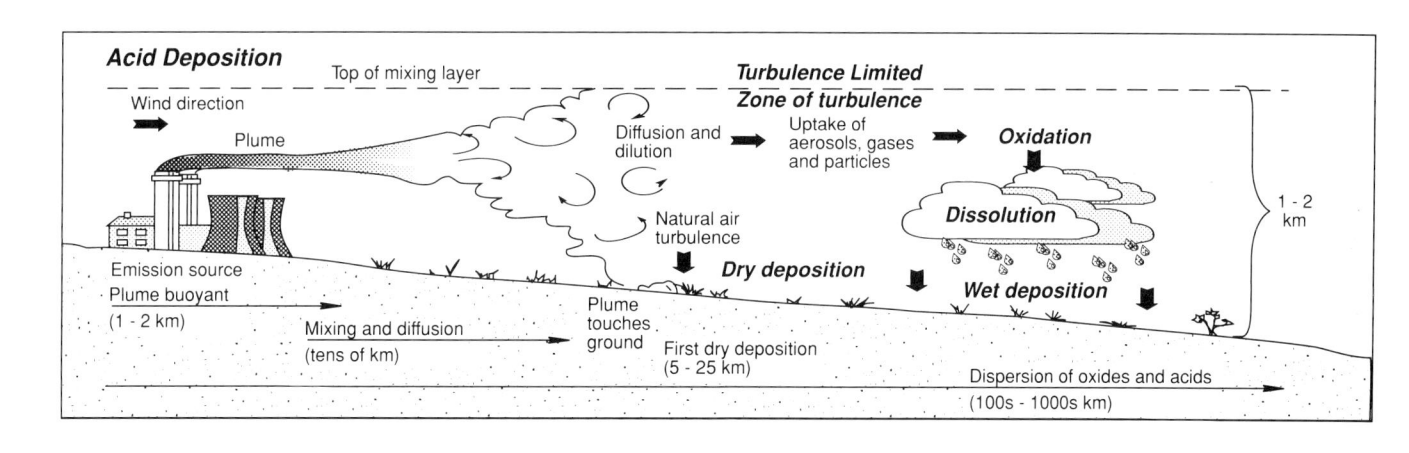

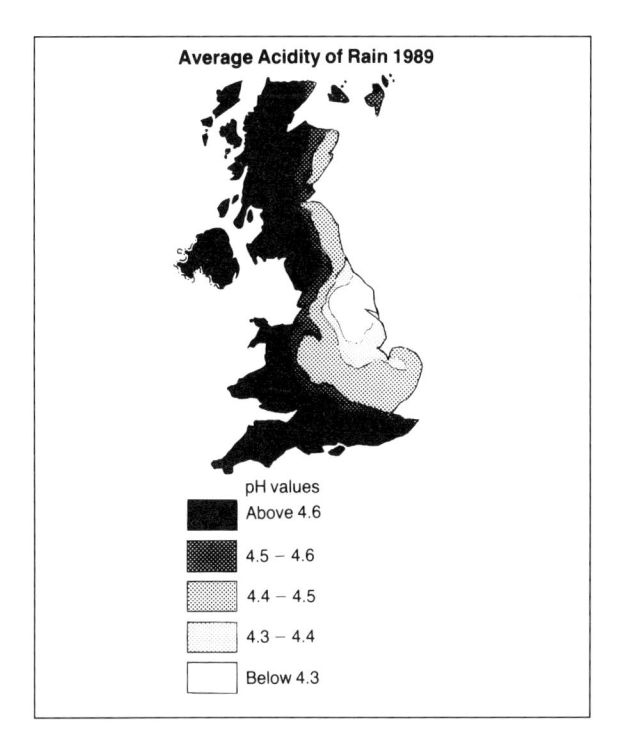

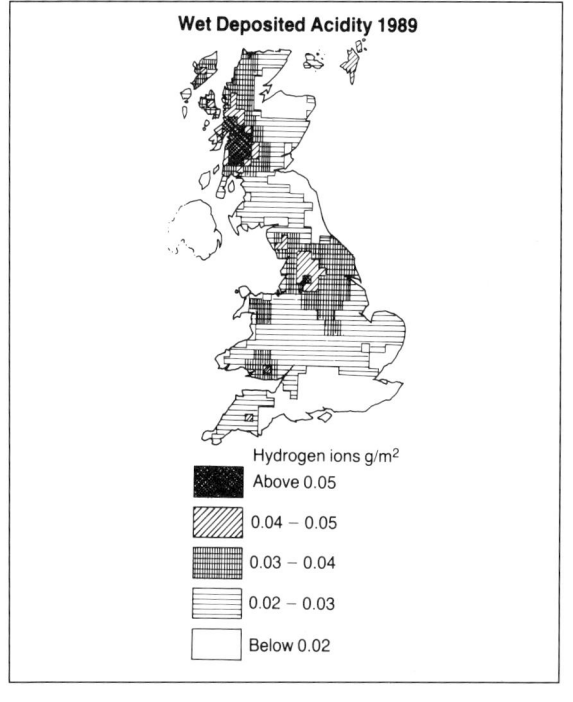

Fig. 2.7.15

Fig. 2.7.16
Independent 26
March 1990

Areas with Soil and Rocks which cannot Resist Acidification

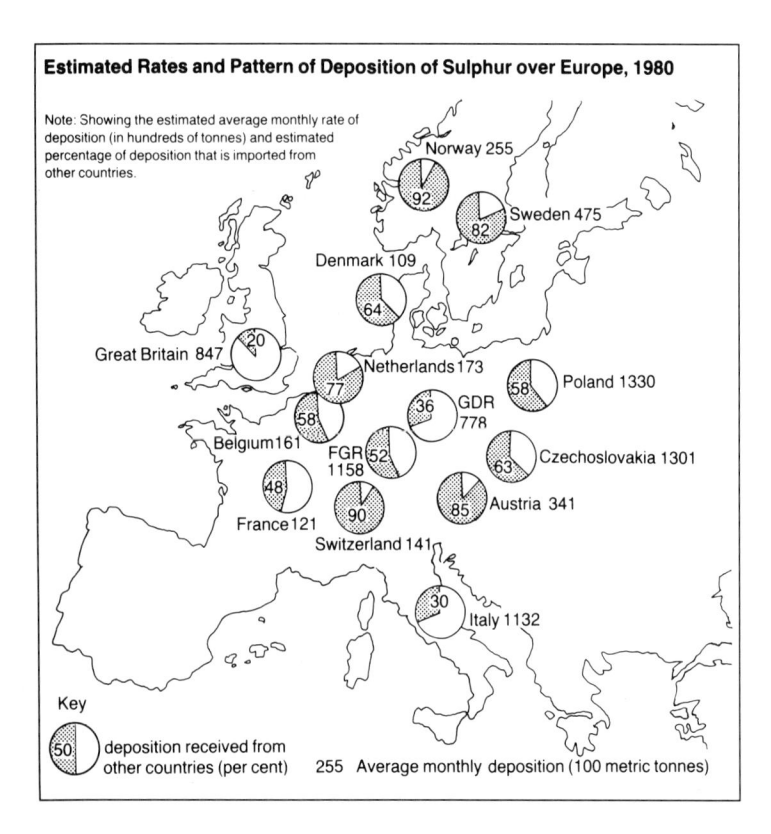

Evidence from cores taken from sediments in lakes in Scotland show that acidification first began about 150 years ago. By June 1988 nearly 80 British waters were acidified or very vulnerable to acidification. In Scotland, Wales and Northern England 25 acidified lakes were found and 64 per cent of the trees were suffering from the effects of acid rain.

The map (Fig. 2.7.17) shows the amount of acid deposition in European countries and the estimated percentage that is imported from other countries.

1   What are the shortcomings of representing the data in this way?

2   After studying the chart (Fig. 2.7.18) showing the main countries responsible for emissions of sulphur dioxide, comment on the relationships that may exist between the distribution of acid deposition and the emission of sulphur dioxide.

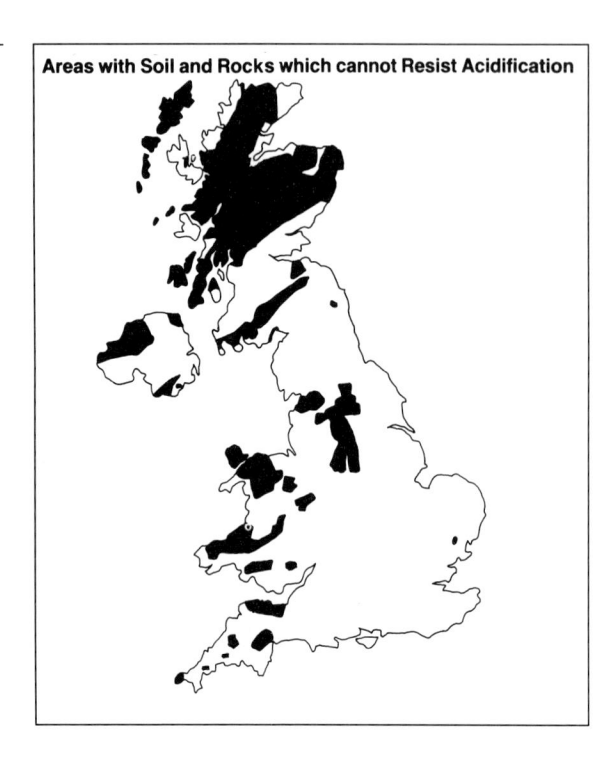

1   What factors influence the two different distributions shown?

2   Figure 2.7.16 shows those areas with soils and rocks that cannot resist acidification. What implications does this map have for the acidification of lakes and rivers?

Fig. 2.7.17

Widespread concern has been expressed about the international implications of sulphur dioxide emissions. International action was clearly necessary if the damage caused by acid deposition was to be reduced.

Controlling acid deposition

Various initiatives have been tried in the 1980s. In 1979 the United Nations (UN) Economic Commission for Europe drafted a series of resolutions concerning reduction of sulphur dioxide levels, but it was not legally binding and therefore was not very effective. The Ottawa conference of 1984 resulted in much firmer commitments to cut emissions. Britain made no promises to cut emissions in 1984. In 1986 however, it gave a commitment to fit all new coal-fired power stations with flue-gas de-sulphurisation equipment (FGD) and fit FGD equipment to some existing power stations. In 1988 European Community directives were issued which required all EC countries to successively reduce their emissions in three stages by 1993, 1998 and 2003. Britain agreed reductions of 20 per cent 30 per cent and 60 per cent respectively. In 1989 measures were announced to fit six existing power stations with FGD equipment. Electricity privatisation would appear to have led to some modification of these schemes. National Power is fitting equipment to its Drax power station, and Powergen is fitting Ferrybridge C and Ratcliffe on Soar. Both companies favour the import of low sulphur coal as an option in reducing emission levels. Although it is cheaper, it will not reduce emissions by the same amount.

Estimated Rates and Pattern of Deposition of Sulphur over Europe, 1980

Note: Showing the estimated average monthly rate of deposition (in hundreds of tonnes) and estimated percentage of deposition that is imported from other countries.

- Norway 255 — 92 (82%)
- Sweden 475 — 82
- Denmark 109 — 64
- Great Britain 847 — 20
- Netherlands 173 — 77
- Poland 1330 — 58
- Belgium 161 — 58
- GDR 778 — 36
- FRG 1158 — 52
- Czechoslovakia 1301 — 63
- France 121 — 48
- Austria 341 — 85
- Switzerland 141 — 90
- Italy 1132 — 30

Key:
50 — deposition received from other countries (per cent)
255 — Average monthly deposition (100 metric tonnes)

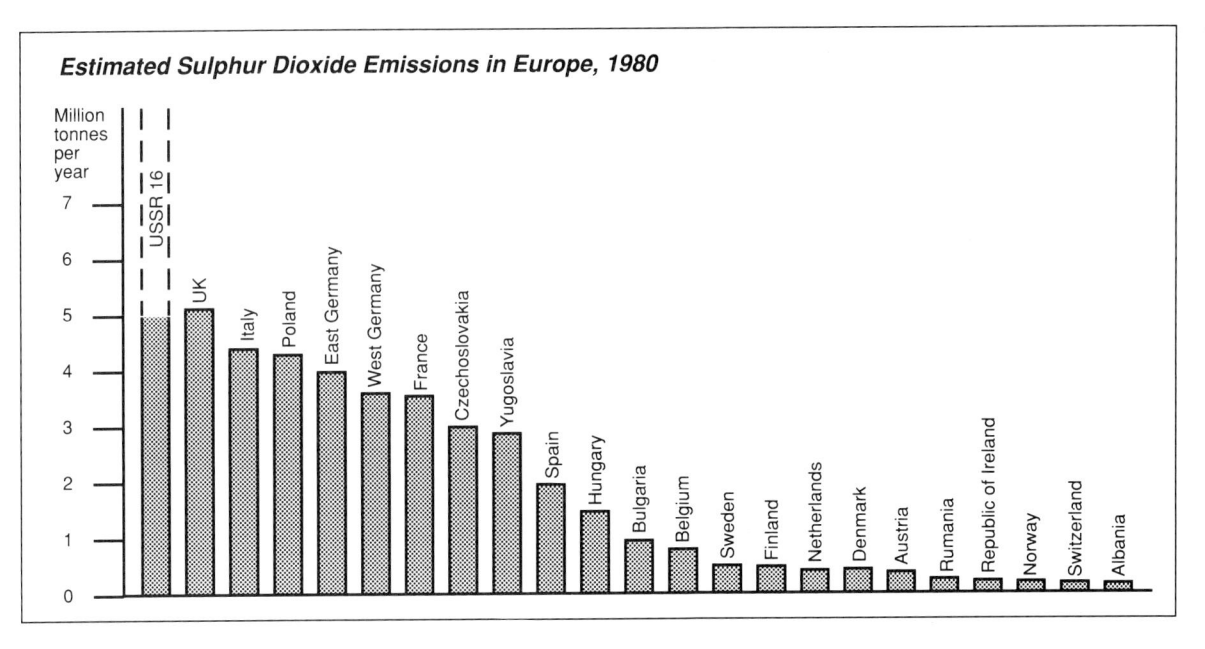

INDUSTRIAL POLLUTION

Estimated Sulphur Dioxide Emissions in Europe, 1980

Fig. 2.7.18

Figure 2.7.19 shows the level of sulphur dioxide emissions in Britain in the 1970's and the 1980's.

1 Discuss the pattern of emissions shown on this diagram.

2 What other measures, apart from the ones mentioned above could lead to the reduction of emissions in Britain?

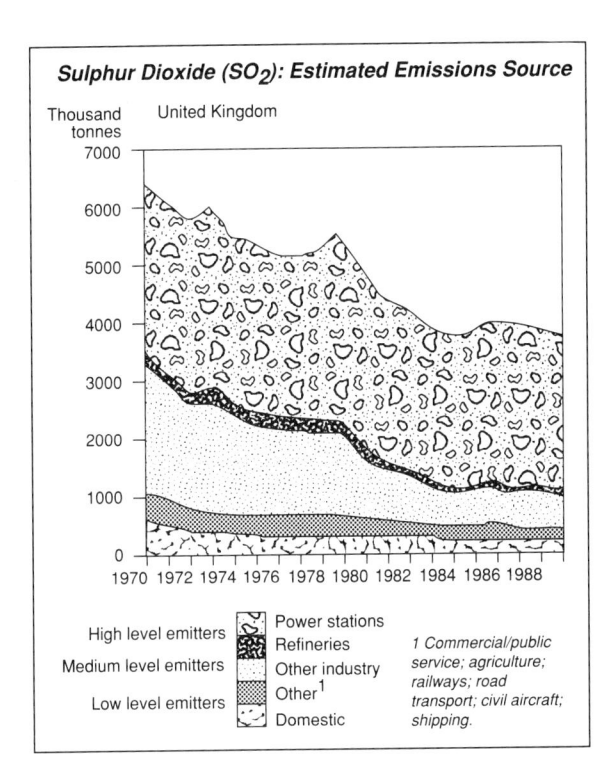

Sulphur Dioxide (SO_2): Estimated Emissions Source — United Kingdom

High level emitters: Power stations, Refineries
Medium level emitters: Other industry, Other[1]
Low level emitters: Domestic

[1] Commercial/public service; agriculture; railways; road transport; civil aircraft; shipping.

Fig. 2.7.19

WASTE LAND

Derelict land

De-industrialisation has resulted in the creation of considerable amounts of derelict land, to be added to the extensive areas that already exist as a result of mineral extraction. The map (Fig. 2.7.20) shows the principal changes that took place in the creation and restoration of derelict land in the period 1969–82 in England.

1 Discuss the main trends that are apparent from the map. How might these trends be explained?

2 What changes may have occurred in the last decade (1982–92) as a result of further de-industrialisation?

3 Which organisations would make funds available for reclamation of derelict land in the 1980s?

Reclamation of derelict land encounters a number of problems, including lack of adequate funding, lack of expertise at the local level, difficulties over land ownership and lack of suitable equipment. Problems may sometimes arise when derelict land is contaminated, as is often the case with derelict land resulting from industrial closures.

Fig. 2.7.20

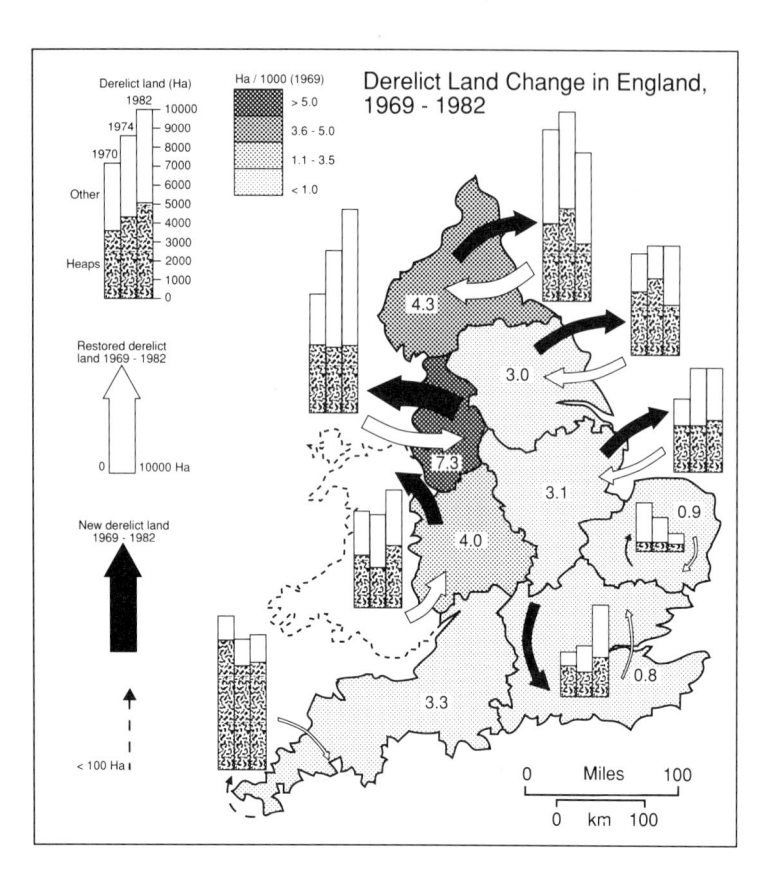

Fig. 2.7.21
The Ebbw Vale Garden Festival site

The Garden Festivals

One of the most interesting and innovative ideas for the reclamation and redevelopment of derelict land has been the series of Garden Festivals held in the 1980s and the early 1990s. Liverpool began with its festival in 1984 and others followed in Stoke-on-Trent, Glasgow and Gateshead. Ebbw Vale is the site of the latest, and last Garden Festival.

The Garden Festival of Wales opened in May 1992 on the reclaimed site of one of the sturdiest reminders of the industrial past in the valleys of South Wales, the Ebbw Vale steel works. The steelworks closed in 1978, with the loss of 14000 jobs.

'No-one could have imagined in those days that steel making in Ebbw Vale would ever come to an end. Since the old steel works reopened in 1938, Ebbw Vale steel workers had shown that they could beat the world.... Ebbw Vale had the lowest unemployment of any industrial centre in Wales: never higher than two per cent and at times nothing at all.'

Michael Foot, on the beginning of his by-election campaign in Ebbw Vale in 1960.

'But the day came when we had to admit defeat... we realised that we had to be content with half a loaf, or we would lose the lot. So we settled for a brand new finishing plant, which is still going strong, a £50 million development of the new Rassau estate, and the promise, when the chance came, we would have full national assistance to clear up the mess left by the old steel masters and the coal owners.' Michael Foot, in the foreward to the Garden Festival brochure, 1992.

The Garden Festival site (Fig. 2.7.21) occupies some 57 hectares to the south of Ebbw Vale. The nearby village of Waun Rhyd once had the reputation of being the most polluted in Britain: the air was thick with red dust from the steel works; great mounds of waste slag filled the valley sides; and the river was reddish-brown with polluting liquids from the plant. To produce the festival site, 1.9 million cubic metres of waste sludge and slag have been moved and landscaped creating thousands of temporary jobs. Total cost of the reclamation project was estimated at £20 million. It was completed by the Welsh Development Agency within three years.

At present the 57 hectare site is occupied by a variety of attractions (Fig. 2.7.22). These include a number of gardens, landscaped backdrops with artificial waterfalls leading to a series of lakes and promenades, a range of exhibitions and entertainments and a shopping complex. All of these are linked together by an integrated transport system. It is hoped that two million people will visit the festival before it closes in October 1992. What of the future? Much of the landscaping will remain but a completely new village and technology campus will be built on the site.

Fig. 2.7.22
Attractions of the Garden Festival of Wales

Included will be:

- A new village centre (5 hectares) focusing a High Street and village square.
- Residential areas (13.5 hectares) with a variety of housing for rent or for sale, which will provide homes for 500 people.
- A technology complex (9.5 hectares) which will house new high-tech industries in a series of purpose built units and provide potential employment for 700 people.
- The business area (5 hectares) which will provide another 450 jobs in new office buildings.
- The Festival Park (25 hectares), which will retain the essential features of the Garden Festival.

Is it all worth it?

Four Garden Festivals preceded the one at Ebbw Vale:

- 1984 Liverpool: derelict dockland was landscaped and attracted three million visitors
- 1986 Stoke-on-Trent: the derelict site of the old Shelton steelworks was reclaimed and the festival attracted three million visitors
- 1988 Glasgow: derelict dockland in the Govan district attracted 4 million visitors
- 1990 Gateshead: industrial wasteland was redeveloped and brought in three million visitors.

Their onward development after the initial festival has not been entirely successful.

Liverpool: one third of the 100 hectare site was scheduled for commercial development but it failed to attract businesses and was eventually turned over to housing. Eight hectares still remain undeveloped. The 35 hectare Festival Park has been beset with business difficulties: the first operator went bankrupt, and the Merseyside Development Corporation was forced to intervene; in 1992 a private leisure company began to invest £11 million on a new complex.

Gateshead: the site had to be fenced off after joy-riders started to dump cars on it.

Glasgow: in 1992 the Glasgow Development Agency took back 14.5 hectares from the house builders Laing, with the aim of reclaiming the land for business.

Although the five festivals will have attracted 15 million visitors by the end of 1992, critics point out that:

- inner city redevelopment, manifested in the festivals, slipped down the political agenda in the 1980s (although there are signs now that this trend has been reversed);
- on the continent Garden Festivals have been a longer lasting success because they have been located in more promising locations (not in inner city areas) and have a much longer lead-in time (up to ten years to reclaim, plan and plant);
- too much of the investment goes "underground" i.e. is spent on reclamation, with less left for other purposes;
- the initial greening is followed too often by 'hard reclamation' i.e. with housing, offices and industrial units.

1   Carefully review the arguments for and against the reclamation of derelict land through the medium of the Garden Festival.

2   What lessons could have been learnt from the first four Garden Festivals?

3   How far do these lessons appear to have been heeded in the development of the Ebbw Vale site?

4   Write a short introduction to a brochure designed to attract small high-tech companies to the technology complex at the festival site at Ebbw Vale.

RURAL BRITAIN

A PERSPECTIVE OF RURAL BRITAIN

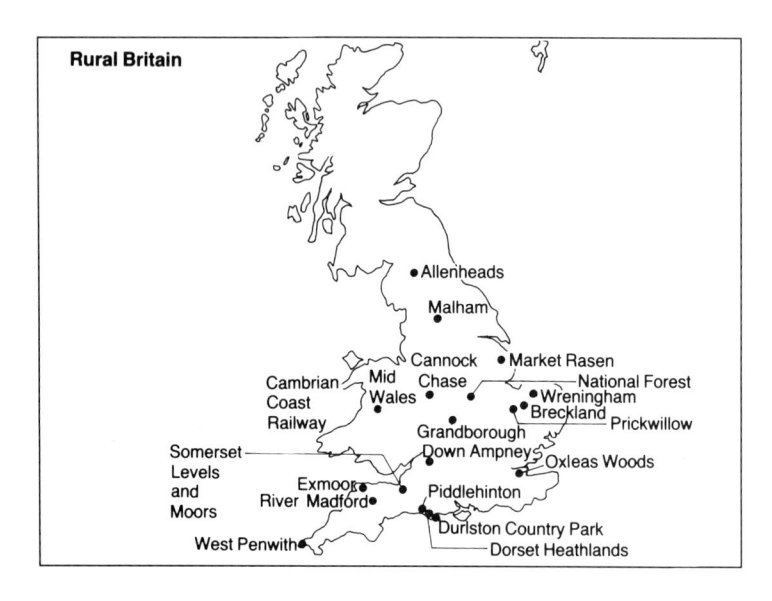

'Rural Britain is no longer agricultural Britain'

H. Newby – *'The Countryside in Question'*

Fig. 3.1.1
The English rural landscape: the Otter Valley in East Devon

Since World War Two a remarkable transformation has taken place in rural Britain. Much of this change results from a fundamental reorganisation of farming itself. Agribusiness, where farming is operated on strictly commercial principles, is becoming increasingly widespread. Associated modern farming methods necessarily embrace science and new technology. Government intervention has increased, with measures aimed at prosperity on the farm and adequate supplies of food in the shops. In order to become more efficient, farms have subsequently grown larger, through amalgamations, bankruptcies and retirements. More efficient farming has meant more mechanisation, a search for greater productivity, and hence a steady diminution of the farm labour force. Changing objectives and methods on the farm have also resulted in striking changes in the rural landscape.

The chequerboard landscape of rural Britain (Fig. 3.1.1) in the early twentieth century remains part of our perception of the countryside idyll. Landscape change has been quite dramatic: the cosy image of farmstead and field, hedgerow and coppice, has disappeared in many areas. The draining of meadows, the removal of hedgerows and copses, the enlargement of fields, the erection of fences and the construction of silos and winter-feeding sheds have created new, perhaps alien, vistas in the countryside.

Reaction against these changes has prompted a strong campaign for the conservation of the countryside, for protection of endangered species, and for a more balanced approach to future rural landscapes.

The decline in numbers working on the land has led to a significant change in the nature of rural communities themselves. Less work available on the land has meant that many of the rural workforce have had to seek alternative employment elsewhere in neighbouring towns. In many cases this has meant a move from the countryside to the town, although the theme of rural depopulation is not a new one. To a considerable degree the exodus to the towns has become increasingly balanced by an opposite movement of town-dwellers into the country (Fig. 3.1.2). Except in the most remote areas, rural communities have an increasingly large proportion of people that, whilst living in the country, are able to enjoy many of the advantages of urban living.

Fig. 3.1.2
New commuter homes in the country: Alton Pancras, Dorset

Private transport has enabled the new rural dwellers to work, shop and seek their entertainment in the nearest town, rather than where they live. The inevitable, consequent rise in rural house prices had disadvantaged the less well paid that still work on the land, the young, and the old. An increasing problem is the difficulty of providing affordable housing for these groups. The provision of services for rural communities has suffered in many cases too, with the loss of village shops, post offices and schools, and the closure or curtailment of bus and rail services. Thus villages within easy reach of sizeable towns have increasingly become commuter settlements, whilst the more remote areas have seen population decline bolstered only by the influx of the retired, and the second-home owner. Rural deprivation, be it a result of inadequate housing, poor services or lack of mobility is a reality in many of Britain's rural areas.

Change in the countryside continues. The 1991 census shows that many rural areas have gained population within the last decade. These areas are not confined to those nearest the heartland of urban Britain. Areas such as West Wales, north Norfolk and much of Cornwall, previously regarded as remote are now showing a healthy increase in population. This is not a reflection of people of retirement age moving into these areas, but embraces a much wider spectrum of ages. Modern small scale industry or the telecommuter, working from a computer terminal at home may well find these areas attractive. Population growth may encourage the need for improved services and rural Britain may be entering a new phase.

RURALITY, RURAL IDYLL AND RURAL CONFLICT

The map (Fig. 3.1.3) shows the index of rurality (the extent to which areas may be regarded as rural) for England and Wales. It was calculated from the 1981 census, using a range of socio-economic indicators, and others related to spatial remoteness. Areas that are extremely rural are identified, together with another group that is described as intermediate rural.

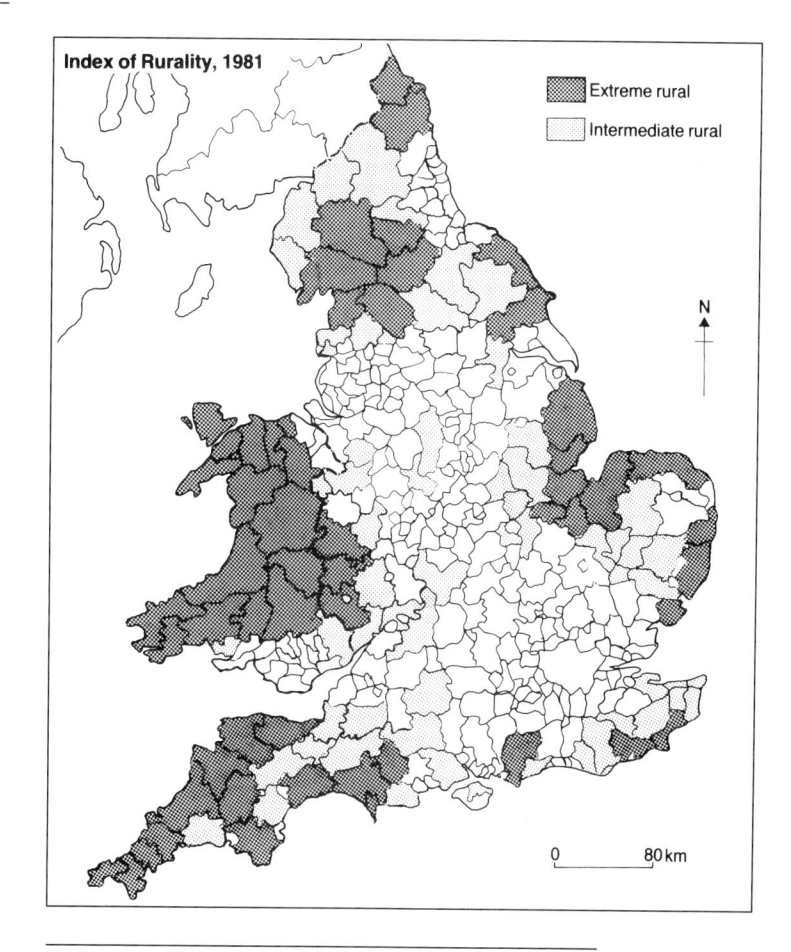

Fig. 3.1.3

Happy as crickets amid the oil-seed rape

AMATEUR COUNTRYMAN

By Philip Oakes

I BECAME an amateur countryman when we left London two and a half years ago, exchanging the Highgate borders (agent speak for the less salubrious Archway) for the flatlands of Lincolnshire. House prices were peaking and the exodus was at its height. Many wanted a country cottage, preferably a second home. But our move was wholesale; a clean break with no looking back. On a wet winter afternoon we arrived at our destination, mid way between Lincoln and Grimsby, in the dead centre of what a local newspaper columnist called "the forgotten lands".

They grow a lot of sugar beet in Lincolnshire. And wheat. And potatoes. It is the heart of Britain's agribusiness. The wolds rise gently to the east. There are small woods, slow rivers, a lot of redundant churches. Scenically, it is less than spectacular. For anyone contemplating a new way of life, free of stress but just as remote from any kind of rural idyll, it is the acid test. This is the real country, where people earn their bread by working the land. As non-farmers, we are seen – tolerantly, I hope – as amateurs, otherwise engaged.

What we wanted was clean air, quiet and a change of pace, and the country has given us all three. We live a mile from the village, surrounded by pasture and plough, in what was a farm labourer's cottage and is now a much-expanded house with a conservatory in which bougainvillea climbs to the roof. One of our neighbours has a small dairy herd. Another has a flock of 4000 free-range hens. Another grows wheat and oil-seed rape on his 400 acres.

London lies 150 miles to the south, but for most people it's only a rumour. The truth is that the provinces are oblivious to London. Customs are different here; so are values. As incomers we disagree with the locals on hare coursing, crop-spraying, factory farming. There is a sense of continuity which is fiercely local. Opinions are slow to form, slower to change. People take their time, but when it matters they're courteous and prompt. The garage sends a mechanic 15 minutes after a phone call. A neighbour brings his chainsaw to a tree which is shadowing the vegetable plot.

As amateurs we arrived believing the horror stories about poor provincial shopping (nothing but sliced bread) but they have proved entirely false. We buy most of our supplies in Market Rasen, six miles away. Starbuck's bakes wonderful bread; Tasker the butcher sells superb meat. Grimsby (15 miles away) has an Asda and a Sainsbury's where we buy wine and groceries in bulk. We have two deepfreezes, one packed with our own vegetables, the other filled with meat. Last year Hedley, the dairy farmer, supplied us with half a sheep and half a pig. We paid around 50p a pound and every ounce was delicious, even the pig's head, which we made into brawn. I can't imagine doing that in London.

There are other pluses. Market Rasen has at least one excellent dentist, so good that after years of paying privately I'm once again a National Health patient. There is also a first-rate health centre.

On the debit side we live in a cultural desert. There are few cinemas and even fewer theatres. We have become passionate gardeners. We read a lot. We watch a good deal of TV. We enjoy having friends to stay. By focusing on what appears to be small and immediate things, we end up taking the longer view.

On a dark winter day the sense of isolation can be intense; but the compensations multiply all year long. We live surrounded by birds and beasts. The air is tonic. People say "please" and "thank you".

Fig. 3.1.4
Independent

1 Suggest a range of simple socio-economic indicators that could be used to index rurality.

2 Comment on the distribution of the areas of extreme and intermediate rurality.

3 What does the map (Fig. 3.1.3) suggest about the socio-economic conditions over the remaining area that is left unshaded?

Study the newspaper article by Philip Oakes, 'Happy as crickets amid the oil-seed rape' (Fig. 3.1.4).

1 Discuss the writer's views on the advantages and disadvantages of rural living.

2 To what extent are these determined by a sense of urban values that appear in the article?

3 How far do you agree with this view of living in the country?

Much controversy exists over the role of the farmer as custodian of the rural landscape. Farmers have often been castigated for turning the much-loved British landscape into a characterless prairie, all in the name of greater efficiency and higher profits. Conflict with those that wish to conserve the landscape is inevitable.

Nature would ruin the landscape

There is a widespread notion that Britain's landscape is a gift of nature. Farmers are presently cast as the ogres of the piece, eating hedgerows for breakfast, spewing out nitrates for lunch and supping on a heady pudding of minced ramblers, seasoned with a sprinkling of conservationists.

If only nature were allowed to take its course, people say. Freed from the grasp of the wicked agriculturists, how much more beautiful the landscape would be. We could picnic among cowslips and wild orchids. Clouds of butterflies would fill the summer pastures and the hedgerows would be alive with the sound of birdsong.

The greatest fallacy here is the idea that our landscape is a natural phenomenon. Nature provided the raw material, the underpinning, the geology of the countrytside. The climate controls the way that the raw material can be handled, but the views that we croon over and write about, with too many adjectives, have almost entirely been shaped by farmers.

The process has been going on so long that, even as far back as the Domesday Book, only 20 per cent of the country was covered in wild wood. It did not look as it does today, but nor was it natural. The landscape we know, the hedges, fields, the mixture of pasture and plough, has been claimed, lost and reclaimed by the agricultural activities that for centuries occupied the vast proportion of people living in Britain.

The second fallacy is that this landscape, left to its own devices, would become a more beautiful place than it is now. Within 15 years, one dairy farmer estimates, his pasture would become impenetrable scrub: thorn, elder, bramble, hazel, with docks and nettles, the bully boys, swarming through the ditches and along the verges.

The oldest generation of farm labourers can still remember what some agricultural land looked like in the depression before the last war. It was, they say, not a pretty sight. "Set-aside land', the present Government's answer to overproduction in the corn sector, does not grow carpets of wild violets or sheaves of poppies. it grows poisonous ragwort, much on the increase this year, and thistle. The poppy is a flower of arable land, succeeding only where earth has been freshly turned over.

To retain the diversity at the core of its appeal (the Fens are a special case), landscape has to be managed. Eighty per cent of Britain's countryside is now farmed, with less than 10 per cent designated as nature reserve. The real power to determine the face of the countryside rests with farmers.

Since the Second World War, yield has been the driving force behind all farming. During that cataclysmic time, farmers were heroes, not ogres, wringing their land for the last pint of milk, the last ounce of corn. Government subsidies subsequently continued to steer farming in this same direction, where success was measured in output per acre. Farmers took pride in running well-organised, productive farms.

It is not surprising then, after two generations of brainwashing, that they should now find it difficult to come to grips with the role that is being thrust upon them: the farmer as conservator, rather than producer. Nor is it surprising that ardent new converts to the conservation cause, quick to tell farmers what they should do with their land, have sometimes been sent packing with several fleas in their ears. On this score, sympathy has drifted away from the farmers. But few ordinary householders have the experience of a stranger moving into view and telling them that they should be planting pink roses not red, and must not cut the lawn for the next two months.

The conservation lobby may argue that the best solution would be for them to hold more land and the farmers less. The danger with this approach is that we slide even further down the road of country as theme park.

The Government is now offering grants to farmers to do the opposite of what they were urged to do before. Instead of being paid to bulldoze hedges, they are now encouraged to keep them. In July, Michael Heseltine, the Secretary of State for the Environment, and the Ministry of Agriculture announced their Countryside Stewardship Scheme. Over three years, £13m is being put aside to enable farmers to protect and enhance five kinds of marginal landscape: chalk and limestone grassland, lowland heath, waterside areas, coastal land and uplands.

This is only one of the schemes now on offer. Within Environmentally Sensitive Areas there are grants available to persuade farmers to continue with habitat-friendly, but time-consuming, practices. The Farm Woodland Scheme helps towards planting trees on former agricultural land. The Nature Conservancy Council, now English Nature, can assist in the management of Sites of Special Scientific Interest. But, as farmers are quick to point out, a grant to be spent on a specific project is not an income.

Some of the most effective work on farms is being done through Farming and Wildlife Advisory Groups, now well established in most of the counties of England, though struggling in Wales. These are led by farmers, chairing a mixed body of other interested parties – English Nature, the RSPB, the Worldwide Fund for Nature and others..

Because they are working from within, rather than bossing from without, they have found it easier to persuade members that ponds, hedges, trees, field margins, scrub and verges can be managed to benefit wildlife without detracting from the viability of the farm.

We do not have to pass on to our children exactly the same landscape as the one we know now. Change is a symbol of vitality. Agriculture cannot and should not be set in aspic. We can ensure, though, that the landscape suffers no loss of quality. For this, we depend on the farmers.

Fig. 3.1.5a
Independent 12 November 1991

Read the article by Anna Pavord, 'Nature would ruin the landscape' (Fig. 3.1.5a).

1 Summarise the comments on the evolution of the rural landscape.

2 What is the nature of the conflict between the farmer and the conservationist?

3 How might the conflicting viewpoints be reconciled?

Now read the letters to *The Independent* that the article stimulated (Fig. 3.1.5b).

Construct a matrix to show the extent to which each of the three correspondents reflects the following values positions. (For each of the values positions, score each of the correspondents on a scale of 1 to 5. 1 = weakly supports, 5 = strongly supports.)

Fig. 3.1.5b
Independent 16 November 1991

- The landscape is best left to nature.
- Farmers are the best custodians of the landscape.
- Current farming practices not only damage the landscape, but also put ecological sustainability at risk.
- Conservation is essential for the maintenance of the rural landscape.

LETTERS

Wilderness and farmland in England's green and pleasant land

From Mr J.J. Putnam
Sir: Anna Pavord's article ("Nature would ruin the landscape", 12 November) on the horrors of leaving the landscape to itself may be a trifle pessimistic.

She refers to inter-war recollections of the agricultural depression and fields full of weeds, and quotes a dairy farmer's estimate of 15 years for dereliction to take over on set-aside land.

Happily, we have an interesting record of what does happen to set aside land when really left to its natural devices.

When the Rothamsted Experimental Station, Hertfordshire, was set up some 120 years ago, its founder, while exploring different cultivation patterns for wheat, also wondered what would happen to a field of wheat if left unharvested. He grew an acre of wheat and allowed it to stand; the grains fell to the ground and the field remained untouched. After several years no wheat was visible. Thorns, brambles, weeds and bushes grew. So much for the experiment? Not at all. Leave it alone; the experiment has only just begun, he told his staff.

My recollection when being shown round Rothamsted a couple of decades ago is that this experiment, then more than 100 years old, had produced a mixed coppice; predominantly oak and other deciduous trees. I feel sure it is still there, as a living memorial and a continuing experiment.

So we just need to look a little longer ahead than 15 years in our countryside planning. Incidentally, ragwort, though a weed in pasture – even a poisonous one, as emotively described by Anna Pavord – can be nourishing. I vividly recall as a child walking on the South Downs and seeing ragwort smothered in orange-and-black-striped caterpillars which clearly enjoyed devouring the leaves.

Certainly, land left to revert to nature would be different; but maybe a little more wilderness is not a bad balance to the burgeoning theme parks ...

What would the world be, once bereft
Of wet and of wilderness? Let them be left,
O let them be left, wilderness and wet;
Long live the weeds and the wilderness yet.

... wrote Gerard Manley Hopkins 100 years ago.
Yours truly,
J.J. PUTNAM
Blackawton, Devon
12 November

From Mr Mark Comwell
Sir: Anna Pavord's article made a refreshing change. It was reassuring to read a well informed, reasoned approach to the agriculture-and-the-environment debate.

Ms Pavord states that after two generations of government "brainwashing" (to achieve maximum output), the farmer is having difficulty adjusting to a role more akin to conservator than producer. To a certain extent, this is true, but common ground is being sought and realised, between farming and conservation groups. This, in turn, is enhanced by a gradual realisation that the farmer is better qualified than any alternative in land husbandry.

However, Ms Pavord is right to warn against dangers of "country as theme park". While farmers are the best custodians of the countryside, they must remain primarily farmers if they are to remain effective.

It is also true that the biggest threat to the countryside is posed by dereliction. Agriculture and the countryside are synonymous, and the interests of all, particularly of our rural communities, would be best served by encouraging more people to go into farming.

Finally, with reference to what these farmers should be producing, government money and research facilities should be directed towards developing replaceable fuel sources, instead of consuming fossil fuels at ever increasing rates.
Yours sincerely,
MARK CROMWELL
Half Yard Farm
Barcombe Mills, East Sussex
13 November

From Mr Roger Martin
Sir: Farmers certainly deserve sympathy these days, but it is a pity you could not find a more sophisticated defender than Anna Pavord (12 November). She treats the conservation case purely in terms of landscape. She does not address the real issue of ecological sustainability. Rural areas produce not just food and "landscape", but the air we breathe, the water we drink, the soil that grows plants, and the species-base that keeps the system resilient.

In these terms, modern, intensive agro-industry is not sustainable. It has been a purely post-war historic mistake, driven by misguided government policies. It must now end.

Quarrying the soil for its nutrients, wiping out species, chemical distortion in land and water eco-systems, the negative energy-input balance of modern farming – all these must give way to more sustainable farm management. Reform of European agricultural policies should be primarily about this.

The countryside will doubtless look different; but this is vastly less important than its ability to sustain life.
Yours faithfully,
ROGER MARTIN
Director
Somerset Trust for
Nature Conservation
Broomfield, Somerset
13 November

CONTEMPORARY ISSUES IN BRITISH FARMING

Farming in Britain has seen many changes since World War Two, but the 1980s have witnessed some of the most important. If present trends continue, farming will have a very different face at the turn of the century from that of 20 years ago. It is necessary, first, however, to examine some of the trends that have characterised change in farming since the beginning of World War Two.

The tables and diagram (Fig. 3.2.1a–c) show:

- Changes in the farm size structure of Britain 1939–87
- Change in the farm labour force 1939–89
- Change in the production and yield of selected crops in the UK 1939–89.

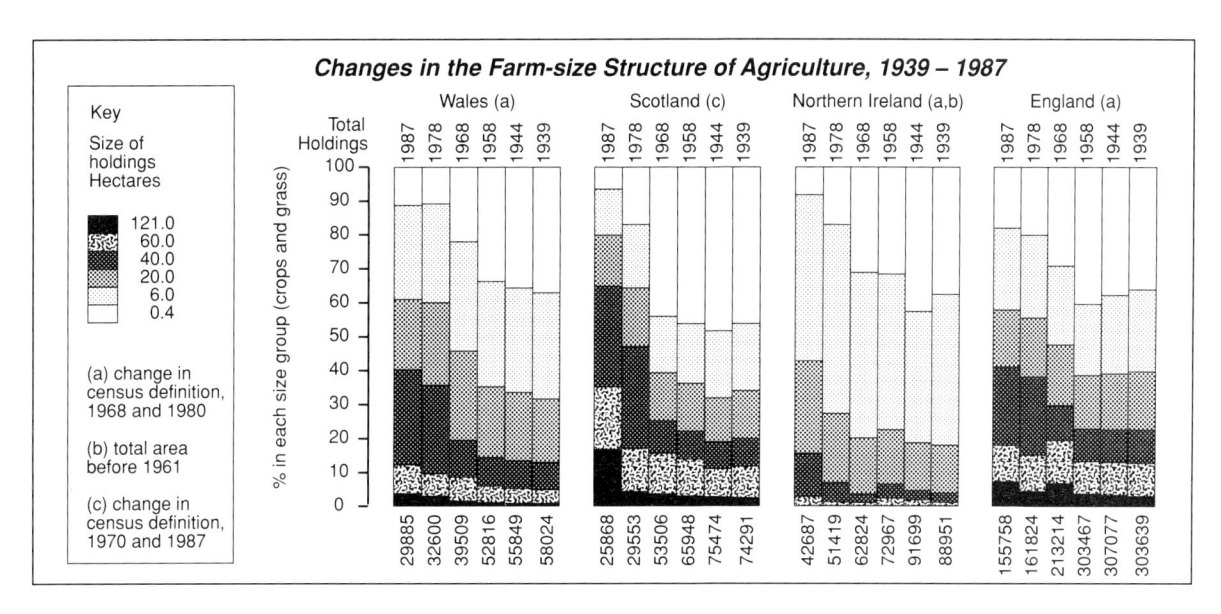

Fig. 3.2.1a

Changes in the Farm-size Structure of Agriculture, 1939 – 1987

Key
Size of holdings Hectares
- 121.0
- 60.0
- 40.0
- 20.0
- 6.0
- 0.4

(a) change in census definition, 1968 and 1980
(b) total area before 1961
(c) change in census definition, 1970 and 1987

Fig. 3.2.1b

CHANGING TYPES OF WORKERS EMPLOYED ON AGRICULTURAL HOLDINGS IN THE UK, 1939–1989 (PER CENT TOTAL WORKFORCE IN EACH YEAR)

TYPE OF WORKER	1939	1944[1]	1958	1968	1977	1987	1989
Total full-time	82.5	77.2	73.5	72.0	56.4	47.6	46.7
male	73.7	63.4	66.7	65.7	51.0	42.7	41.3
female	8.8	13.8	6.8	6.3	5.4	4.9	5.4
Total part-time	17.5	22.8	11.0	11.0	18.6	20.4	21.2
male	11.2	13.3	6.5	5.9	8.6	10.6	11.1
female	6.3	9.5	4.5	5.1	9.0	9.8	10.1
Total seasonal and casual[1]	–	–	15.5	17.0	25.0	32.0	31.9
male	–	–	9.2	9.3	14.0	19.2	19.5
female	–	–	6.3	7.7	11.0	12.8	12.4
Total workers	803 526	902 117	730 253	450 102	37 0523	297 257	276 000

Notes: [1] Excludes Womens Land Army and prisoners of war
[2] Part-time workers in Northern Ireland included with casual workers

Fig. 3.2.1c

UK PRODUCTION (000, tonnes) AND YIELD OF SELECTED CROPS (tonnes per hectare), 1939–1989

CROP		1939	1944	1958	1968	1978	1989
Wheat	P	1672	3189	2755	3469	6613	14030
	Y	2.0	2.2	2.7	3.5	5.3	6.74
Barley	P	906	1780	3221	8270	9848	8070
	Y	2.2	2.2	2.9	3.4	4.2	4.88
Oats	P	2035	3001	2172	1224	706	525
	Y	2.0	2.0	2.4	3.2	3.9	4.45
Sugarbeet	P	3586	3320	5835	7119	6382	8115
	Y	25.8	19.3	33.1	38.1	31.7	41.8
Potatoes	P	5302	9243	5646	6872	7331	6215
	Y	18.6	16.1	17.1	24.6	34.2	35.4

CHANGE IN FARM SIZE STRUCTURE

1 Critically evaluate the diagrammatic method used for presenting these statistics.

2 What are the principal trends revealed by these statistics?

3 Why do you think that there are important variations between the different areas of the UK?

CHANGE IN THE FARM LABOUR FORCE

1 What are the main trends revealed in the table?

2 How might these trends be explained?

CHANGE IN PRODUCTION AND YIELDS OF SELECTED CROPS

1 Is there a clear relationship between production and yield for the selected crops?

2 Suggest some of the factors that are likely to have influenced the trends that are revealed in the table.

Considering all of the statistics together now:

1 What general relationships exist between the three sets of statistics?

2 What other statistics would be useful in seeking a fuller explanation of the changes that have taken place in this period?

CHANGES IN THE 1980S

Subsidies and the EEC

A Cambridgeshire farmer, with 1200 hectares, mostly in cereals, said in the *Farmers' Weekly* 1983: 'One day the tax-paying public is going to wake up with a start and decide that it is fed up with the Common Agricultural Policy (CAP). Until this actually happens, I shall continue to accept subsidies with alacrity and gratitude'.

Quoted in *The Times* 25 November 1986: 'We, as arable farmers, enjoyed a greater level of prosperity between 1972 and 1983 than I suspect any farmer has ever enjoyed. It was impossible to lose money. Even bad farmers made some money, good ones made a lot, and excellent ones made fortunes.... The intervention system, which was supposed to be a safety net, has turned the CAP into a lunatic asylum. I have sold wheat into intervention in preference to selling it to my local merchant.'

Quoted in *The Independent on Sunday* 5 May 1991: 'It is not surprising that the British public yawns. We have an appalling credibility problem brought on by ourselves.' He was referring to, 'the image that has grown more persistent of fat farmers roaring around the despoiled countryside in Range Rovers and making a handsome living off European Community hand-outs.'

Roger Trapp, The Independent on Sunday, 5 May 1991

Before Britain joined the EEC in 1973 a system of subsidies had been in operation since the 1930s which made up the difference between the average price of farm commodities and a guaranteed price. Each year these guaranteed prices were reviewed for

CONTEMPORARY ISSUES IN BRITISH FARMING

Fig. 3.2.2

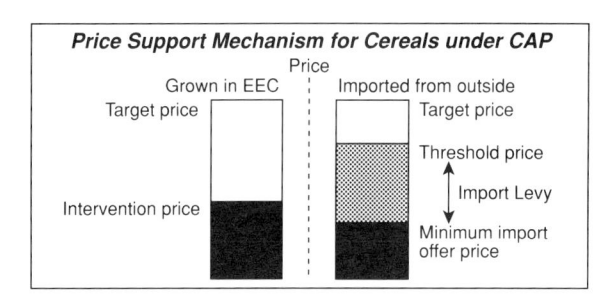

Price Support Mechanism for Cereals under CAP

Many British farmers benefited from the operation of the Intervention price. Farmers, aided by new technological improvements in fertilisers and more efficient farming methods, increased production to levels where it far outstripped demand. They sold at the Intervention price, and surpluses were stored by Governments (see Fig. 3.2.4). Farm incomes and land prices rose to very high levels in the early 1980s as a result of the support mechanism. Not all farmers enjoyed this apparent boom; the main beneficiaries were those that operated the very large farms, run on agribusiness lines (see Fig. 3.2.3).

Fig. 3.2.4

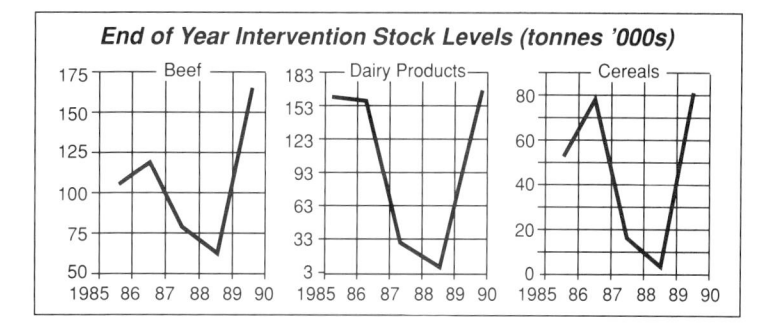

the next harvest or twelve months. On joining the EEC a different system came into operation. There are three basic price elements (Fig. 3.2.2) in the farm support scheme of the CAP.

- The Target price is the desired market price set by the Community or the price that the farmer should hope to get for his produce.

- The Intervention price is the price at which member countries buy in farm produce if the Target price is not reached. Much controversy has arisen over the level of the Target price, since many regard it as being set too high thus encouraging farmers to over-produce.

- The Threshold price is applied to imports. Imports from outside of the EEC that are cheaper than the Threshold price are not allowed into the Community until a levy has been paid that raises the price of the commodity to the Threshold price. This has been criticised on the grounds that it encourages inefficient farming in the EEC.

End of Year Intervention Stock Levels (tonnes '000s)

With high incomes farmers were encouraged to invest in new and sophisticated machinery, a range of storage facilities, and many acquired or leased additional land. Inevitably this led to increased borrowing, and high interest payments. Towards the end of the 1980s a different outlook became apparent amongst farmers:

Jonathan Dixon-Smith sits in the study of his large detached house and says: '*If I take out an agricultural wage (about £130 a week) I could make a profit this year*'. When you consider that this is the centre of a holding of about 460 hectares in the rich cereal lands of north Essex, and there is a tennis court and a swimming pool out back, it is difficult to believe him.... But a couple of miles away, David Macmillan also has a well-appointed house, and he, too, says he is feeling the pinch. He currently operates a 400 hectare spread, but is currently trying to sell the 80 hectare farm that he bought after the last bumper harvest in 1984 in order to wipe out his borrowings. Land in this part of East Anglia is now selling for about £3000 a hectare, compared with about £5000 a hectare in the mid-eighties.... But he is prepared to sustain the loss to clear his debts. '*I'll still have inheritance taxes and the likes, but I can take care of that by selling farm cottages. Reducing costs is the key. We don't talk about increasing production now. We talk about reducing costs.*'

Fig. 3.2.3
Agribusiness: Farms on Salisbury Plain, near Longbridge Deverill: note the large fields and the lack of hedgerows

Roger Trapp, Farmers lose ground, The Independent on Sunday 5 May 1991

1 The cost of the price guarantee fund for the CAP in 1991 was £3.5 billion ecus (£19 billion); in 1975 it was 4.6 billion ecus (£2.77 billion). On what social and economic grounds can this increase be justified?

2 Suggest some of the environmental consequences of the use of the guarantee fund to boost farm incomes.

In the 1980s various measures were introduced in order to reduce the surplus that accumulated through the buying in of farm produce at the Intervention price. Two important schemes were the introduction of milk quotas in 1984, and the set-aside scheme in 1988 (to be discussed on pages 114–117). The milk quota scheme had the effect of reducing the size of the national dairy herd by 15 per cent from 1984–9, and cutting the production of milk by 18 per cent. The buying and selling of milk quotas between farmers has, however, become a very lucrative business, with 42 per cent of dairy farmers involved in lease or transfer of their milk quota in 1988–9. The milk quota has therefore become a very important commodity. The set-aside scheme pays farmers for leaving some of their land fallow in order to reduce production.

In the early 1990s controversial proposals to reduce the level of intervention have been put forward by Ray McSharry, the EC Agricultural Commissioner. Some of the main points of the proposals are as follows:

- Reductions in the Intervention price for cereals (35%), beef (15%) and milk (10%). The levels of intervention stocks for the late 1980s for these three commodities are shown in Fig. 3.2.4.

- Compensation payments for loss of income from the reduced payments but tapered in such a way as to favour small farmers.

- Producers farming more than 20 hectares will be forced to set aside 15 per cent of their land. Compensation will be paid for loss of subsequent income, but only up to a 50 hectare limit.

- Bigger grants for afforestation.

- Early retirement benefits for full-time farmers over 55.

In the final package settled on 21 May 1992 the price reductions agreed were 29% for cereals, 15% for beef and poultry and 15% for dairy products. The other proposals remain broadly the same.

Fig. 3.2.5

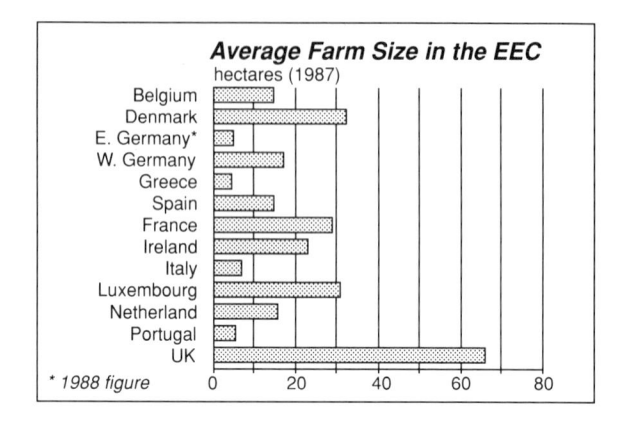

Average Farm Size in the EEC
hectares (1987)

Fig. 3.2.6

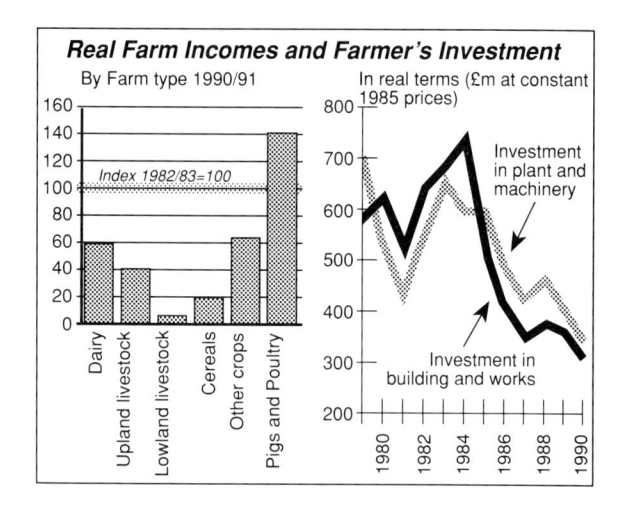

Real Farm Incomes and Farmer's Investment
By Farm type 1990/91 — Index 1982/83=100
In real terms (£m at constant 1985 prices) — Investment in plant and machinery; Investment in building and works

1 Study Figure 3.2.5 which shows the average size for farms in the EEC and Figure 3.2.6 which shows levels of farm income and investment in farming in Britain. Why have the McSharry proposals been received with so much dissent in Britain?

2 John Gummer, Minister of Agriculture is reported as saying (*The Independent* 22 May 1992) concerning the final agreement on farm support reforms: 'It is the fundamental reform we have all been waiting for, it is a fair deal for Britain and good for the rest of the world'. To what extent do you agree with this view?

The set-aside scheme

One of the options that can be used to cut production of farm commodities is to take land out of cultivation. Agreement was reached among Community members in February 1988 to reduce the area in cultivation through the set-aside scheme.

Participation is entirely voluntary on the part of the farmer, and is in no way guaranteed. The scheme was launched in Britian in June 1988. In order to take advantage of the annual payment of £200 per hectare farmers have to take at least 20 per cent of their arable land out of cultivation for at least five years. Three options are available to farmers, when choosing what to do with their set-aside land.

- It may be kept as fallow, with a green cover top e.g. clover.
- It may be converted to woodland.
- It may be used for non-agricultural purposes (such as farm shops, farm-based accommodation, provision of educational facilities, but excluding residential, industrial and retail functions).

The set-aside agreement is initially for five years, but the farmer can opt out after three years. Farmers who set aside at least 30 per cent of their arable land will be exempted from the payment of the levy on up to 20 tonnes of grain sold. Rates of compensation vary according to the type of option selected and the nature of the land in the region. Less Favoured Areas (LFAs) are mainly in upland Britain. Variations are shown in the table (Fig. 3.2.7).

Fig. 3.2.7

COMPENSATION PAYMENTS PER HECTARE OF LAND SET-ASIDE (£)

TYPE OF LAND	NON-LFA LAND	LFA LAND
Permanent fallow	200	180
Rotation fallow	180	160
Non-agricultural use		
Woodland		
– set-aside only	200	180
– with farm woodland scheme	190	150 (disadvantaged areas)
		100 (severely disadvantaged areas)

Compensation is lower in the Less Favoured Areas because of the lower yields that are obtained in these regions. Applications of pesticides and fertilisers in the set-aside areas are in general prohibited. Payments are based on the farmers' crop pattern in 1987–88. The base year is 1987–8, even for farmers thinking of entering after 1989.

CASE STUDY: WRENINGHAM, NORFOLK

Dennis Long set aside 80 hectares in the initial scheme, and this gives him an income from this land of £16 000 per annum. He has altered the options on his farm over a period of years; until 1962 he was mainly involved with dairy cattle, then switched to beef and later to cereals. Profits from the wheat and oil seed rape that he was growing fell markedly in the mid 1980s.

'We grain farmers had been doing very well in cloud cuckoo land. Eventually it had to have repercussions. The EEC were trying to maintain a price above the market level. I've heard my father 50 years ago arguing that supply and demand will always rule the market, eventually. I think this came home to roost in the mid-1980s.' He reviewed his options:

- Increase yields – unlikely since the land is quite difficult to work.
- Retire and let his son operate the farm.
- Rent out the farm.

None of these options were really viable, but the set-aside scheme seemed to offer something better.

'I didn't go into it as an easy option to do nothing. With no fertiliser used, at the end of the five year period it could be eligible to be organic. I was unhappy about the use of chemicals – not that I'm saying that it did any damage – we just don't know.'

The set-aside land has been planted with clover and grass seed, in order to improve the land so that it is in better shape than when it was first set-aside. Sixteen hectares were set aside to pasture with limited public access as an amenity, qualifying for an extra £120 per hectare.

Is it right for the country to subsidise farmers to do nothing? Mr. Long's reply: *'That depends on whether it achieves the objectives it was set out to achieve, to reduce production.'*

1 Analyse the conflicting sets of values that a farmer has to consider in making the decision to set-aside land.

2 Consider the other options (apart from set-aside) that were open to this farmer. Why were they unviable? What were the main reasons for his decison to set-aside arable land?

Arable set-aside in English counties:

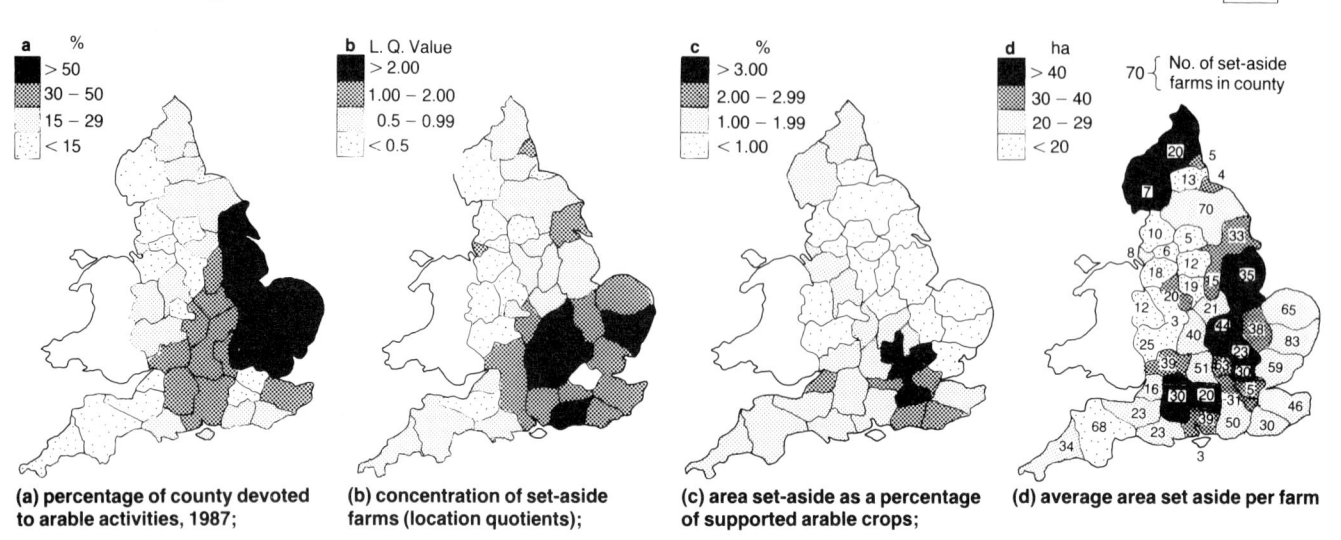

(a) percentage of county devoted to arable activities, 1987;
- > 50 %
- 30 – 50
- 15 – 29
- < 15

(b) concentration of set-aside farms (location quotients);
- > 2.00 L.Q. Value
- 1.00 – 2.00
- 0.5 – 0.99
- < 0.5

(c) area set-aside as a percentage of supported arable crops;
- > 3.00 %
- 2.00 – 2.99
- 1.00 – 1.99
- < 1.00

(d) average area set aside per farm
- > 40 ha
- 30 – 40
- 20 – 29
- < 20
- 70 { No. of set-aside farms in county

Levels of adoption of set-aside

The four maps above (Fig. 3.2.8) show details of the potential and actual level of set-aside land in England.

1 Study map A which shows the percentage of arable land in each county in 1987. Describe the distribution that the map reveals. To what extent does it reflect the relatively high levels of intervention that encouraged farmers to plant their land to cereals?

2 Map B shows the location quotients (L.Q.) for set-aside land in England. L.Q. of >1 indicates counties with more than their fair share of set-aside land; L.Q. of <1 indicates counties that have less than their fair share of set-aside land. Comment on the distribution of the counties with a L.Q. of >1. Are there any reasons why this should not coincide with the counties with the biggest percentage of arable land?

3 Map C shows the area set aside as a percentage of the area devoted to supported arable crops in each county. What does this map tell us about the response to set-aside within the country? Why should there be a slightly better response in the Home Counties to the west of London?

4 Map D shows the average area set-aside per farm. What factors are likely to be responsible for the relatively high areas in Northumberland and Cumbria?

5 Taking all four maps together, what conclusions can be drawn concerning the take up of the set-aside scheme?

Fig. 3.2.8
Geography, Volume 76 (Part 1), 1990

In the third year of set-aside (1990) various changes to the regulations were introduced. Plant cover on set-aside fallow land had to be cut twice a year instead of once. A wider range of environmental features will have to be protected by farmers participating in the scheme e.g. stone walls, vernacular buildings, unimproved grassland, moorland and heath. A limited grazing option was introduced on fallow land. In 1991 a new one year set aside was introduced. Farmers would have to set aside at least 15 per cent of their 1991 arable area for 1991–2 and also reduce their cereals area by 15 per cent in that period.

Fig. 3.2.9
The Field

Farm Diversification Scheme
WALKS IN THE LONG GRASS
£1.00

Other schemes aimed at encouraging farmers to diversify away from traditional arable farming are:

- The Farm Woodland Scheme (which operates in tandem with the set-aside scheme). It was introduced in October 1988, and allows the planting of trees on a maximum of 36 000 hectares over a three year period.

- The Farm Diversification scheme (Fig. 3.2.9). This acknowledges the fact that only 15 per cent of farmers earn all of their income from agriculture. From January 1988 grants to cover capital investment towards the cost of the development of non-agricultural enterprises were to be made available.

Environmentally Sensitive Areas

An Environmentally Sensitive Area (ESA) is 'an area of national environmental significance. The conservation of the area depends on the adoption, maintenance, or extension of a particular form of farming practice, in which changes have occurred (or will occur), in farming practices which pose a threat to the environment. ESAs represent a discrete and coherent unit of environmental interest which would permit the economical administration of appropriate conservation aid.'

The location of the first two sets of ESAs are shown on the map (Fig. 3.2.10). The original twelve were announced in 1986, followed by a further seven in 1987. The latest announcement in November 1991, designated a further twelve in England (six for 1992, and six for 1993). Four new areas were announced for Wales and two for Northern Ireland: further ESAs would be designated in Scotland. The new proposals are indicated on the map; precise boundaries have yet to be decided. The ESAs cover a wide variety of farming and landscape types, from the granite moorlands of West Penwith in Cornwall and the grazing marshes of the Broads to the hills of the Shropshire Borders and the Pennine Dales of Yorkshire. In many of these areas the main purpose of the designation is to maintain the high ecological quality of the grassland, and to preserve other features, such as walls, hedges, ponds and barns. All farmers with eligible land are invited to enter the scheme, which is entirely voluntary. In some areas, farmers who opt to enter land must include all their eligible land (as in the Pennine Dales), while in others, such as the Broads, they can enter as much, or as little, as they wish.

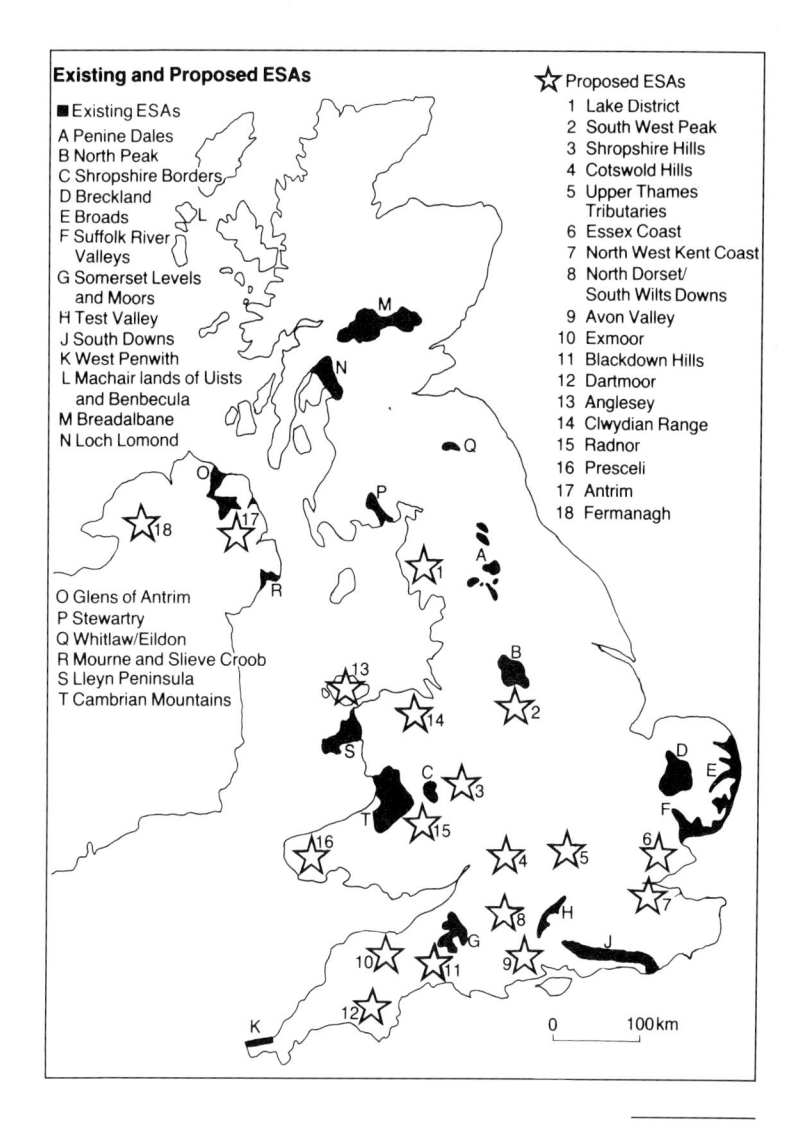

Existing and Proposed ESAs

■ Existing ESAs
A Penine Dales
B North Peak
C Shropshire Borders
D Breckland
E Broads
F Suffolk River Valleys
G Somerset Levels and Moors
H Test Valley
J South Downs
K West Penwith
L Machair lands of Uists and Benbecula
M Breadalbane
N Loch Lomond

O Glens of Antrim
P Stewartry
Q Whitlaw/Eildon
R Mourne and Slieve Croob
S Lleyn Peninsula
T Cambrian Mountains

☆ Proposed ESAs
1 Lake District
2 South West Peak
3 Shropshire Hills
4 Cotswold Hills
5 Upper Thames Tributaries
6 Essex Coast
7 North West Kent Coast
8 North Dorset/ South Wilts Downs
9 Avon Valley
10 Exmoor
11 Blackdown Hills
12 Dartmoor
13 Anglesey
14 Clwydian Range
15 Radnor
16 Preseli
17 Antrim
18 Fermanagh

Fig. 3.2.10

DECISION-MAKING EXERCISE: EVALUATING THREE ESAs

You are to assume the role of a MAFF Inspector, required to evaluate the suitability and efficacy of three ESAs (Figs. 3.2.11, 3.2.15–17). Your report should be structured as follows:

1 An assessment of the respective claims of the three areas to the status of an Environmentally Sensitive Area.

2 An evaluation of the conservation aims of the three schemes.

3 A review of the response to the schemes by farmers within each area.

4 A reasoned ranking of the three areas as ESAs based on your findings.

gorse and bracken provide, with the small fields and the steep granite cliffs a wide range of habitats for wild life.

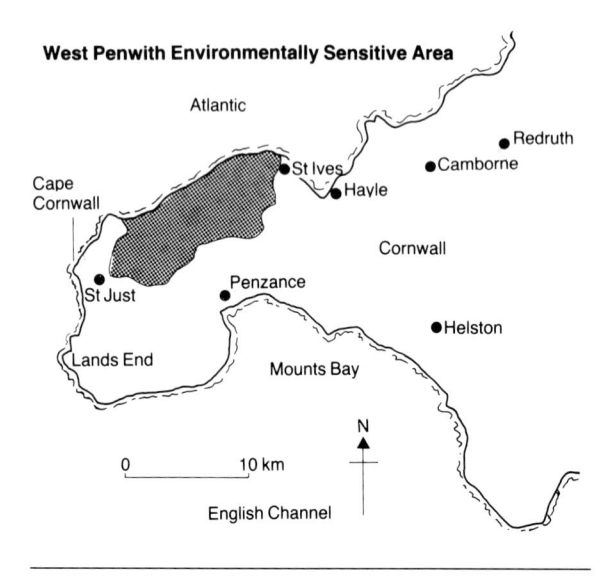

Fig. 3.2.12

Fig. 3.2.11
(a) Granite moorlands in the Environmentally Sensitive Area of West Penwith, Cornwall
(b) The Somerset Levels (West Sedge Moor) Environmentally Sensitive Area
(c) Breckland, Environmentally Sensitive Area in East Anglia, with large areas of heathland and forest

The main guidelines issued to farmers wishing to join the scheme may be summarised as follows.

- On improved land the farmer must keep the hedges up and in good condition, but there will be no restrictions on improvements to the land e.g. no restrictions on fertiliser.

- On rough land the farmer must graze, but not poach (trample) the land with cattle and must not carry out any improvements such as drainage, fencing, applications of lime fertiliser, herbicides or pesticides.

- The farmer shall seek written advice before constructing farm buildings or roads and shall farm in such a way which protects historic features, gateways, boulders, ponds and streams.

Farmers enter into an agreement with MAFF for five years and are paid £60 per hectare.

West Penwith – designated area 7200 hectares. The area has been farmed for over 5000 years, and is noted for its pattern of small, intricate, stone enclosed fields and its ancient granite farmsteads. On the granite moorlands there are many sites of prehistoric settlements. Open expanses of heather,

Somerset Levels and Moors (Fig. 3.2.13) – designated area 26 970 hectares. The region is valued both for its distinctive landscape and the wealth and variety of its wildlife.

The landscape is open low-lying and almost flat, characterised by extensive tracts of almost treeless grassland, small rectangular fields separated by many miles of rhynes (ditches) and lines of pollarded willows on ditch banks and roadside verges. Winter flooding is a

regular occurrence over much of the area, and while giving it its character also contributes to the ornothological value by encouraging overwintering of birds e.g. waders, swans, geese. The exceptional wildlife interest is associated not only with the birdlife but also the many flowering herbs and grasses in the grazing meadows and the aquatic plants and animals of the rhynes. The area has considerable archaeological interest including remains from Roman, Medieval and post Medieval times. 'MAFF' leaflet.

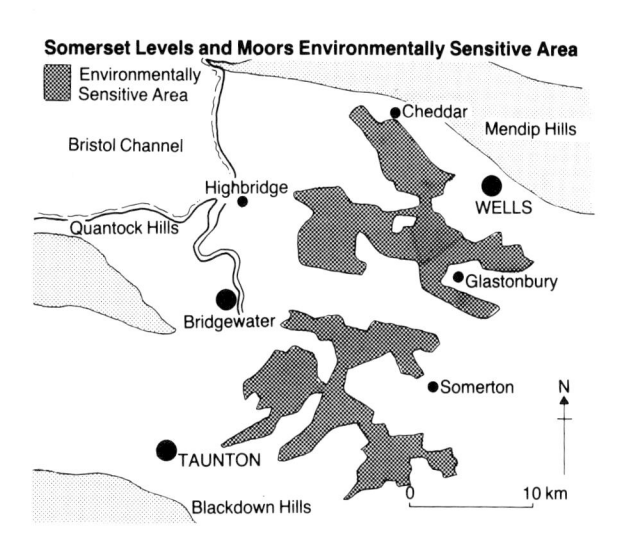

Fig. 3.2.13

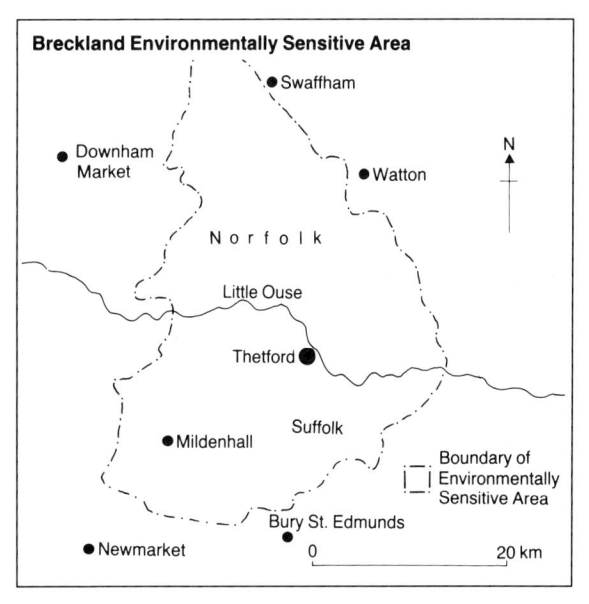

Fig. 3.2.14

The main guidelines for farmers wishing to join the scheme are as follows:

- maintenance of livestock farming with retention of land under permanent pasture

- limits on the application of fertiliser, herbicides, pesticides and lime

- no new field drainage; maintenance of rhyne water levels

- non-chemical maintenance of ditches, rhynes (Somerset for ditch), field gutters, rig and furrow (ancient ridges and furrows in fields)

- maintenance of hedges, trees and pollarded willows, and no damage to features of historical interest.

Farmers enter into an agreement with MAFF for five years and receive payments varying from £82 per hectare to £120 per hectare according to the type of agreement (conditions are much stricter for the higher payment).

Breckland (Fig. 3.2.14) – designated area 88 480 hectares. Breckland is an area of open and grassland landscape based on sheep raising, rabbit warrening and rye cultivation. Heathland communities still make up important areas of Breckland, but considerable areas have now been planted with Scots Pine. The heathland has been ploughed up in the past (broken, hence Breckland) for the cultivation of cereals. When it was returned to fallow it was grazed by sheep, and in more recent times by warrened rabbits, often resulting in serious soil erosion and even the formation of sand dunes in places. Shelter belts of pine trees were planted to restrict soil erosion, and form a distinct part of the mosaic of the Breckland landscape. The region is seamed by shallow river valleys and contains a number of shallow ponds or meres, that occupy hollows formed in periglacial times in the area. Within the ponds, rivers and streams there is a wide and varied range of aquatic life.

The main guidelines for farmers are as follows:

- maintain heathland and dry grassland and do not plough, level, reseed, cultivate or apply fertilisers

- revert arable land to heathland or dry grassland. Two successive cereal crops may be grown, but no liming, limited use of fertiliser and all straw to be removed, but not by burning, after the second cereal crop the land reverts to heathland or dry grassland

- maintain wet grassland, graze with livestock excluding pigs or poultry, but avoid poaching, undergrazing or overgrazing; no use of pesticides and limited use of herbicides

- conserve wildlife strips at the edge of cultivated areas; no applications of fertiliser, lime, herbicides or pesticides

- conservation of headlands (a six metre wide strip at edge of cereal crops), limited use of fungicides, insecticides and herbicides.

Farmers enter into agreement with MAFF for five yeas and receive payments varying from £100 per hectare for maintaining dry grassland and heathland to £300 per hectare for conservation of the wildlife strips.

Fig. 3.2.15

ESA DETAILS

SCHEME	TOTAL DESIGNATED AREA (ha)	PURPOSE	ANNUAL PAYMENT (£/ha)	WHOLE/PART*
West Penwith	7200	Maintain enclosed land and heath	60	Whole
Somerset Levels and Moors	26 790	Retain permanent grassland	82–120	Part
Breckland	88 480	Maintain heath and dry grassland	100	Part
		Maintain wet grassland	125	Part
		Arable strip	300	Part
		Conservation headlands	100	Part

* Whole/part refers to the requirement to enter either the whole or part of the eligible area.

Fig. 3.2.16

LEVELS OF RESPONSE

ESA	NUMBER OF AGREEMENTS/APPLICATIONS			HECTARAGE		
	1987	1988	1989	1987	1988	1989
West Penwith	139	149	160	5500	5850	5945
Somerset	523	671	716	7700	9150	9630
Breckland	–	n.a.	101	–	3023	3535

Fig. 3.2.17

AREAS AND HOLDINGS AS A PERCENTAGE OF THAT WHICH IS ELIGIBLE (hectares)

ESA	AREAS TAKEN UP AS % OF THOSE ELIGIBLE		HOLDINGS AS % OF THOSE ELIGIBLE	
	1987	1988	1987	1988
West Penwith	79	84	77	83
Somerset	48	57	29	37
Breckland (but almost certainly lower than the other two)	n.a.	n.a.	n.a.	n.a.

FARMING AND WATER POLLUTION

Intensification of agriculture, using a range of modern methods, has led to greater risk of water pollution. The demand for higher productivity on cropland has seen increasing use of fertilisers, pesticides and herbicides. Livestock enterprises produce increasing amounts of farm slurry and silage liquor which both pose serious threats of pollution. Although agriculture only accounts for 12 per cent of all water pollution incidents, this is thought to be only a small proportion of what actually occurs. Put in perspective, farm livestock produces about three times the amount of human waste and nearly all of the farm waste is spread on to the land. It is regarded as being much 'stronger' in terms of its polluting potential, so, if only two per cent of it reaches water, it is equivalent to the load from all treated human waste.

PERCENTAGE OF ALL FARM POLLUTION INCIDENTS 1985–1989

POLLUTANT	PERCENT
Cow slurry	55.2
Silage	20.5
Pigs	9.9
Poultry	1.6
Miscellaneous:	
oil	3.0
pesticides	2.0
nutrients	0.8
minor pollutants	7.0

Fig. 3.2.18

Distribution of total pollution incidents from organic waste (1985–1989)

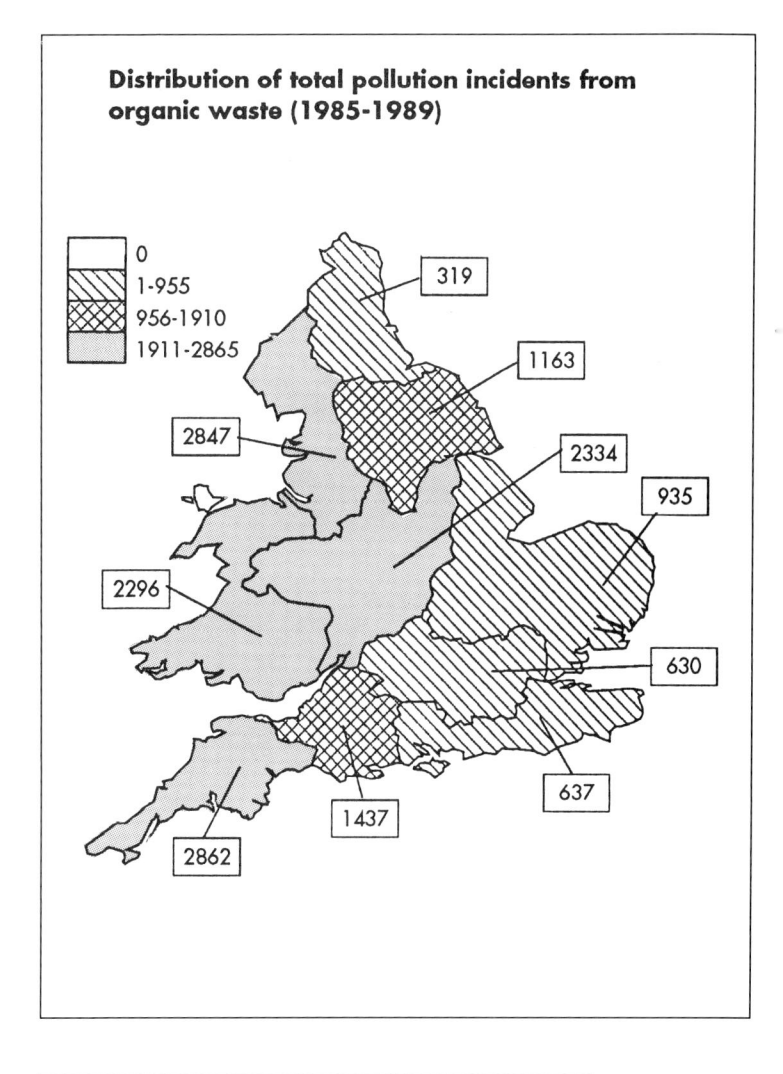

Key:
- 0
- 1–955
- 956–1910
- 1911–2865

Values on map: 319, 1163, 2847, 2334, 935, 2296, 630, 637, 1437, 2862

Fig. 3.2.19

The table (Fig. 3.2.18) shows the percentage contribution of the various polluting elements to the total reported incidents. The map (Fig. 3.2.19) shows the distribution in England and Wales of the total pollution incidents from organic waste (1985–9).

1 Comment on the regional distribution of these pollution incidents.

2 Suggest reasons for the greater concentration on the western side of the counties.

CASE-STUDY OF ORGANIC WASTE POLLUTION

- **July 1989,** about 455 000 litres of pig slurry escaped from a storage tank and entered the River Madford, a tributary of the River Culm in Devon. The cause was poor operation of the storage facility. Ammonia concentrations in the river rose to 140 mg/l and approximately 1500 fish were killed.

Shortly after this incident, in spring 1990, farms in the catchment and its sister stream, the Bolham, were subject to a farm survey, carried out by the NRA (National Rivers Authority). The results of the survey are shown on the map (Fig. 3.20). The majority of the farms in the polluting and high risk category were involved in dairying, with the problems being caused by parlour washings and yard run-off. The survey showed that 64 per cent of the farms relied solely on a dungheap for waste storage, which was often unwalled. During the period of the campaign only six incidents were reported by the public, which represented 24 per cent of the incidents discovered by NRA staff.

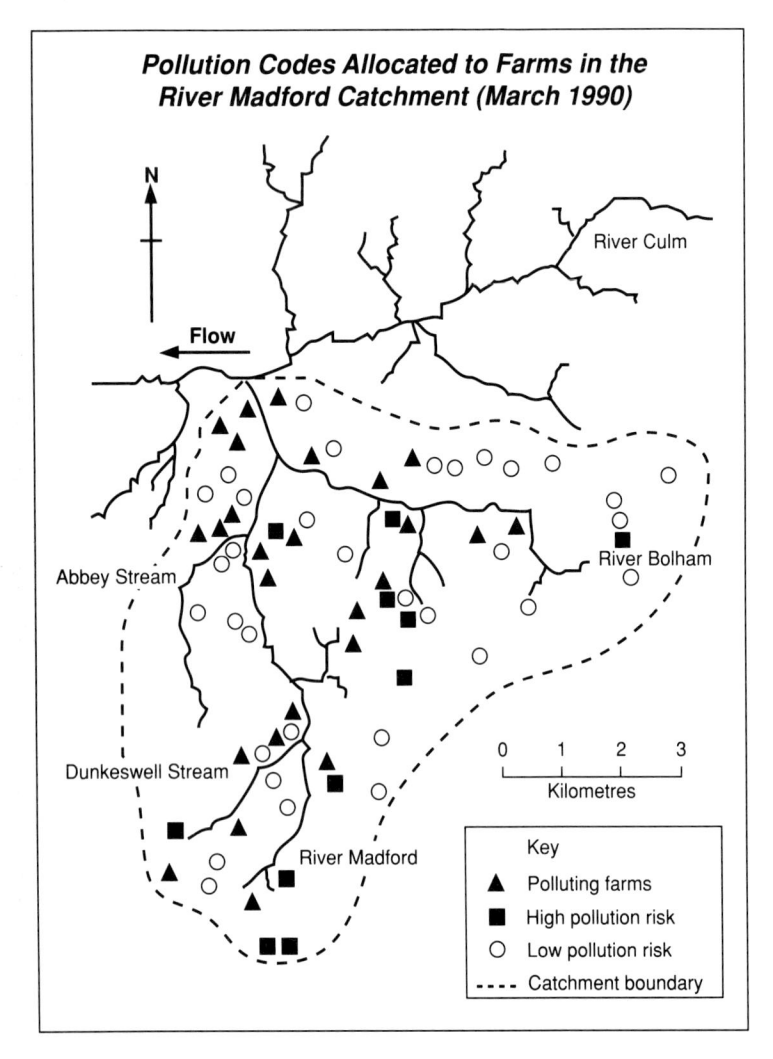

Fig. 3.2.20

1 Comment on the distribution of the polluting farms on the map.

2 Along which stretches of river would you expect pollution to be highest?

3 What measures need to be taken to reduce pollution in this area?

The nitrate problem

Nitrate pollution has probably received more publicity than any other source of pollution. Study the newspaper article 'Nitrates worst in East Anglian water', *The Guardian* 20 September 1989. (Fig. 3.2.22).

1 What are the main causes for concern in the high concentrations of nitrates in water supply in East Anglia?

Now Study the article, 'Nitrates order threatens rural economy' from *The Guardian* September 1989 (Fig. 3.2.23).

1 Examine the nature of the conflict between EEC directives and the farming lobby.

2 Are the measures suggested adequate to resolve the controversy?

Nitrogen, phosphorus and potassium are all nutrients that are essential for plant growth. The main sources of these nutrients are the soil itself, fertilisers, animal manures, rainfall, sewage and silage liquor. Each year 2.5 million tonnes of mineral fertilisers are applied as plant food and smaller quantities of other nutrients, such as animal slurry and sewage sludge are spread to assist plant growth. A popular misconception is that the high levels of nitrates in river water are the result of excessive applications of nitrogenous fertiliser, surpluses being washed off the land and into streams and rivers. Research has shown that remarkably little nitrate is left over from the fertiliser to be leached away in the autumn.

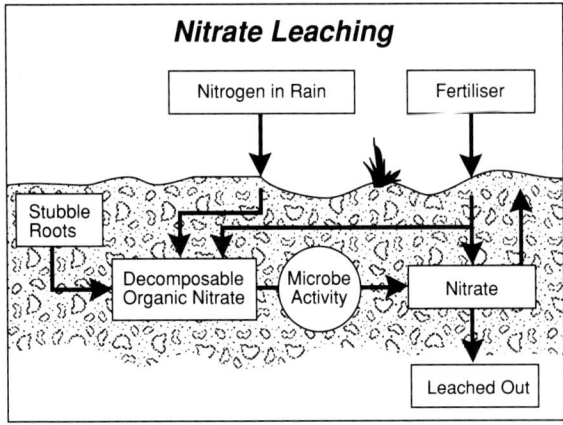

Fig. 3.2.21

Nitrates worst in East Anglia water

Andrew Culf

EAST ANGLIA'S water supply has the highest nitrate concentration in the country. Anglian Water, the authority which covers an area stretching from Essex to Lincolnshire, has 35 water sources which exceed the European Commission's nitrate limit of 50 milligrams a litre.

In some areas, consumers, alarmed by health warnings about the effects of nitrates, regularly filter their household supplies, despite protestations from the authority that the water is perfectly safe to drink.

The authority is embarking on a programme to reduce nitrates to acceptable levels and work on its first nitrate removal plant, costing about £2 million, should be complete by the end of the year.

Ms Iris Webb, Friends of the Earth's East Anglian water and toxics campaigner, said yesterday: "Routinely over the region, nitrates are in excess of safety levels. Nitrates are a bigger problem for Anglian Water than any of the other 10 water authorities."

Ms Webb said that although European surveys had revealed that blue baby syndrome and stomach cancer were associated with nitrates, there was little evidence of this in East Anglia. However, she said that other health risks such as diarrhoea and thrush could be caused by the poor water quality. Ms Webb regularly advises worried mothers who bottle-feed their babies to use bottled rather than tap water supplies to make up the feed.

Like many people in the region she treats her water with a jug filter. In the worst-affected areas families pay about £400 to have a filter plumbed in.

Ms Webb claimed that three factors have contributed to East Anglia's unenviable nitrate problem . Firstly, a large quantity of water comes from underground sources and the water table is heavily loaded with nitrates created by years of intensive agriculture, encouraged by Common Market farm policies.

Secondly, fertilisers running off the land have added to the problem, she said, with some 50 per cent poorly applied by farmers. High nitrate levels are also caused every winter by the natural breakdown of roots and vegetation in the soil, but the process causes more problems in East Anglia because of the intensive use of the land.

Nitrates order 'threatens rural economy'

Alan Travis   Political Correspondent

MORE than half of the most fertile farmland in East Anglia will have to be taken out of production if the Government is to comply with the European Commission's directive on nitrate levels in drinking water, a House of Lords committee warns today.

The peers on the Lords European Communities Committee believe that the Commission's proposals to cut nitrate levels were drawn up without taking account of the scientific evidence of the impact they would have.

They say in a report: "The restrictions on agriculture which the Commission seems to have in mind would have a drastic effect on output and would bring extensive social and economic changes to rural society in the most intensively-farmed areas in the United Kingdom.

"It appears that intensive arable farming is simply incompatible with the Commission's target."

The report warns that in most of East Anglia and parts of the Midlands and Kent the, EC's nitrate target of below 50 parts per million in drinking water sources can only be met by turning extensive areas into unfertilised land for grass or trees.

"As these areas are the most productive in the country, this would have a major effect on arable output and on the rural economy generally."

Expert witnesses told the committee that the vulnerable areas of East Anglia and Lincolnshire currently grow two-fifths of the UK's wheat crop, 29 per cent of its potatoes, 73 per cent of its sugar beet and 61 per cent of its field vegetables.

The Lords report agrees that some form of restriction on the quantities of nitrate entering water sources from agricultural land is essential if the tap-water supplies are to be safe-guarded. But they question the effectivensss of the EC's proposals.

In particular, the peers think the commission misguided when it says that relatively minor restrictions on fertiliser applications will be enough to cut nitrate pollution levels and that their impact on arable farmers' profits will not be great.

The peers argue that, on the contrary, many of the farmers involved in intensive agriculture will suffer substantial losses.

They prefer the more cautious government approach, based on pilot programmes in selected areas. But they urge ministers to give clear advice on how and when to use fertilisers. Farmers should also be encouraged to make better use of organic manures.

Fig. 3.2.22&23
The Guardian, 20 August 1989

Most of the nitrate that is leached out is derived from the breakdown of organic material in the soil by microbe activity in the autumn. At this time of the year the soil is either bare or carrying a growth of a newly sown crop. Increased rainfall in the autumn results in a flow of water down through the soil and the microbe-produced nitrate is leached away into streams and rivers (Fig. 3.2.21). It is likely, however, that the fertiliser applied increases the amount of organic nitrogen in the soil that the microbes can break down.

Study the following statistics, which show levels of nitrate concentration in three English rivers (Fig. 3.2.24).

1   Draw line graphs based on running means to show the main trends in nitrate concentration in the period between 1972 and 1988.

2   Why is it necessary to use running means as a method of recording this data?

3   What are the main trends revealed and what factors could have caused the irregularities that the running means have eliminated?

The map (Fig. 3.2.25) shows nitrate trends (positive = increase in nitrate levels, negative = decrease) in selected rivers in England and Wales (two from each regional Water Authority).

Study the map and then use it to evaluate the evidence that it provides to test the validity of the following statements.

a   Rivers in the main arable districts show the most positive trends.

b   Rivers in areas of higher rainfall show positive trends.

c   Much higher levels of nitrate pollution are found in the upland rivers of England and Wales.

d   Nitrate levels remain much the same in the rivers on the eastern side of the country.

On a more local scale the two graphs (Fig. 3.2.26) show the concentration of nitrates in groundwater (in Jurassic Limestones) at Swell, and the increase in land devoted to arable cultivation.

Fig. 3.2.24

ANNUAL MEAN NITRATE (NO_3-N) CONCENTRATIONS (MG L^{-1}) FOR THREE ENGLISH RIVERS, FOR THE WATER YEARS 1972–1988

YEAR	SLAPTON WOOD (DEVON) (AREA=1 KM^2)	WINDRUSH (GLOUCS.) (AREA=300 KM^2)	THAMES AT OXFORD (AREA=3400 KM^2)
1972	5.9	n.a.	6.5
1973	5.6	5.1	5.9
1974	5.9	6.0	6.7
1975	6.4	6.1	6.4
1976	6.9	4.5	5.4
1977	8.3	7.7	10.7
1978	7.7	7.6	7.8
1979	6.8	7.7	8.8
1980	6.9	7.3	7.4
1981	6.7	7.6	7.9
1982	7.8	7.2	7.8
1983	6.8	7.7	8.6
1984	6.9	7.3	7.9
1985	8.3	9.1	9.4
1986	9.2	8.5	7.7
1987	9.2	8.5	7.7
1988	8.8	9.0	8.0

Fig. 3.2.25

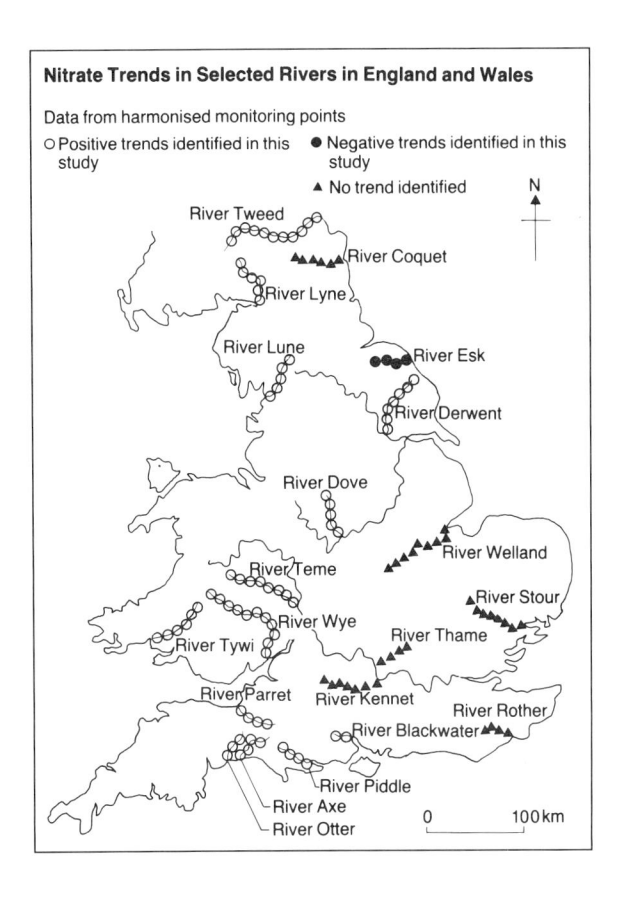

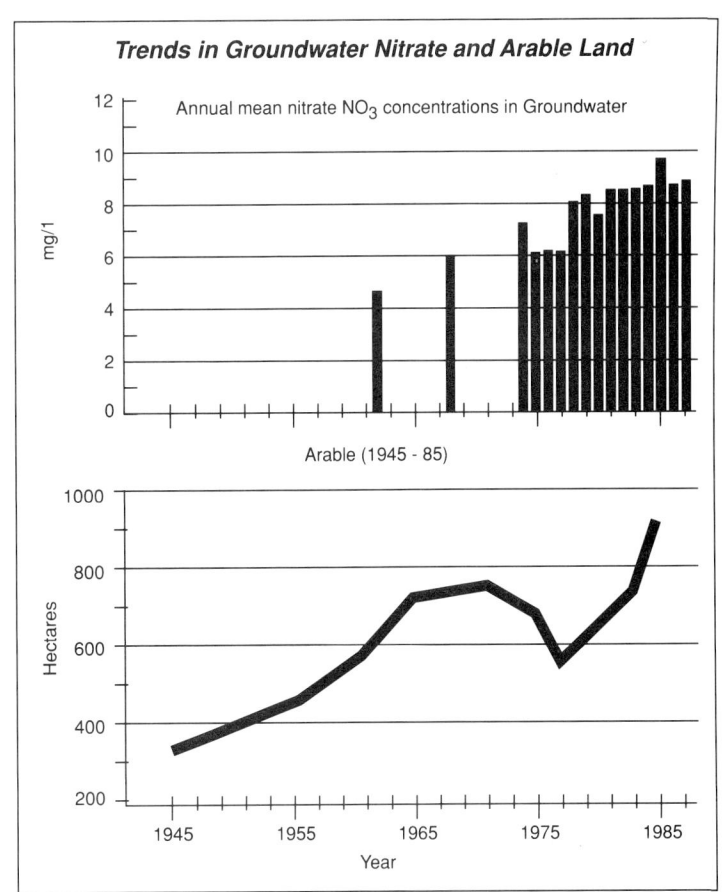

Nitrate Trends in Selected Rivers in England and Wales

Data from harmonised monitoring points
○ Positive trends identified in this study
● Negative trends identified in this study
▲ No trend identified

Trends in Groundwater Nitrate and Arable Land

Annual mean nitrate NO_3 concentrations in Groundwater

Arable (1945 - 85)

Fig. 3.2.26

1 What simple conclusion can be drawn from a comparison of the two graphs?

2 Why would it be unwise to generalise from your findings from the two graphs?

Clearly nitrate pollution is a growing problem. The following code of conduct for good farming practice draws together some of the themes explored above:

- do not apply nitrogen fertiliser in autumn
- do not leave the soil bare in the autumn
- sow winter crops early in the autumn
- if a spring crop is to be sown, grow a winter crop, such as ryegrass, which can then be ploughed in before the spring crop is sown
- use animal manures sparingly
- do not plough up too large a proportion of the grassland in any one area at any one time
- plough in straw; this locks up some nitrate in organic matter, but may lead to increased nitrate pollution in the long run.

THE FUTURE FOR FARMING

Charlie Pye-Smith and Richard North published 'Working the Land' in 1984. Their last chapter was entitled 'Epilogue: Rural Britain in the year Two Thousand'. Read the short extract from this chapter below (Fig. 3.2.27).

1 Based on your work in this section on farming, how far do you think they have been correct in their prophecies so far?

2 What do you consider to be the most promising trends displayed in British farming in the early 1990s?

There were, of course, no longer great fortunes to be made in farming. By the beginning of the 1990s farmers were having to dispose of surpluses as best they could on the world market and there were many problems associated with indebtedness. The price of land in 1995 was roughly the same as it was in 1985; in real terms this meant its value had halved. The farmer who had borrowed heavily found himself in trouble with the banks but De-escalation Grants helped many to weather the storm. Where farmers who had over-extended themselves wished to sell a portion of their holdings the county committees established an imaginative scheme to buy their land and set up family farms. County councils, rather than selling off their small-holdings as they had once been encouraged to do, were now doing the opposite; and large landlords were given a package of incentives which encouraged them to relet tenancies which fell vacant and to create new ones. The decision that no one should be eligible for development grants if he farmed over 300 acres [120 hectares] of lowland, 1500 acres [600 hectares] of marginal land or 3000 [1200 hectares] of mountain land led to the break-up of some of the large estates. The part-time farmer, for so long a despised element of the rural community, became increasingly common towards the end of the century. The success of part-time farming, particularly in the uplands, owed much to the shift of small industries into rural towns and villages.

Changes in attitude were even more pronounced than changes in the farming landscape. The process of de-escalation was welcomed by all but a handful of farmers. The sales representatives of the big chemical companies were no longer as welcome as they had been in the halcyon days of the seventies when their produce was slapped on the land with gay abandon. It was with relief that the farmers came off the chemical treadmill: few became strict organic farmers but nearly all relied heavily on their skilful manipulation of rotations and stock to curb pests and disease. There were some long-established traditions, ridiculed in the post-war boom period, that came back into fashion. Interestingly, much of the impetus for change came from the new generation of farm students who had little sympathy for the reckless ways of many of their parents.

Fig. 3.2.27
Extract from Working the Land, C. Pye-Smith and R. North

RURAL LANDSCAPES

Changes in farming practice are reflected in the rural landscape. Those who knew the rural scene before World War Two would find it difficult to recognise the same landscape today. Many familiar features no longer appear as landmarks, and their absence marks the progress of modern farming. Where whole ecosystems, such as ancient woodlands, or hay meadows have been lost, they will be difficult to replace. However much we lament the passing of the old countryside image, it is as well to remember that the features of the modern landscape which some find so jarring, are equally transient. Already new woodlands darken the farm horizons, and new hedgerows give a more substantial look to field boundaries. Meadows in danger from the plough still retain their ecological richness and diversity, and new ponds and meres attract the wild life that suffered in the draining of the old wetlands. Trees and woodlands were cleared in the post-war years, but in the 1990s there are plans for new recreational forests and even a new National Forest in the English Midlands. Heathland, one of the richest and most grievously damaged of Britain's habitats, now has a better chance of survival than it did in the last few decades. To imagine the rural landscape of the twenty-first century is as difficult as it was for others to envisage the changes of the post-war years.

THE AGRICULTURAL LANDSCAPE

To examine changes in the agricultural landscape three case studies have been selected.

- **Prickwillow:** located in the Cambridgeshire fenland, with intensive arable and horticultural cultivation on a highly fertile soil.

- **Piddlehinton:** located on the Dorset chalklands; in the twentieth century there has been a dramatic swing away from the traditional sheep rearing on the downland turf to extensive cereal cultivation.

- **Grandborough:** located on the Midland clay plain; mainly under grass with livestock enterprises dominating. Flat or slightly rolling enclosure landscape with large numbers of hedgerow trees.

Details on these three landscapes are available for 1920, 1970 and 1980. No survey has been carried out since 1980, but the changes monitored over 60 years reveal some striking changes in all three landscapes.

Fig. 3.3.1 English rural landscape in late summer: Vale of Wardour, west of Salisbury Plain from Wyn Green (Wiltshire)

PRICKWILLOW

CROPS	PROPORTION OF IMPROVED FARMLAND (%)		
	1920	1970	1980
Wheat	29	34	36
Other cereals	19	10	11
Total cereals	48	44	47
Potatoes	19	16	12
Sugar Beet	–	20	21
Mangolds	4.5	–	–
Dry peas	–	4.5	2
Vegetables	–	6.5	13.5
Other crops	11	3.5	1.5
Total other crops	34.5	50.5	50
Temporary grass	3.5	1.5	–
Permanent grass	14	4	4
Total improved grass	17.5	5.5	4

The three sketches below represent the Prickwillow landscape in 1945, 1972 and 1992 (Fig. 3.3.2).

Fig. 3.3.2

1945

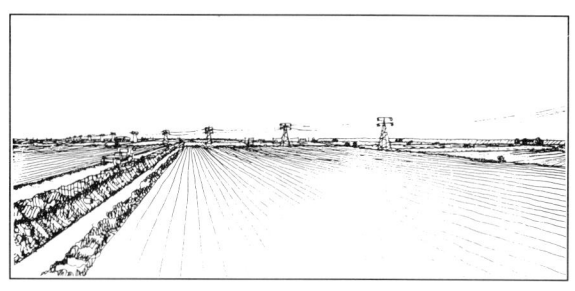

1972

1992

HORIZON TYPES (1970): AVERAGE PERCENT OF HORIZON MADE UP BY DIFFERENT LANDSCAPE FEATURES.

VIEW OVER 1 KM	TREES	HEDGEROWS	ROADSIDE HEDGES	BARE GROUND	BUILDINGS
52	8	0	0	20	17

The three sketches below represent the Piddlehinton landscape in 1945, 1972 and 1992 (Fig. 3.3.3).

Fig. 3.3.3

1945

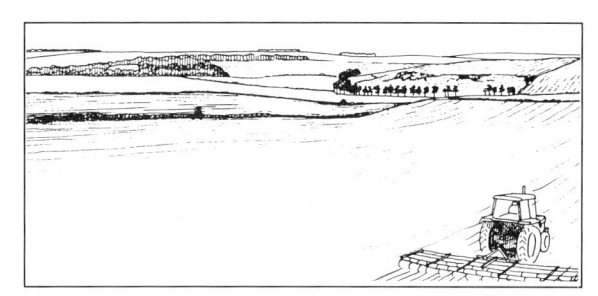

1972

1992

PIDDLEHINTON

CROPS	PROPORTION OF TOTAL FARMLAND (%)		
	1920	1970	1980
Wheat	4	13.5	22
Barley	9	31	25
Oats	10	3	1
Other cereals	1.5	0.5	0
Total cereals	24.5	48	48
Temporary grass	11	21.5	25.5
Permanent grass	41	21	19
Rough grazing	6.5	3	3.5
Total grass	58.5	45.5	48
Total other crops	18	6.5	4.5

HORIZON TYPES (1983): AVERAGE HORIZON MADE UP BY DIFFERENT LANDSCAPE FEATURES.

VIEW OVER 1 KM	TREES/WOODS	HEDGES	ROADSIDE HEDGES	CROPS BARE GROUND	BUILDINGS
34	9	13	24	17	3

GRANDBOROUGH

CROPS	PROPORTION OF IMPROVED FARMLAND (%)		
	1920	1970	1980
Total cereals	10	30	36
Total other crops	5	5	3
Temporary grass	3	20	13
Permanent grass	82	45	50
Total improved grass	85	65	62

Three sketches represent the Grandborough landscape in 1945, 1972 and 1992 (Fig. 3.3.4).

You are to assume the role of a landscape consultant. You are to give a presentation of your research in these three areas to a group of colleagues, and have to prepare a series of overhead transparencies for your talk.

1 Use a series of diagrammatic techniques to show the changes that have taken place in the crop and grassland percentages.

2 Use a technique to show the horizon composition (there are different possibilities here e.g. bar graphs, pictograms).

3 Devise an evaluation matrix to produce an assessment of the landscape changes that have taken place over the sixty-year period in the three locations.

4 Produce a flow diagram to show the factors that have been responsible for landscape change. List the main changes first, then list the factors responsible for these changes. Finally construct the diagram to show how the two sets are related.

Fig. 3.3.4

1945

1972

1992

HORIZON TYPES (1983): AVERAGE HORIZON MADE UP BY DIFFERENT LANDSCAPE FEATURES (%).

VIEW OVER 1 KM	TREES/WOODS	HEDGES	ROADSIDE HEDGES	CROPS BARE GROUND	BUILDINGS
41	14	22	15	5	3

TREES, HEDGEROWS AND WOODLANDS IN THE RURAL LANDSCAPE

The thirty years since 1945 have been a time of unprecedented destruction of ancient woodland, in contrast to the active conservation or slow decline of the previous thousand years. This sort of thing has often been said before, but this time it is real. Earlier damage was usually reversible. This time it is not a matter of felling trees which will grow again, but of converting sites to other uses. It is an uncanny experience to trace an identifiable wood through five or seven centuries, and on going to the spot to be just in time for the dying embers of the bonfires in which it has been destroyed.

Oliver Rackham, 'Trees and Woodland in the British Landscape'.

Fig. 3.3.5
Ancient beech woodland: Ashmore, Dorset

Ancient woodlands (see Fig. 3.3.5) are those that existed in or before the Middle Ages, and are particularly important for nature conservation.

Such forests are inevitably under threat since the stock that was inherited in 1600 was a finite resource and could only diminish. The maps below (Figs 3.3.6a and b) show:

- ancient woodlands as percentages of county areas in England and Wales
- percentages of ancient woodlands that have been cleared for agriculture and other uses in the last 50 years.

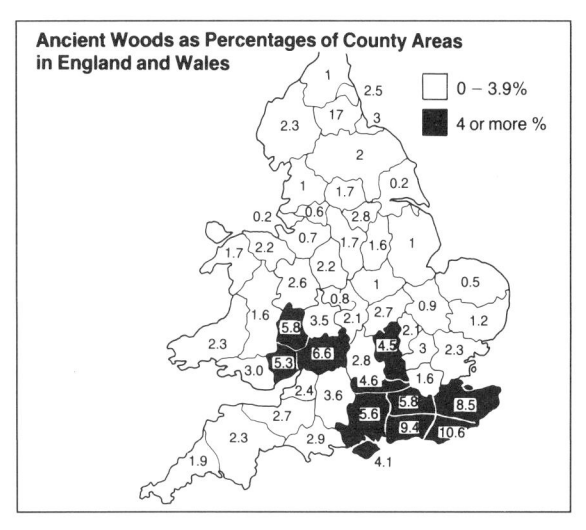

Fig. 3.3.6a

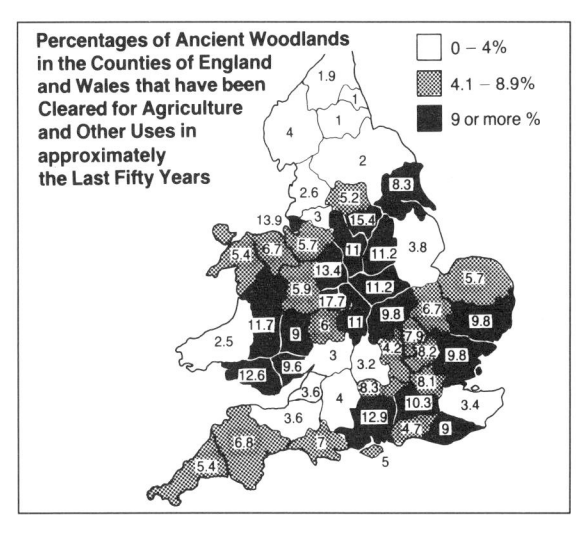

Fig. 3.3.6b

1 Comment on the distributions shown on the two maps.

2 Discuss the factors that might have been responsible for the variations in loss of ancient woodland.

Fig. 3.3.7
Oxleas Woods

One of the most recent publicised cases of a threatened piece of ancient woodland is that of Oxleas Woods in south-east London (Fig. 3.3.7). The woods are thought to be 8000 years old and are oakwoods rich in bird, plant and insect life. They survived a threat from a proposed housing development in 1934, after a long protest campaign, led by *The Observer* newspaper. The woods were under threat from the proposed East London River Crossing that would have resulted in a four lane highway being driven through the woods (see Fig. 1.7.4). The proposed cutting would have divided the woods in two and would have resulted in the felling of at least 500 mature oaks, hornbeams and sweet chestnuts and many other species of smaller trees and shrubs. Most of the 33 breeding bird species would have been lost. Pollution from exhaust gases and fumes would have had a serious effect on the ecological balance in the woods.

After a long protest campaign alternative proposals were put forward by the objectors:

- a 1200 metre tunnel that would add £32 million to the cost
- a cut-and-cover tunnel that would allow the wood to regenerate at a cost of £23 million.

The Inspector of the enquiry could not allow either of these alternatives, since they would cancel out the 'value for money' calculated for the new road. He proposed an alternative 400 metre cut-and-cover tunnel that would only add £10 million to the cost and would avert the worst of the environmental damage. This was rejected by the two Government ministries involved on the ground that the scheme initially proposed was environmentally acceptable.

In October 1991 the European Community Commissioner, Carlo Ripo di Meana warned that legal action could be taken by the Community if the construction of the East London River Crossing, and the building of the Oxleas Woods dual carriageway went ahead, on the grounds that Britain had failed to carry out adequate environmental impact assessments. The Government abandoned its plans in July 1993 (see page 39).

David Black, 'P.A.R.C. (People against the River Crossing),' *'How can Britain criticise the Brazilians for cutting down the rain forest, when our primeval forest is being destroyed by the Government'*.

1 How do the values of Government ministries which were in favour of the road differ from the values of those opposed to the destruction of the wood?

2 Why do you think the Government changed its mind?

Hedgerows

The most obvious effect of modern British farming, particularly in eastern England, is the disappearance of the hedgerows. From the windows of our farmhouse near Huntingdon we cannot see a single hedge except for the one around our garden. Until 1945 tall hawthorn hedges lined both sides of the road, and surrounded every field. They have all gone.

Kenneth Mellanby, 'Farming and Wildlife', 1981.

Fig. 3.3.8
Hedgerow near Abbotsbury, Dorset

Hedgerows were an intrinsic part of the British countryside as portrayed by the great landscape painters such as Constable and Turner. Their traditional functions as stock barriers, field boundaries and providers of shelter and shade have been complemented recently by their value as a wildlife habitat, forming corridors within intensively farmed areas linking separate pieces of copse and woodland and other semi-natural habitats. In a survey, farmers in seven different areas of the country were asked to give their reasons for hedge removal; their replies were as follows.

Cambridgeshire (dykes or ditches instead of hedges, which are largely absent, but perform the same function); field amalgamation and benefits of allowing large scale mechanisation.

Huntingdonshire; efficiency of mechanised working and intensification of production.

Dorset; field sizes were traditionally large and thus few hedges had been removed, removal was primarily to increase the efficiency of mechanised production.

Somerset; improves the efficiency of mechanical operations and facilitates grassland management.

Herefordshire; improves the efficiency of mechanical operations.

Yorkshire; untidiness and loss of agricultural function – the hedges in this area are of very low quality.

Warwickshire; poor state of the hedgerows, amalgamation of very small fields, and to achieve uniform size in fields for rotation purposes.

AREA	AVERAGE FARM SIZE IN 1983 (HA)
Cambridgeshire	68
Huntingdonshire	158
Dorset	140
Somerset	45
Herefordshire	72
Yorkshire	105
Warwickshire	57

The diagrams (figs. 3.3.9 to 3.3.11) show:

- hedges removed 1945–83 and hedges remaining 1983
- average field size 1945, 1972 and 1983
- quality of remaining hedgerows 1983.

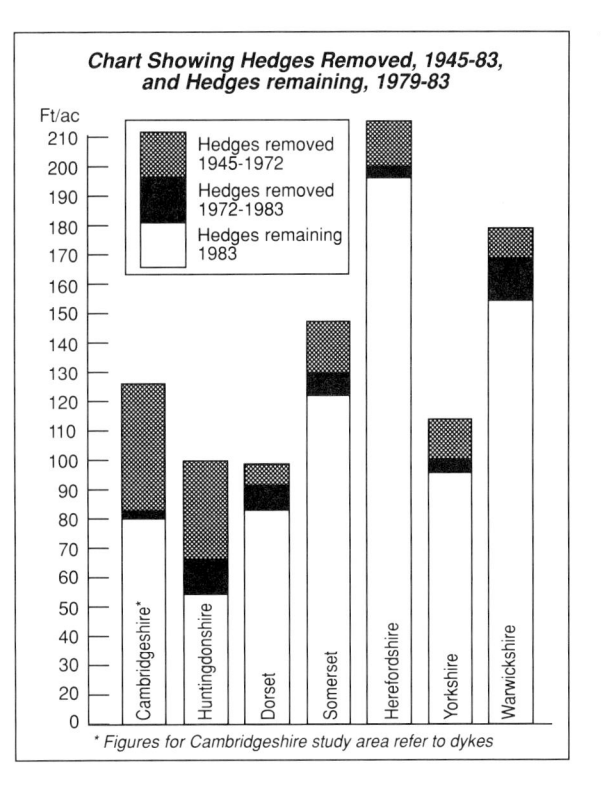

Fig. 3.3.9 — Chart Showing Hedges Removed, 1945-83, and Hedges remaining, 1979-83

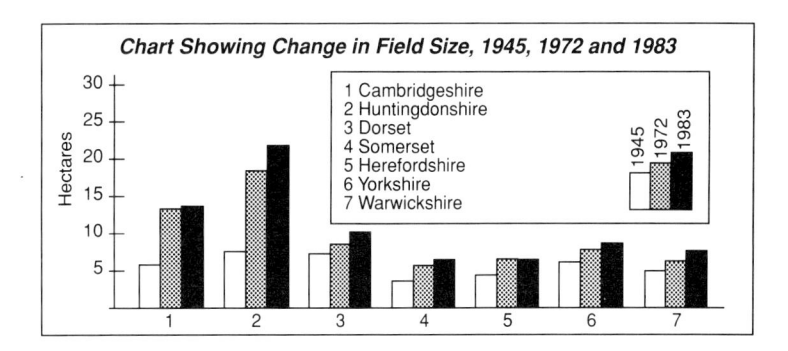

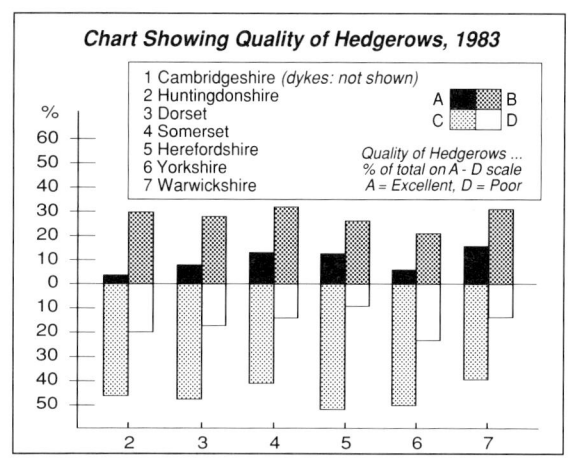

Fig. 3.3.10 — Chart Showing Change in Field Size, 1945, 1972 and 1983

Fig. 3.3.11 — Chart Showing Quality of Hedgerows, 1983

Using all of the resources on hedgerows write a summary report on hedgerow loss in the seven areas, suitable for presentation to a group of conservationists.

New forests for the next century

In the last five years important new initiatives have been launched to create new forest areas within Britain. New planting in much of the twentieth century has taken place in the British uplands (see section on Upland Britain) and some specific locations in lowland Britain, such as Breckland in East Anglia. However, the dreary coniferous plantations have received much criticism for their drab appearance and their ecological poverty. Pressures for planting in lowland Britain have been growing for a number of reasons. Tree losses from Dutch Elm disease, acid rain, and recent storms in 1987 and 1990 called for replacement plantings. Surplus arable land now has to be put to other uses, and the Farm Woodland scheme aims at achieving new plantings, although the scheme has so far had a disappointing response. Global concern over the loss of tropical rain forest has prompted fresh thinking about tree cover in middle latitude areas such as Britain. In comparison with its continental neighbours Britain is poorly endowed with woodland and forests. Only 10 per cent of the UK is devoted to forests, in France the figure is 27 per cent and in Germany 30 per cent.

Fig. 3.3.12

Community forests

Plans for a series of community forests were announced in 1989 and the national programme was launched jointly by the Countryside Commission and the Forestry Commission in 1990. The main aims of these new forests were to create places on the edge of towns and cities where major environmental improvements will create well wooded landscapes, and will provide opportunities for:

- a thriving forestry and farming industry with increased scope for diversification
- recreation, walking, riding and sports
- education – as an outdoor classroom
- new habitats for wild-life

Three community forests were announced at the launch of the scheme, and a further nine were added in February 1991 (see fig. 3.3.12). No major changes in land ownership are likely as a result of the new initiative and there will be no compulsory purchase of land. The main aim is to improve parts of the urban fringe that have become neglected, through encouraging landowners to plant tracts of woodland that can be used by the wider community.

The new National Forest

The broad principles that underly the idea of the community forests also apply to the concept of the new National Forest. The idea embraces multi-purpose forestry, with the New Forest in Hampshire acting as a model. It would be a mix of land uses, with some wooded areas, open farmland interspersed with small towns and villages. It would also fulfil a number of roles, such as a producer of timber, a recreational resource, and a habitat for wild-life. In 1989 the Countryside

Fig. 3.3.13

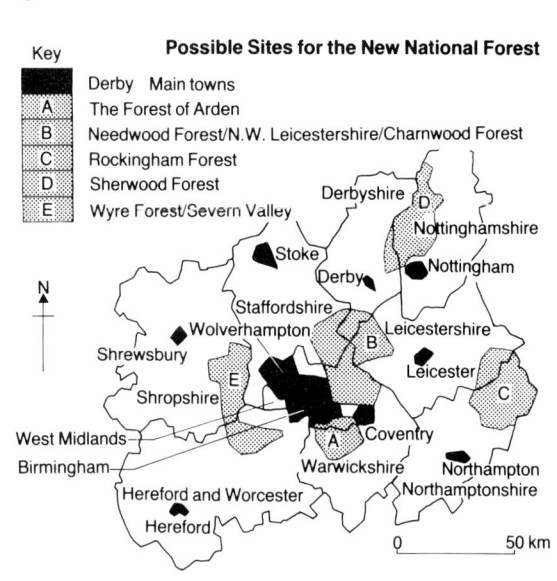

Fig. 3.3.14

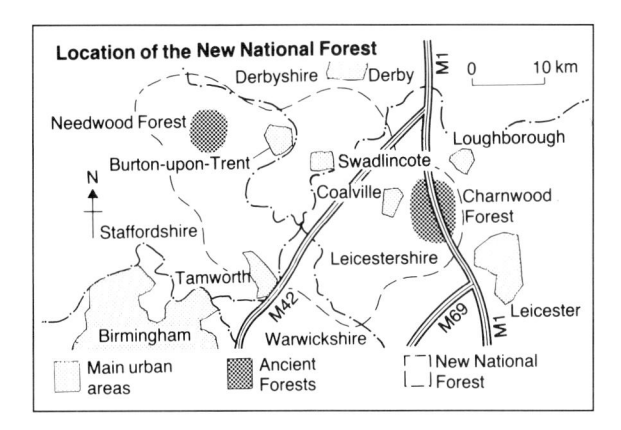

Location of the New National Forest
[Map showing: Derbyshire, Derby, Needwood Forest, Burton-upon-Trent, Loughborough, Swadlincote, Coalville, Charnwood Forest, Staffordshire, Leicestershire, Tamworth, Leicester, Birmingham, Warwickshire, M1, M42, M69. Legend: Main urban areas, Ancient Forests, New National Forest]

Commission announced a shortlist of five possible locations, each of which was centred on fragments of ancient woodland; these locations are shown on the map (Fig. 3.3.13). The final decision on a location for the National Forest was made in October 1990. The selected option was the Needwood/Charnwood site, shown on the map (Fig. 3.3.14). The Countryside Commission chose this location for a variety of reasons, not the least of which was that the opportunities for environmental improvement were considerable in an area that was much scarred by old mineral workings, particularly in the north-west Leciestershire coalfield.

Compulsory purchase will not be employed in the development of the forest. The Commission will seek to persuade farmers and other owners that timber production, and other options can yield worthwhile profits. The trees will be predominantly native broadleaved species, such as oak, ash, beech and willow, although conifers will also play their part. The forest is likely to be funded from a variety of sources, both public and private. Total investment of some £90 million is needed over the next 30 years. Initial pilot planting began in late 1990, and the main woodland planting programme is scheduled for late 1992.

1 Study the map which shows the location of the new National Forest. Environmental improvement of the area has been suggested as one reason for choosing this area. What locational advantages does it possess?

2 What major arguments could be deployed to persuade a sceptical farmer of the advantages of planting trees on land within the designated area?

3 It has been suggested that the fund-raising for the new forest could be organised through a New National Forest Trust – a charitable organisation. Suggest a suitable job description for the new chairman of this trust.

THE HEATHLAND LANDSCAPE

To recline on a stump of thorn in the central valley of Egdon, between afternoon and night, as now, where the eye could reach nothing of the world outside the summits and shoulders of heathland which filled the whole circumference of its glance, and to know that everything around and underneath had been from prehistoric times as unaltered as the stars overhead....
(Thomas Hardy, 'The Return of The Native').

Fig. 3.3.15
The Dorset heathlands, near Furzebrook, Isle of Purbeck: note the invasion of heathland by gorse and Scots Pine

No other English writer has captured the image of heathland better than Thomas Hardy, writing about Egdon Heath or the Dorset Heathlands (see Fig. 3.3.15). Yet Hardy would find it difficult to recognise the same heathland today. Within the area he knew as Egdon, there is now an atomic energy research establishment, the largest on-shore oilfield in western Europe, the main producing open cast pits for high quality ball clay in Britain, a maze of gravel workings, and a rash of firing ranges where armoured vehicles exercise. Although the Dorset heathlands are perhaps under more pressure than the other remnants of lowland heath in England, the heathland landscape and habitat in general is a fast disappearing one. Heathland in England appears to have originated from the clearance of woodland, and has been maintained by grazing, burning and removal of any invading trees. Today it possesses a unique ecology, dominated by a rare and distinctive flora and fauna (for example, on the Dorset Heathlands are found the Dorset heath, the Dartford Warbler, the sand lizard and the smooth snake).

Use the 1:25 000 Tourist map of the Isle of Purbeck on the inside of the front cover.

1 Consider the area bounded by eastings 86 and 90 and northings 81 and 85. Draw a sketch map of the area to show the main types of land use.

2 Discuss the main conflicts arising from the close proximity of such different types of land use.

Fig. 3.3.16

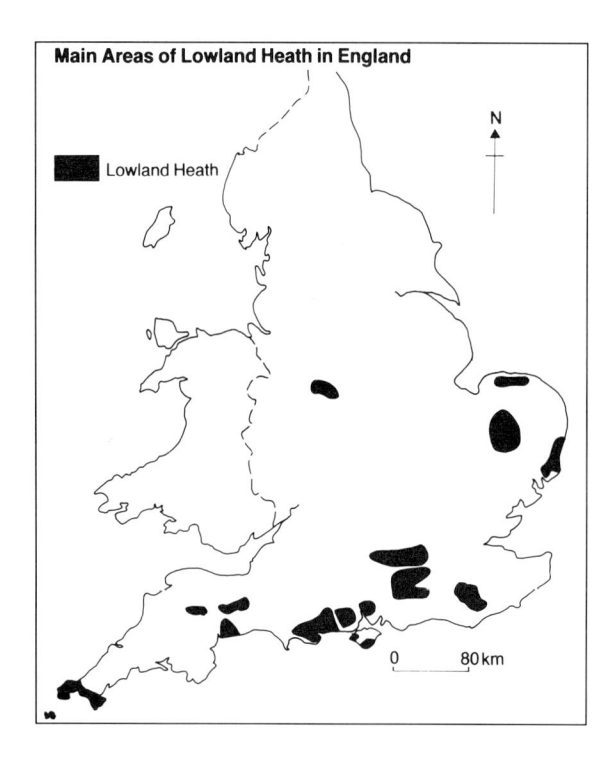

Main Areas of Lowland Heath in England

The map (Fig. 3.3.16) shows the distribution of lowland heathlands in England, and the diagrams (Fig. 3.3.17) shows the decline in the areas of heathland in the period between 1800 and the mid-1880s.

Fig. 3.3.17

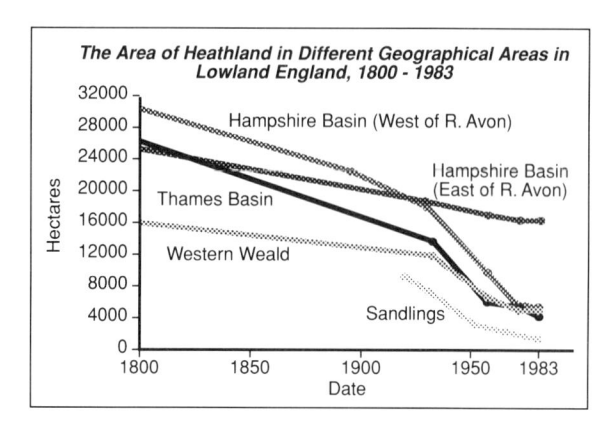

The Area of Heathland in Different Geographical Areas in Lowland England, 1800 - 1983

1 With the help of a good atlas, discuss the reasons for the present vulnerability of the heathland landscape in England.

2 What are the broad trends that are apparent in the series of graphs of heathland decline?

3 Suggest reasons for the slowing down of the decline in the period since 1950.

The Dorset heathland

1759

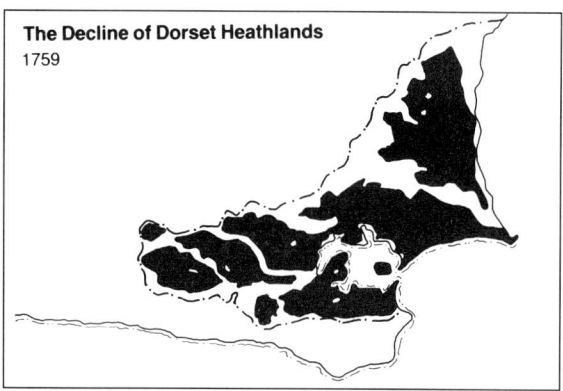

The Decline of Dorset Heathlands
1759

1934

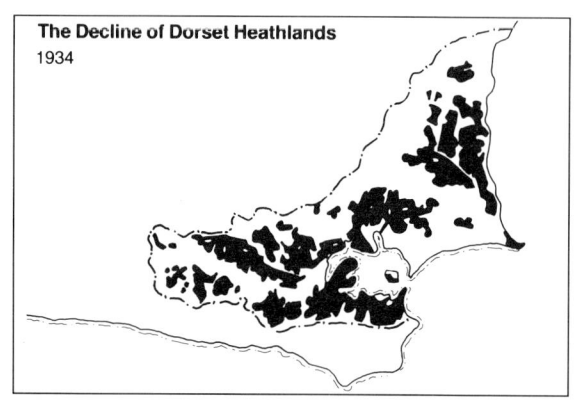

The Decline of Dorset Heathlands
1934

RURAL LANDSCAPES

1960

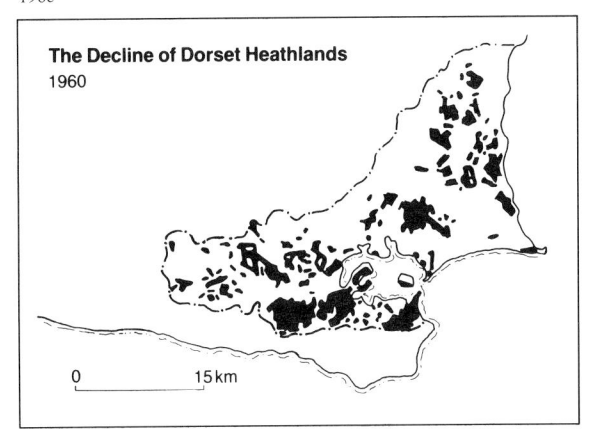

The Decline of Dorset Heathlands
1960

0 15 km

1987

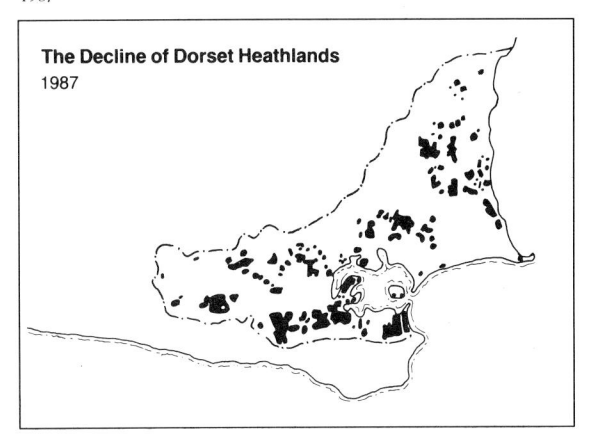

The Decline of Dorset Heathlands
1987

Fig. 3.3.18

The four maps (Fig. 3.3.18) show the progressive decline in the area of the Dorset heathlands. The map (Fig. 3.3.19) shows the range of heathland nature reserves in the country. The creation of these reserves has been essential to the preservation of the remaining parts of the heath. The pressures on the heathlands have been referred to above.

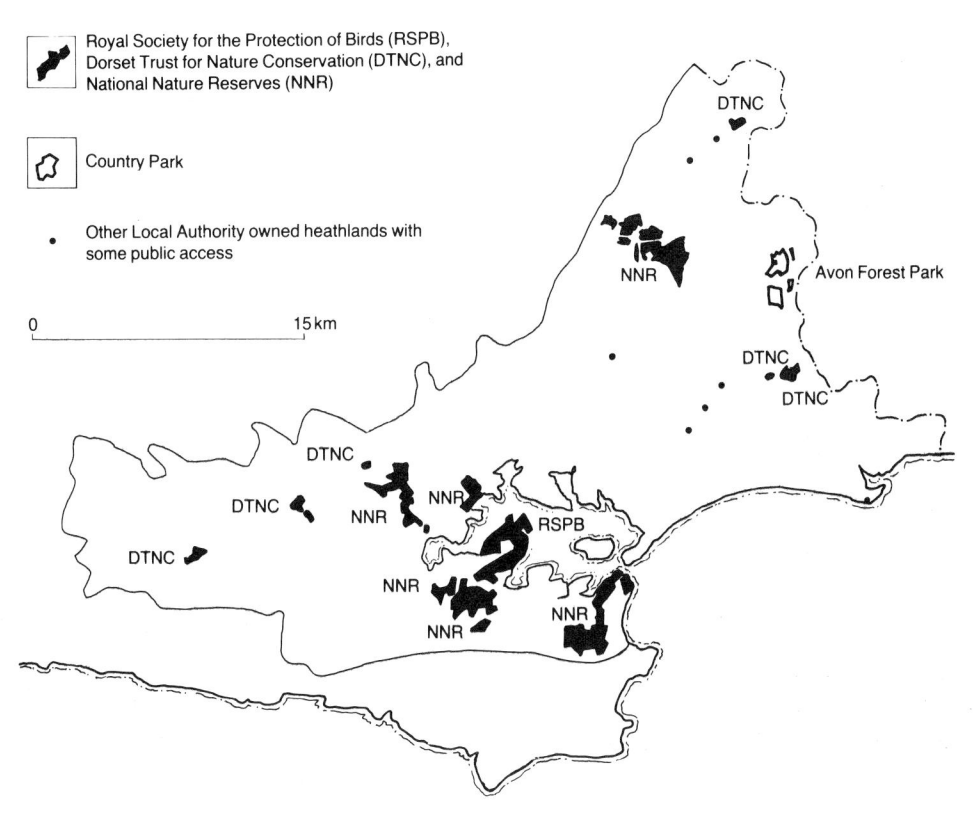

Read the Extract from *The Observer*, June 1991 'Natives rage as Hardy's Heath waits to be blasted' (Fig. 3.3.20).

1   What are the main conflicts indicated in the article?

2   How can the extraction of gravel from this area be justified, bearing in mind its high ecological status?

3   'The not-so-pretty areas of Dorset have all been worked out. Now the prettier ones must be worked.' Is this a value judgement that can be justified?

Fig. 3.3.19

Heathland Nature Reserves in South-East Dorset

Royal Society for the Protection of Birds (RSPB), Dorset Trust for Nature Conservation (DTNC), and National Nature Reserves (NNR)

Country Park

Other Local Authority owned heathlands with some public access

0 15 km

Natives rage as Hardy's heath waits to be blasted

EGDON Heath, Thomas Hardy's wild, brooding tract of wilderness that has been 'from prehistoric times as unaltered as the stars overhead', according to the Wessex novelist, is about to meet what may be its final crisis, **writes Sarah Lonsdale**.

So frightened was the young Hardy by stories of witches and bogeymen hiding out in the heath's dark swallow holes that he dared not cross it at dusk.

Now prospecting on the heath, which is near the Dorset villages of Turnerspuddle and Affpuddle, has revealed gravel and sand deposits just a few feet below the surface. These could be excavated cheaply.

English China Clays has submitted the first of what locals fear will be a series of planning applications to start excavations on parts of the heath.

The application, to dig 250 000 tonnes of gravel from a 27-acre [11 hectare] field, is at the particularly sensitive Throop Corner, where. Clym Yeobright and tragic Eustacia Vye used to walk on summer evenings in *The Return of the Native*.

Tess of the d'Urbervilles walked over Egdon Heath to reach the Valley of the Great Dairies. The meadows were backed by the 'swarthy and abrupt slopes' of the heath, on which stood 'clumps and stretches of firs, whose notched tips appeared like battlemented towers crowing black-fronted castles of enchantment'.

The haulage route, over which 60 32-tonne lorries a day will travel from the site to ECC's Warmwell processing plant, six miles away, overlooks Clym and Eustacia's Brickyard Cottage, still standing and occupied by Lucy Woollen, who remembers Hardy – 'a funny little man' – from when she was a girl. The route crosses Diggory Venn, the reddleman's trade path over the heath.

The site is adjacent to the Piddle Valley Conservation Area and includes three Sites of Special Scientific Interest. But the results of other planning applications in the area offer little comfort to the local action group, Rage (Residents Against Gravel Extraction).

ECC's last application to dig gravel in Dorset, from near Wareham on the Purbeck coast, was turned down by the county council. But at the appeal the conservation status of the area – part of the Dorset Heritage Coast, an Area of Outstanding Natural Beauty and an SSSI – was disregarded and the application was waved through.

During the past 10 years not a single planning appeal by ECC has been rejected. Rage chairman Paul Badcock said: 'This area is full of history. Virtually every tree, every clump and hollow of the heath has a story to tell. We refuse to bear witness to the devastation of our countryside without a fight.'

But ECC spokesman Ian Lanyon said: 'The not-so-pretty parts of Dorset have all been worked out. Now the prettier parts must be worked.'

The wild, untameable heath, which while 'the sea changed, the fields changed, the rivers, the villages and the people changed', remained the same, appears to be facing its 'one last crisis – the final overthrow', ominously referred to in Hardy's *Return of the Native*.

Fig. 3.3.20
Observer, June 1991

Fig. 3.3.21
Povington Heath: Army Firing Ranges on the Dorset Heathland

RURAL SETTLEMENTS AND SERVICES

If Thomas Hardy would find it difficult to recognise his much-loved Egdon Heath, it is equally true that he would find that the villages and hamlets of southern England today bear little resemblance to those of his native Wessex. Gone are the village workers, the furze-cutter, the hurdle-maker and the reddleman: few shepherds and dairymen remain. Most of the village shops will have gone too, and the revered schoolmistress would no longer have enough pupils to keep the school open. He would be baffled by the large number of people that leave the village every morning to work in offices or business parks, often over twenty miles away, and in some cases nearly a hundred miles distant. He would notice that some houses were empty, and only showed signs of life at the weekends or for a longer period in the summer. He was used to reaching distant places, such as the Cornish villages where he did advisory work as an architect, on the emerging railway network. Many of these rural branch lines have long since been taken up, their routes used as walkways or cycleways. He would find the filling station at the end of the village a new feature, with its newspaper stall and cold cabinets, often a surrogate for the village shop. But, most puzzling of all would be the lack of identity of the place: village gossip and chatter is of less substance, since fewer people know their neighbours any more and the sense of community has gone in some villages.

TWO VILLAGES COMPARED: DOWN AMPNEY AND ALLENHEADS

Down Ampney

Down Ampney has cedars tall,
Ornate chimneys – Tudor Hall:
Warplanes flew from nearby fields,
Now splendid crops our 'Airfield' yields.

All Saints Church with graceful steeple
Memorials has to valorous people.
Pheasants strut in every copse –
No huntsmen now pursue the fox.
Enthusiasts at the vicarage peer;
Yes – Ralph Vaughan Williams was born here!

Winifred Dickens

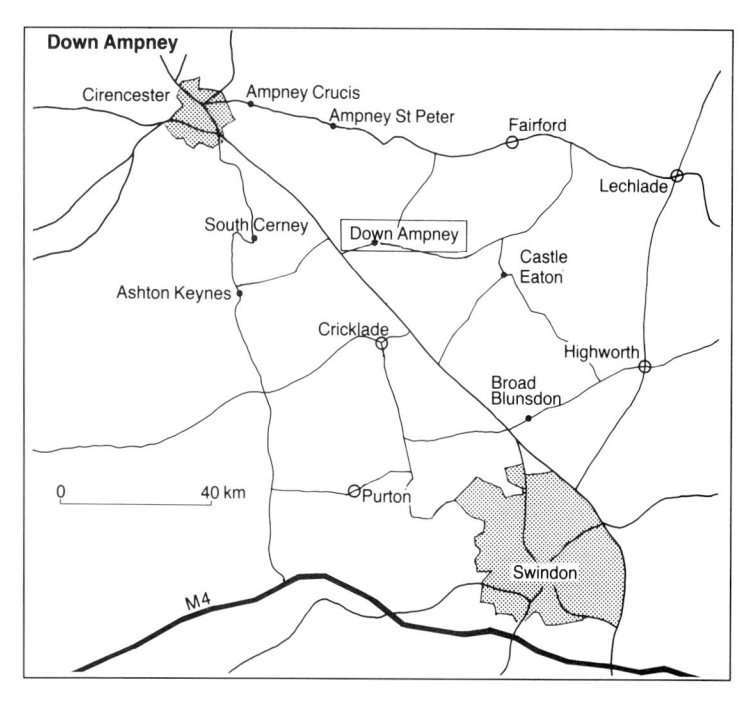

Fig. 3.4.2

Fig. 3.4.1
The village of Down Ampney in the Cotswolds

The map (Fig. 3.4.2) shows the position of Down Ampney, in the Cotswolds, within relatively easy reach of Cirencester, and Swindon. The charts and diagrams (Fig. 3.4.3) show the main demographic and occupational structure of the village. Of the population who are in work few work in the village itself. The following percentages indicate the population's place of work.

Fig. 3.4.3 Down Ampney: some sociological perspectives

PER CENT	PLACE OF WORK
23	Circencester
21	Swindon
17	Fairford or Lechland
12	Down Ampney or within 5 miles
5	Cheltenham or Gloucester
2	London
20	Variety of location more than 5 miles

DISTANCES PEOPLE TRAVEL TO WORK

42	work from home
22	work in the village
106	within 10 miles
56	11–30 miles
24	30 miles

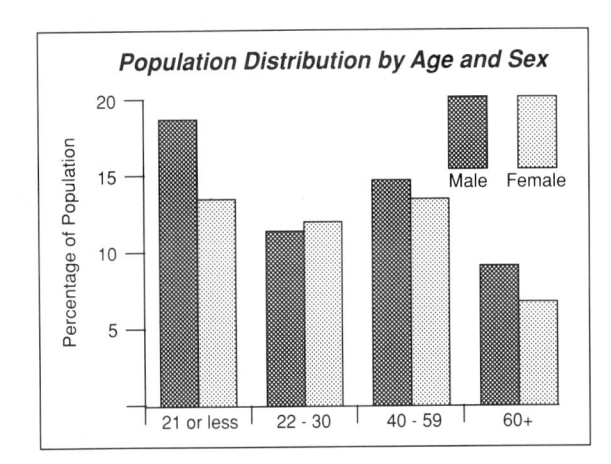

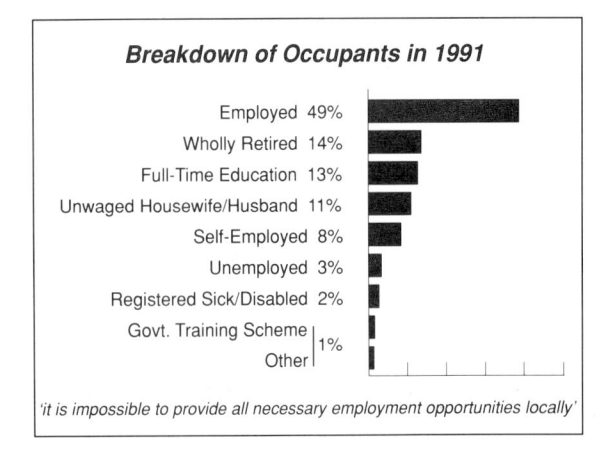

The village has retained its school, and has a Post Office and general stores. Surprisingly there is no pub in the village. Significantly 86 % of the households possess a car, 42 % have two cars, 9 % have three cars, 2 % have four cars and 2 % have five cars! Within households 80 % of the men have daytime access to a vehicle, but the figure is only 60 % for women. Of those people that work 70 % use their own car and 13 % travel by bus. However 53 % of the village do not use the bus service.

Some comments on Down Ampney

Headmistress of the village school 'The significant strength of the village school is the great sense of security, identity and natural feeling of belonging which makes the children strong and secure characters'.

'The co-op houses are a disgrace to the village – they should be renovated or sold off to those that would do it.'

'Only small houses should be built until balance be restored as larger houses would make Down Ampney a "Yuppie" village.'

'Too many heavy traffic problems which could be redressed by weight and speed restrictions and a police presence.'

'More frequent bus service or minibus to Cirencester, Lechlade and Fairford.'

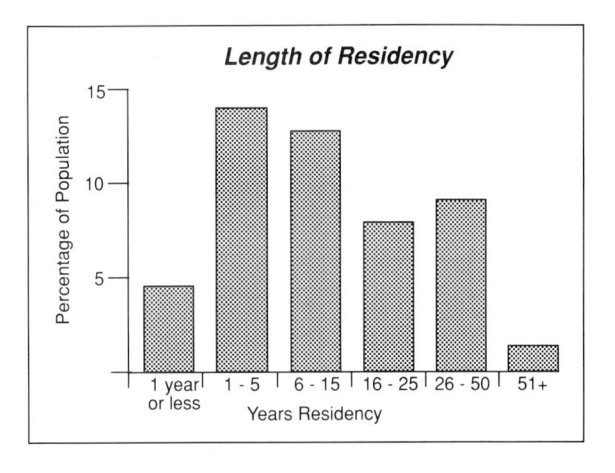

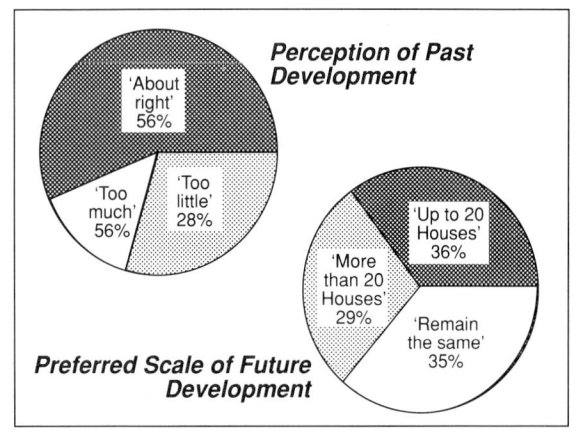

Population Distribution by Age and Sex

Length of Residency

Breakdown of Occupants in 1991

Employed 49%
Wholly Retired 14%
Full-Time Education 13%
Unwaged Housewife/Husband 11%
Self-Employed 8%
Unemployed 3%
Registered Sick/Disabled 2%
Govt. Training Scheme 1%
Other

'it is impossible to provide all necessary employment opportunities locally'

Perception of Past Development

'About right' 56%
'Too much' 56%
'Too little' 28%

Preferred Scale of Future Development

'Up to 20 Houses' 36%
'More than 20 Houses' 29%
'Remain the same' 35%

RURAL SETTLEMENTS AND SERVICES

Allenheads

Fig. 3.4.4
(a) Allenheads village in the northern Pennines, one of England's highest villages
(b) the Heritage Centre, Allenheads village

Fig. 3.4.5

The map (Fig. 3.4.5) shows the remote position of Allenheads in the northern Pennines. Similar data to those on Down Ampney are not available, since no Village Appraisal was carried out in Allenheads.

Fig. 3.4.6
Observer, 3 December 1989

A village pulled back from the dead
Tim Walker

MRS Elizabeth Ridley did not expect to enjoy the final years of her life. The village she had lived in for the best part of the century was starting to die a long and painful death around her.

The decline of Allenheads, a small upland community on the borders of Northumbria, Cumbria and Durham, had seemed irreversible ever since the lead mine that had sustained it from the seventeenth century was shut down a decade ago.

Many other villages in the area had come to the same sad, ignominious end. Mrs Ridley, like just about everyone else, had thought that nothing could be done to save it. 'I felt that it would just be me pottering about the ruins,' she said last week. 'I don't think it would have given me much to live for.'

Today, Allenheads looks as if it has come back from the grave. Work has begun on the first six new homes to be built in the village for more than 100 years. A trout farm is being established nearby. There is a new community centre with exhibition rooms in the middle of the village. The post office, pub and general store, all of which had looked likely to close, are enjoying a healthy trade.

The population of the village was 2500 in its heyday. It reached an all-time low of 120 shortly after the mine closed and is slowly starting to rise again. Employment is running at almost 100 per cent and the future suddenly looks promising.

The change in the village's fortunes was brought about by the efforts of a small but determined group of local people – initially just a pensioner, a publican, a gamekeeper, a housewife and an unemployed man – who had a bold new vision for the village.

They set up the Allenheads Trust to relaunch the village as a tourist destination, using its natural resources in a new role. It is the highest village in the country, some 1,400 feet above sea level, and commands spectacular views of the surrounding countryside. Its situation so far off the beaten track, once its greatest disadvantage, became an asset as ramblers and naturalists were urged to use it as a base.

Nevertheless a full account of life in Allenheads appeared in *The Observer* on 3 December 1989 and it is reproduced in Fig. 3.4.6.

1 How does the quality of life differ in Down Ampney and Allenheads?

2 Is there any evidence for a greater sense of village identity in Allenheads than in Down Ampney?

3 What predictions could be made about life in the two villages in twenty years time?

RURAL SETTLEMENT PLANNING

The planning of rural settlements is the responsibility of the county and district councils established under the 1972 Town and Country Planning Act. Each county is charged with the production of a Structure Plan (covering the whole county), and each district council has to produce a Local Plan (covering its own particular area).

Fig. 3.4.7

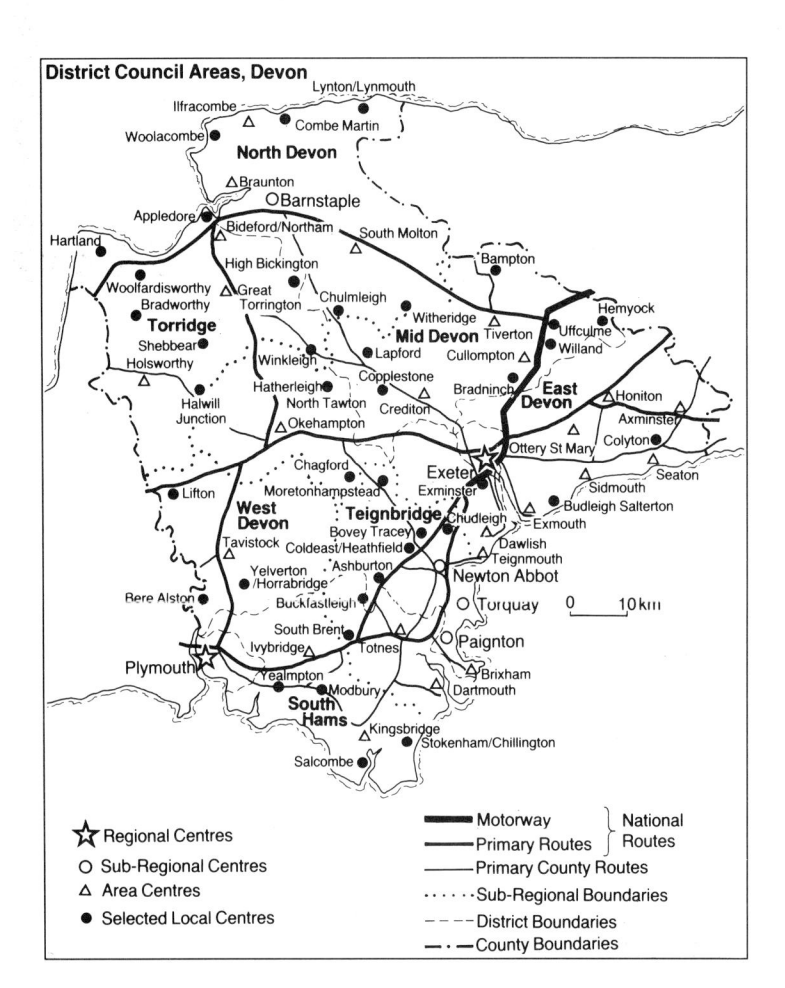

The map (Fig. 3.4.7) shows that local government in Devon is organised through seven district councils: East Devon; Mid Devon; North Devon; South Hams; Teignbridge; Torbay; Torridge and West Devon. Rural settlements are classified as Area Centres and Selected Local Centres. Most of the area centres are small towns which offer a range of higher order services such as secondary schools, hospitals, a local newspaper and some comparison shopping (such as clothing and furniture). It can be argued that the real linch-pins for the maintenance of the rural fabric in the county are the selected local centres.

Fig. 3.4.8

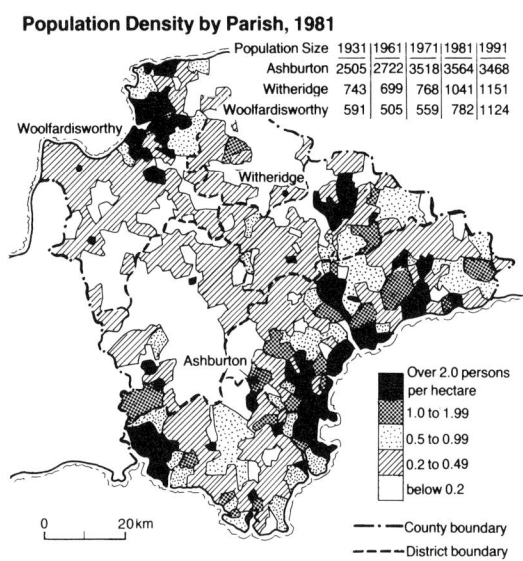

Population Density by Parish, 1981

Population Size	1931	1961	1971	1981	1991
Ashburton	2505	2722	3518	3564	3468
Witheridge	743	699	768	1041	1151
Woolfardisworthy	591	505	559	782	1124

1 Suggest the range of criteria that would have to be considered in designating these Selected Local centres.

2 Study the map (Fig. 3.4.8). Suggest why the range of services offered in the three Selected Local Centres of Woolfardisworthy, Witheridge and Ashburton is likely to show some considerable differences. Their respective populations for 1931–91 are shown below the map.

Rural services

The range of rural services will have been considered in assembling the criteria for choosing the Selected Local Centres in the previous section. The village shop acts as a community focus in most villages, and is seen to have more than commercial significance.

Village shops put up their shutters

Peter Freedman on the fate which awaits even the best rural stores

THE THREE nearest villages to Steeple Aston in north Oxfordshire have all lost their village stores in the last two years. The Post Office store in Steeple Aston itself, owned and run by John and Julie Wain, carried on to win the 1989 award for the Best Village Shop in Buckinghamshire, Berkshire and Oxfordshire. Now it is up for sale – and unlikely to find a buyer.

The Wains did everything right from the moment they bought the shop in 1984. They invested £30 000 to double its size. They opened a delicatessen counter, stocked wholefoods and offered courgettes and asparagus.

They drew up cash flow forecasts and used the techniques of industrial engineering time-study, learnt by Mr Wain during his previous job as a management consultant. They worked 84 hours a week and, for two years, lived on food that had passed its sell-by date.

The judges of the Best Village Shop Award held them up as a model of how to run a business.

But the Wains made a loss from the start. Village stores – even the best-run ones – look set to follow trams, galoshes and threepenny bits into history.

The Rural Development Commission revealed last week that the number of grocery outlets in Britain had fallen to fewer than 39 000 from 147 000 in 1961. Only about 9000 shops are left in rural areas; according to the commission, they are closing faster than ever because of supermarket competition, high property prices, rising overheads and – last, but many feel most – the thumping increase in costs caused by the introduction of the Uniform Business Rate in April.

Proposals for saving village shops – put forward by academics from the Institute of Agricultural History in Reading last week – include turning them into branches of the social services departments and giving them state income supplements. But even if such improbable solutions were adopted, they would come too late to save the Wains. The shop has been on the market for 18 months. 'People have advised us now to market it for alternative use." said Mr Wain.

The 1300 people in the village – itself an award-winner which was twice named Oxfordshire's Best Kept Village – are a mixture of local workers and London commuters. They have already lost another general store, which has become an architect's office. The local authority has advised the Wains that a similar option would be open to their premises.

Village store prices are not the problem. When the Wains compared the price of 30 lines in their shop with those in the multiples earlier this year, the difference was just fourpence in the pound. Once you buy petrol for the 20-mile round trip the supermarket, the local shop may well be cheaper.

The Wains know they are up against larger cultural forces – Sainsbury and Tesco have both opened stores in Banbury, 10 miles away, in the last two years. 'The new Tesco's – it's more a hypermarket than a supermarket – has got this magical attraction,' Mr Wain said. A visit is a leisure activity, a day out.

Mr Wain says closure of village shops has "a devastating effect". The village store and post office are a lifeline of the elderly and young mothers.

Fig. 3.4.9 Independent

Often it is combined with a Post Office, its community function then becomes doubly important, with the payment of pensions, its savings bank and range of mailing facilities. The village shop, like its corner counterparts in towns, is under threat – 'Use it or lose it' has become a common slogan.

Read the article 'Village shops put up their shutters' from *The Independent*, (Fig. 3.4.9).

1   Why is this article particularly pessimistic concerning the future of the village shops?

2   Why does the slogan 'Use it, or lose it' have something of a hollow ring about it?

Several surveys have shown that, after the cost of travel to town has been taken into account, there is little difference in the cost of groceries in the village shop and in supermarkets in the nearest town!

The six maps (Fig. 3.4.10) show the distribution of a range of rural services in Devon.

Trace blank maps of Devon and then use separate maps to:

1   plot all the villages that have more than 11 shops

2   plot all the post offices, banks, surgeries and primary schools that have closed

3   plot all of the dispensing chemists and dentists.

Mark in on an overlay map the areas of rural Devon that appear to be the least well served for this range of services.

4   What factors would be likely to be responsible for this distribution?

A NEW VIEW OF BRITAIN

Fig. 3.4.10

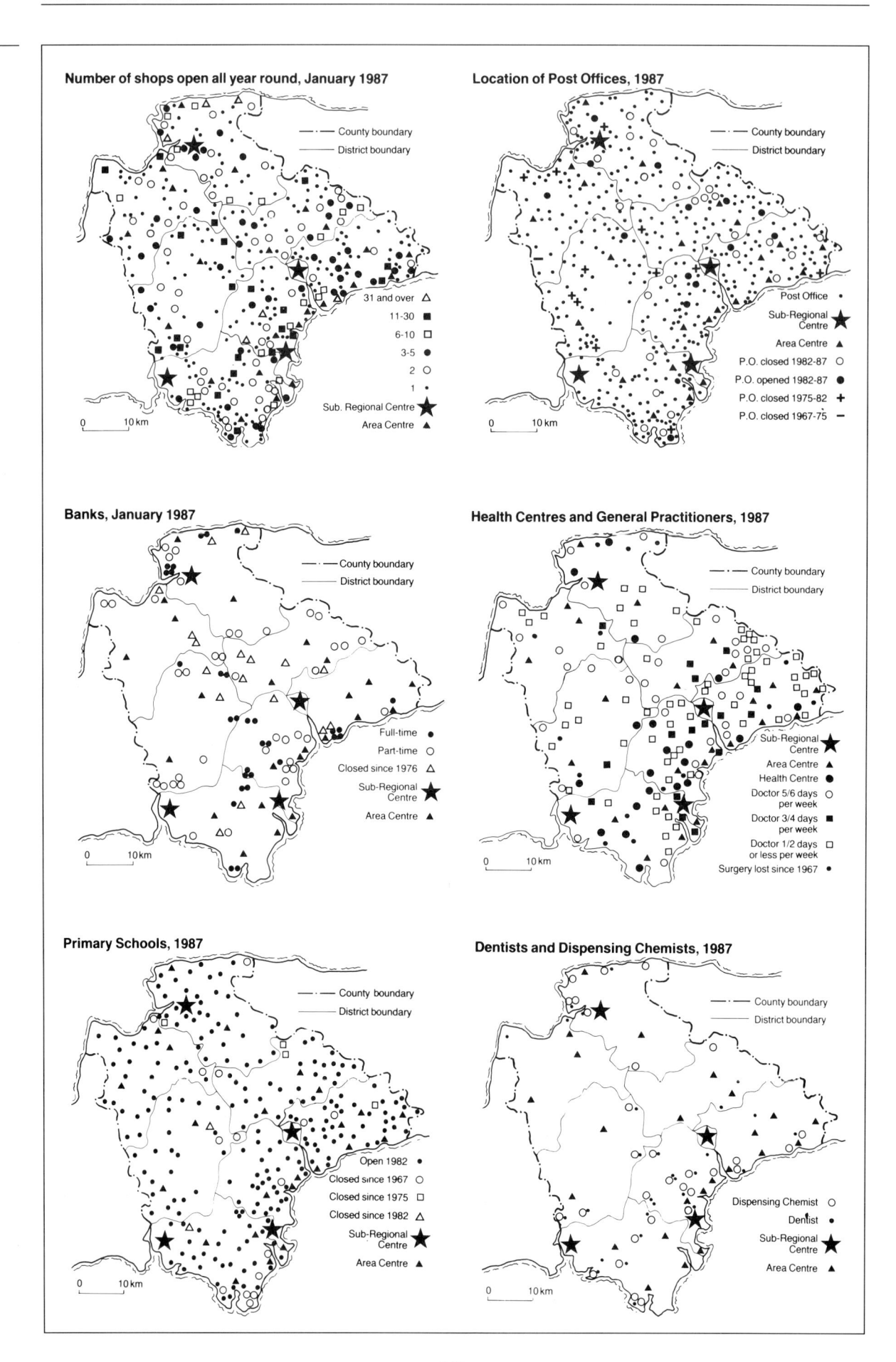

Number of shops open all year round, January 1987

- — · — County boundary
- ——— District boundary
- 31 and over △
- 11-30 ■
- 6-10 □
- 3-5 ●
- 2 ○
- 1 ·
- Sub. Regional Centre ★
- Area Centre ▲

Location of Post Offices, 1987

- — · — County boundary
- ——— District boundary
- Post Office ·
- Sub-Regional Centre ★
- Area Centre ▲
- P.O. closed 1982-87 ○
- P.O. opened 1982-87 ●
- P.O. closed 1975-82 +
- P.O. closed 1967-75 —

Banks, January 1987

- — · — County boundary
- ——— District boundary
- Full-time ●
- Part-time ○
- Closed since 1976 △
- Sub-Regional Centre ★
- Area Centre ▲

Health Centres and General Practitioners, 1987

- — · — County boundary
- ——— District boundary
- Sub-Regional Centre ★
- Area Centre ▲
- Health Centre ●
- Doctor 5/6 days per week ○
- Doctor 3/4 days per week ■
- Doctor 1/2 days or less per week □
- Surgery lost since 1967 ·

Primary Schools, 1987

- — · — County boundary
- ——— District boundary
- Open 1982 ●
- Closed since 1967 ○
- Closed since 1975 □
- Closed since 1982 △
- Sub-Regional Centre ★
- Area Centre ▲

Dentists and Dispensing Chemists, 1987

- — · — County boundary
- ——— District boundary
- Dispensing Chemist ○
- Dentist ·
- Sub-Regional Centre ★
- Area Centre ▲

RURAL SETTLEMENTS AND SERVICES

Loss of village shops, schools and public houses is often exacerbated by the reduction or closure of local public transport services. For those without their own private transport this can mean a sudden deterioration in the quality of life, and a feeling of deprivation (as serious in its own way as the deprivation discussed in the section on urban Britain). Many of the rural branch lines of the railways were closed as a result of the Beeching cuts in the 1960s, but other closures have followed, and some of the remaining rural railways are still under threat, particularly those in the more remote parts of Wales and Scotland. Rail closure means more reliance on the bus services in rural areas, but there is considerable argument over whether the latter are an adequate substitute.

One interesting study chose to look at this. Ten railway lines were chosen (see Fig. 3.4.11) in different parts of Britain, and passengers were asked to compare the convenience of rail travel and bus travel. Figure 3.4.12 gives details of the routes and the findings of the passenger survey.

Fig. 3.4.12
Railway Closures: Customer Survey

CHARACTERISTICS OF THE TEN CLOSED RAILWAY LINES AND THE SURVEY AREAS

RAILWAY LINE CLOSED TO PASSENGERS	MINSTER'S CONSENT	DATE OF CLOSURE CLOSURE	STATIONS CLOSED	APPROX. MILEAGE	TOURIST USE	LINK OR BRANCH
1 Haltwhistle-Aston	Jan 1973	May 1976	5	13	high	branch
2 Maiden Newton-Bridport	Dec 1974	May 1975	3	9	high	branch
3 Paignton-Kingswear	Jan 1972	Oct 1974*	3	7	high	branch
4 Alton-Winchester	Aug 1971	Feb 1973	4	19	low	link
5 Exeter-Okehampton	Nov 1970	Jun 1972	4	24	high	branch
6 Penrith-Keswick	Nov 1971	Mar 1972	5	18	high	branch
7 Cambridge-St. Ives	Feb 1970	Oct 1970	5	15	low	branch
8 Bangor-Caernarfon	Oct 1969	Jan 1970	1	8	high	branch
9 Dundee-Newport-on-Tay East	Aug 1968	May 1969	3	5	low	branch
10 Edinburgh-Hawick-Carlisle	Jul 1968	Jan 1969	24	98	low	link

SURVEY AREA	REFERENCE IN TABLES	POP. (000'S)	POSITION ON LINE	TOWN/CITY AT END OF LINE: NAME	POP. (000'S)	TRAVEL BEYOND BRANCH LINE
1 Alston, Cumbria	A1	2	far end	Haltwhistle	7	high
2 Toller/Powerstock, Dorset	TP	0.5	middle	Bridport	7	high
3 Kingswear, Devon	Kg	1.5	far end	Paignton	35	low
4 New Alresford, Hants	NA	4	middle	Winchester	31	high
5 Okehampton, Devon	Ok	4	far end	Exeter	96	low
6 Keswick, Cumbria	Ks	4	far end	Penrith	11	high
7 St. Ives, Cambs	SI	7	far end	Cambridge	99	low
8 Caernarfon, Gwynedd	Cn	9	far end	Bangor	15	high
9 Wormit, Fife	Wo	1.5	middle	Dundee	182	low
10 Newtown St. Boswells, Borders	NB	1	middle	Edinburgh	453	low

*Effective date of closure. British Rail withdrew the service in Dec 1972 and it was replaced by a service run by the Dart Valley Railway Company until Oct 1974, after which date there were no more winter services, a summer tourist service with steam trains being run from May to October each year.

Fig. 3.4.11

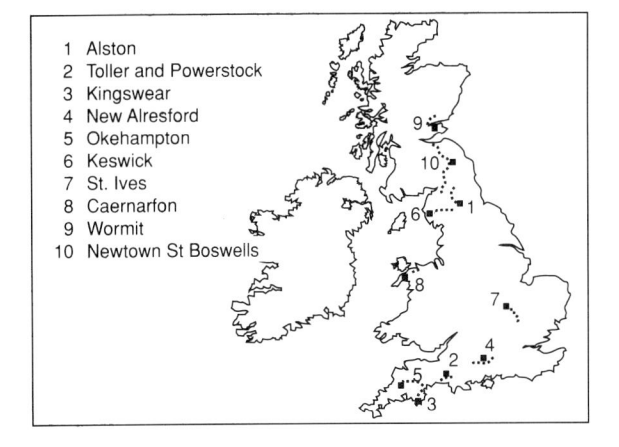

1 Alston
2 Toller and Powerstock
3 Kingswear
4 New Alresford
5 Okehampton
6 Keswick
7 St. Ives
8 Caernarfon
9 Wormit
10 Newtown St Boswells

ADJUSTMENT TO CLOSURE AMONG FREQUENT AND INFREQUENT RAIL USERS

	PER CENT OF FORMER USERS WHO PREVIOUSLY USED RAILWAYS:				
	SEVERAL TIMES A WEEK	ONCE OR TWICE A WEEK	SEVERAL TIMES A MONTH	ALL THESE AT LEAST ONCE A MONTH	LESS THAN ONCE A MONTH
No change in former journeys (except travel method)*	20	15	22	19	47
former journeys reduced*	65	75	70	71	35
former journeys stopped*	15	10	8	11	18
some or all former journeys continued:†					
by bus/coach	51	57	56	55	31
as car passenger	31	30	30	30	24
as car driver	28	20	29	25	39
partly by train	18	21	20	20	11
n (people in sample)	136	215	153	504	293

* Includes some reorientation in travel.
† More than one answer possible.

Assume that you are chairman of an action group fighting the proposed closure of a well-used rural branch line that serves a market town and a number of villages along its ten mile length.

You are to give evidence at a public enquiry and wish to use examples from previous closures to show that bus services are not an adequate replacement for rail.

Use all of the resources in the tables to produce a case against closure, illustrated by appropriate diagrams and charts.

COMPARISON OF DOOR-TO-DOOR JOURNEY TIMES BEFORE AND AFTER CLOSURE, ACCORDING TO TRAVEL METHOD USED TO REACH THREE TYPES OF DESTINATION AFTER CLOSURE

PER CENT OF RESPONDENTS WHO CONSIDERED FORMER RAIL JOURNEY	RESPONDENTS TRAVELLING TO:		
	LOCAL DESTINATIONS	END OF LINE	BEYOND END OF LINE
faster than bus	78	76	77
of which:			
much faster than bus	19	30	31
faster than car	45	51	58
of which:			
much faster than car	7	16	12

CHANGES IN TRAVEL TIME FOR JOURNEYS BY BUS NOW COMPARED WITH JOURNEYS ON FORMER RAIL SERVICES[1]

FROM	TO	APPROX. TIMES IN MINUTES[2] TRAIN	BUS
Haltwhistle	Alston	34	41–47
Carlisle	Alston	68	75–81
Hexham	Alston	60	67–73
Maiden Newton	Bridport	22	36
Dorchester	Bridport	40	37
Dorchester	Powerstock	25	57
Toller	Bridport	16	28
Paignton	Kingswear (non-stop)	18	25
Paignton	Kingswear (summer only)	18	32
Paignton	Kingswear (via Brixham)	18	45–90
Alton	Winchester	35	41–60
New Alresford	Winchester	15	15–29
Exeter	Okehampton	42	50–67
North Taunton	Okehampton	10	25
Crediton	Okehamtpon	30	55
Penrith	Keswick	35	55
Carlisle	Keswick	60	70–95
Cambridge	St. Ives	27	35–79
Swavesey	St. Ives	5	26
Bangor	Caernarfon	15	25–32
Dundee	Wormit	8	15[3]
Edinburgh	Hawick	74–93	135
Newcastleron	Carlisle	32–40	60
Galashiels	Newton St. Boswells	12	28
Hawick	Newtown St. Boswells	13–19	55–68

[1] Excluding time to walk to railway station or bus stop.
[2] Shorter travel times are for the fast/limited stop services.
[3] Travel time often doubled owing to congestion on road bridge.

COMPARISONS OF TRAIN AND BUS FOR VARIOUS CHARACTERISTICS OF SERVICE, IN TEN AREAS

	Al	TP	Kg	NA	Ok	Ks	SI	Cn	Wo	NB	ALL AREAS
	Per cent of respondents considering train better than bus										
comfort[1]	83	74	50	86	79	91	60	93	72	94	78
evening service[2]	48	59	71	40	67	47	45	43	46	64	53
day service[3]	62	66	67	58	62	67	54	55	52	64	60
frequency	49	54	73	39	41	57	48	36	48	51	50
reliability	51	50	41	72	56	60	49	60	47	48	54
access to stop[4]	22	14	14	13	14	36	16	17	22	8	18
	Per cent of respondents considering bus better than train										
comfort[1]	3	2	7	2	5	0	3	0	2	0	2
evening service[2]	3	9	2	4	2	14	3	16	12	4	7
day service[3]	3	6	9	12	4	5	5	26	11	5	9
frequency	10	11	2	17	7	13	5	36	13	11	13
reliability	0	5	6	3	3	4	2	12	2	6	5
access to stop[4]	35	16	47	21	56	32	20	52	41	21	34

[1] And ease of using the service.
[2] Running at convenient times in the evening.
[3] Running at convenient times in the daytime.
[4] Getting to the bus stop or station

'Bustitution' became a new word in transport terminology and the last word should go to the Cambrian Coast Line Action Group who successfully fought the closure of the line that runs from Machynlleth to Pwllheli.

'The idea was ludicrous to anyone with the remotest knowledge of the area concerned, and that perhaps was the root of the problem: distant, alien and bureaucratic government trying to implement a secret masterplan under the veneer of consultation and democracy. But they were coming across totally unforeseen opposition and strength of feeling.'

Today British Rail can promote the Cambrian Coast Line as 'one of Britain's most scenic railway lines (Fig. 3.4.13) providing its passengers with a succession of mountains and seascapes. It winds its way leisurely for over 50 miles, serving 28 stations or quiet rural halts on the shores of Cardigan Bay (see Fig. 3.4.14). It offers unique panoramas of the outstandingly beautiful Dovey, Mawddach and Glaslyn estuaries. For 40 of its miles it skirts the Snowdonia National Park opening up breathtaking views of the Snowdon Mountain Range and Cader Idris'.

Fig. 3.4.13
The Cambrian Coast railway: crossing of the Mawddach estuary at Barmouth

Fig. 3.4.14

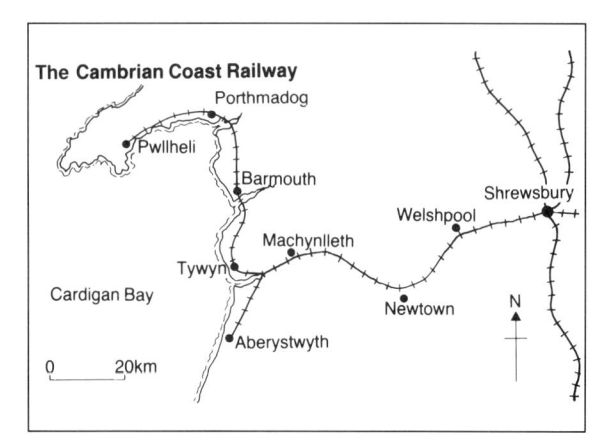

1   British Rail obviously promoted the line for its tourist value. Use a good atlas map of Wales to suggest what other social uses of the railway (particularly out of the tourist season) could be cited to justify its retention.

2   What likely economic reasons could be given for its closure?

INDUSTRY AND RURAL DEVELOPMENT

'Businesses located in rural areas form a valuable and growing part of our national economy – and high technology and engineering companies are as likely to be found in rural areas as traditional craftspeople. For these businesses to flourish, a thriving community is needed to provide the workforce and the services on which rural communities depend. Twenty per cent of the population of England live in rural areas. They need someone to look after their interests.'

(Rural Development Commission)

The urban-rural shift in manufacturing industry first appeared as a trend in the 1960s and can be traced into the late 1980s, although growth has tended to be uneven over the thirty year period. Even in the early eighties, when most areas were recording losses in manufacturing jobs, counties such as Cambridgeshire, Clywd and Shropshire were still showing increases (see Fig. 3.5.1). A number of possible reasons have been put forward for this shift.

- Factory floorspace is much more readily available in rural areas than in congested urban areas. As machinery replaces labour the additional floorspace required is simply not available in crowded urban locations, so industries cannot expand and take advantage of new technologies. Such problems are not encountered in rural areas.

- Production costs are much higher in urban areas than in rural ones – differences of up to 50 per cent of gross profits exist between the two locations. Wage and salary costs are lower in rural areas, and rent and rates are similarly beneficial to the industrialist.

- Quality of life for managers and staff is far superior in rural areas to that experienced in towns.

- Rural locations are far less likely to suffer from the serious traffic congestion that now affects so many urban areas.

- It has also been suggested that large companies may open branch plants in rural areas in order to reduce production costs. Although this may happen in some cases, the large number of small independent manufacturing companies that are starting up in rural areas suggests that this element in rural industrial growth is of limited application.

There is evidence also of the continued growth of service industries in the rural areas during the second half of the century. Figure 3.5.2 shows the structure of employment change in rural areas in the north and south of Britain in the 1981–7 period.

Fig. 3.5.1

Manufacturing Employment Change, 1981-4

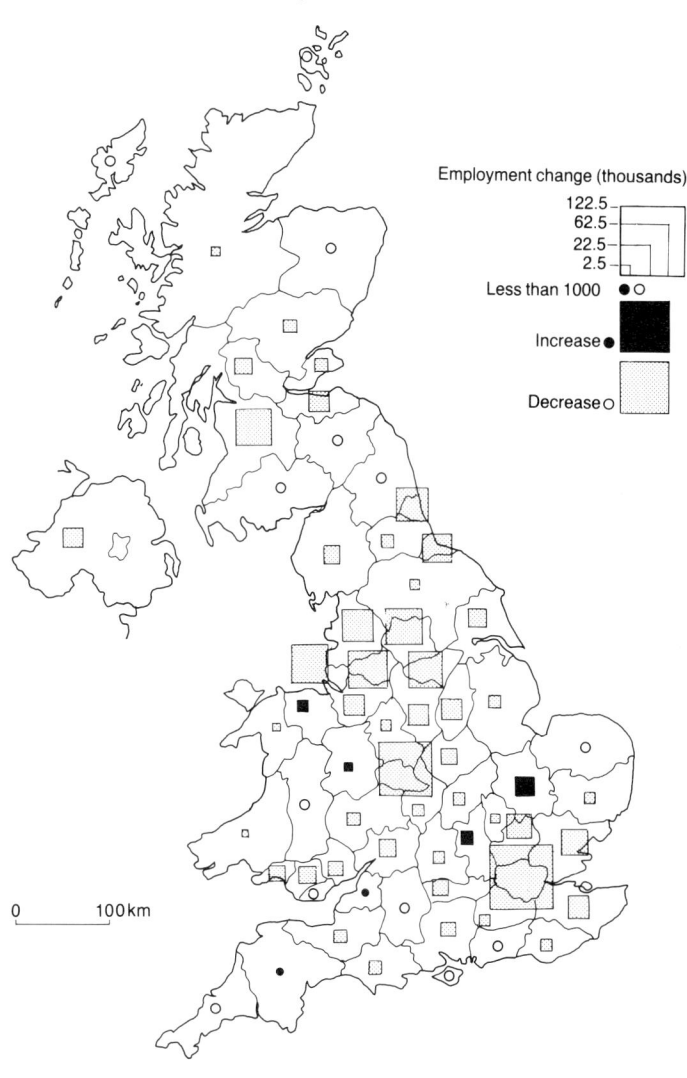

Fig. 3.5.2

THE STRUCTURE OF EMPLOYMENT CHANGE IN REMOTER, MAINLY RURAL, DISTRICTS, 1981–1987, BY NORTH/SOUTH GROUPS (EMPLOYEES IN EMPLOYMENT)

	SOUTH		NORTH	
	TOTAL CHANGE		TOTAL CHANGE	
	000s	%	000s	%
Agriculture, forestry etc	−9.7	−10.6	−13.1	−17.1
Energy and water supply	−3.3	−20.8	−4.7	−20.1
Other minerals, etc	−2.4	−7.5	−4.4	−15.6
Metal goods, engineering	−7.8	−7.1	−0.4	−0.8
Other manufacturing	+1.8	+1.3	+1.1	+1.3
Construction	−2.3	−4.1	−8.7	−15.3
Distribution, hotels, etc	+27.3	+12.0	+4.8	+2.9
Transport etc	+4.4	+7.7	+0.5	+1.2
Banking, finance etc	+21.9	+42.2	+6.9	+8.3
Other services	+44.3	+17.4	+34.1	+15.8
Total, of which	+74.0	+7.2	+16.1	+2.1
Males	+2.8	+0.5	−20.5	−4.5
Females	+71.1	+16.8	+36.7	+11.3
Full-time	+25.0	+3.2	−18.7	−3.1
Part-time	+48.9	+20.4	+34.9	+19.6
Tourism-related industries	+12.4	+18.8	+9.7	+15.7

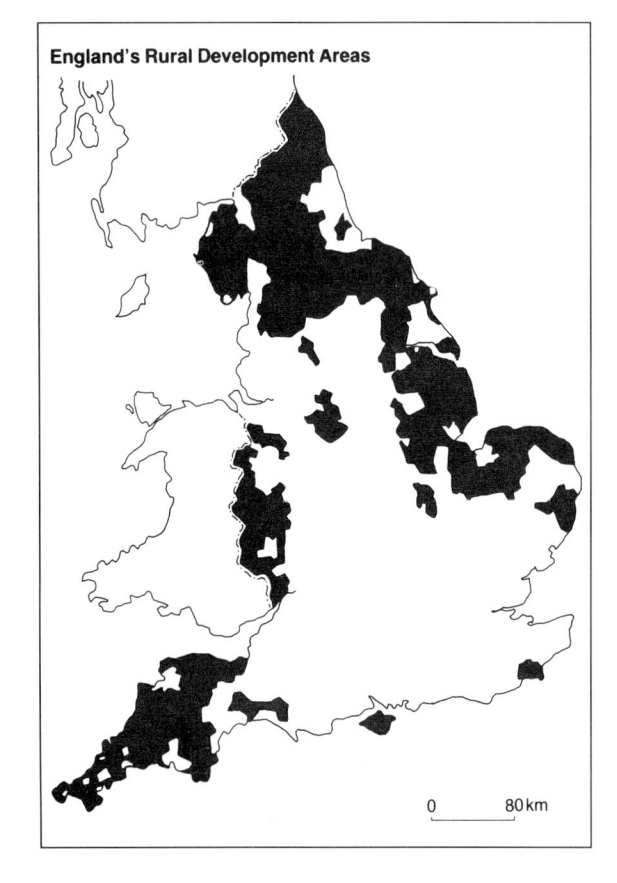

England's Rural Development Areas

The move from urban to rural areas has been much encouraged by assistance given from various organisations. In England the Rural Development Commission (RDC) is responsible for promoting rural industries and community development. Until 1988 when it merged with the Development Commission, COSIRA (Council for Small Industries in Rural Areas) was responsible for the grant funding of rural industry. A series of priority Rural Development Areas have been identified by the Rural Development Commission and are shown on the map (Fig. 3.5.3). Similar bodies to the RDC exist in both Wales and Scotland. In Wales, the Development Board for Rural Wales was created in 1977 to promote the economic and social development of Mid-Wales, which includes 40 per cent of Wales, but only has 7 per cent of its population. In Scotland, the Highlands and Islands Development Board (now renamed Highlands and Islands Enterprise) was formed in 1965, with very much the same aims and objectives as the Development Board for Rural Wales.

Fig. 3.5.3

INDUSTRY AND RURAL DEVELOPMENT

CORNWALL: OPPORTUNITIES FOR INVESTMENT

'My vision of Cornwall's future is of a special place, where people want to live, where they can enjoy an improved standard of living and a high quality of life. The County Council is committed to working with all those who shared this vision, to achieve the economic conditions necessary to realise this'.

(Joan Vincent, Chairman of the Planning and Economic Development Committee, Cornwall County Council.)

DECISION-MAKING EXERCISE

Lying at the extremity of the south-west peninsula, Cornwall is one of the more remote parts of rural Britain. The county is keen to attract more industry.

Fig. 3.5.4
Four aspects of Cornwall
(a) The coast south-west of Portreath
(b) Sandy Beach near Godrevy lighthouse: St. Ives Bay
(c) St. Ives Harbour
(d) The village green at Blisland, North Cornwall, on the fringe of Bodmin Moor

ECONOMIC INDICATOR	CORNWALL	UK
Unemployment rate (%)	10.0	6.9
Gross value added in manufacturing £ per employee Index UK = 100	97.8	100.0
Gross domestic product Index UK = 100 (1981)	78.4	100.0
(1987)	76.0	100.0
Net capital expenditure in manufacturing £ per employee Index UK = 100	123.2	100.0
Earnings (April 1990) Average gross weekly earnings (£)		
male	241.9	295.6
female	166.8	201.5
Household disposable income Index UK = 100	94.4	100.0

Fig. 3.5.5

149

Findings from a survey of Cornwall by Cambridge Consultants:

- 80 % rated skill recruitment above average
- 100 % had experienced no labour disputes in the previous year
- 90 % found an above average response to training

Absenteeism was lower than elsewhere

- 84% had expanded since their initial location
- 80 % rated profitability higher than other plants in their group.

Some comments from employers

Ranco Controls: *'We have never regretted our move to the south-west 27 years ago. The workforce has grown with us. Their co-operation is a major factor in our success.'*

AVERAGE HOUSE PRICES FROM THE THIRD QUARTER, 1989 (£)

	CORNWALL	SOUTH-EAST	UK
Terraced	56 807	66 322	50 951
Semi-detached	62 294	83 121	61 650
Detached	106 653	161 433	112 152
Bungalow	72 997	100 288	69 086

J.I. Case: *'Since moving to Cornwall, our productivity has risen by 89 per cent. In Cornwall you can achieve more with your workforce than anywhere else in the United Kingdom.'*

Spectra Brands plc: *'There is a total flexibility in jobs. An indication of this adaptability is the fact that our turnover has gone from £4 million to £12 million in three years with only a 15 per cent increase in labour.'*

Assume the role of the Chairman and Managing Director of a small electronics company wishing to move from congested premises in the Brentford area of West London. You are thinking of moving to the West Country, and probably to Cornwall.

1 Outline the advantages of relocating in Cornwall.

2 What might be the disadvantages of relocating in Cornwall?

3 Consider the four site options within Cornwall. Using all the resources (Figs. 3.5.4–3.5.8), give a reasoned ranking of these sites in order of suitability for your company.

Four possible industrial sites are on the market in Cornwall:

Truro: Property contains several self-contained units on the ground and first floors. Freehold: £100 000

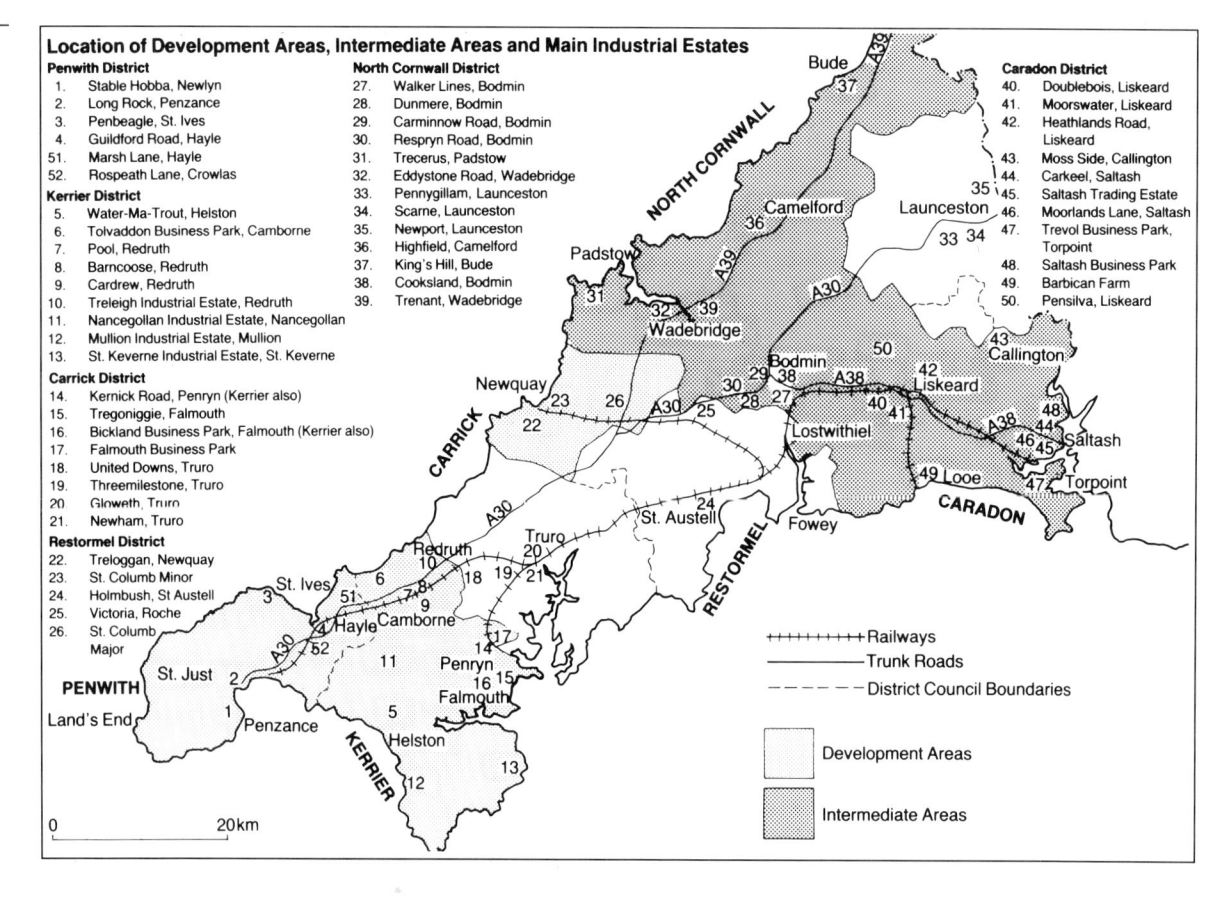

Fig. 3.5.6

Location of Development Areas, Intermediate Areas and Main Industrial Estates

Penwith District
1. Stable Hobba, Newlyn
2. Long Rock, Penzance
3. Penbeagle, St. Ives
4. Guildford Road, Hayle
51. Marsh Lane, Hayle
52. Rospeath Lane, Crowlas

Kerrier District
5. Water-Ma-Trout, Helston
6. Tolvaddon Business Park, Camborne
7. Pool, Redruth
8. Barncoose, Redruth
9. Cardrew, Redruth
10. Treleigh Industrial Estate, Redruth
11. Nancegollan Industrial Estate, Nancegollan
12. Mullion Industrial Estate, Mullion
13. St. Keverne Industrial Estate, St. Keverne

Carrick District
14. Kernick Road, Penryn (Kerrier also)
15. Tregoniggie, Falmouth
16. Bickland Business Park, Falmouth (Kerrier also)
17. Falmouth Business Park
18. United Downs, Truro
19. Threemilestone, Truro
20. Gloweth, Truro
21. Newham, Truro

Restormel District
22. Treloggan, Newquay
23. St. Columb Minor
24. Holmbush, St Austell
25. Victoria, Roche
26. St. Columb Major

North Cornwall District
27. Walker Lines, Bodmin
28. Dunmere, Bodmin
29. Carminnow Road, Bodmin
30. Respryn Road, Bodmin
31. Trecerus, Padstow
32. Eddystone Road, Wadebridge
33. Pennygillam, Launceston
34. Scarne, Launceston
35. Newport, Launceston
36. Highfield, Camelford
37. King's Hill, Bude
38. Cooksland, Bodmin
39. Trenant, Wadebridge

Caradon District
40. Doublebois, Liskeard
41. Moorswater, Liskeard
42. Heathlands Road, Liskeard
43. Moss Side, Callington
44. Carkeel, Saltash
45. Saltash Trading Estate
46. Moorlands Lane, Saltash
47. Trevol Business Park, Torpoint
48. Saltash Business Park
49. Barbican Farm
50. Pensilva, Liskeard

Railways
Trunk Roads
District Council Boundaries
Development Areas
Intermediate Areas

0   20km

Penzance: New industrial unit 1092 sq ft (102 sq m). Freehold: £41 000

Helston: Workshop unit 1000 sq ft (93 sq m). Leasehold: £4.00 per sq ft (£10 per sq m) per annum

Launceston: Unit comprising yard with security fence and two storey workshops and offices. Leasehold: £3650 per annum.

Department of Trade and Industry – Regional Initiative – Regional Selective Assistance

Description: *Discretionary project grants based on fixed capital costs and on the number of jobs created or safeguarded. The level of grant will be negotiated as the minimum necessary to enable the project to proceed.*

Eligibility: *Most manufacturing and some service sector projects located in Development and Intermediate Areas, which can demonstrate viability, need and a Regional/National benefit. Projects that have already started will not normally be assisted.*

Department of Trade and Industry – Regional Initiative – Regional Enterprise Grants

Description: *Discretionary grants for investment projects in most manufacturing and some service sectors; 15 per cent of eligible fixed assets (plant, machinery, buildings, land purchase, site preparation, vehicles used solely on site) of the project up to a maximum of £15 000.*

Eligibility: *Available to individuals, partnerships and companies with under 25 employees (full time equivalents). Projects must be in a Development Area or in Plymouth Travel to Work Area. Projects that have already started will not normally be assisted.*

Fig. 3.5.7 Government assistance to industry in Cornwall

Fig. 3.5.8

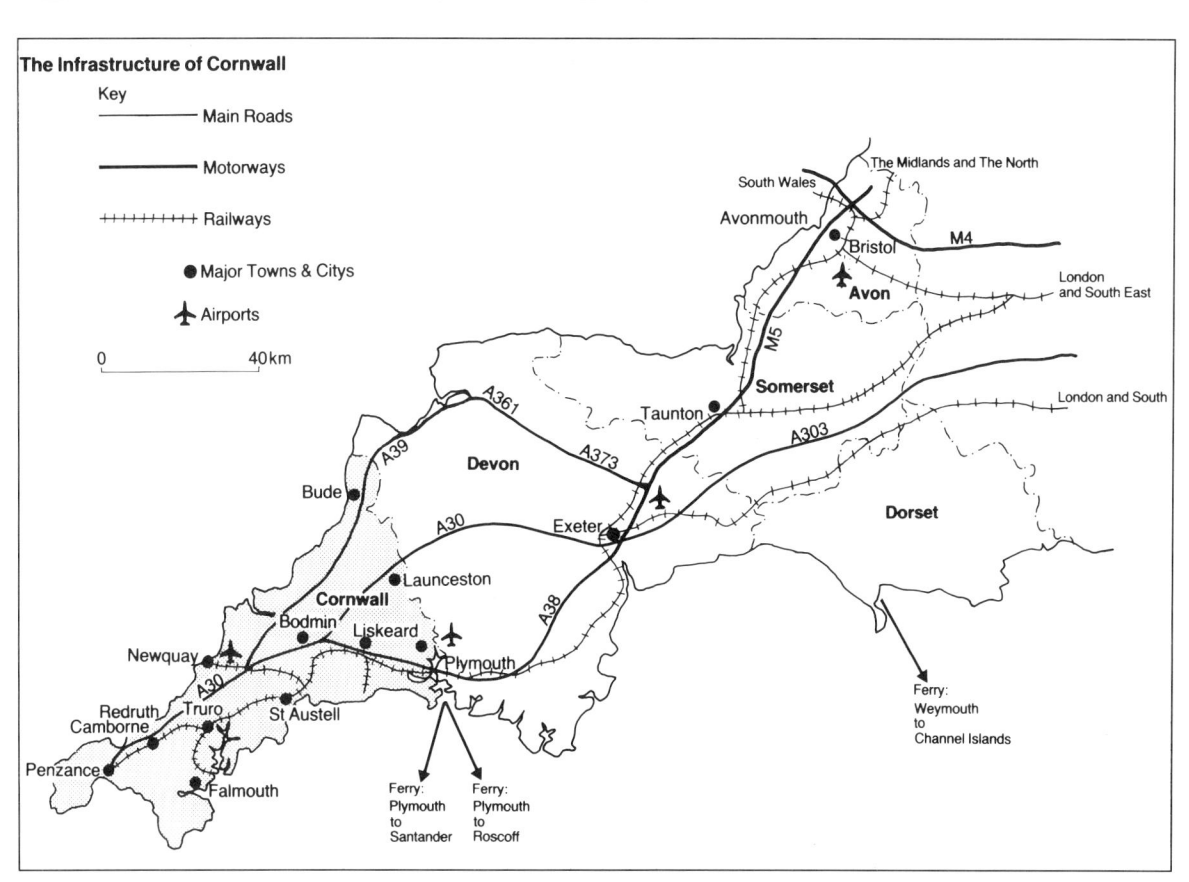

The Infrastructure of Cornwall

A NEW VIEW OF BRITAIN

MID-WALES: THE NEW WALES!

The area administered by the Development Board for Rural Wales lies between the more densely populated areas of the south Wales industrial and mining area and the coastal strip of north Wales. The statistics in the table (Fig. 3.5.10) show the changes in population in the Rural Wales Development area.

1 Use graphs and charts to show absolute and relative rates of change.

2 What common trend emerges from the study of the graphs and charts?

Now examine the map (Fig. 3.5.11) which shows relative and absolute employment change in the Development Board area for the period 1976–87 (the first twelve years of the existence of the Board).

1 Comment on the regional variations in employment change in the area.

2 It is sometimes asserted that the western part of the Board's area is disadvantaged compared to the east. Does the data on the map confirm this view?

3 Is it possible to establish any relationship between the changes in population, and the changes in employment structure?

Fig. 3.5.9

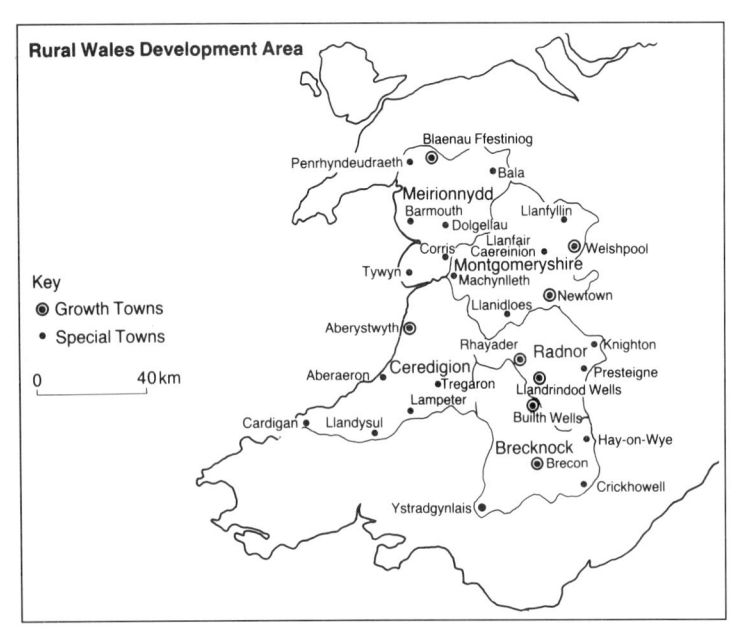

Fig. 3.5.10
Changes in population: Rural Wales Development Area

DISTRICT	1961	1971	1981	1991
Brecknock	39 599	37 750	40 700	41 300
Ceredigion	53 644	54 878	57 400	63 600
Meirionnydd	34 393	31 512	32 100	33 400
Montgomery	44 154	43 107	48 200	52 000
Radnorshire	18 512	18 321	21 600	23 200

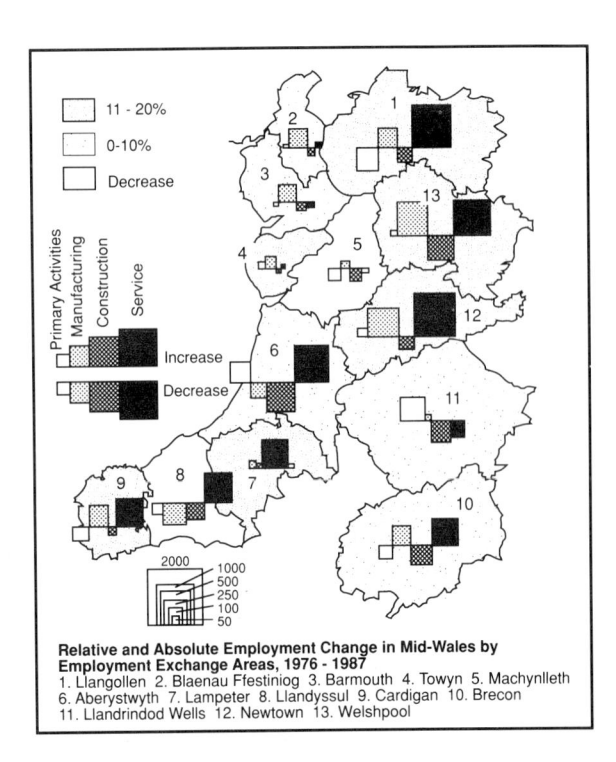

Relative and Absolute Employment Change in Mid-Wales by Employment Exchange Areas, 1976 - 1987
1. Llangollen 2. Blaenau Ffestiniog 3. Barmouth 4. Towyn 5. Machynlleth 6. Aberystwyth 7. Lampeter 8. Llandyssul 9. Cardigan 10. Brecon 11. Llandrindod Wells 12. Newtown 13. Welshpool

Fig. 3.5.11

INDUSTRY AND RURAL DEVELOPMENT

The map (Fig. 3.5.12) shows the variety of financial assistance that is available to companies wishing to locate within the Rural Wales Development area.

One of the main problems in attracting industry into areas such as mid-Wales is to persuade prospective clients that the area is not remote, and is within relatively easy reach of the main centres of population in Britain.

Using a good atlas draw an annotated map suitable for a business brochure to show that mid-Wales is not as remote as might be imagined. Use your map to support such contentions as:

- it only takes three hours from mid-Wales to the M25
- mid-Wales is only an hour from the West Midlands
- good rail connections link mid-Wales into the main network
- Newtown is a key focus for mid-Wales.

Fig. 3.5.12

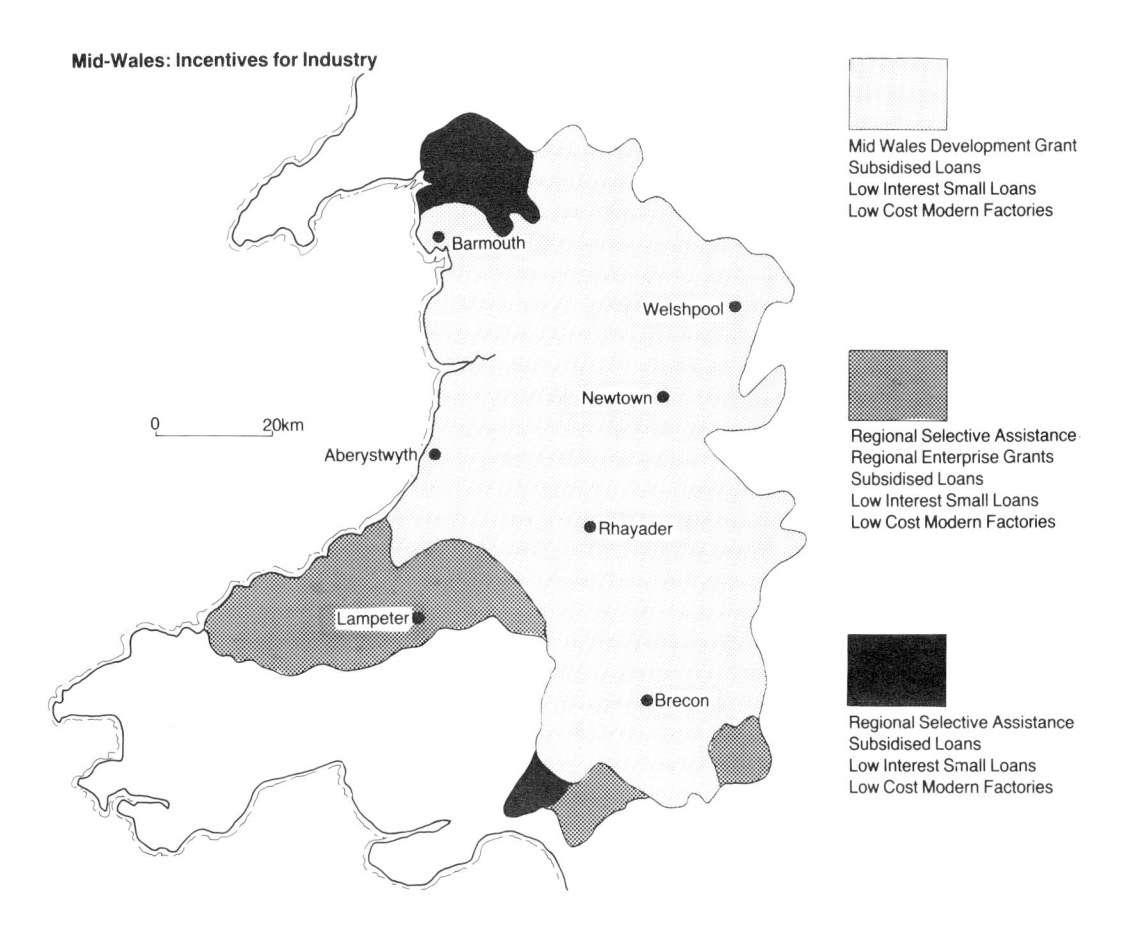

Mid-Wales: Incentives for Industry

Mid Wales Development Grant
Subsidised Loans
Low Interest Small Loans
Low Cost Modern Factories

Regional Selective Assistance
Regional Enterprise Grants
Subsidised Loans
Low Interest Small Loans
Low Cost Modern Factories

Regional Selective Assistance
Subsidised Loans
Low Interest Small Loans
Low Cost Modern Factories

Fig. 3.5.13
New factories for mid-Wales on new technology park at Newtown

RECREATION IN RURAL BRITAIN

If you are looking for some peace and quiet on a bank holiday or a summer's weekend, the countryside is not necessarily the best place to go. On a typical summer Sunday about 18 million of us can be found there, hoping to 'get away from it all'. But getting away from it all is often more difficult than its seems. Having struggled through the traffic jams and found a suitable parking place, where is the harassed visitor to go? For the countryside is private property, not an extended municipal park, and the right of public access is restricted. Those wishing to get away from it all might more easily find a place to relax in the peace and quiet of the city centre.

(Howard Newby, The Countryside in Question)

Fig. 3.6.1
Independent 4
January 1992

Conflicts in the urban fringe, we discovered in the section on urban Britain, are principally concerned with controversies over land use. Pressures to convert amenity and recreational uses in this zone are considerable, and usually meet with healthy resistance, which may, or may not be successful. Recreational demands of the urban population, however, extend well beyond the perceived urban fringe. As Howard Newby points out, most of the countryside is privately owned and access is thus restricted. Providing the right kind of facilities for those seeking relief from the pressures of urban living has exercised rural authorities for much of the post-war period.

Study the article (Fig. 3.6.1) from *The Independent* which considers some of the conflicts that have arisen over the provision of recreational facilities for visitors in the Malham area of the Yorkshire Dales National Park.

On the sandwich-board trail
Picturesque Malham, in the Yorkshire Dales, is prosperous and popular. Perhaps too popular, as **Peter Dunn** reports

JOHN DYSON, a stockbroker in West Yorkshire, must have had Malham very much in mind when he wrote a report called *The Northern Playground Britain's Most Prosperous Area*.

The stone-built village, tucked into the southern folds of the Yorkshire Dales, epitomises the new-found economic confidence Mr Dyson discovered across a wide swathe of northern England, from the Lake District in the west, over the spine of the Pennines and crossing the Vale of York to Pickering and the North York Moors.

Mr Dyson's report, written for his firm, B W D Rensburg of Huddersfield, painted glowing panoramic views of the Playground's low unemployment and its prosperity underpinned by tourism, which flourishes even in midwinter. His report says: 'This rapidly emerging picture is challenging some long-held misconceptions about the relationship between North and South. The main one is that the decline of the North is inevitable and inexorable and will continue once the recession is over.'

Malham's villagers made a good living from tourism during the Eighties. The old farming families stopped fleecing sheep and instead spun gold out of trinket shops, camping sites and barn conversion. Sharp elbowed market forces, exemplified by a long running ice-cream war between two local entrepreneurs, turned Malham into a microcosm of Mrs Thatcher's giddy capitalism.

But even Mr Dyson concedes that there is a down side to prosperity in the Northern Playground. When a visitor to Malham was heard saying this year, 'Where's the chip shop, then?' conservationists in their posh conversion began to twitch with concern. The Dales National Park, increasingly under fire for its encouragement of mass tourism, is now conducting an inquiry into the effect of mounting pressure on the Dales. Even in Malham some are suggesting that the park keepers' priorities must switch from populism to conservation before it is too late.

Others, such as David Howard, the warden of Malham youth hostel 50 years ago, have walked away in despair.

Three years ago Mr Howard returned to the village as guest of honour at the hostel's fiftieth anniversary celebrations. The transformation of a once-insular farming community appalled him. "I made a speech saying I never wanted to go near Malham again," he says. "I was horrified by it all: the crowds, the shops, the barn conversions, the sheer commercialisation. The car's wrecking everything, isn't it?"

Malham is under the protection of the Dales National Park, which has a car-park and information centre behind the Methodist chapel. The number of visitors to the centre has almost trebled in the past 10 years. Motorists who do not want to pay £1 to use the official site cram the village approach roads. Their metallic jam blocks the village bus and competes with a kerb-side clutter of sandwich boards advertising shops, cafés and pub meals. One of the sweet stalls now sells Malham rock.

Controlling the influx has left the park's governing body with an impossible dilemma. Faced with severe damage to the landscape around the village, its honourable policy of "access to all" has remained uncompromised. The park has installed mini highways complete with two-way wall stiles the size of decorators' ladders, to ease the crush and reduce destruction to the dry stone walls. The result has been to increase pressure on local beauty spots.

Julie Boocock, a local farmer's daughter who has run a fancy goods shop next to Malham's chapel for eight years, is chairman of the community's parish meeting. She says that without the business and its modest B&B sideline, she, her husband, Granville, and their three young daughters would have been driven away from the area.

'Funnily enough,' she says, 'I think the real locals take everything in their stride. The people who complain most are those that moved in and expect it to be a sleepy little village.'

She was furious when she went to a meeting and heard a park official suggest that Malham might cut back on its tourism. 'I went home with steam coming out of my ears. Anyone who takes their living from tourism thinks the park would like to see Malham pickled. In a peculiar way, we're treated like we shouldn't be here.'

Jonathan Ditchfield, 30, and his brother Andrew, 32, ran a café in Manchester before taking over the Listers Arms in Malham three years ago. Their first act was to stop the traditional after-hours drinking. Locals such as Tommy Clark, a farmer, took their trade off to the Victoria in Kirkby Malham, a mile away.

Jonathan says: 'It probably made us very unpopular the first 12 months. People round here don't know how to cater for tourists. There's no originality of ideas. We realised from the start it was more than just a local pub. There's a big market out there that hasn't been tapped.'

Christine Langrish, clerk to the parish council for eight years, has watched tourism mushroom in Malham in the 19 years that she and her retired husband have lived there. 'The national park, having encouraged tourism, now says, "Oops. Look what we've done," she says. 'It's made us rather mad. Even I've said you can't let people come into a village, buy a guest house and then tell them" 'We're cutting down on tourism.' It's just not on.

The park has become obsessed with this idea of making the natural beauty of the countryside available to everybody. They've virtually put down wheelchair tracks to the Cove, and everyone's been appalled with the Janet's Foss development. 'I've been saying for a long time now that if we're not careful we'd attract the wrong sort of people to Malham. By wrong people, I mean those who aren't really interested in the Dales for the peace and beauty of the walking,' she says.

Fig. 3.6.2
Busy Whitsun bank Holiday at Malham, Yorkshire Dales National Park

DAY TRIPS TO SCOTLAND'S COUNTRYSIDE

During the 1987–9 period a study was carried out for the Countryside Commission for Scotland and other bodies on leisure day trips. The survey was run bi-monthly from February to December in each year, with an additional survey in July to cover the summer peak. The sample size was 1000 people in each month, selected to be representative of the Scottish population at 40 points throughout the country.

Some results of the survey are summarised in the tables and charts (Figs. 3.6.3 and 3.6.4).

1 What are the perceived consequences of the National Park's policy for tourism in the area?

2 How may such a policy be justified?

3 Discuss the ways in which the conflicts in the area may be related to different sets of values?

1 Summarise, in a short report, the main findings of the survey.

Fig. 3.6.3

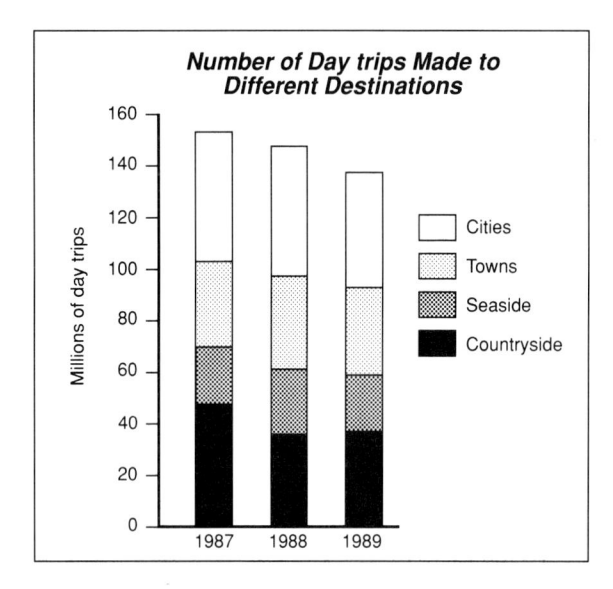

Number of Day trips Made to Different Destinations

[Bar chart showing millions of day trips for 1987, 1988, 1989 broken down by Cities, Towns, Seaside, Countryside]

Fig. 3.6.4

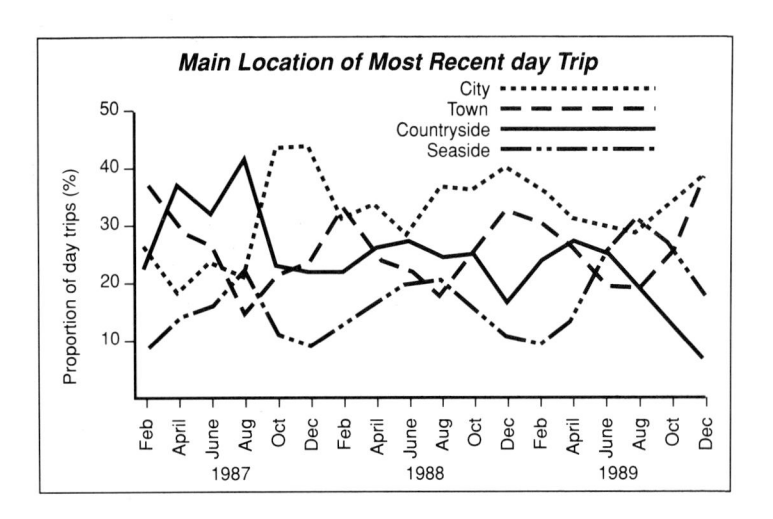

Main Location of Most Recent day Trip

[Line graph showing proportion of day trips (%) by City, Town, Countryside, Seaside from Feb 1987 to Dec 1989]

COUNTRY PARKS

The idea of country parks was first suggested in the White paper 'Leisure in the Countryside'. Many existing beauty spots (or honeypots as they became inevitably known) were already becoming overcrowded, and the new country parks were proposed to relieve the pressures. They had three principal aims:

- they would make it easier for town dwellers to enjoy leisure in the countryside without having to travel too far

- they would ease the pressure on the more distant and solitary locations particularly those of outstanding landscape and ecological value

- they would reduce the risk of damage to the countryside, which occurs when people spend an hour or two at one location.

The map (Fig. 3.6.5) shows the distribution of country parks in the UK, which are managed either by local authorities or non-public bodies.

Fig. 3.6.5

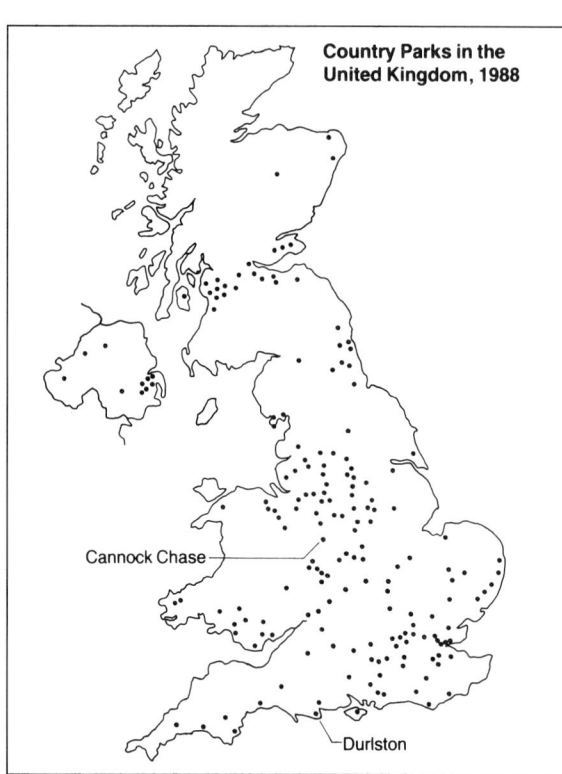

[Map of United Kingdom showing Country Parks in the United Kingdom, 1988, with Cannock Chase and Durlston labelled]

CANNOCK CHASE AND DURLSTON: TWO COUNTRY PARKS COMPARED

The locations of the two country parks are shown on the map (Fig. 3.6.5). Cannock Chase Park lies within the Cannock Chase Area of Outstanding Natural Beauty (AONB), and occupies an area of 1113 hectares (the largest in Britain) out of a total of 6880 hectares in the AONB (Fig. 3.6.6). By way of comparison, the Durlston Country Park on the Dorset coast is one of the smaller ones occupying 106 hectares within the AONB of the Isle of Purbeck (Fig. 3.6.7).

Fig. 3.6.6
Cannock Chase Country Park

Fig. 3.6.7
Durlston Country Park

Cannock Chase Country Park (see Fig. 3.6.8): The AONB is a low plateau rising to a height of 245 m at Castle Ring hill fort, and bounded on the valleys of the Sow to the north, the Penk to the west, and the Trent to the east. Valleys leading into these rivers give the topography of the Chase an immediate appeal to visitors. Forestry Commission land covers nearly a third of the AONB, agriculture about 25 per cent, and the Country Park some 16 per cent. The park has the largest area of open lowland heath in the Midlands, which is of particular interest because of its mix of southern and northern species. In places the heathlands grade into birch and pine woods and the mosaic is completed by areas of wetland in the valleys. Over four million people live within a radius of 50 km. Records show that over half a million people visit the park every year, usually on Sunday or afternoon half day visits. Car parking is organised in some 60 sites across the park. The majority of people use the park for informal recreation. Horse-riding and illegal motor cycle scrambling, orienteering and informal walking have been responsible for some erosion problems, particularly on the steeper slopes. Mountain-biking is causing increasing problems, particularly where those involved in the sport use the same tracks as walkers.

Fig. 3.6.8

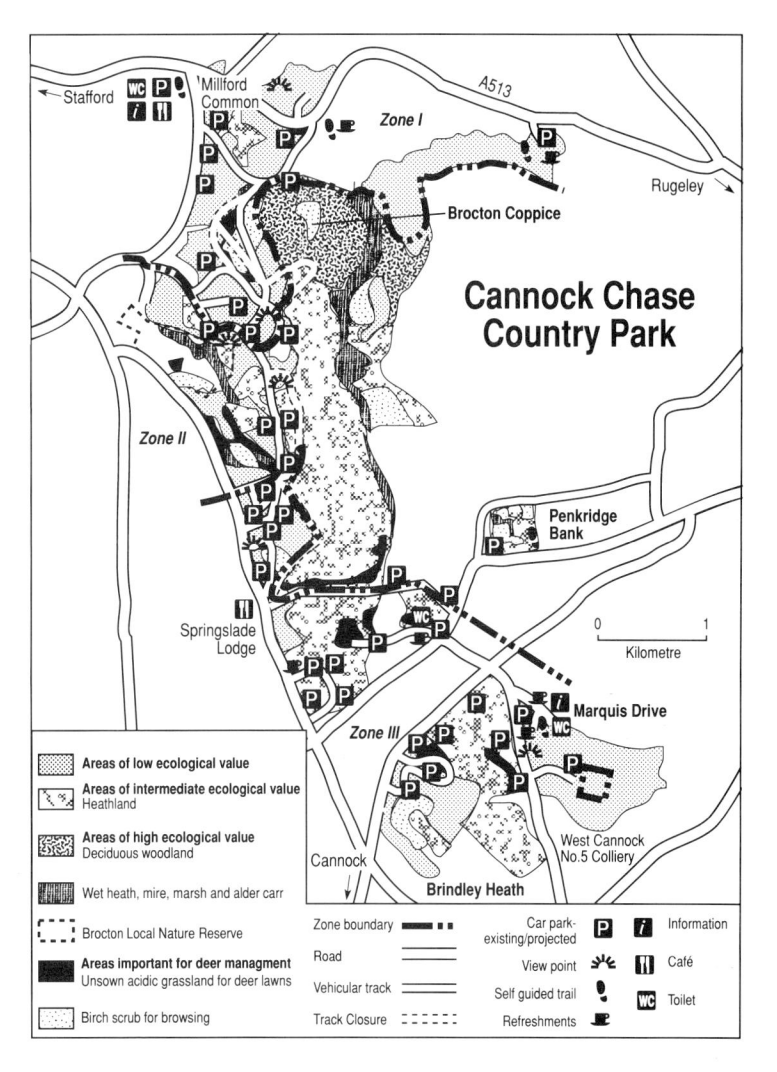

Durlston Country Park (see Fig. 3.6.9) is much smaller than Cannock Chase. It lies on the south coast of the Isle of Purbeck, and enjoys additional status and protection through its location on a stretch of Heritage Coast, and its importance as a SSSI. The coast is spectacular, with its cliffs of Portland and Purbeck limestone dropping sheer into the sea. Inland the open limestone country is enclosed within a framework of stone walls, and is devoted to a combination of farming and conservation uses. The flora and fauna of the park

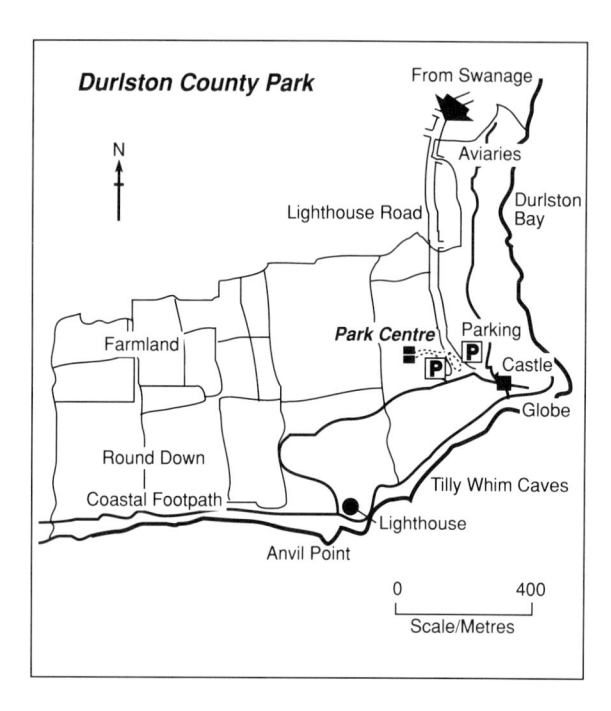

Fig. 3.6.9
Durlston Country Park

are of particular note, with the rich and varied bird life along the coast, and the rare and sensitive limestone plants on the downland behind. Part of the cliffs between Tilly Whim Caves and Durlston Bay has been declared a sanctuary to enable sea-birds to breed undisturbed, including guillemots, fulmars and razorbills. Within a 50 km radius there are nearly 750 000 people, but the park receives many summer visitors who are staying in the resorts of Swanage and Bournemouth and in the many camping and caravan sites along this part of the coast. To emphasise the difference in scale between the two parks, Durlston has just one paved car park, which is often full to capacity. More than 250 000 people visit the park every year. Apart from walking along the Heritage Coast the main recreational activity is rock climbing on the limestone cliffs, but this is now restricted since the creation of the sanctuary.

1 Discuss the different physical and ecological backgrounds to the two parks.

2 What would have been the main reasons for designating the two parks?

3 What are the different pressures that will be experienced in the two parks? How will the management plans for the two parks differ as a result of these pressures?

FROM FARMLAND TO GOLF COURSE

One of the interesting spin-offs from the set-aside scheme is the proliferation of applications from farmers to build golf courses! During the first three years of set-aside 70 farmers took set-aside to build golf courses; 8 used the land for car-parking; 59 set up livery stables; 30 established commercial game shoots; 15 set up caravan parks and 36 went into turf production.

Dennis Craggs, who farms 84 hectares at Knotty Hill Farm, near Sedgefield, Co, Durham will receive a subsidy of £12 000 a year for the next five years from set-aside grants, *'The subsidy is not to build the golf course. It is to take 44 hectares of my cereal land out of production. I am building the golf course, but I would get the grant anyway, I don't really want the taxpayer to pay for subsidies, but if you want to reduce grain mountains without paying for them, farms would go out of business, the banks would move in and land would be left derelict....'*

Eric Roylance used the set-aside grant on 24 hectares of his 48 hectares that he turned into a golf course in 1988. Now there is no agricultural activity on his farm, he receives £4000 a year in grants, which help to pay the interest on his investment. *'I can understand why set-aside is such an emotive issue. I don't think it is a good scheme and I don't think it is well administered by the EEC. But farmers aren't the ogres that people think they are. There is a system and they use it....'*

Such developments have, inevitably created a great deal of controversy:

Tony Burton, Senior Planner at the Council for the preservation for Rural England (CPRE), *'We are talking about the change of use of vast areas of farmland. There has been no other kind of development of this kind in the countryside for decades. Many proposals constitute major urban development that just happen to have 18 flags around it.'* (The Independent 5 December 1991)

Andy Wilson, Assistant Secretary CPRE, *'We do not see how golf club houses, farm shops and stables can readily be returned to good agricultural condition nor would there appear to be any intention of doing so'.* (The Times 4 December 1991)

1 What are the major issues involved in this controversy?

2 What is your view on set-aside land being used for golf-courses?

UPLAND BRITAIN

LIFE IN UPLAND BRITAIN

'Upland Britain has often been described as the better half of Britain.'

(Roy Millward and Adrian Robinson, Upland Britain.)

Fig. 4.1.1
Upland Britain: Beinn Eighe, Wester Ross, one of the high peaks of Torridon, now part of a National Nature Reserve

It is interesting to explore this value judgement on Britain's uplands. Are the uplands better because their scenery is more striking, their vistas more extensive and embracing? Or are they superior because many of the pressures and stresses of urban and industrial Britain are largely absent in the uplands? Many would argue that the uplands possess some of the finest and most awe-inspiring scenery in Europe, from the ice-scarred and lochan-speckled lands of the north-west of Scotland to the sweep of the moorlands of the south-west of England, from the high granite summits of the Cairngorms to the red escarpments of the Brecon Beacons and the Black Mountains of Wales. It is equally the case that traffic levels on the roads in the uplands are much lower, that despite acid rain, pollution is less damaging to the environment, and the high crime rates of our towns and cities are absent. Quality of life is one of the uplands most prized attributes, and few would deny that the upland farmer or forestry worker has real benefits here.

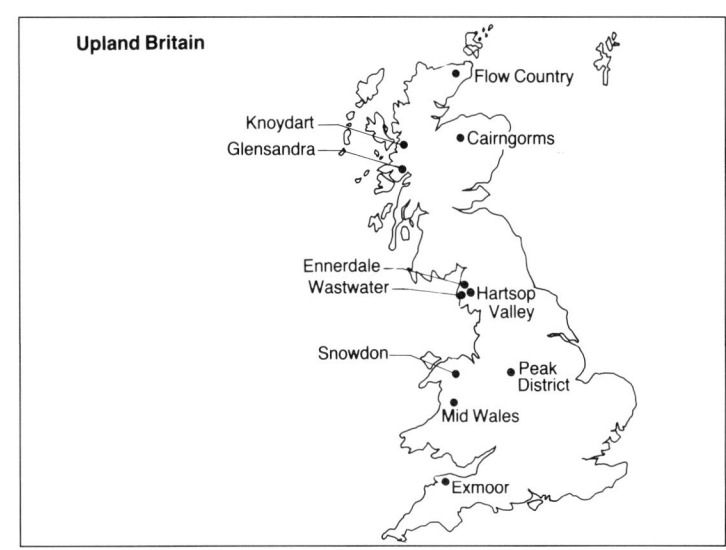

On the other hand, the physical environment in the uplands is a harsh one, and a perennial challenge to those that live there. Farming struggles and adapts to weather that is often damp or cold, weather that can bring isolation in winter: farmers learn to cope with grazing damaged by the spread of bracken, slopes that defy machinery, and soils that need frequent enrichment, and finances that are finely balanced. Much greater distances have to be travelled to enjoy the full range of services that are taken for granted by urban dwellers: the nearest general hospital may be an hour's journey away, and few schoolchildren in remote areas will find a secondary school within easy reach. So the judgement on the uplands is indeed one based on a sense of values, balancing landscape and quality of life against possible hardship and remoteness.

LIVING IN REMOTE BRITAIN: A FUTURE FOR KNOYDART?

On a single December day in 1954 259 mm of rain fell at the western end of Loch Quoich. In an average year over 3000 mm of rain are recorded here, making this rugged territory – the Rough Bounds of Knoydart – one of the wettest in our islands…. The narrow hill road continues westward beyond Loch Quoich, twisting this way and that to avoid ice-worn slabs and silent peaty lochans before suddenly dropping to the head of our finest fjord, Loch Hourn, with neighbouring Loch Nevis and Loch Scavaig on Skye, comes nearest to our idea of the genuine ice-worn, sea-flooded inlet of the Norwegian and south-west New Zealand coasts. For seven miles the loch goes down to the Sound of Sleat in three, sharp-angled sections. Rock walls and jagged steeps rear from the black waters of this 'hell's loch, dark-shadowed and formidable…. Suddenly we could look over the nearby bracken ridge, down the sinister length of Loch Hourn to Ladhar Bheinn, inaccessible King of Knoydart.

(Roger Redfern, A Country Diary The Guardian)

(a)

(b)

Fig. 4.1.2
(a) Knoydart: Loch Hourn 'dark-shadowed and formidable'
(b) Ladhar Bheinn 'inaccessible King of Knoydart'

1 Knoydart is one of Britain's last wildernesses. What 'wilderness qualities' give Knoydart its distinctive physical nature? (Use the passage from 'A Country Diary', and the photographs (Fig. 4.1.2) to help you determine these qualities)

2 Use the Ordnance Survey extract: Sheet 33 Loch Alsh and Glenshiel 1:50,000 Second Series for an evaluation of the scenery of this part of Knoydart. Each Grid Square needs to be given two scores, one for landforms, and one for land-use. The landform scores are given below.

- Lowlands: below 150 m in height.

- Rolling countryside: between 150 m and 450 m in height with available relief of less than 120 m.

- Upland plateau: exceeding 450 m in height, but with less than 120 m of available relief.

- Hill country: either between 150 m and 450 m in height, and with available relief exceeding 120 m, or between 150 m and 600 m with available relief between 120 m and 250 m.

- Bold hills: either exceeding 600 m in height, with an available relief of between 120 m and 250 m or between 450 m and 600 m with available relief exceeding 180 m.

- High hills: exceeding 600 m in height with more than 250 m of available relief.

Available relief is the vertical distance between the highest points of the dissected land surface and the valley floors of the local rivers.

The categories are scored successively from 0 (lowlands) to 5 (high hills).

The land-use scores are given below.

- Urban areas: all built-up land and land within 1.5 km of such land.

- Agricultural land: crops and grass are the dominant land use and no other use is significant.

- Woodland: woodland is the dominant use and no other use is significant.

- Moorland: moorland is the dominant use and no other use is significant.

- Diversified use: at least two of the categories, crops and grass, woodland and moorland, make significant contribution.

- Water: squares where water makes a significant contribution.

The categories are scored successively from 0 (urban) to 5 (water). Each square can have a maximum score of 10.

Complete the scores for each square on the map and then choose an appropriate choropleth method to show the variations in landscape quality in the map area. Comment on the pattern that your map displays. Are there any shortcomings of this method of evaluating landscape?

The position of Knoydart estate on the north-west coast is shown on the map (Fig. 4.1.3). Before the Highland Clearances, when crofters were evicted from these estates in order to make way for large scale sheep-rearing, there were four hundred crofters in Knoydart. Lady Josephine MacDonnell, who was responsible for their removal, predicted, when she sold the estate that none of the future owners would ever make a profit in the running of the estate. Her prophesy has proved all too correct. Knoydart was put on the market in 1983, with an asking price of £1.9 million. It took a long time for a buyer to emerge: during that period several alternative proposals were put forward for the future use of this mountainous, rain-swept peninsula, much of which can only be reached by sea from Mallaig. The sale brochure described the deer stalking as excellent and the salmon fishing as superb. The four alternatives were:

- use as a military training ground (this was shortly after the Falklands War, when the need for such training and facilities had been sharply focused)

- conservation of the physical and ecological environment under a body such as the National Trust for Scotland

- use by a sporting consortium, with shareholders investing a large sum in return for a time-sharing arrangement which guaranteed hunting, shooting and fishing at certain times of the year

- maintaining the status quo, with opportunities to sell off some of the estate to buyers who could develop their land with a view to encouraging some modest economic growth, whilst stimulating the intrinsic community culture of Knoydart.

Location of Knoydart Estate

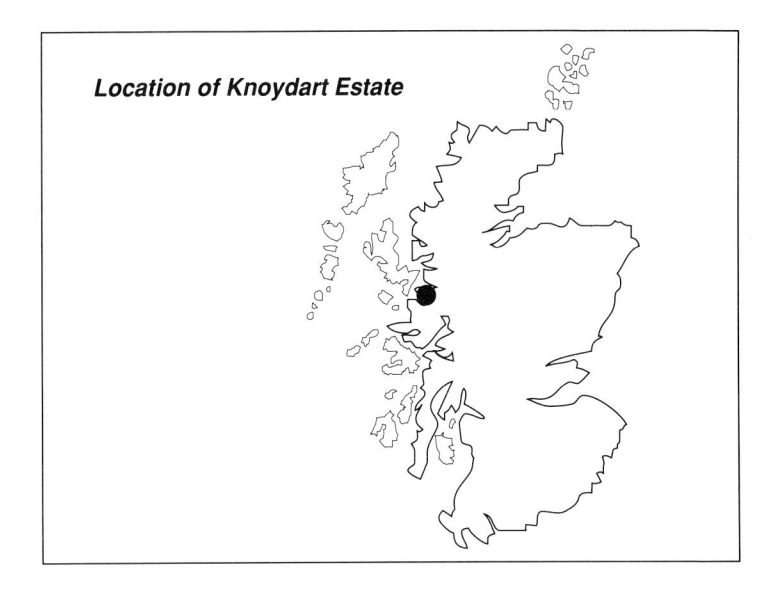

Fig. 4.1.3

Some comments

Dave Smith, crofter, 'I came to escape from London, but there is much more to it than that for me – it is the real world here'.

Knoydart resident, 'Some of those higher up think that we are simple, ignorant, with straw coming out of our ears, and that angers us. We have as much to contribute here as anybody, we want to make a go of Knoydart, make it a success'.

Present owner: 'I bought Knoydart because it is stunningly beautiful. I would have been pleased if we could have got some grants for forestry, but nothing came about'.

Ministry of Defence official: 'The services at present are suffering from a serious shortage of training grounds'.

Sporting consortium, 'The sporting possibilities for Knoydart are unlimited. A proposal such as ours could bring a much-needed injection of capital'.

1 Each of these proposals reflects a certain set of values. Discuss what these values are, and draw an annotated diagram to show how there is conflict between the different sets.

2 What are the merits of the different proposals?

3 What, in your opinion, should be the attitude of such bodies as the National Trust for Scotland, or the Countryside Commission for Scotland to the future of Knoydart, which in 1992 is again for sale? The estate has now been reduced, through sales of land by the present owner, to 6500 hectares. The asking price is £1.7 million.

THE PROSPECT FOR FARMING IN THE UPLANDS

'In wind and driving rain, Dr. Domenico Berardelli plods over the mountainous acres of the Great Glen near Fort William to tend his blackface sheep. Life often seems a constant battle against the elements for the Italian chemical engineer turned Highland farmer. 'It's not an easy life and I don't expect high financial rewards. The attraction is the countryside and the open air – even in this climate. I expect to earn a living – that's all.'

Fig. 4.2.1
Sheep farmer in Scottish Highlands

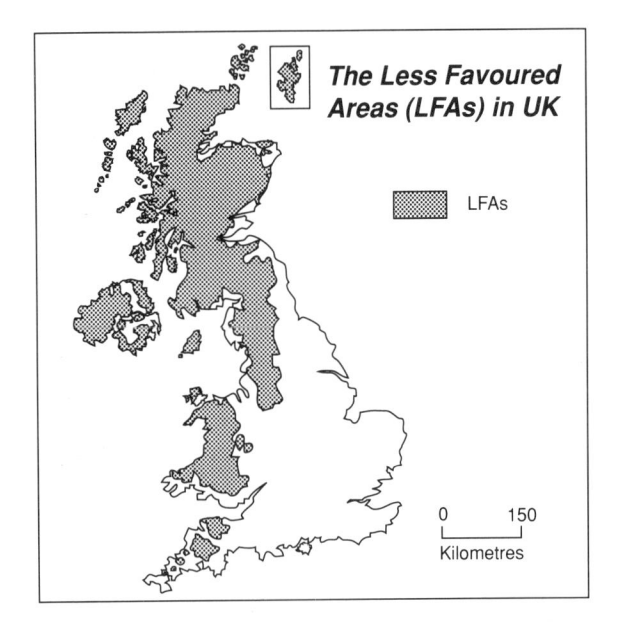

Fig. 4.2.2

CONSTRAINTS ON FARMING IN THE UPLANDS

The map (Fig. 4.2.2) shows the distribution of the Less Favoured Areas (LFAs) in Britain. The LFAs are defined by the European Community as 'areas in danger of depopulation: farming areas regional in character, having permanent natural handicaps, essentially those of poor land where the potential for production cannot be increased except at excessive cost. To qualify these areas must also show low productivity as reflected in crop yields or stocking of livestock and have a low population (or dwindling one) whose economy is extensively dependent on agricultural activity'. It can be seen that most of Scotland and Wales qualify as LFAs, considerable parts of the Pennines in Northern England, and the moorlands of the south-west Peninsula are also included.

A desk study of upland areas in England and Wales (Fig. 4.2.3) sought to classify different zones in the upland areas according to their physical characteristics. Two main groups were recognised, known as marginal and central uplands, corresponding to classes 1–4, and 5–8, both shown on the map (Fig. 4.2.3). The general characteristics of the different classes are shown in the Figure 4.2.4a.

Comment on the distribution of all eight classes in the uplands of England and Wales. Which two classes represent the most severe conditions? Comment on their particular distributions.

It is also useful to establish the percentage of each of these upland classes that is occupied by the various grades of land quality as determined by MAFF. The five different categories are shown below.

Grade 1: Land with very minor or no physical limitation to agricultural use (little is found in the uplands and therefore it is included with class two).

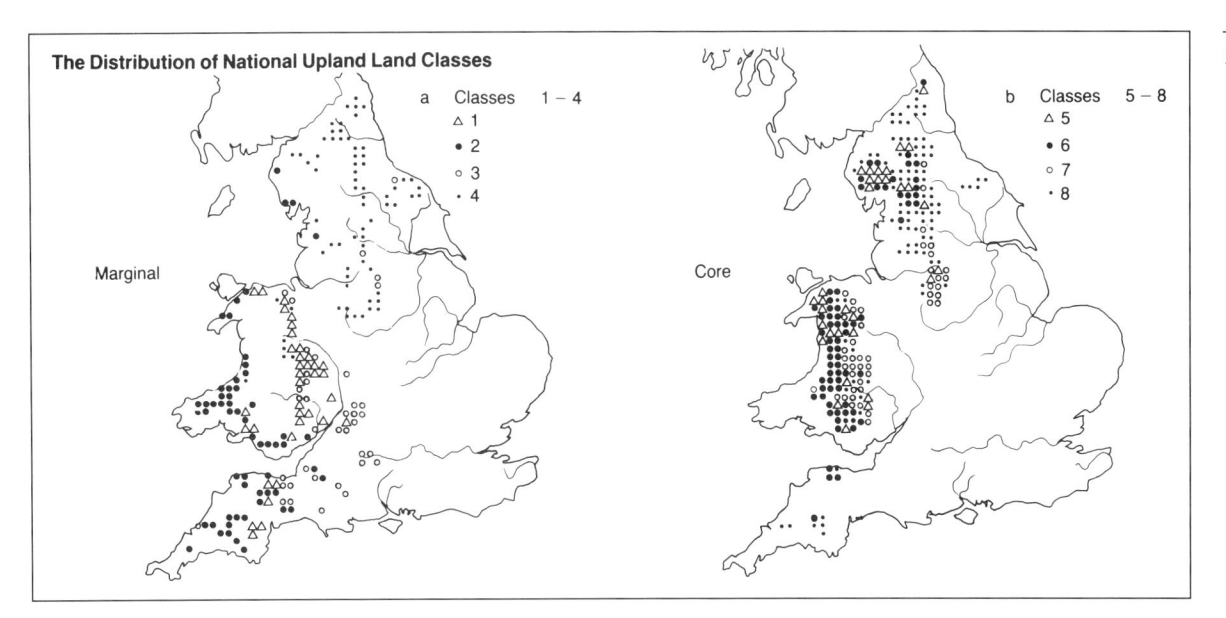

Fig. 4.2.3

The Distribution of National Upland Land Classes

a Classes 1 – 4
△ 1
● 2
○ 3
· 4

Marginal

b Classes 5 – 8
△ 5
● 6
○ 7
· 8

Core

DOMINANT RANGE IN CHARACTERISTIC

Fig. 4.2.4a

CLASS	ALTITUDE	RELIEF	SLOPES	RAINFALL	SOIL
1	Low	Moderate	Moderate	Low	Brown earth
2	Low	Low	Gentle	Moderate	Brown earth
3	Very low	Low	Gentle	Low	Brown earth and calcareous soils
4	Low	Low	Gentle	Low	Gley, brown earth and peaty gley
5	Very high	Very strong	Very steep	Very high	Podzol, brown earth and peaty gley
6	High	Strong	Steep	High	Brown earth, podzol, peaty gley and peat
7	Moderately high	Moderate	Steep	Moderate	Brown earth, gley and peaty gley
8	Moderately high	Moderate	Gentle	Moderate	Peaty gley, gley, brown earth and podzol

Fig. 4.2.4b
Upland Britain: Landscape classes
1. North Yorkshire Moors (overlooking upper Rye Dale) – Class 4
2. Exmoor's rolling farmland – Class 2
3. Snowdonia: Tryfan, the Ogwen valley and the Carneddau – Class 5
4. Lakeland: the central Fells in winter – class 5/6

Grade 2: Land with some minor limitations, particularly in soil texture, depth and drainage.

Grade 3: Land with moderate limitations due to soil relief and climate.

Grade 4: Land with severe limitations due to adverse soil relief and climate.

Grade 5: Land of little agricultural value with severe limitations due to adverse soil, relief or climate.

The distribution of these land grades in the various upland classes in England and Wales is shown in the table (Fig. 4.2.5).

Similar tables (Figs. 4.2.6a-e) are included to show the percentages of various physical constraining factors in the various upland classes (altitude, slope, average annual rainfall, temperature and soils).

1 Use all of these tables to summarise the variety and level of constraints on farming in the uplands.

2 Use a good atlas to decide how constraints in the Scottish uplands compare to those shown in England and Wales.

Fig. 4.2.5

AGRICULTURAL LAND CLASSIFICATION: PERCENTAGE OF LAND IN THE AGRICULTURAL LAND CLASSIFICATION GRADES

CLASS	GRADE				NON-AGRICULTURAL USES	
	2	3	4	5	URBAN	OTHER USES
1	8*	33	32	12	4	12
2	1	29	42	15	3	10
3	14*	51	13	3	8	11
4	1	28	35	17	9	10
5	0	1	18	69	1	11
6	0	1	21	61	1	16
7	0	5	48	34	2	10
8	0	8	26	54	3	9

*Less than 2 per cent Grade 1 land included.

Fig. 4.2.6a–e
Some physical characteristics of Upland Britain

(a)
ALTITUDE: PERCENTAGE OF LAND IN ALTITUDE BANDS (in metres)

SPECIFIC PHYSICAL FEATURES OF THE EIGHT UPLAND LAND CLASSES

CLASS	OVER 240	OVER 420	OVER 600	OVER 900
1	28	1	0	0
2	16	<0.5	0	0
3	9	0	0	0
4	20	<0.5	0	0
5	76	34	9	<0.5
6	73	27	3	0
7	71	10	<0.5	0
8	62	14	1	0

(b)

SLOPE: PERCENTAGE OF LAND IN THREE SLOPE BANDS (in degrees)

CLASS	0–11	12–22	OVER 22
1	40	56	4
2	63	35	2
3	79	21	0
4	79	21	0
5	14	54	32
6	26	63	11
7	28	68	4
8	72	27	1

Notes: Fig. 4.2.6b: a track laying tractor operates up to a limit of 20 degrees, a wheeled tractor up to 15, and a combine harvester up to 4.

(c)

RAIN: PERCENTAGE OF LAND IN BANDS OF AVERAGE ANNUAL RAINFALL
(in millimetres)

CLASS	610–759	760–1014	1015–1524	1525–2284	2285–3174	3175–5079
1	9	61	24	6	0	0
2	0	5	80	15	0	0
3	21	73	6	0	0	0
4	10	75	15	1	0	0
5	0	1	30	41	22	6
6	0	1	18	74	7	0
7	0	16	79	5	0	0
8	0	23	58	19	0	0

(d)

TEMPERATURE: PERCENTAGE OF LAND IN TEMPERATURE BANDS
(Annual sum of day degrees above 5.6°C)

CLASS	LESS THAN 825	825–1099	1100–1374	1375–1649	1650–2000
1	0	0	9	46	45
2	0	0	4	31	65
3	0	0	1	28	71
4	0	1	36	58	5
5	4	26	48	19	3
6	2	20	43	28	7
7	0	5	62	30	3
8	1	15	50	30	4

Fig. 4.2.6d: day degrees represent the accumulated total of degrees by which the daily temperature exceeds the minimum for plant growth.

(e)

SOIL: PERCENTAGE OF LAND WITH MAJOR SOIL TYPES

CLASS	BROWN EARTH VARIANTS	CALCAREOUS SOILS	GLEY SOILS	BROWN PODSOLIC SOILS	PODSOLS AND PEATY PODSOLS	PEATY GLEYS	PEAT SOILS
1	80	2	10	4	2	2	0
2	75	0	14	1	5	4	1
3	50	31	19	0	0	0	0
4	24	0	52	0	5	18	1
5	27	0	4	0	45	16	8
6	37	0	4	0	28	17	14
7	58	0	12	0	10	16	4
8	17	0	25	1	17	29	11

Fig. 4.2.6e:

- *brown earths are freely draining soils and are the most productive found in the uplands.*

- *calcareous soils are only important in class 3, since that is the class in which most limestones fall.*

- *gley soils and peaty gley soils occur on gentle slopes where the lower soil horizons are almost entirely water-saturated: they are difficult and expensive to improve for agriculture.*

- *podsols are typical of heather moors and require intensive fertilisation after they have been reclaimed.*

- *brown podsolic soils are better, and are suitable for improvement for grassland.*

- *peat soils are usually found in plateau sites, or in shallow basins: their agricultural use is very limited.*

Fig. 4.2.7
Hartsop Valley

CHANGING PATTERNS OF LAND USE IN THE UPLANDS

TWO CONTRASTED CASE STUDIES OF FARMING IN THE BRITISH UPLANDS: THE HARTSOP VALLEY AND EXMOOR

The farming systems of these two contrasted areas of upland Britain raise different issues. The Hartsop Valley (Fig. 4.2.7) lies on the eastern side of the Lake District. Most of the land is upland class 5 land, and is farmed under some of the most severe physical conditions in England. The second case study examines a farming environment, which is still distinctly upland, but is much more favourable. Here the land falls into the upland classes of 1, 2, 6 and 8, with only a quarter of the land in the harsh category 6. The Exmoor study (Fig. 4.2.8), covers an area of 1200 sq km compared to the 41 sq km of the Hartsop Valley. On Exmoor the main issue is the reclamation of moorland for additional grazing land, a subject that has caused much controversy in the area.

Fig. 4.2.8
Moorland in the central part of Exmoor

The Hartsop Valley

The tiny settlement of Hartsop lies at 180 m on the side of a deep glacial trough that bites deep in the craggy country formed by the Borrowdale Volcanic series on the eastern side of the Lake District. Within the hamlet and in the surrounding valley floor are seven farms whose land extends from the flat, moraine-strewn floor to the high fells some 500 m above. Farming such land is difficult and unrewarding, and making a living is a precarious matter. In such an environment careful management of farming operations is essential if such an agricultural system is to survive. Compounding the economic pressure on farmers is the additional awareness that they are operating a farming system in an area of high landscape value, which has considerable recreational potential. A number of different interest groups are likely to be in conflict in such an area. Farmers need financial support to survive, and use this support to invest in new technology. This, in turn, may lead to a clash of interests with local bodies who have a responsibility for landscape conservation, and others who may wish to promote the recreational potential of the area. From the diagrams which show altitudinal variations in relief, land use and soils (Fig. 4.2.9) and the map of a typical farm holding (Fig. 4.2.10) it can be seen that there are three main altitudinal zones:

- the in-bye land on the valley floor (relatively high quality land)

- the allotment (land enclosed by stone walls) which extends from approximately 200–270 m where the fields are much larger and contain rocky outcrops. The land is steeper and has a good sward of fescues and bent grasses

- the high fell, some of which has been enclosed, the remainder retained as open grazing.

All of the farms in the Hartsop valley are involved in the rearing and selling of both sheep and cattle, with the main emphasis on sheep.

1 Suggest the main farming uses of the three types of land indicated above.

2 In what ways might each of these types of land be improved, and what environmental conflicts might result from such improvements?

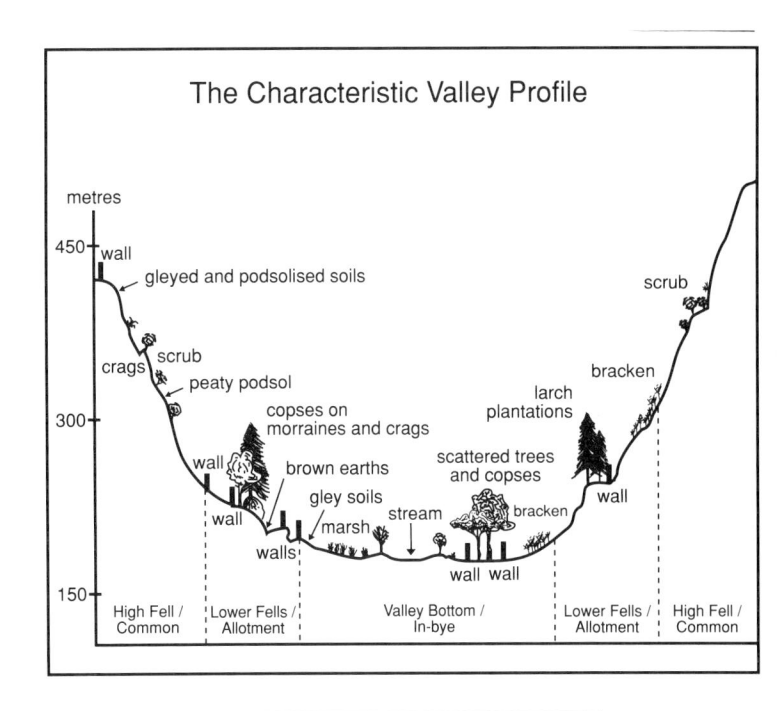

Fig. 4.2.9

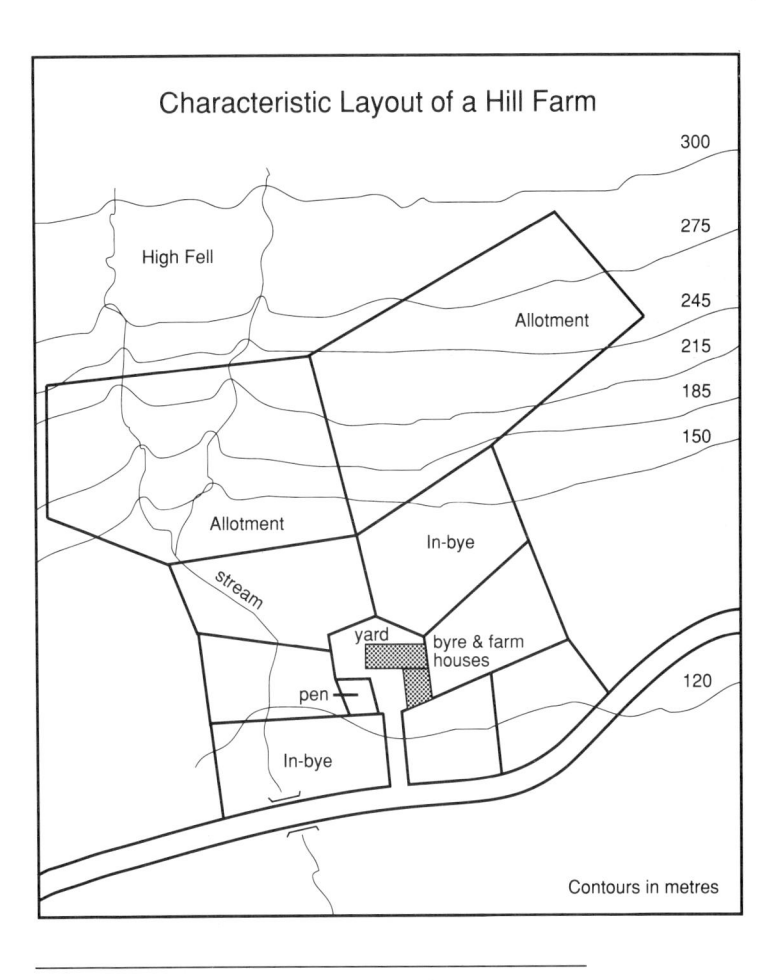

Fig. 4.2.10

HARTSOP VALLEY FARM SYSTEM

INPUTS: Livestock Costs; Fixed Input (Machinery Repairs Rates Rental); Crops; Contractors fees Fertilisers

Farm Enterprises (Swaledale) Sheep, Friesian Cattle, Galloway Cattle

OUTPUTS: Sheep; Wool; Store Cattle

Subsidies (HLCA)

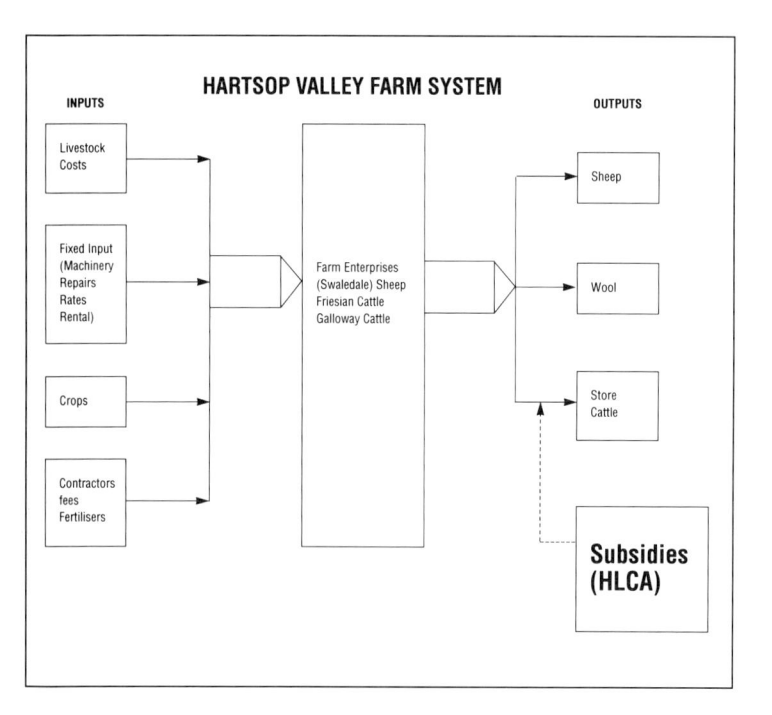

Fig. 4.2.11

Most of the farms in the Hartsop valley are relatively small, with an average size of 52 hectares. The diagram (Fig. 4.2.11) shows the farm system in the Hartsop valley. It will be apparent from the diagram that farmers are heavily dependent on the Hill Livestock Compensatory Allowance (HLCA), which is a subsidy paid to all farmers in the LFAs. Without these subsidies few would survive financially. Farmers constantly need to review their options for the future: these include:

- reviewing farm operations in order to reduce costs

- adopting new techniques of reclamation, drainage and reseeding, with a view to pasture improvement

- improving the level of livestock care, particularly in the winter

- introducing a greater degree of co-operation in bulk-buying and marketing

- broadening the range of activities on the farm to increase sources of income from activities other than farming.

Grant aid is available (up to 50 per cent in some cases) for improving the status of soils, for the control and eradication of bracken, and the draining and reseeding of pastures. It is also available for the construction of new farm buildings.

Evaluate each of these options, and discuss how they could be integrated to produce a greater sense of security and stability on these upland farms.

Exmoor: reclamation or conservation: the moorland issue

Exmoor, like the people who live there, has alternately resisted and accepted change, but over the past 30 or 40 years the pace has quickened. Each of the individual changes in the landscape and pattern of farming has affected only a small section of the moor, but taken as a whole they have given it a very different aspect from the early years of this century. The 'hills of great height covered with heather' of which Richard Jefferies wrote in 1883 have lost much of their purple colouring. The heather, though still plentiful in parts, has retreated, and on Winsford Hill, for example gorse or 'vuzz' has begun to invade. Why?

(Glyn Court, Exmoor National Park)

Exmoor suffers less severe conditions than many of Britain's uplands, partly because of its lower altitude, partly because of its position in the mild south-west. Nevertheless much of the moor has rainfall amounts of 1000–2300 mm and severe falls of snow are not unknown in winter. Opportunities for plant growth are much better, with many moorland parishes having well over 1500 day degrees above 5.6°C suggesting a much improved supply of forage compared to other more northerly uplands. Brown earths mantle much of the southern slopes, and it is only on the high windswept and damp plateaux, where leaching is much in evidence, that podsols occur. Poor, ill-drained peaty soils occur in some hollows on the plateau surface and at the heads of some valleys. Much of this central area is covered by purple moor grass, with deer sedge dominant across the Chains (Fig. 4.2.12). The grass moorland gives way to fragmented heather moorland on most sides,

Fig. 4.2.12
The Chains: Exmoor's most desolate expanse of moorland

THE PROSPECT FOR FARMING IN THE UPLANDS

cut into by the steep valleys whose sides are covered by bracken and gorse, but retain their oak woodlands. It is an environment which has promoted a healthy livestock rearing system, soundly based on the Exmoor Horn sheep and Devon cattle.

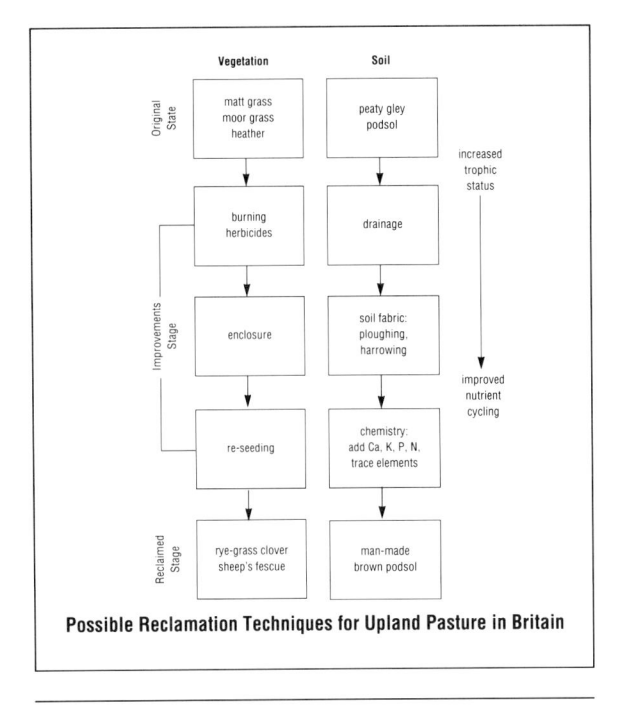

Possible Reclamation Techniques for Upland Pasture in Britain

Fig. 4.2.13

Reclamation of moorland has become an increasingly controversial issue over the last 30 years (see Fig. 4.2.13). Once moorland has been ploughed, drained and fenced it is planted with a green crop in the first year, which is eaten on the land by sheep. A second year's crop is also eaten on the land, and a final selected grass mixture is sown in the third year for a long ley, which will be left as pasture for a number of years before reseeding. Reclamation grants were payable, with up to 50 per cent of the costs of ploughing and reseeding allowable, and up to 70 per cent of the cost of drainage. Once the land has been reclaimed, Hill Livestock Compensatory Allowances are payable on the extra livestock pastured on the reclaimed land. Attitudes became polarised. On the one hand there were the farming and improvements interests, very much supported by the National Farmers' Union and the Country Landowners association, and encouraged by the MAFF. On the other hand there were the amenity and conservation interests, who wished to see the moorland preserved for scenic and ecological interest. These interests were supported by the National Park Authority, the former Nature Conservancy Council, and a number of voluntary bodies such as the Exmoor Society.

Study the maps which show the changes consequent on reclamation in the central part of Exmoor just north of the Chains (Fig. 4.2.14).

1 Comment on the changes in field pattern that have taken place in the 90 year period.

2 What main landscape changes result from the reclamation?

Reclamation of Exmoor (Lynton Parish)

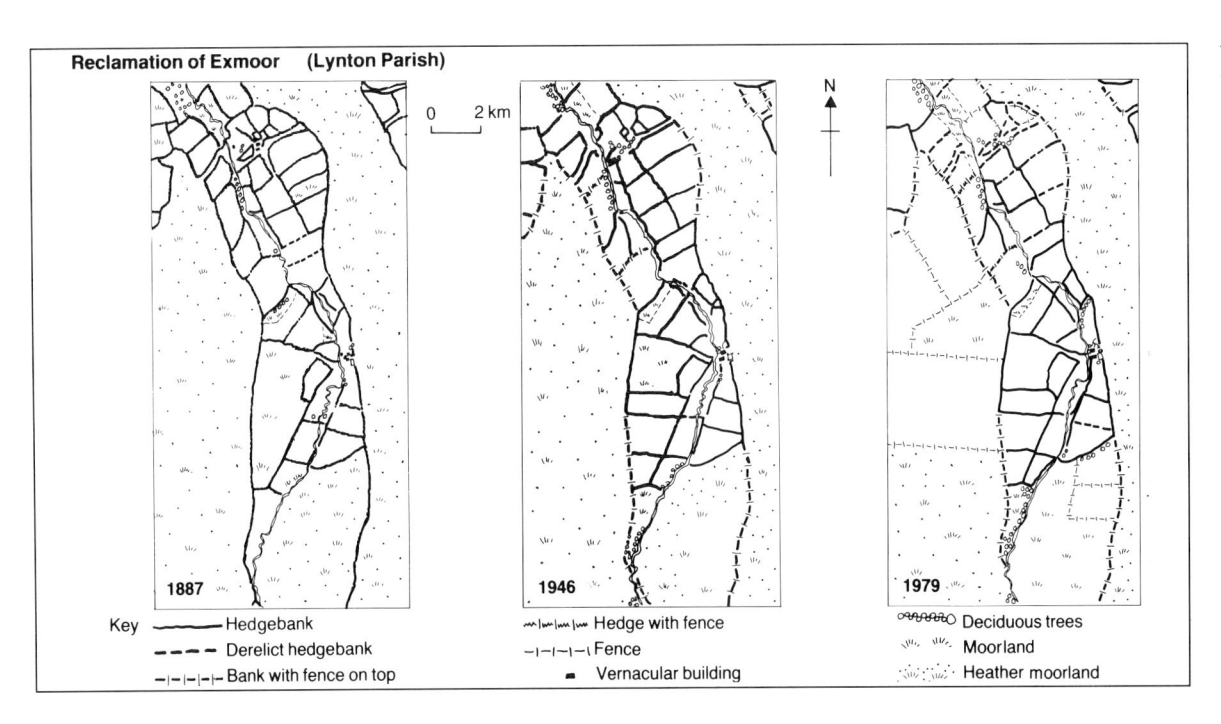

1887 | 1946 | 1979

Key
— Hedgebank
- - - Derelict hedgebank
-|-|-|- Bank with fence on top
⌒⌒⌒⌒ Hedge with fence
-·-·-·- Fence
▪ Vernacular building
○○○○○ Deciduous trees
ᴡᴡ Moorland
∴∴∴ Heather moorland

Fig. 4.2.14

Porchester Map

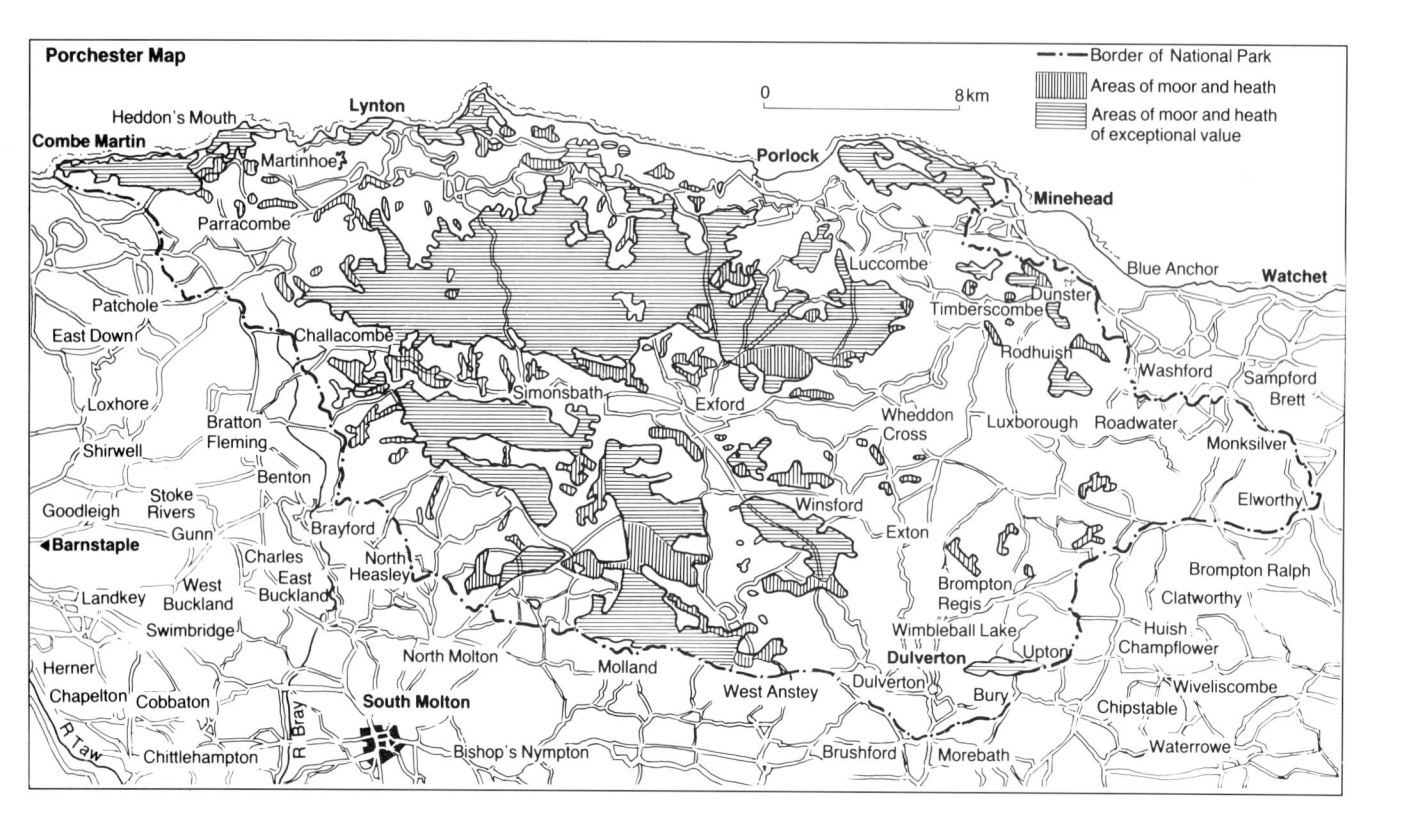

Fig. 4.2.15

The dispute became so acrimonious in 1977 that the Porchester enquiry was set up to attempt to find a solution. The report produced two maps (combined on one in Fig. 4.2.15) Map 1 showed the total area of heath and moorland on Exmoor, Map 2 showed 'those particular tracts of land that the National Park Authority would wish to see conserved for all time.'

1 Study the map (Fig. 4.2.15) which shows Porchester Map 1 moorland and Porchester Map 2 moorland. Comment on the distribution of the two types of land.

2 Why was the publication of the maps very controversial?

Porchester recommended that reclamation grants should cease to be paid and that farmers should enter into management agreements in which they were compensated for not reclaiming moorland. The mangement agreement involved the following:

- payment to be made to the farmer would be equivalent to the profits that he would have made if he had reclaimed the land

- payment could either be a lump sum, in which the farmers were bound for all time, or an agreement over 20 years

- farmers would agree to keep the unreclaimed moorland in good shape.

1 Identify the different sets of values that are relevant to this controversy.

2 Are upland farmers justified in reclaiming moorland?

3 How could the system of management agreements be abused?

4 Farmers must agree to keep the unreclaimed moorland in good shape. What problems are likely to be encountered here?

The management agreement seems to have worked reasonably well. Since 1979 only 88 hectares have been reclaimed, all of these outside of the map 2 areas. An example of a successful management agreement is the Glenthorne Estate, which lies on the northern side of Exmoor, several kilometres to the east of Lynmouth (see Fig. 4.2.16). The estate consists of two moorland farms of about 400 hectares, together with about 200 hectares of cliff and woodland along the coast. The area was particularly rich in wildlife, with some 200 plant species and 47 species of birds identified so far. Approximately 120 hectares were considered suitable for reclamation. However, the land was already used for informal recreation, and was obviously of high conservation value. The management agreement sought to achieve some real sense of balance between production, recreation and conservation. In the end this was resolved by devoting 40 hectares to each of the three different uses (see Figs. 4.2.16 and 4.2.17).

- 40 hectares: Yenworthy Common was to be fenced and reclaimed.

- 40 hectares: Cosgate Hill was to be sold to the National Park Authority. The existing cottage would be converted into an Information and Interpretive Centre.

- 40 hectares: North Common would be retained as unreclaimed moorland.

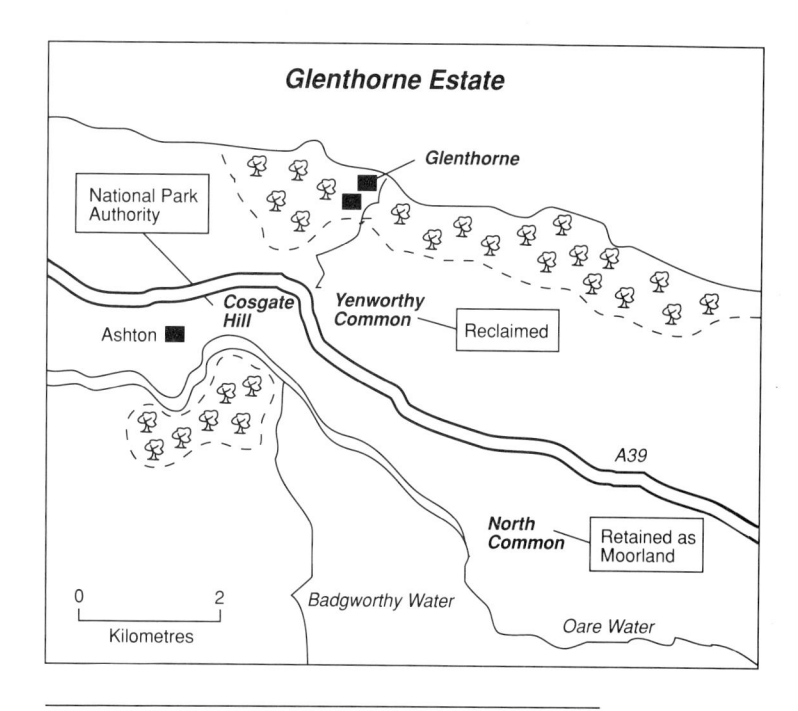

Fig. 4.2.16

(a)

(b)

(c)

Fig. 4.2.17
The Glenthorne Estate (see Map 4.2.16)
(a) Yenworthy Common: reclaimed for pasture with heather moorland in the foreground
(b) Cosgate Hill: recreational area with interpretive centre
(c) North Common: moorland to be left unreclaimed

As an agreement this seems to have worked well and has achieved the balance that it set out to achieve. In the mid-1980s the Government withdrew all grants for land reclamation, whether in the National Parks or elsewhere. At the same time it offered grants to intensify production on grassland and to improve the management of grass and heather moorland.

Summarise the ways in which farming on Exmoor differs from that in the Lakeland Valleys, and discuss the ways in which such differences are likely to be reflected in the landscape of farming in the two areas.

SUBSIDIES AND GRANTS IN UPLAND FARMING

From the two case studies it is apparent that the availability and payment of subsidies and grants to farmers have a fundamental role to play in the sustaining of farming in areas like Hartsop, and in the maintenance of the farming system on Exmoor. The livestock subsidies should be seen not only as a vital source of income for many hill farmers, but in a much broader sense, as a means of maintaining a more stable rural population. Payment systems do, however, tend to favour the larger farms, and work against those that operate much smaller holdings, often in much more disadvantaged circumstances.

Study the table (Fig. 4.2.18) which shows the payment of HLCAs to farmers in England and Wales.

1 Comment on the levels of payments to different categories of livestock unit.

2 Make some suggestions for a fairer system.

Fig. 4.2.18

DISTRIBUTION OF HEADAGE PAYMENTS (HLCAs) IN THE LESS FAVOURED AREAS OF ENGLAND AND WALES

LIVESTOCK UNITS PER FARM ELIGIBLE FOR HLCAS	NO OF FARMS IN LFA	% OF LFA FARMS	% OF ELIGIBLE LIVESTOCK	HLCA TOTAL AT 1981/2 RATES (£M)	HLCA PER FARM (£)
0–50	11 213	53.7	14.8	6.6	590
51–100	4243	20.7	19.4	8.7	2043
101–150	2460	12.0	19.0	8.5	3452
151–200	758	3.7	8.3	3.7	4895
201–250	615	3.0	8.3	3.7	6033
251–300	472	2.3	7.8	3.5	7387
300+	759	3.7	22.4	10.0	13192
Totals	20 520	44.7	100.0	44.7	4591

Notes: 1. Estimated on the basis of 300 farms in the Upland Landsdcapes Study representative sample.
2. At present all farms are entitled to a flat rate HLCA of £44.50 for beef cows, £6.25 for 'hill ewes' and £4.25 for 'upland' ewes.
3. Calculations have been made on average HLCA payments per livestock unit.

Fig. 4.2.19

ATTRACTIONS OF UPLANDS TO FARMERS

ATTRACTION	NUMBER	% OF RESPONSE	% OF RESPONDENTS CITING RESPONSE
Scenery/physical features	149	38	52
Vegetation	47	12	16
Peace and quiet	46	12	16
Variety	23	6	8
Open/unspoilt	19	5	7
Fresh air	19	5	7
Wildlife	15	3	5
Familiarity	9	2	3
Well-farmed	8	2	3
Cannot define	7	2	2
None	27	7	9
Total	392	100	136

WHY DO THEY FARM IN THE UPLANDS?

When all of the difficulties of farming in the uplands are taken into account, the above question seems to be a pertinent one. One of the most important pieces of literature on upland farming is the Upland Landscape Study edited by Geoffrey Sinclair. It involved the study of 12 upland parishes selected from all the upland regions of England and Wales. Farmers were questioned on a wide range of topics in the study. One important issue was the attraction of their area for the farmers. Responses are shown in the table (Fig. 4.2.19).

1 Use diagrams and charts to express the results in a form suitable for presentation to a conference on the future of the uplands.

2 To what extent do all of the views recorded seem to reflect the same set of values?

3 How well do these views accord with that expressed by the Highland sheep farmer, Domenico Berardelli at the beginning of the section on upland farming?

FORESTRY IN THE HILLS

There were the little farmsteads of the lower slopes, the huge sweep of the yellow-green turf, the slightly more verdant green blocks of the bracken neatly cut up by narrow sheep-tracks like a gigantic jig-saw puzzle, the wooded cleft of the Dulas gorge, the bare summit of Cusop Hill where we were yesterday. And further to the right – the great dark angular slabs of the Forestry Commission plantations spreading over the hills like a disease....

(A.L. Le Quesne, After Kilvert. OUP 1978)

While pursuing its main objective of wood production the Forestry Commission seeks to achieve an acceptable balance between efficient and timely forest operations and good landscape design. New forests are designed to blend with the landscape as far as possible, and opportunities are taken to improve the pattern of existing forests so that they too blend with the form of the land. The intrusive effects of the landscape caused by any forestry operations are kept to a minimum.

(Forestry Commission, The Forestry Commission and Landscape Design)

However good they may become as wildlife habitats in their own right, these forests are not, and never can be a compensatory substitute for such open ground ecosystems (the original moorlands), which are highly valued in their original condition.

(English Nature, Nature Conservation and Afforestation in Britain)

It is scandalous that vast tracts of upland Britain lie barren awaiting productive development.

(Forestry and British Timber Magazine, editorial May 1978)

Farming in the uplands will have to be intensified if land is to be made available for afforestation without loss of agricultural production and loss of livelihoods in hill farming.

(Ramblers' Association, Afforestation: The Case against Expansion)

Britain cannot afford the luxury of a countryside as a free park for the visual delights of ramblers and 'eco-freaks'.

(Forestry and British Timber Magazine editorial, May 1980)

Modern foresters are not insensitive: they have learned from mistakes of the past. They are already conscious of the need to avoid hard edges to plantations.

(Forestry and British Timber Magazine, 1986)

Fig. 4.3.1a Forestry Commission Plantations (North Wales looking east from slopes of Rhinog Fawr): note the angular boundaries and their effect on landscape

Fig. 4.3.1b
Coniferous plantations; Rhinns of Kells, southern Scotland: note the abandoned shepherd's hut, now used as a walkers' bothy

Later, with the advent of burning and clearance of land for farming, and grazing of animals, further diminution of the forests took place. Today natural woodlands occupy only a small part of the uplands. Natural pine forests are restricted to a few areas in the central and north-eastern parts of the Grampian Mountains in Scotland – the remnants of the ancient Caledonian forest (Fig. 4.3.2) showing ancient pine forest near Loch Maree. Further south some remnants of oak, ash and birch and, particularly in Wales, the alder, remain. Oaks appear to heights of 450 m in sheltered locations on Dartmoor (Fig. 4.3.3), ashwoods appear up to 300 m in the northern Pennines and alders reach similar heights in Snowdonia.

Fig. 4.3.2
Caledonian Forest: part of one of few remaining native coniferous forests in Scotland on shores of Loch Maree

Forestry in the uplands has been a controversial topic for several decades. The above quotes represent some of the different views on afforestation in the uplands.

1   What are the two polarised sets of values that are represented here?

2   Is a compromise between the two represented?

3   How can the two polarised views be justifed, and is the compromise a valid one?

Fig. 4.3.3
Wistman's Wood, Dartmoor: primeval oak forest

NEW FORESTS FOR THE UPLANDS

Evidence from buried tree remains suggests that woodland was formerly quite extensive in the uplands, up to heights of about 600 m. In some sheltered locations it even reached heights approaching 700 m in the north of England and even 900 m in some parts of Scotland. Once the ice sheets had vacated land at these heights, birch, and later pine began to invade these areas about 11 000 years ago. Around 7000 years ago the climate became wetter, and encouraged the growth of sphagnum mosses, and other peat-forming plants, which checked the regeneration of the pine forests.

New planting in the uplands is in the hands of the Forestry Commission or private forestry concerns. The Forestry Commission was formed in 1919, with one of its specific aims the provision of strategic supplies of timber, particularly in time of war. Since that time there has been a steady increase in the area planted by the Forestry Commission. The original target set in 1919 was 0.75 million hectares, but this has been steadily revised upwards. Today the area of Forestry Commission plantations alone stands at 0.9 million hectares. Annual target rates for afforestation have increased from 20 000 – 25 000 hectares at the beginning of the 1980s to 33 000 in the late 1980s. In the UK as a whole the area under woodland has increased from 5 per cent to nearly 10 per cent over the last 50 years, and the proportionate increases in the various categories are shown in the diagram (Fig. 4.3.4). There has been a significant increase in private forestry planting in the 1980's (in 1985/6 planting was running at a level of 19 000 hectares a year compared to only 4000 hectares by the Forestry Commission) The graph (Fig. 4.3.5) shows trends in planting in the early and mid–1980s. One of the main reasons for the increase in private planting was the very considerable tax advantage to be had from investment in private forestry. This advantage ceased to operate after the 1988 budget, but this was offset to a certain extent by the considerable grants made available for planting under the Woodland Grant Scheme.

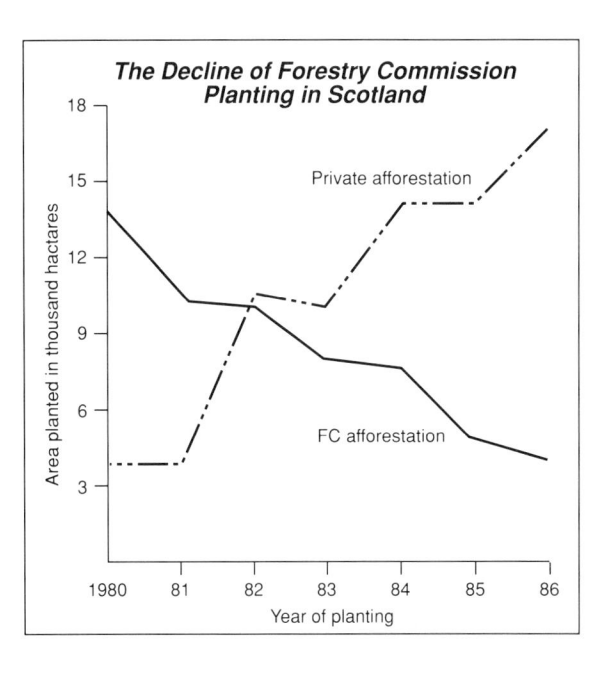

Fig. 4.3.4

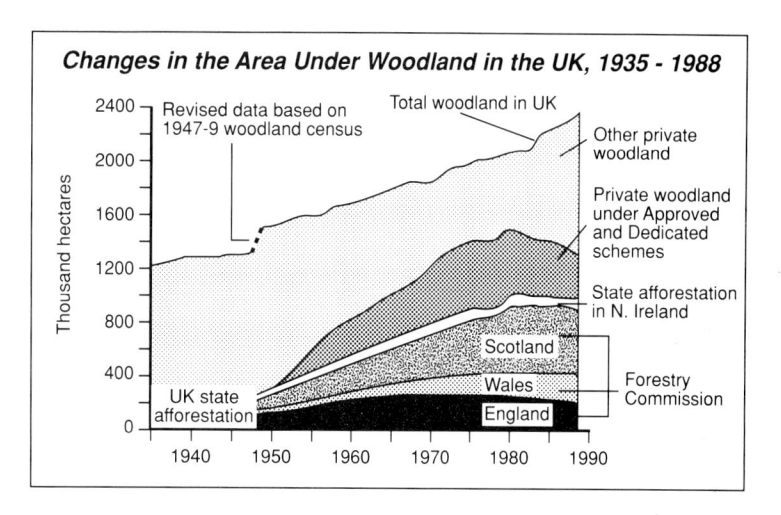

The Environmental impact of blanket afforestation in the uplands

Fig. 4.3.6
New coniferous plantations in the Flow Country

Fig. 4.3.5

Blanket afforestation has caused increasing concern among the many organisations involved with the status and stability of the upland environment and its habitats. The main objections that have been raised are:

- semi-natural habitats in the uplands are being replaced by artificial habitats, which lack both variety and richness

- the loss of rare, often unique ecosystems such as the blanket bogs of the Flow Country
- the loss of specialised upland bird communities which have a high proportion of waders
- the initial increase in run-off that follows the deep ploughing and draining of the areas to be planted
- the increase in sediment levels carried into streams
- the effect of evergreen foliage in filtering existing atmospheric pollution responsible for acid rain. Pollution is trapped by the foliage, and is eventually washed into streams
- the washing of pesticides and fertilisers from the plantations into streams in a part of the country where this type of pollution has hitherto been absent
- the effects of the above pollution on aquatic life
- the effect of angular shapes of plantations on the natural outlines of the upland landscape.

1 Which, in your opinion, are the most serious criticisms?

2 How might the forestry industry reply to each of the above criticisms?

PLANTING FOR LANDSCAPE

Planting on land which has not carried trees for centuries can have a major impact on the appearance of the countryside.... The design of new forests, therefore, is very important and involves an analysis and understanding of the essential character of the landscape where trees are to be planted. Almost all new planting takes place on infertile sites in the uplands and it must be recognised that only a limited number of species, all of them coniferous, are capable of economic growth here.

Wherever possible the boundaries of new forests are designed to follow the natural landform. Every effort is made to avoid hard unnatural lines, and to prevent obscuring prominent landmarks and other interesting features such as crags, streams, waterfalls, and gullies.

Broadleaved trees are retained where they are important for visual or nature conservation reasons, especially near water and bordering farmland.

(Extract from Forestry Commission and Landscape Design; Forestry Commission Policy and Procedure Paper No. 3)

The above extract indicates the broad general principles of landscaping new forest planting.

LANDSCAPE PLANNING AND FORESTRY EXERCISE

You are to assume the role of a landscape consultant, seconded to the Forestry Commission, and working in partnership with a team of foresters. You have been asked to prepare landscape designs for an area in mid-Wales in which new planting is to take place.

You have been provided with the following resources:

- Fig. 4.3.8 showing physical controls on options for planting in mid and north Wales
- Fig. 4.3.9 showing the distribution of soil types on a hillside profile
- Fig. 4.3.10 showing the vulnerability of tree species to grazing animals.

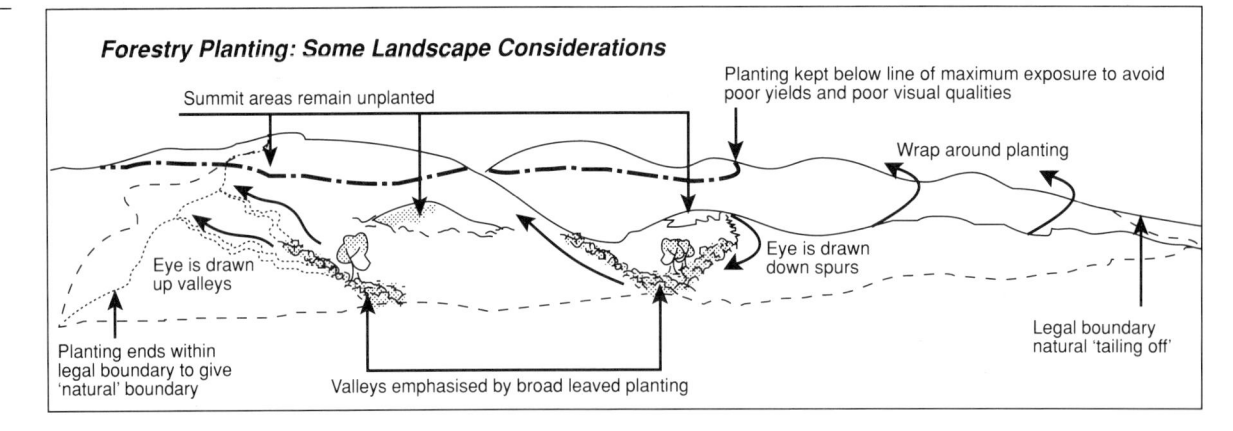

Fig. 4.3.7

Forestry Planting: Some Landscape Considerations

- Summit areas remain unplanted
- Planting kept below line of maximum exposure to avoid poor yields and poor visual qualities
- Wrap around planting
- Eye is drawn up valleys
- Eye is drawn down spurs
- Planting ends within legal boundary to give 'natural' boundary
- Valleys emphasised by broad leaved planting
- Legal boundary natural 'tailing off'

FORESTRY IN THE HILLS

PHYSICAL CONTROLS ON OPTIONS FOR PLANTING IN NORTH AND MID-WALES

ALTITUDE	200–400 METRES										400–600 METRES										MORE THAN 600 METRES									
							DEEP PEATS										DEEP PEATS										DEEP PEATS			
SOIL	Brown Earths		Peaty podsols		Gleys		Flushed peats		Unflushed peats		Brown Earths		Peaty podsols		Gleys		Flushed peats		Unflushed peats		Brown Earths		Peaty podsols		Gleys		Flushed peats		Unflushed peats	
	S	E	S	E	S	E	S	E	S	E	S	E	S	E	S	E	S	E	S	E	S	E	S	E	S	E	S	E	S	E
Norway Spruce	1	2	1	3	1	2	1	2	3	–	1	3	3	–	2	3	3	–	–	–	–	–	–	–	–	–	–	–	–	–
Sitka Spruce	1	1	1	1	1	1	1	1	2	2	1	2	1	2	1	2	1	2	2	3	3	3	3	3	3	–	3	–	–	–
Scots Pine	1	1	1	2	1	2	1	2	2	–	1	3	2	3	2	–	3	–	–	–	3	–	3	–	–	–	–	–	–	–
Corsican Pine	1	2	1	3	1	–	3	–	–	–	3	–	3	–	–	–	–	–	–	–	–	–	–	–	–	–	–	–	–	–
Lodgepole Pine	1	1	1	1	1	1	1	1	1	1	1	2	1	2	1	2	1	2	1	2	3	–	3	–	3	–	3	–	3	–
Japanese Larch	1	1	1	2	1	2	1	2	2	–	1	2	2	3	2	3	3	–	–	–	3	–	–	–	–	–	–	–	–	–
European Larch	1	2	3	–	1	3	3	–	–	–	1	3	3	–	3	–	–	–	–	–	–	–	–	–	–	–	–	–	–	–
Other conifers	2	–	2	3	2	2	2	2	3	–	–	–	3	–	3	–	3	–	–	–	–	–	–	–	–	–	–	–	–	–
Oak	1	3	3	–	1	3	3	–	–	–	3	–	–	–	–	–	–	–	–	–	–	–	–	–	–	–	–	–	–	–
Birch	1	2	1	3	1	2	1	3	3	–	2	3	3	–	2	–	3	–	–	–	3	–	–	–	–	–	–	–	–	–
Alder	1	3	1	3	1	2	1	3	3	–	3	–	3	–	2	–	3	–	–	–	–	–	–	–	–	–	–	–	–	–
Other broadleaves	2	3	3	–	3	–	3	–	–	–	3	–	–	–	–	–	–	–	–	–	–	–	–	–	–	–	–	–	–	–

S = Sheltered. E = Exposed.
1 = strong probability of producing crops more than 10 metres high.
2 = probability of producing such crops.
3 = possibility of producing such crops.
– = probability unproductive.

Fig. 4.3.8

Distribution of Soil Types on Hillside in Mid-Wales

Deep Peats | Gleyed and Podsolised Soils | Peaty Podsols | Leached Brown Earths | Brown Earths
600m

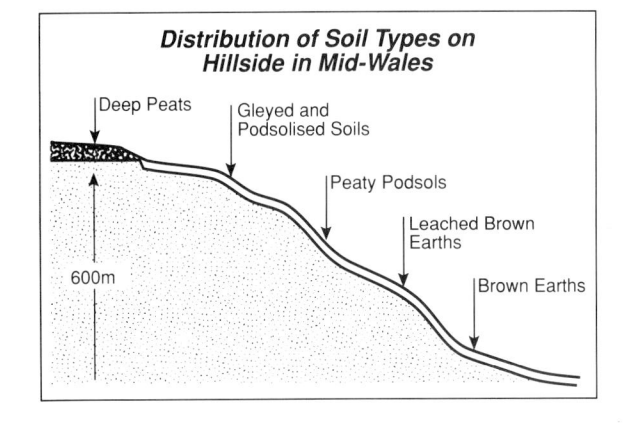

Fig. 4.3.9

VULNERABILITY OF TREE SPECIES TO GRAZING ANIMALS

Very vulnerable*	lime, elm, birch, willow, alder
Fairly vulnerable	sycamore, silver fir, oak, rowan, ash, beech, red cedar
Not so vulnerable	pines, holly, hawthorn, larch, yew, hemlock, Norway spruce
Least vulnerable	sitka spruce

*i.e. most palatable

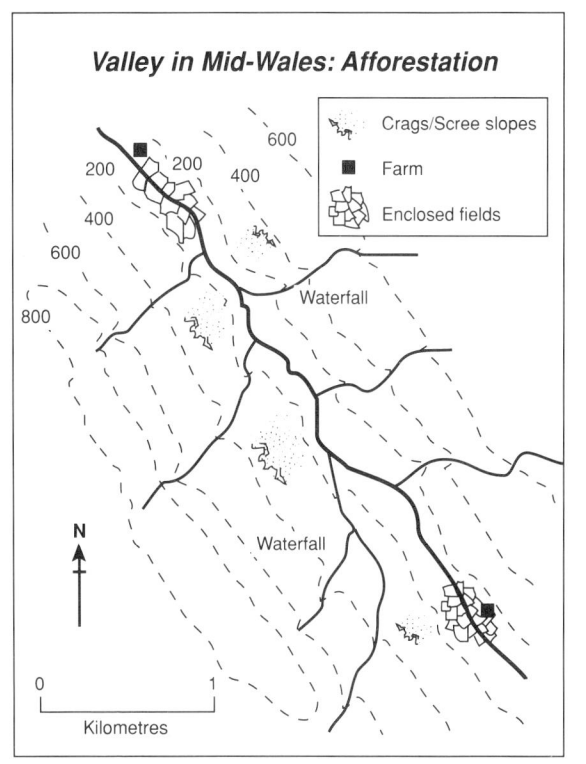

Fig. 4.3.10

Valley in Mid-Wales: Afforestation

Legend: Crags/Scree slopes, Farm, Enclosed fields

Fig. 4.3.11

You have to submit plans for the afforestation of a section of valley approximately 3 km long. Details of the valley are shown on the sketch map (Fig. 4.3.11).

1 Outline the main priorities that will govern your approach to landscape planning in this valley.

2 What are the main physical and environmental constraints that will have to be taken into account when planning your allocation of different areas to different types of trees?

3 Draw a block diagram showing your plans for treeplanting and landscape design in this valley section.

A NEW VIEW OF BRITAIN

Forestry in the Flow Country: a threatened wilderness

The Flow Country is our largest and possibly last primeval landscape; but, despite its vast size, it is extremely fragile and vulnerable to change. From the ground, the area looks flat and featureless, but once you have gained even a few feet, you can begin to appreciate the beauty of the place – vast skies, scattered pools reflecting the light, the silence broken only by the distant rippling call of the curlew, or the eerie wail of the red-throated diver. There is no doubt that the Flow Country is a magical place, which we are lucky to have on our doorstep.

(Michael Scott, Paradise Ploughed. Anglia Television)

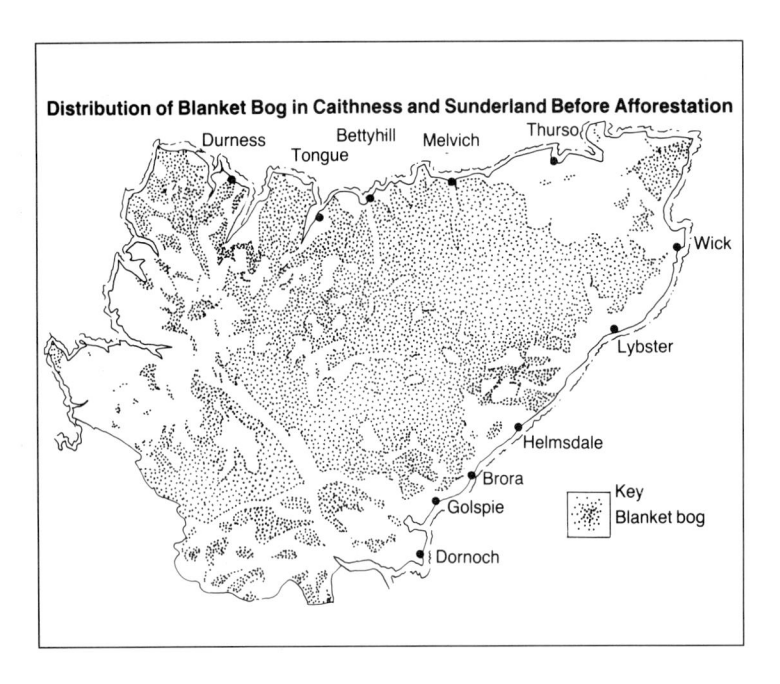

Fig. 4.3.12

The map (Fig. 4.3.12) shows the extent of the Flow Country of Caithness and Sutherland. Treeless tracts of bare peat-covered moorland, broken by the occasional distant hills, extend over an area of some 400 000 hectares in the counties of Caithness and Sutherland. This blanket bog is patterned with pools and small peat-stained pools – the dubh lochans, and is regarded as one of the last true wildernesses in Europe. Similar expanses of such a rare habitat are found only in Ireland, Norway and Iceland, but in these locations it is of much more limited extent. It has been recognised by international experts as 'unique and of global importance, equivalent to the African Serengeti, or Brazil's tropical rainforest.' Its tundra-like moorland serves as a breeding ground for distinctive and rare Arctic birds. Britain thus has an international responsibility to protect birds such as the dunlin, greenshank and the black-throated diver and their habitat.

The Flow Country has increasingly attracted forestry interests over the last 15 years. The Forestry Commission owns some 21 500 hectares in the Flows, but has greatly reduced its planting in recent years. On the other hand a private forestry company, Fountain Forestry began to purchase land in the late 1970s, and now holds over 40 000 hectares. Wealthy investors are encouraged to purchase blocks of this land, and then pay the forestry company to plant trees on the acquired land. Although the tax incentives for investors have now ceased after the 1988 Budget, investment in commercial forestry still continues. Fountain Forestry claims to have created some 200 jobs in Caithness and Sutherland, but it is estimated that this is at an annual cost of £37 000 per job. It is argued by some that this money could create more jobs in the area if it were spent in other ways. Some estimates put the loss of land to forestry at 50 hectares per week since 1980.

The landscape issue in the Flow Country

Study the two aerial photographs (Fig. 4.3.13) which show two locations within the Flow Country, the untouched Dubh Lochs of Shielton and a section that has been afforested. Use the bi-polar semantic framework to evaluate the landcape contrasts between the two areas.

+						−
3	2	1	0	1	2	3
Beautiful						*Ugly*
Interesting						*Dull*
Varied						*Monotonous*
Flowing						*Regimented*
Spectacular						*Ordinary*
Stimulating						*Depressing*
Undeveloped						*Spoilt*
Pleasing						*Harsh*
Welcoming						*Hostile*
Serene						*Disturbed*

Comment on the results of your assessment. What criticisms of the form of landscape assessment can you offer?

Fig. 4.3.14
Sphagnum moss

Fig. 4.3.15

The Peatland Ecosystem

Fig. 4.3.13
Two aerial views of the Flow Country

The habitat and wildlife issue in the Flow Country

The peatland ecosystem

The peatlands are an ecosystem that has developed over 7000 years. In the lowest horizons of the many accumulated layers of the peat, evidence of the former cover of birch and scattered pines can be found, in the form of pollen grains and the remains of tree stumps. With the advent of wetter conditions, which favoured the growth of the sphagnum mosses (see Fig. 4.3.14), successive layers of peat have been built up from the dead mosses, and the other plants of the bogland community. The functioning of the peatland ecosystem is shown in the diagram, (Fig. 4.3.15). Sphagnum mosses vary in the degree of moisture they require, therefore the surface layers of the ecosystem tend to develop something of a hummocky appearance, with the sphagnum varieties that can tolerate least water occupying the highest parts of the hummock. The insectivorous sundews are another important species that grow in the peatlands, depending on the insects that they have trapped for their supply of nutrients. The diagram (Fig. 4.3.16) shows a cross-section through one of the hummocks. The dubh lochans (black pools) are an essential component of the peatland environment. They result either from the development of hollows between the peat hummocks, or from the fracturing of the peat surface, and their subsequent formation.

1   Using the diagrams and the text write a concise summary of the factors controlling the formation and functioning of the peatland ecosystem.

2   Comment on the changes that appear to have taken place in the peatlands as a result of drainage (a pre-requirement of the preparation of the land for planting conifers) as shown in the diagram (Fig. 4.3.17).

A NEW VIEW OF BRITAIN

Fig. 4.3.16

Cross-Section: Peat Hummock – Dubh Lochan

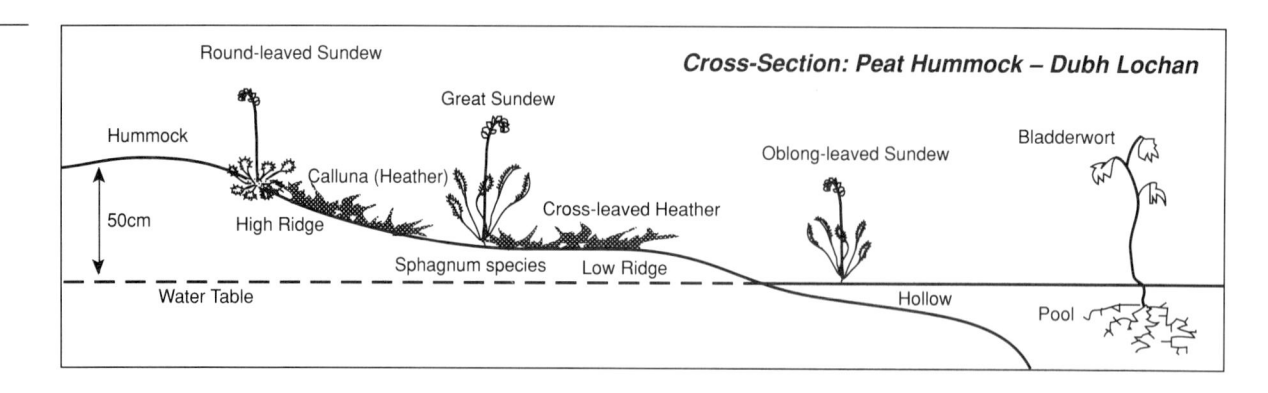

Fig. 4.3.17

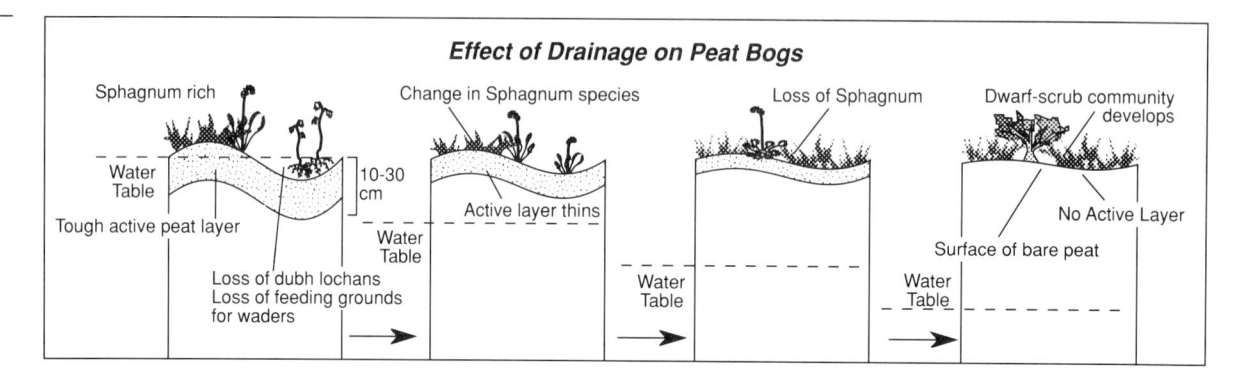

Effect of Drainage on Peat Bogs

The birdlife of the peatlands

Fig. 4.3.18
(a) Golden plover
(b) Dunlin
(c) Greenshank
(d) Black-throated diver

(a)

(b)

(c)

(d)

The bird fauna (Fig. 4.3.18) of the peatlands is regarded as being of outstanding international interest. The moorlands are noted not only for their variety of bird life, but also, because of their extent, for the very large numbers of breeding populations. Waders, waterfowl, and a range of birds of prey are particularly important in the peatlands, and the arctic affinities of many of the species give the Flow Country its distinctive ornithological character.

Forestry and bird habitat in the Flow Country

Afforestation of the Flow Country will adversely affect many of the bird species that use the peatlands as breeding grounds. Birds such as the dunlin, the greenshank and the black-throated diver depend on the semi-aquatic environment of the dubh lochans for feeding, and the surrounding blanket bog for breeding. The larger birds of prey, such as the golden eagle nest on the higher craggy hills, but depend on the peatlands for feeding on the moorland birds (Fig. 4.3.19b). The changes brought about in peatland vegetation after afforestation are shown in diagram (Fig. 4.3.19a) and the newly established coniferous ecosystem is shown in the diagram (Fig. 4.3.20).

1 How are the changes in the peatland vegetation likely to affect their bird populations?

2 Contrast the peatland ecosystem with the coniferous forest ecosystem. Why is the latter regarded as being poorer ecologically?

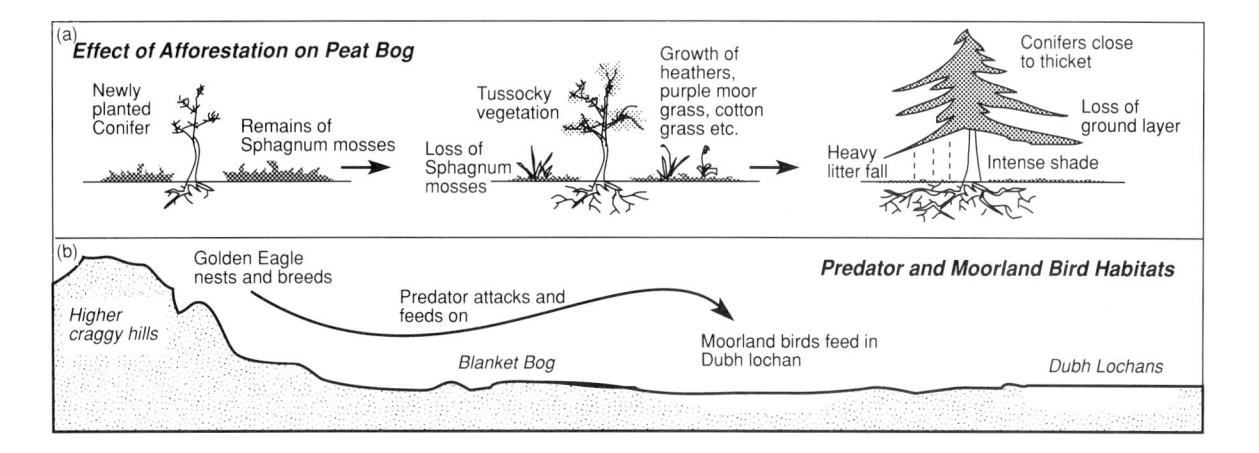

Fig. 4.3.20

CONIFEROUS WOODLAND ECOSYSTEM

Fig. 4.3.19a and b

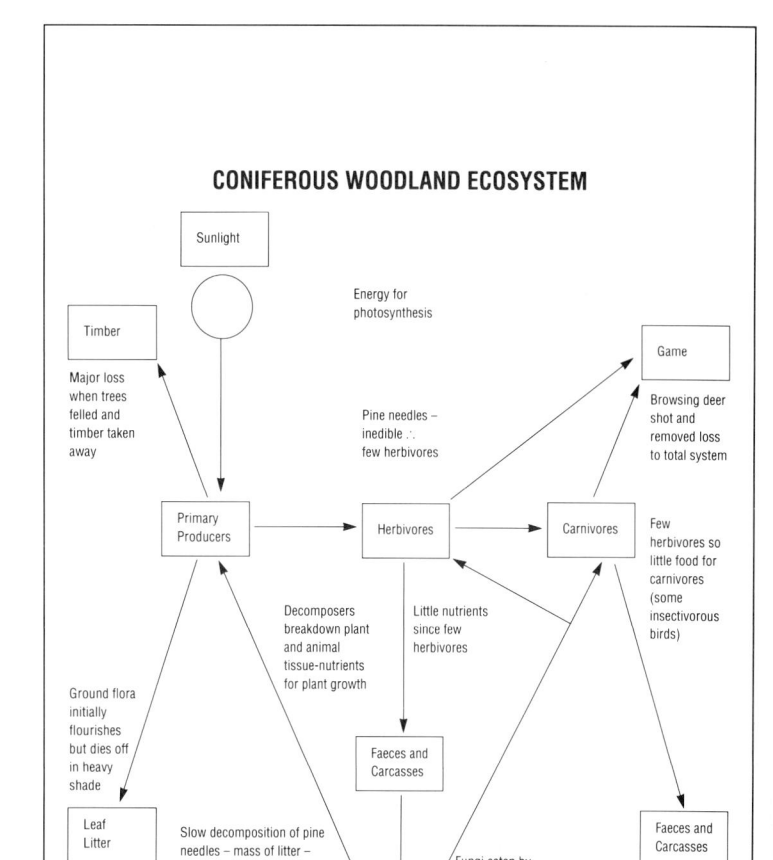

A NEW VIEW OF BRITAIN

Environment Essential to the Maintenance of the Birdlife of the Moorlands

Extent of provisional key peatland areas in Caithness and Sutherland including associated hydrological units and 1 km buffer zones
Key: Provisional key peatlands

Key unafforested hydrological systems in Caithness and Sutherland
The shaded areas represents 23 individual systems which had less than 6% cover of conifer plantations in January 1986
Key: Key hydrological systems

Extent of high quality peatland wader habitat in Caithness and Sutherland including associated 1 km buffer zones
Key: High quality wader habitat

River catchments (including loch systems) of exceptional importance for black-throated divers
Key: River catchments

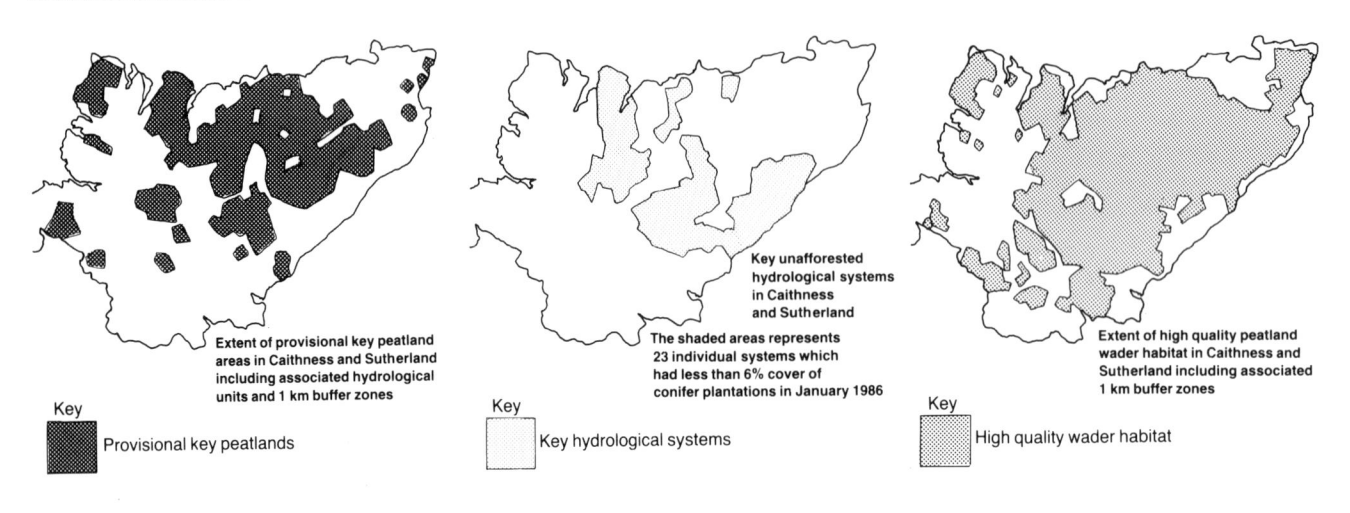

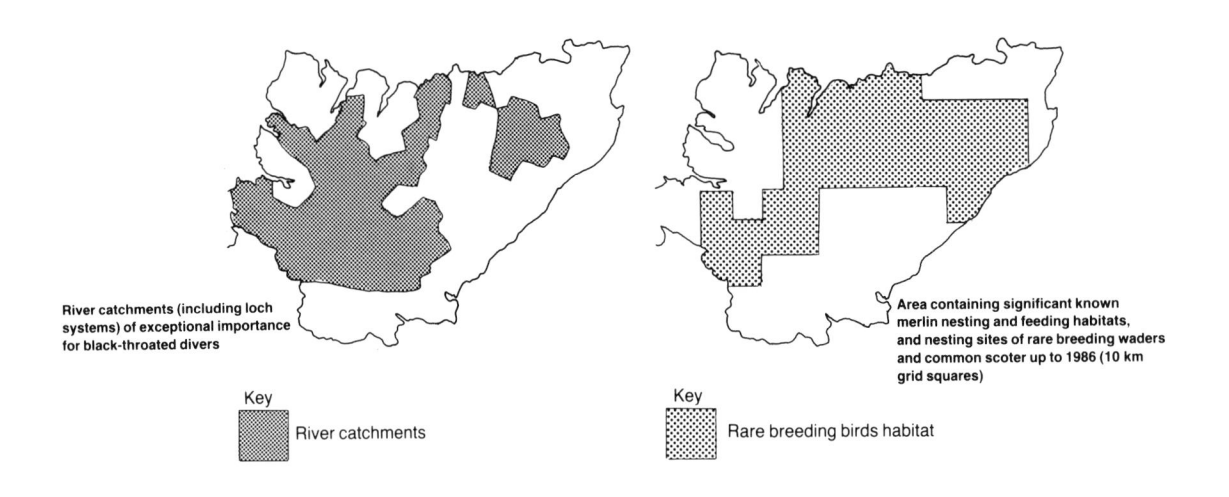

Area containing significant known merlin nesting and feeding habitats, and nesting sites of rare breeding waders and common scoter up to 1986 (10 km grid squares)
Key: Rare breeding birds habitat

Fig. 4.3.21

The Distribution of Plantations on Blanket Bog

Key:
- Blanket bog
- Plantations on blanket bog
- Plantations on other soils

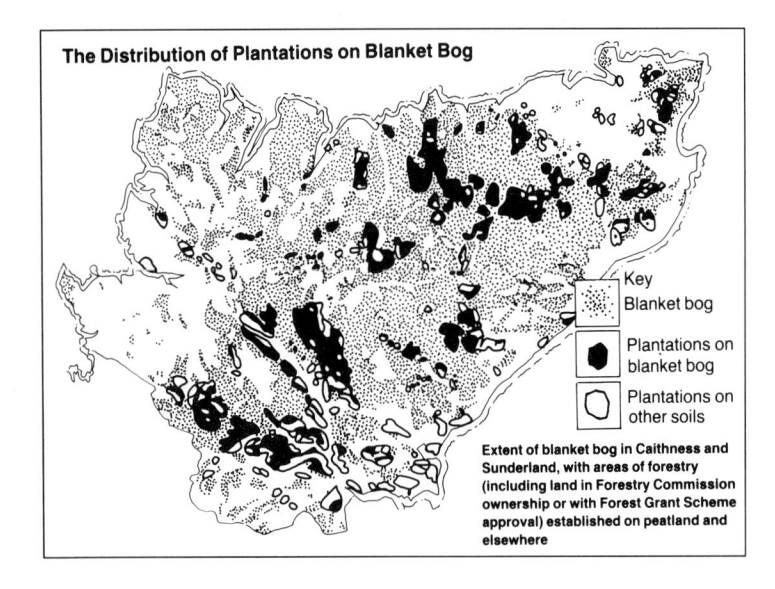

Extent of blanket bog in Caithness and Sunderland, with areas of forestry (including land in Forestry Commission ownership or with Forest Grant Scheme approval) established on peatland and elsewhere

Fig. 4.3.22

The five maps (Fig. 4.3.21) show key environments essential to the maintenance of the birdlife of the moorlands.

1 Use a sieving technique to produce a map showing the areas that are most at risk from afforestation i.e. those that combine the largest number of the key environments.

2 Now compare the map that you have produced with the map (Fig. 4.3.22) which shows the distribution of plantations on blanket bog. What conclusions can be drawn concerning the selection of these areas for afforestation from the point of view of bird ecology?

The future of the Flow Country

'My client has been advised that it is much better for the environment of northern Scotland that trees should be planted than it should be left as open moorland.'

Financial Consultant

'Broadly speaking, afforestation replaces the threatened and vulnerable with the commonplace and adaptable.'

(Royal Society for the Protection of Birds)

By the late 1980s 17 per cent of the original peatland area had been planted or programmed for planting. Most of the main peat-dominated river catchments in Caithness and Sutherland now contain some plantation forest. Actual or predictable losses of breeding birds as a direct effect of afforestation are:

- 19 per cent of golden plovers
- 17 per cent of dunlins
- 17 per cent of greenshanks.
- Similar levels of loss are thought to apply to other species.

The peatlands may be regarded of international value for the following reasons.

- One of the largest and most intact known areas of blanket bogs.
- A northern tundra-type ecosystem in a southerly geographical and climatic location.
- Development of unusually diverse systems of patterned surfaces on blanket bog.
- A floristic composition of blanket bog and associated wet heath vegetation unique in the world.
- A tundra-type breeding bird assemblage showing general similarity to, but specific differences from that occurring on arctic-subarctic tundra.
- Significant fractions of the total populations of certain breeding bird species in Europe.

(The Flow Country, Scottish Nature)

This assessment by Scottish Nature needs to be set against the criteria established by the World Heritage Convention for sites that will qualify for 'natural heritage' status.

- Natural features consisting of physical and biological formations or groups of such formations, which are of outstanding universal value from the aesthetic or scientific point of view.
- Geological and physiographical formations and precisely delineated areas which constitute the habitat of threatened species of animals and plants of outstanding universal value from their point of view of science, conservation or natural beauty.

(Natural Heritage Criteria, World Heritage)

1   Using the World Heritage Criteria, summarise the case for the award of World Heritage status to the Flow Country.

2   Outline the case for a limited extension of commercial forestry in the Flow Country.

MINERAL EXTRACTION IN THE UPLANDS

He told the Lee Moor enquiry at Exeter yesterday that his company was conscious that there was a clash of interests over Dartmoor and the minerals lying under it. But both the minerals and the open space were needed, and the development at Lee Moor was a coupling of modern methods of working with a sympathetic landscape treatment, which would produce an ever-diminishing scale of exposure.

(Alan Dalton, Managing Director of English Clays, Lovering and Pochin & Co. Ltd., quoted in John Blunden, Mineral Resources and their Management)

(a)

The hey day of Lakeland Mining was past; a new factor was beginning to appear and it had nothing to do with the richness or the poorness of the veins or the prices of the minerals. This was the effect of the mining works on the scenery.

(W.T. Shaw, Mining in The Lake Counries)

(b)

In the final analysis, and in the real perspective, there is only enough copper in Capel Hermon to satisfy current world demand for that mineral for four weeks; there are other qualities in Capel Hermon and places like it, which are more difficult to quantify but of no less value. The choice between them must be made. Is Britain rich enough to sell Snowdonia? Are we poor enough to need to?

(Graham Searle, Copper in the Snowdonia National Park in Peter J. Smith, Politics of Physical Resources)

(c)

Fig. 4.4.1
(a) China clay pit: Lee Moor, south-west Dartmoor
(b) Old Lakeland Mine
(c) Capel Hermon; landscape threatened in the early 1970s through Rio Tinto Zinc's proposals for extensive copper mining in Snowdonia

The three quotes refer to mineral working in three different areas of the British Uplands. Lee Moor in south-west Dartmoor has long been important for the extraction of china clay from the large quarries in the altered granite that yields this valuable mineral. Mining in the Lake District has a very long history, going back to Roman times. There have

MINERAL EXTRACTION IN THE UPLANDS

been few large scale workings, but many valleys are scarred with spoil heaps and other mining debris. In the early 1970s Rio Tinto Zinc, the international mining company, was granted permission to carry out exploration drilling to prove the extent of copper ores around Capel Hermon in the south of the Snowdonia National Park. Although no permission for mining the ores was given, a large scale open cast pit in the National Park was a real possibility. Adverse publicity was one factor that caused Rio Tinto Zinc to abandon its proposals in 1973.

Using the quotes and text suggest the main issues likely to be raised by proposals to work minerals in the British uplands.

LIMESTONE QUARRYING IN THE PEAK DISTRICT

Limestone is used mainly for roadstone, concrete aggregate and in the manufacture of cement. High purity limestone is used in the manufacture of sodium carbonate, in the manufacture of glass, and as a metallurgical flux. Other specialised uses are in water treatment and in the ground limestone powders used in paints and rubbers. Carboniferous Limestone is one of the most valuable sources of this mineral, and its distribution is shown on the map (Fig. 4.4.2). This should be compared with a map of the National Parks of England and Wales (Fig. 4.4.3).

1   Trace the outlines of the National Parks to use as an overlay on the map of the Carboniferous Limestone outcrops.

2   In which National Parks is limestone quarrying likely to pose the biggest threat?

The two maps (Figs. 4.4.4 and 4.4.5) show the geology of the Peak District and the distribution of the major quarries within, and close to the Peak District National Park.

Comment on the distribution of quarries in relation to

a)   the outcrop of the Carboniferous Limestone and

b)   the boundary of the Peak District National Park.

Fig. 4.4.2

Fig. 4.4.3

Fig. 4.4.4

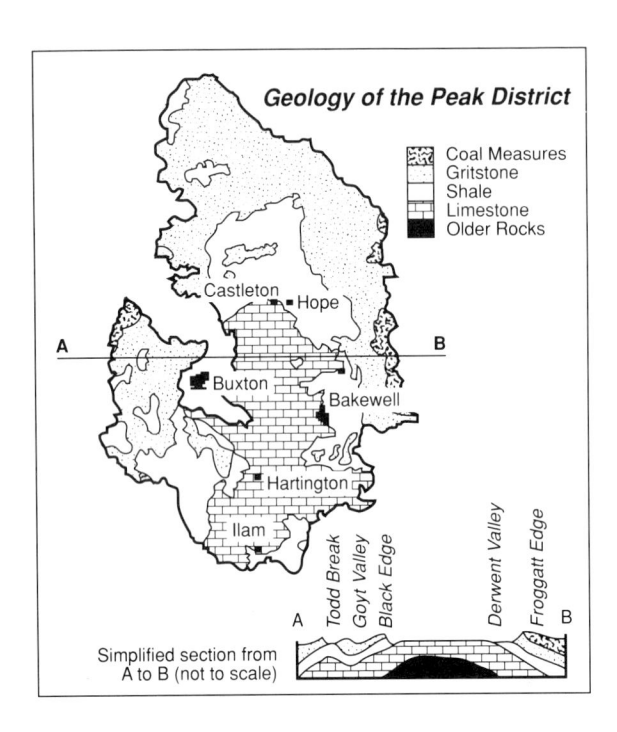

Geology of the Peak District

Fig. 4.4.5

Mineral Workings Peak District National Park

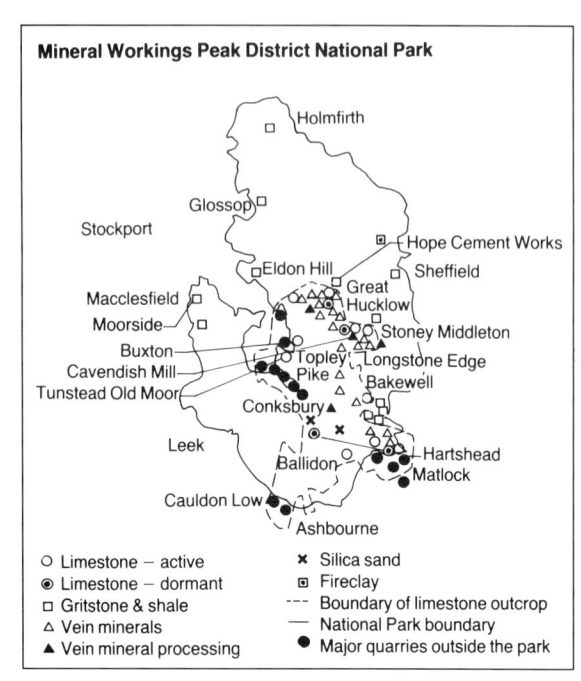

Fig. 4.4.6

TABLE 1 LONG TERM TRENDS (MILLION TONNES PER ANNUM)

1951	1961	1971	1981	1990
1.5	2.2	5.4	4.0	8.5

TABLE 2 SHORT TERM TRENDS (MILLION TONNES PER ANNUM)

	1985	1986	1987	1988	1989	1990
Aggregate	2.4	3.1	3.5	4.4	4.1	4.5
Non-aggregate	1.6	1.9	2.8	3.1	3.3	4.0
Total	4.0	5.0	6.3	7.5	7.4	8.5
Permitted reserves	385	373	373	362	350	340

TABLE 3 LIMESTONE QUARRIED IN THE PEAK DISTRICT NATIONAL PARK, 1989

CONSUMER	PER CENT
Aggregate (roadstone)	56
Cement	23
Chemicals	17
Iron and Steel	4
Agriculture	0.2

The tables (Fig. 4.4.6) shows the output of limestone in the Peak District.

1   Comment on the long and short term trends in limestone output revealed in the tables (you should consider relative as well as absolute growth).

2   What implications do they have for the future of the industry?

Planning and permission for limestone quarrying in the Peak District

A browse through the last two years' issues of Mineral Planning *will leave the reader in no doubt that mineral extraction in National Parks has become an increasingly important national issue.*

(Mineral Planning December 1987)

For many years applications for mineral working were subjected to the Silkin Test, first set out in 1949. In this examination all proposals were to be subjected to rigorous scrutiny by reference to four criteria:

MINERAL EXTRACTION IN THE UPLANDS

- need to conserve environment of Park and extent to which the proposal would be damaging
- effect of traffic on safety and character of Park
- need nationally and locally for the minerals to be worked
- lack of practical alternative sources.

The Silkin Test has undergone a series of modifications, and the present policy in the Peak District may be summarised as: Proposals for mineral extraction, mineral processing, waste disposal, or ancillary activity, will not be approved unless the applicant has demonstrated that the proposals are justified in the national interest. In assessing whether a proposal is justified the Board will consider whether:

a) the production or disposal capacity is essential to meet a national or regional need which overrides the need to protect the national Park

b) any alternative source site or means of disposal is available.

Permission will not normally be granted for proposals which would either:

(i) form a significant landscape intrusion;

or (ii) have a detrimental impact on the wildlife, cultural heritage, geological or archaeological interest or the character of the Park;

or (iii) cause significant nuisance or disturbance to local residents;

or (iv) cause a hazard to traffic safety or generate excessive heavy traffic on unsuitable roads;

or (v) result in a site being left in an unsatisfactory state on cessation of operations;

or (vi) cause disruption to the network of public rights of way or severely reduce public rights of access to 'open country';

or (vii) result in excessive amounts of limestone being removed from vein mineral workings.

Fig. 4.4.6a
Peak Park criteria for refusing permission

TOPLEY PIKE AND DARLTON QUARRIES

Both Topley Pike and Darlton Quarries lie within the Peak District National Park: their locations are shown on the map (Fig. 4.4.7). Both submitted applications in the 1980s for extensions to the area being quarried, one was approved and the other refused. The background to the two quarries is fully set out below, together with the details of the proposed extensions.

Fig. 4.4.7

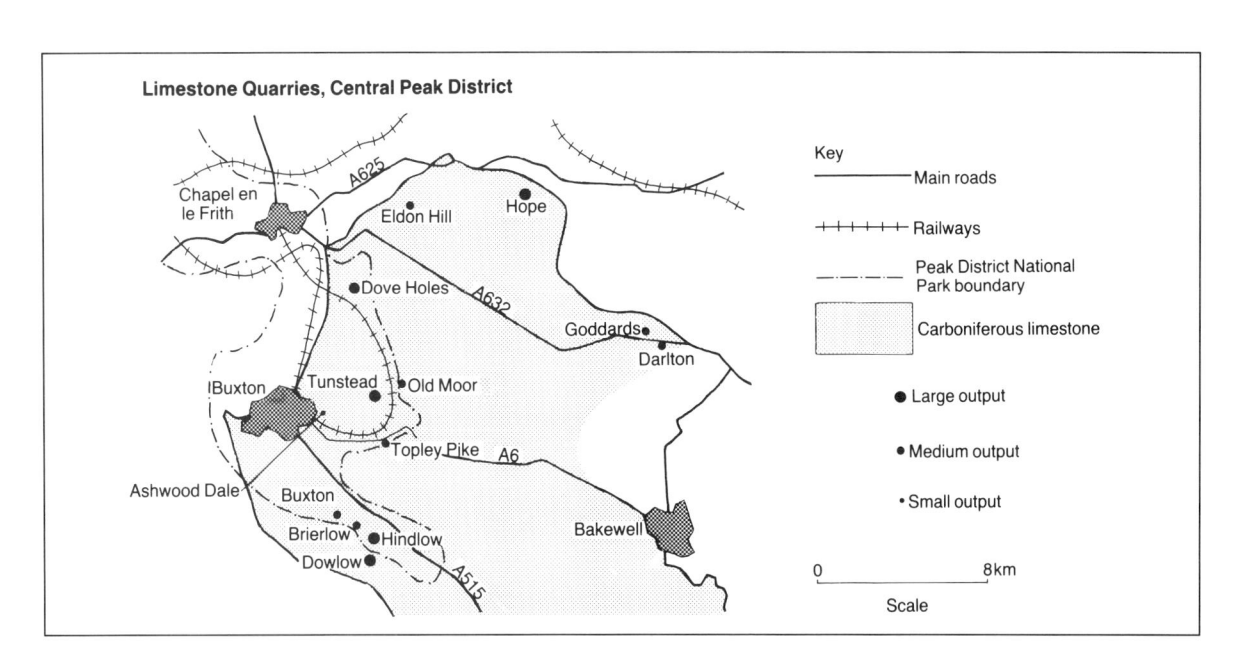

Topley Pike Quarry

The details of the location of the quarry are shown on the map (Fig. 4.4.9) and in the photograph (Fig. 4.4.8).

Location: Topley Pike lies just inside the Peak District National Park, bounded to the north by the Wye Valley, and to the south by Deep Dale, a SSSI, with the hamlet of King Sterndale just to the west.

Area: the present quarry, plant and tipping areas occupy 31 hectares.

Current Output: 500 000–750 000 tonnes per annum for coated roadstone or graded dry materials.

Market: main market is in Greater Manchester, Merseyside, Cheshire and Derbyshire.

Employment: 49 people, with another 60 contract lorry drivers dependent on the quarry.

Fig. 4.4.8 *Topley Pike Quarry, Peak District*

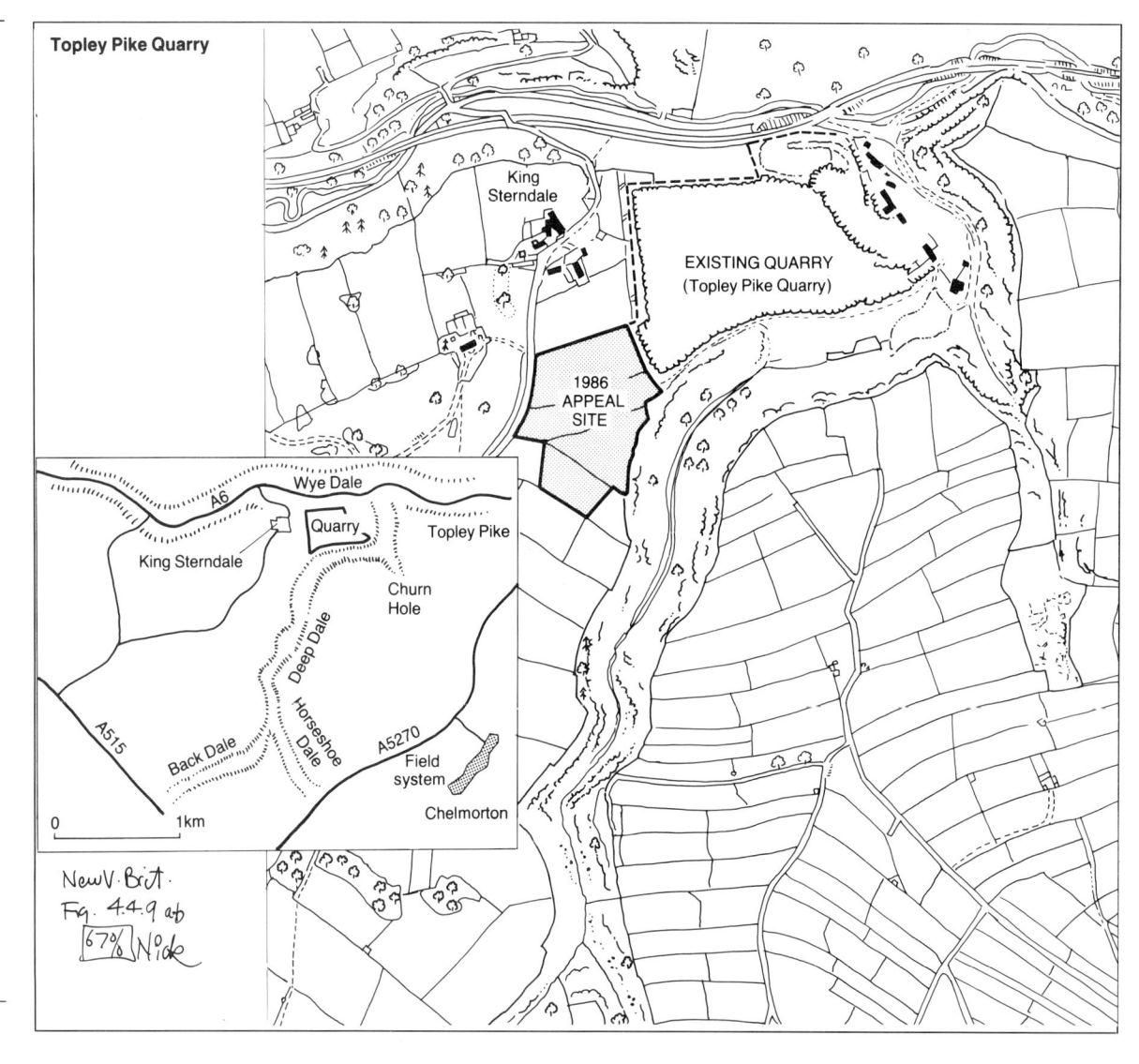

Fig. 4.4.9

Proposal: to extend the quarry workings to the area shown on the map (Fig. 4.4.9). Some 8 million tonnes of high grade limestone would be extracted from the site, together with some 6.2 million tonnes from the existing site. With an extraction rate of some 750 000 tonnes per annum, the life of the extended quarry would be 16 years.

Environmental Safeguards

- Prior to commencement of quarrying in the extension, Tarmac would plant screens of trees to ensure that properties in King Sterndale would have no sight at ground level of the workings.
- New dry stone walls and bankings of soil and waste would obscure the quarry workings from local footpaths and the King Sterndale Road.
- The floor of the new quarry would provide a place for new silt ponds to replace those in Deep Dale. The latter would be regenerated into a more visually attractive state.
- Blasting within 150 metres of King Sterndale Road would necessitate its closure for two to three minutes up to 60 or 70 times a year.
- At the end of its working life Tarmac would undertake ameliorative work to create an artificial limestone dale. Market Assessment by Tarmac: important source of aggregate for the north-west. Topley Pike has one of the two rapid-load rail facilities in Derbyshire. Alternative sources of supply would lead to increased transport costs. Alternative sources from within the Peak Park would lead to additional lorry traffic in the Park.

Assessment by the Peak Park Planning Board

Market: stone at Topley Pike is of high chemical purity and its marketing as an aggregate does not make use of its unique qualities. Remaining reserves of high grade limestone at Tunstead (outside of National Park) are adequate for industrial end uses. Other sources outside of National Parks are adequate for aggregates.

Environment: the proposed extension would form a major intrusion into the views to the east from King Sterndale because of the quarry itself, the new screening, quarry warning signs and diverted power lines. The present quarry has major visual impact along the line of the Wye valley, and further disfigurement would result from the extension of the quarry. The extension would damage the views from the lanes and paths through the Chelmorton Field System (see Fig. 4.4.8). Noise levels in King Sterndale would rise above present ones, but would be well below the recommended 65db. Dust from the quarry extension could possibly affect pH values in the soils in the Deep Dale SSSI, with a possible deterioration of the nature and variety within the ecosystem there.

Darlton Quarry

The location of the quarry is shown on the map (Fig. 4.4.10) and the photograph (Fig. 4.4.11).

Location: the quarry lies on the south side of the A623 in Stoney Middleton Dale, about 1.5 km west of Stoney Middleton village and about 1 km south of Eyam village.

Area: the approximate area of the existing quarry, with plant and tippings is 21 hectares.

Current output: 650 000 tonnes per annum for coated roadstone, concrete blocks, and dry graded materials.

Market: market is mainly in South Yorkshire and east Midlands, and in times of high output, the north-west.

Employment: 90 employed within the quarry, with another 50 employed as lorry drivers.

Proposal:

- to extend the face of Eyam Quarry south-eastwards and extend quarry extraction in that direction by 6.2 hectares.
- to construct new crushing facilities in the existing Darlton Quarry. The existing crusher would be dismantled. The new crusher would be on the lowest floor level and thus completely screened from the Eyam hillside. This is regarded as a substantial 'planning gain.'

Environmental Safeguards:

- topmost bench of the existing quarry to be treated in a roll-over style, and the resultant slope to be resoiled, grassed and returned to grazing within four years of the start of the project.
- to reshape the fields south of the quarry, near Moisty Lane, to create low mounds to help screen the extension.
- to carry out extensive tree planting and general landscaping especially in a block indicated on the map.

Fig. 4.4.10

Darlton Quarry

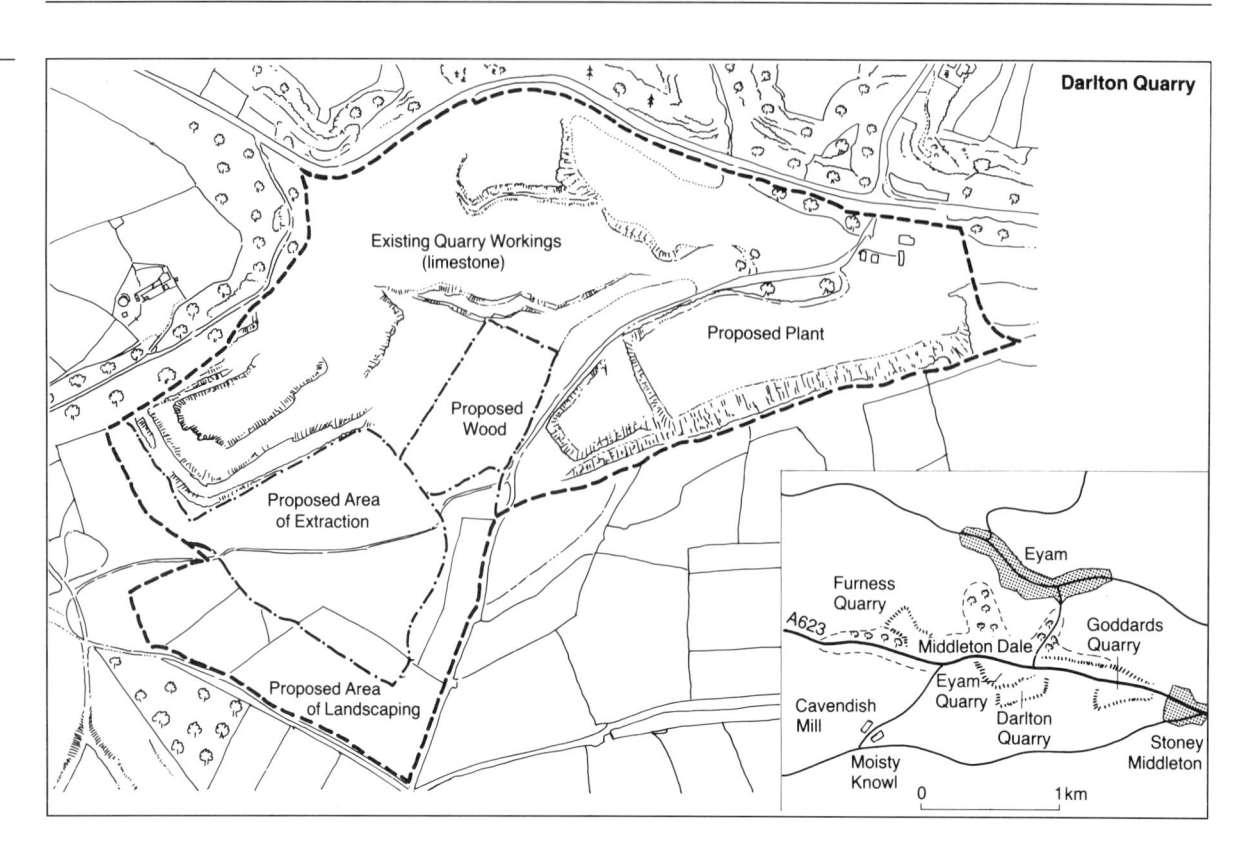

Fig. 4.4.11 Darlton Quarry, Peak District

Market Assessment by Darlton Quarry: important source of roadstone for South Yorkshire and east Midlands and when demand is high, for the northwest. Intended increase in production to 1 million tonnes.

Assessment by the Peak Park Planning Board

Market Assessment: limestone reserves in the National Park suitable for the South Yorkshire aggregates market are confined to the Stoney Middleton area. There is likely to be increased pressure on this area in relation to the forecast aggregate demand. There is no national need for an increase in supply of this type of aggregate stone, but there is a local case for the maintenance of aggregate to the South Yorkshire market, especially for coated ashphalt. The facing blocks, 'Darlstone' produced in the quarry uses stone that could have been used for low-grade products, or have gone to waste.

Environment:

- the quarry and its extension would be visible for a considerable area to the north of Middleton Dale (see Fig. 4.4.10). Landscaping will screen the quarry from many viewpoints, and the crusher and related new plant will be sited in a well-screened position.

- Noise: main noise impacts are to certain residents in Eyam, across the valley and downwind of the quarry. Levels were of the order of 40–48 dB. It is likely that the noise contours will move south with the extension of the quarry; the new position of the crusher should further reduce the noise levels.

- Appropriate measures to be taken to prevent sub-water table working, water pollution and deterioration in quality of groundwater and local springs.

Traffic: an important issue because of the effect of lorry traffic on villages, particularly Stoney Middleton, but also on other villages on the A623 in both directions. There is, however, a certainty of an upper limit on the volume of road traffic because of the maximum output figure (750 000 tonnes per annum).

1 Use the Environmental Impact Matrix (Fig. 4.4.12) to assess the likely impact of the two quarries.

2 Apply the Silkin Test, OR the Peak Park criteria to both of the quarries in turn. Discuss how each fared under such scrutiny.

3 Write a short summary report on the viability of the two proposals and make a recommendation in each case. Remember one of the proposals failed, and one succeeded. You have to decide for yourself.

ENVIRONMENTAL IMPACT ASSESSMENT MATRIX

Existing Situation \ Proposed Development	Site Use	Labour Requirements	Transport of Equipment	Transport of Products	Transport of Employees	Noise of Equipment	Noise of Blasting	Dust from Working	Vibration from Plant	Water Discharge	Groundwater Change	Transport Raw Materials (Coatings)
Landscape												
Land Use												
Flora												
Fauna												
Soil												
Air Pollution												
Water Pollution												
Dust												
Residents												
Tourists												
Traffic												
Road Surface												
Power Supply												
Sewerage												
Employment												
Housing												
Services												

COASTAL SUPERQUARRIES FOR SCOTLAND

This showed that very considerable stone reserves were available as alternative supplies, especially in the other quarries near Buxton. For the longer term, the mammoth coastal quarry at Glensanda in Scotland is a possible alternative source of seaborne supplies of granite into the Mersey.

(Howard White, Senior Minerals Planning Officer for the Peak District, on the dismissal of the appeal against refusal to grant permission for the Eildon Hill Quarry extension in the Peak District.)

The quotation provides a very suitable link between the two case studies in this section. Glensanda, on the shores of Loch Linnhe, is one of a possible new generation of coastal superquarries on the west coast of Scotland. The extract from the article 'Mission to move Mountains' (Fig. 4.4.13) gives details on Glensanda, and discusses some of the issues involved. The map (Fig. 4.4.14), diagram (Fig. 4.4.15) and photograph (Fig. 4.4.16) give some idea of the scale of the operation.

Fig. 4.4.13
Guardian 13 March 1992

Mission to move mountains

**A new generation of superquarries is emerging in Scotland, destroying mountains and dividing communities.
Rob Edwards** investigates

Standing in a hard hat under 450 million tonnes of granite, Kurt Larson talks enthusiastically about moving mountains. Over the next 50 years, it is his job to break up and take away the 2000-foot [610 m] Scottish mountain of MacArthur that towers above his head.

"Our aim is to make it the largest quarry in the world," he says. He is splashing through the mud at the end of a tunnel that his company, Foster Yeoman, has blasted more than a mile into the heart of the mountain. The plan is to shift at least 10 million tones of rock a year to help build England's Europe's and even America's motorways.

Glensanda, on a remote peninsula north of Oban, is the first of a new generation of "superquarries" that are dividing communities, angering environmentalists and changing the face of Scotland. There are proposals for similarly massive operations at another five sites up the west coast which, if the quarry developers get their way, will turn the Highlands and Islands into stone suppliers to the world.

Once understood, there are two ways of viewing the prospect. Either it is a sensible, modern and environmentally friendly way of exploiting Scotland's mineral wealth to the benefit of local communities. Or it amounts to the desecration of the nation's landscape in pursuit of profits and in furtherance of a redundant transport policy. Either way, the implications are only just beginning to sink in.

Larson, a 33-year-old American-born director of Foster Yeoman, is slightly uncomfortable at the idea of removing Scotland's mountains to make England's roads. He prefers to talk about creating glens to ensure a higher standard of living. Excavation plans at Glensanda will leave a hole one and a quarter miles long, two-thirds of a mile wide and 270 yards deep. And that is only the beginning.

The quarry, which has already provided a lining for the Channel Tunnel and an underlay for the M20 between Folkestone and Maidstone, is situated within the largest mass of granite in the UK, with total reserves in the region of a billion tonnes. When Foster Yeoman has exhausted the 450 million tonnes it currently has planning permission to extract, there is every likelihood that there will be an application to double the size of the quarry.

It is the scale of everything that shocks at Glensanda. Five giant trucks, with tyres eight feet high and five feet wide, move around the mountaintop night and day shifting 75 tonnes of granite a trip. Two huge boats both 270 yards long, regularly sail down Loch Linnhe bearing 75 000 tonnes apiece to destinations in south-east England, Germany, Holland and (in the past) the Caribbean. As the quarry expands, even these behemoths are likely to double their carrying capacity.

"It is a real privilege to have a quarry in a place as beautiful as this, I love it," confesses Mr Larson. "The industry has recognised that there is a requirement for coastal superquarries and the government has come round to the idea. The industry now needs to convince people that it can carry out operations as sympathetically as possible."

Although when Glensanda was proposed in the early 1980s there was little opposition, the Highland regional councillor for the area, Dr Michael Foxley, now says that similar applications in the future should be more carefully and comprehensively scrutinised. He argues that the benefits that have accrued to the neighbouring communities are insufficient. The £880 000 which the company has been obliged to commit to the eventual reinstatement of the land is "grossly inadequate".

Some people certainly stand to make a great deal of money from superquarries, not least the mineral consultant who first suggested the idea, Ian Wilson. His companies have leased the mineral rights for Lingerbay on Harris and for at least three other possible superquarry sites: Kentallen on Loch Linnhe north of Glensanda;

on Loch Glencoul near Inchnadamph; and on Loch Eriboll near Durness. He has also identified the granite cliffs enclosing Ronas Voe on Shetland as a suitable site.

Although he dismisses a recent newspaper allegation that he stands to profit personally by some £25 million as "bloody nonsense", he agrees that his companies might receive that much revenue from royalties over the next 100 years. He says his main aim is to develop ways of maximising the benefits that superquarries could bring to the local communities. He sees them as catalysts for other developments, such as offshore wind turbines and other mineral workings.

According to his calculations the six superquarries could extract some 18 billion tonnes of Scottish rock over 200 years, with a total annual turnover of £360 million and a staff of 1200. By developing a Scottish integrated minerals strategy, he believes that crofting communities could be sustainably regenerated.

Superquarries are legitimised by government-sanctioned forecasts of demand for aggregates, despite a well-established tradition of wild inaccuracy. The current prediction is that demand will rise from 300 to over 500 million tonnes a year by 2010. It is difficult to define development on such a scale, which takes no account of environmental constraints, as sustainable.

"There is an awful symmetry about ripping up huge tracts of Scottish hillside to infill roads gouged through precious English countryside," comments Kevin Dunion, director of Friends of the Earth (Scotland) which is planning a major campaign against superquarries. "If we tried to draw up a programme of environmental vandalism we could hardly do better."

"Everywhere you go in Scotland, developers bludgeon environmental concerns with the promise of jobs. These people would have us keep quiet about the plague on the grounds that it provides jobs for undertakers."

The ultimate symmetry of superquarries is the purpose they could serve when their reserves are exhausted. It does not take a genius to realise that the profits to be made from disposing of domestic or industrial waste could tempt future owners into the refuse disposal business. When pressed Larson said it was an "option", while Wilson agreed it was a "theoretical possibility" at certain sites.

That is not to say that huge holes which were once Scotland's mountains are bound to end up being crammed with English and other foreign waste. But the very suggestion could easily provide a Scottish rebellion.

Fig. 4.4.14

Fig. 4.4.15 opposite

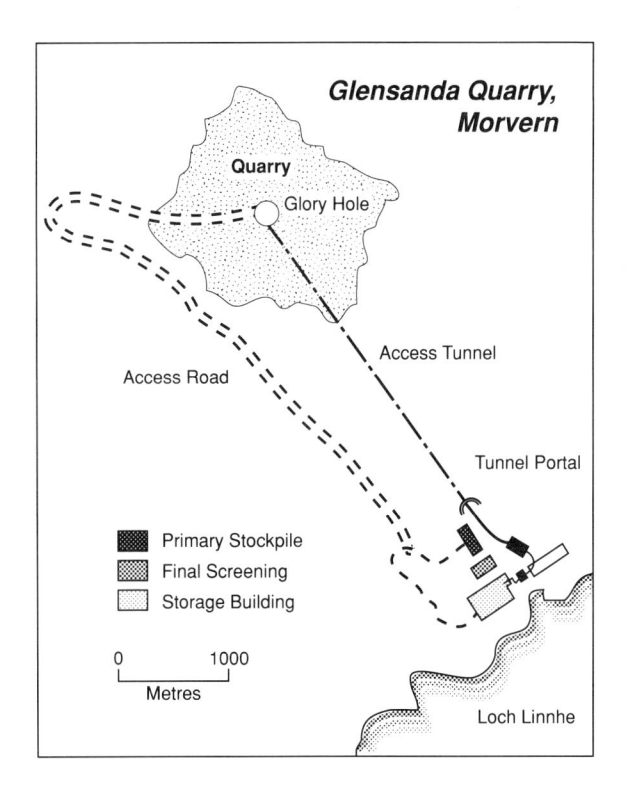

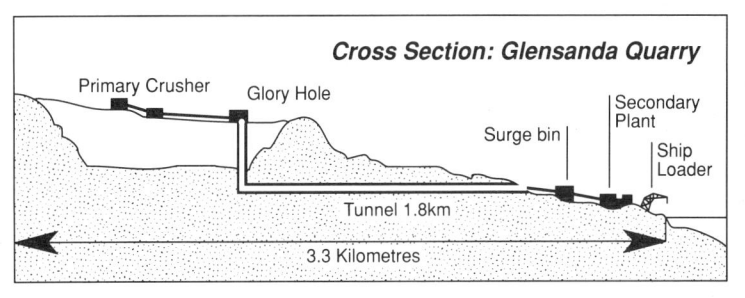

Fig. 4.4.16 Glensanda Quarry, the first of the coastal superquarries

1 What are the commercial advantages of developing coastal superquarries?

2 What are the environmental objections to these developments?

Increasing environmental and land-use pressure have created a dilemma for land use planners, local authorities and users of aggregates (all of us): with this ever-increasing demand for aggregates and ever-increasing conflict on the extraction of minerals in sensitive areas, where are tomorrow's aggregates going to come from? (We) ... recognise this aggregate supply dilemma in the early 1980s and set out to do something about it. We developed Glensanda with a vision to satisfy the needs of civil engineering contractors and help to solve the problems facing land use planners.

(Company brochure)

Fig. 4.4.17

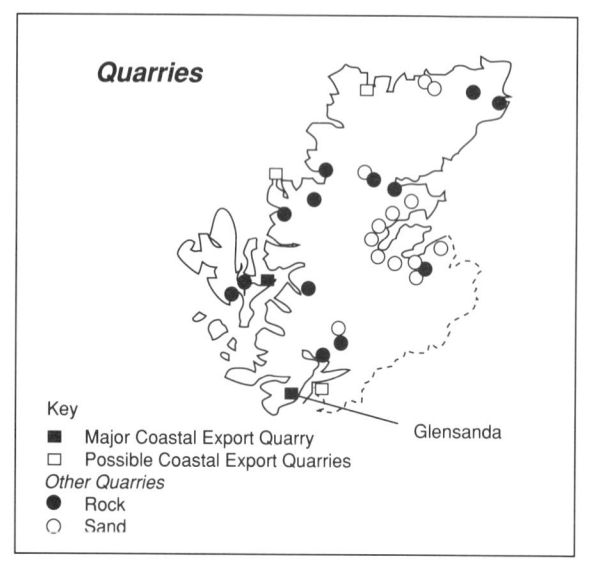

Fig. 4.4.18
Loch Eriboll, Sutherland; site for another superquarry

1   Read the brochure extract carefully. Does it reveal inconsistencies in its argument, or is it just commercial common sense?

2   Study the map (Fig. 4.4.17) which shows the location of other possible coastal superquarries. A photograph of one of the sites is shown in Fig. 4.4.18. Comment on the suitability of this site for superquarry operations.

WATER RESOURCES IN UPLAND BRITAIN

Cities like Birmingham, Liverpool and Manchester now enjoy the legacy of the foresight of early city fathers who invested wisely in the reservoir schemes in the mountain areas of the Pennines, Wales and the Lake District.

(Celia Kirby, Water in Britain)

This was the hamlet of Mardale Green ... but now under sentence of death. There were wild roses in fragrant hedgerows, foxgloves and harebells and wood anemones in the fields and under the trees, all cheerfully enjoying the warmth and sunshine; but there would be no more summers for them: they were doomed ... Manchester Corporation had taken over the valley and built a great dam.... Already the impounded waters were creeping up the valley. Soon the hamlet of Mardale Green would be gone: the church, the inn, the cottages and the flowers....

(A. Wainwright, Fell walking with Wainwright)

Fig. 4.5.1
Talybont Reservoir in the Brecon Beacons

Whilst some of them will never be able to mask their alien, man-made origins, others have, over the years, successfully settled into the landscape. In my own view I would go as far as to say that reservoirs such as Pentewyn and Talybont (Fig. 4.5.1) now enhance their surroundings. ... whatever one's attitude is, it is fair to conclude that, with eighteen in the park, enough is enough.

(Roger Thomas, The Brecon Beacons National Park)

In the Peak Park, an upland of 1404 sq km there are already 55 reservoirs larger than 1.62 hectares and many smaller ones. Within two miles of the Park boundary there are a further 36 reservoirs. Over one third of the total land surface of the National Park is given over in some measure to water catchment. The local river authorities in the last two years have investigated the water storage potential of 11 further sites in the Park.... In view of this level of activity the Board feel that their duty under the National Parks Act to preserve and enhance the natural beauty of the area is being imperilled.

(Peak Park Planning Board, A New Look at Water Resources)

The quotations indicate the perceived need for the water resources of the British Uplands, and the realities with which we are confronted in the National Parks.

1   Using all three of the examples identify the nature of the conflict that stems from the utilisation of the water resources in the uplands.

2   By reference to the preceding sections on Upland Britain, suggest the nature of the conflicts that can develop between the utilisation of the water resources and other forms of land use in the uplands.

ENNERDALE AND WASTWATER: SUCCESSFUL PROTEST

One of the quotations at the beginning of this section indicated the importance of upland water catchments in the water supply of large cities such as Liverpool and Manchester. Water supply to more local consumers, both industrial and domestic, is also drawn from upland sources. One of the most controversial schemes to increase water supplies to local industry emerged in the western part of the Lake District in the late 1970s and early 1980s. The proposals to extract additional supplies of water from Ennerdale and Wastwater for British Nuclear Fuels at Sellafield led to a public enquiry.

Fig. 4.5.2
(a) Wastwater in the Lake District
(b) Ennerdale Water, Lake District

Both Ennerdale Water and Wastwater lie in deep glacial valleys in the west of the Lake District. The streams draining into Ennerdale Water rise on the slopes of Great Gable and Pillar to the south, and on the High Crag – High Stile ridge to the north. Near the western end of the lake, Ennerdale narrows dramatically between Angler's Crag and Bowness Knott. Ennerdale is virtually uninhabited, with much of its valley sides occupied by coniferous plantations. It can justifiably lay claim to being one of Lakeland's wildest valleys. Wastwater has an equally spectacular setting. The delta flats at the head are dominated by the slopes that sweep up to Great Gable, Scafell and Scafell Pike. From its southern shore rise the remarkable screes that lead up to the craggy slopes of Illgill Head. *'To visit Wasdale is to experience the power of the English landscape at its greatest, its constantly changing light, and the varying clarity of its atmosphere, bringing infinite variety to the majesty and wilderness of the scene'.*

(Geoffrey Berry, A Tale of Two Lakes)

The proposals

Ennerdale Water

North West Water proposed to extract more water from the lake in order to provide increased water supply for the Sellafield nuclear complex. The details and impacts of the scheme are shown on the Save Ennerdale poster (Fig. 4.5.3a)

SAVE ENNERDALE

A Campaign to combat the proposal of the North West Water Authority to take more water from Ennerdale.

The scheme would mean:-

1. **Raising** the level of the lake by 4 feet [1.2 m].
2. **Constructing embankments** 6 to 10 feet [1.8 to 3 m] high for considerable distances along the lake shore.
3. **Enlarging the weir** and reshaping the river Ehen at the outflow.
4. **Flooding** agricultural land.
5. **Drowning** of trees and woodland.
6. Providing for a greater degree of **draw-down** which will at times leave exposed large areas of mud and stones.
7. **Disturbance** of shore line vegetation and wild life of the lake.
8. **Jeopardising** the run of salmon and sea trout in the river Ehen by seriously reducing the frequency of spates.

A PLACE OF WILD BEAUTY

Ennerdale is as wild and dramatic a valley as any in England; no building stands upon its shores, no motor road reaches beyond its entrance. Its beauty must be protected.

IN A NATIONAL PARK

Ennerdale is in the Lake District National park. There is a statutory obligation to give it the highest degree of protection and this is a plain duty laid upon all of us.

AN ALTERNATIVE

There is a good alternative, the Derwent Scheme, which would take water from near the mouth of the river Derwent, at Workington. This scheme, at an additional cost of £2 million, would not only provide more water from a vastly greater catchment area but would save Ennerdale.

Funds are essential if we are to put up a worthy fight and be properly represented at a public enquiry.

Please help by sending a donation.

Fig. 4.5.3a

Wastwater

Extra water was required to be drawn from the lake for British Nuclear Fuels (BNFL) at Sellafield. Since World War Two, 18 million litres a day had been drawn from the lake by pipeline. The proposals required a daily take of water from the lake of some three times that amount. The details and impacts of the scheme are shown on the Wastwater Threatened poster (Fig. 4.5.3b).

Ennerdale: the case for North-West Water

North-West Water maintained that it was capable of supplying the extra water to BNFL by raising the

WASTWATER THREATENED

British Nuclear Fuels are seeking permission ot increase abstraction from Wastwater from 4 million gallons [18 million litres] daily to 11 million gallons [50 million litres] daily. At present there is no artificial control of the lake. The new scheme would entail the building of a weir near the outlet of the lake to raise the water level. The increased abstraction would result in a wide variation in the level of the lake and at times of draw-down would expose unsightly areas of mud and stones.

Any control of the lake would open the way to further demands in the future as the huge Windscale complex grows.

England's most dramatic valley in England's premier National Park

Wasdale, with its deep lake, its shapely mountains,
its huge screes, must remain unspoiled
for the enjoyment and inspiration
of the present and future generations.

If you wish to make representations about these proposals yourself you should write to:

The Secretary
Department of Environment
2 Marsham Street
London SW1P 3EB
Quoting reference number WS/5527/764/13.

There is also a scheme by the North West Water Authority to take increased supplies from Ennerdale Water. A separate leaflet about this is available on application to:

Save Ennerdale Campaign
Gowan Knott,
Kendal Road,
Staveley, Kendal.
Cumbria. LA8 9LP.

An Alternative

There is a good alternative, the Derwent Scheme, which would take water from near the mouth of the river Derwent, at Workington. This scheme would not only provide more water from a vastly greater catchment area but would save Wastwater and Ennerdale.

The Wastwater and Ennerdale Schemes are to be the subject of **a public enquiry** in the Autumn. Funds are essential if we are to put up a worthy fight and be properly represented at a public enquiry. Please help by sending a donation.

Fig. 4.5.3b

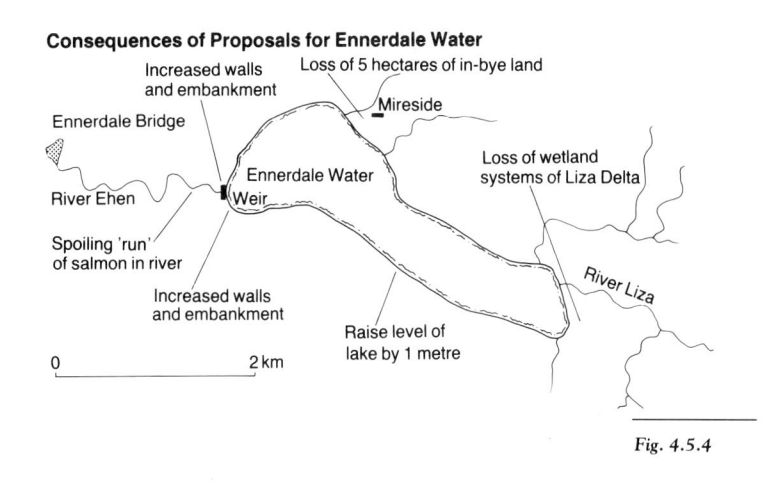

Consequences of Proposals for Ennerdale Water

Fig. 4.5.4

into the lake near its outflow. This scheme would lead to more frequent drawdown of the lake and to a greater degree than at present. The other four schemes involved the construction of weirs of various types to raise the lake level (all of the weirs would be built near the lake's natural bar). Two schemes involved the canalisation of the River Irt towards Lund Bridge (see Fig. 4.5.5).

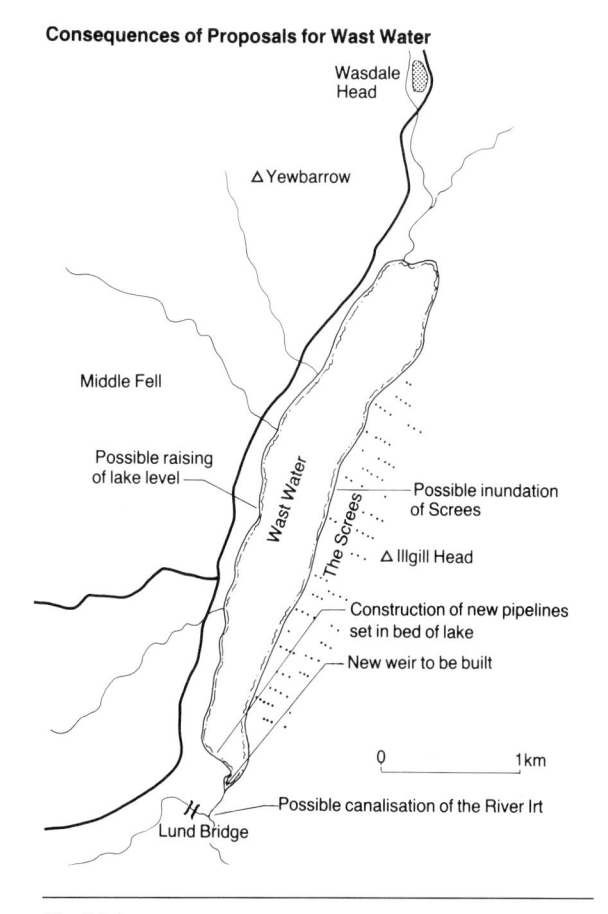

level of Ennerdale Water by 1.2 m (see Fig. 4.5.4). BNFL stated that this scheme was not acceptable to them since it did not provide water of the quality required for their processes. The water, under the proposals of North-West Water would be extracted from the River Ehen at Braystones, 10 miles (16 km) below Ennerdale Water, by which time it would have become heavily polluted by water from the tributary River Keekle. North-West Water compromised by producing another scheme, which would take water directly from the lake to Sellafield, which was more acceptable to BNFL.

Wastwater: the case for BNFL

Water of the highest quality was required for the Sellafield Plant. Only Wastwater could provide water of the quality needed. Five alternative schemes were put forward for the increased abstraction of water from the lake. The preferred and least damaging scheme was No. 1 proposed, in which no raising of lake level by dam or weir would occur. Larger pipelines would be set more deeply

Consequences of Proposals for Wast Water

Fig. 4.5.5

Objections to the proposals

Objections to the proposals were on the grounds of landscape flora, farming and recreation.

Landscape

- The proposals would have more than the minimum adverse effect required by the National Park Authority.
- Raising the lake levels would irretrievably alter the landscape in both Ennerdale and Wastwater.
- Drawdown of the lake would result in unsightly exposure of lake deposits.
- The necessary weir and embankments on Ennerdale Water would be obtrusive and an eyesore, both during construction and after.

Flora

- Wastwater and the Screes were S.S.S.I.s of national and international importance.
- Rare plants around the edge of both lakes would be put at risk. Half of the Ennerdale species would be threatened.

The unique wetland ecosystems of the Liza delta (at the head of Ennerdale Water) and the Mireside mire would be largely destroyed.

Farming

- During construction 15.4 hectares of farmland would be lost at Mireside Farm on Ennerdale Water and 4.6 hectares permanently.
- This land is vital in-bye land, and there would be little land for grazing cattle and none for hay. Cattle on the farm would have to be sold.
- The farm is marginal already, and further loss of land would probably make it unviable.

Recreation

- The heavy construction traffic would detract from the tranquil quality of life in the valleys.
- Children who live in the area would be at risk from the traffic.
- The beauty of both valleys, which attracts many visitors to the area would be spoilt by the new developments.

Hold a class discussion on the merits of the proposals and the objections to them. It could take the form of a mock public enquiry. Roles could include:

The Inspector, who acts as adjudicator

Representative from BNFL

Director of Resource Planning, North-West Water

Representative from Lake District Special Planning Board

Representative from the National Trust, which owns Mireside Farm and extensive areas in Wasdale

Botanist representing Cumbria Trust for Nature Conservation

Farmer at Mireside

Representative from the Fell and Rock Climbing Club.

UPLAND WATER RESOURCES: THE WIDER ROLE

The case study of Wastwater and Ennerdale is essentially a local issue, although it does have some important national implications for similar controversial schemes elsewhere in the uplands, and also in the lowland of Britain. Upland water resources do have a fundamental national significance. The maps (Fig. 4.5.6a-d) show:

- the distribution of annual rainfall in Britain
- the mean annual run off
- the distribution of the main aquifers (water-bearing rocks) in England and Wales
- the frequency of irrigation need in England and Wales (expressed as number of years out of ten).

(Access to a good atlas which shows the distribution of population in Britain is also necessary.)

1   Comment on the level of mismatch beween supply and demand revealed by the study of the above maps, and the map of the distribution of population.

2   What added significance is given to the maps in the early 1990s when ground water supplies in the south-east are low?

WATER RESOURCES IN UPLAND BRITAIN

a) Mean annual precipitation

> 2540
1525 – 2540
1015 – 1524
762 – 1014
635 – 761
< 635

Mean annual precipitation (mm)

b) Mean annual runoff

6 2000
5 1000
4 500
3 250
2 125
1 0

Mean annual runoff (mm)

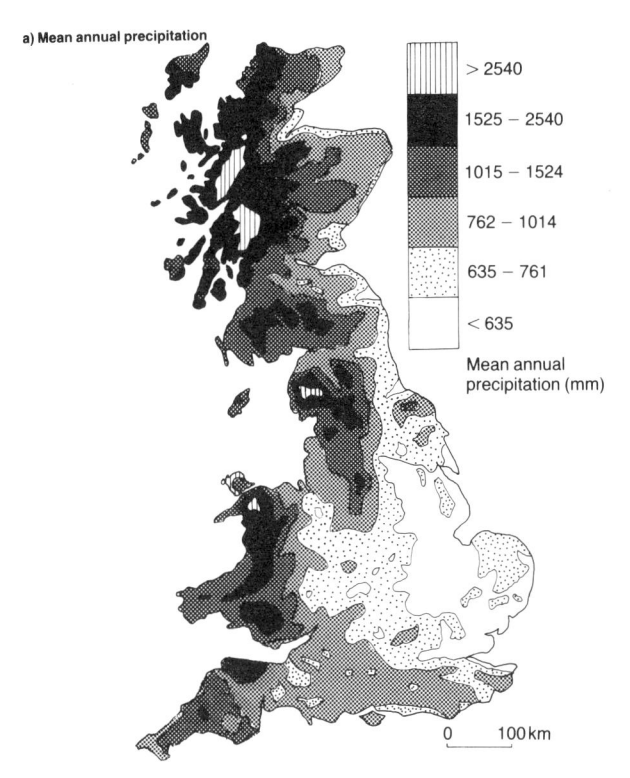

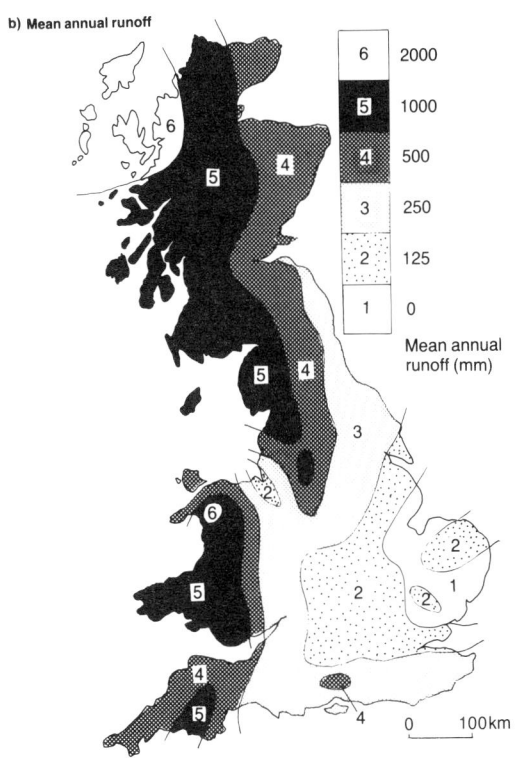

The Geography of Water Supply in England, Scotland and Wales

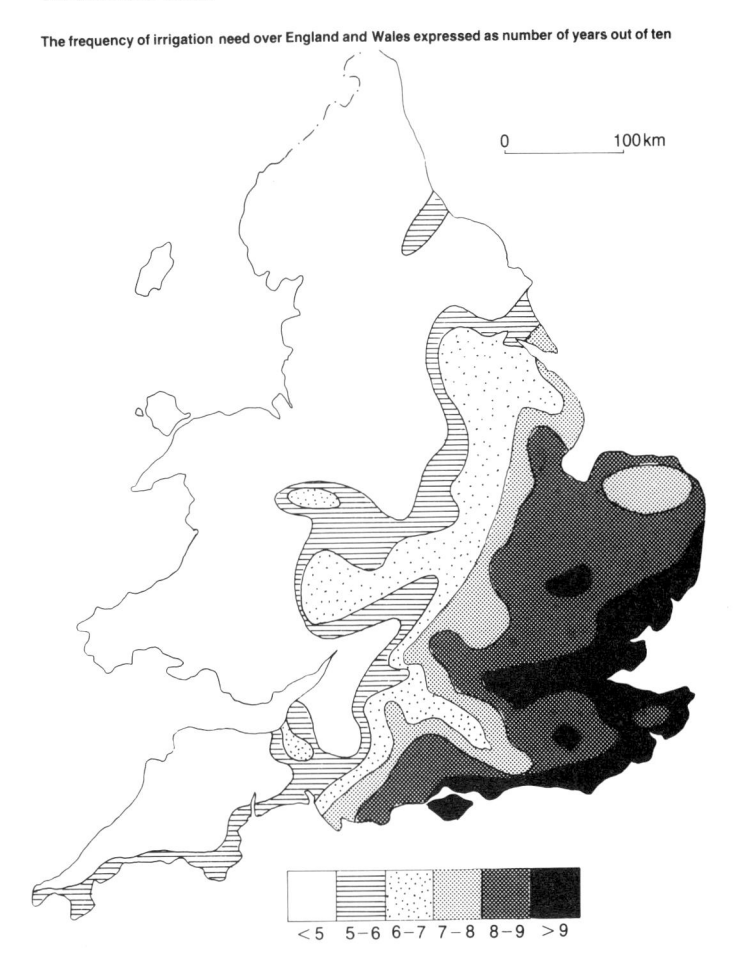

c) Distribution of the main aquifers in England and Wales

Outcrop of principal aquifers
Underground extensions of principal aquifers

The frequency of irrigation need over England and Wales expressed as number of years out of ten

< 5   5–6   6–7   7–8   8–9   > 9

Fig. 4.5.6

A NEW VIEW OF BRITAIN

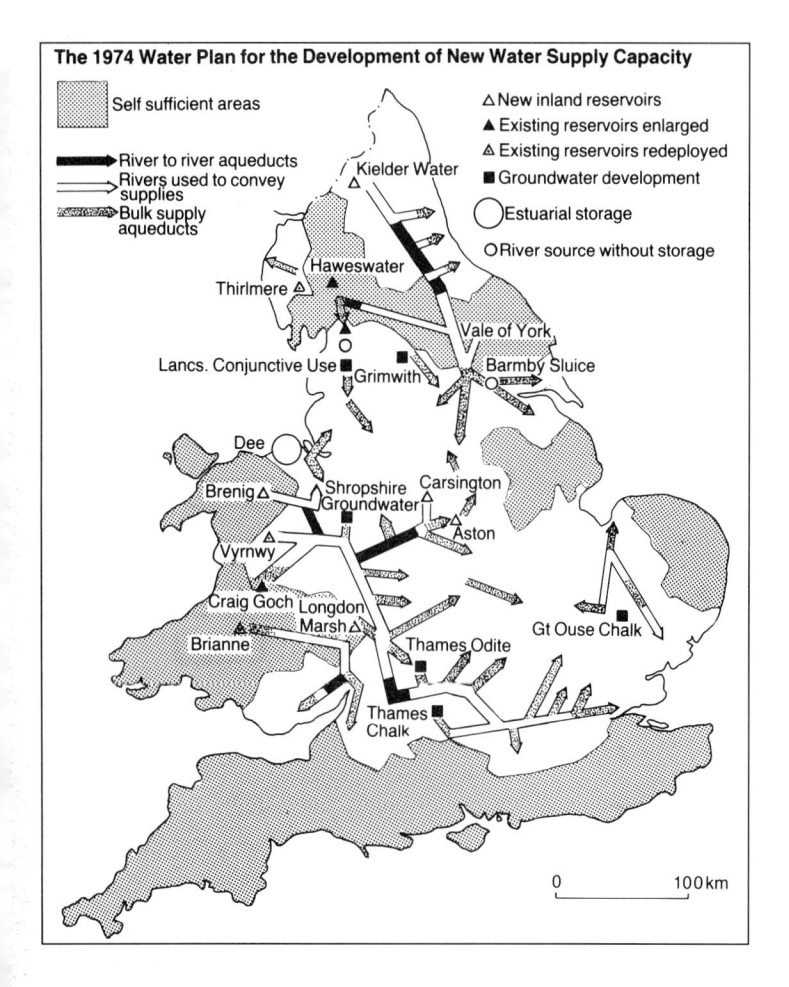

The 1974 Water Plan for the Development of New Water Supply Capacity

Legend:
- Self sufficient areas
- River to river aqueducts
- Rivers used to convey supplies
- Bulk supply aqueducts
- △ New inland reservoirs
- ▲ Existing reservoirs enlarged
- △ Existing reservoirs redeployed
- ■ Groundwater development
- ○ Estuarial storage
- ○ River source without storage

Locations labelled: Kielder Water, Haweswater, Thirlmere, Vale of York, Lancs. Conjunctive Use, Grimwith, Barmby Sluice, Dee, Brenig, Shropshire Groundwater, Carsington, Vyrnwy, Aston, Craig Goch, Longdon Marsh, Brianne, Thames Odite, Gt Ouse Chalk, Thames Chalk

Fig. 4.5.7

The map (Fig. 4.5.7) shows the Water Plan of 1974 for England and Wales. There are four essential elements to the plan.

1 River regulation and inland storage (reservoirs).

2 Groundwater (up to a quarter of expected growth of water in water supplies up to the year 2000.

3 Interbasin transfers.

4 Estuary storage and desalinisation.

Read the article (Fig. 4.5.8) and study the map (Fig. 4.5.9) which give details of a new National Rivers Authority Scheme for water transfer from the uplands.

1 Describe the essential elements of the new scheme.

2 Why is it likely to be more appropriate for the next decade than the 1974 Water Plan?

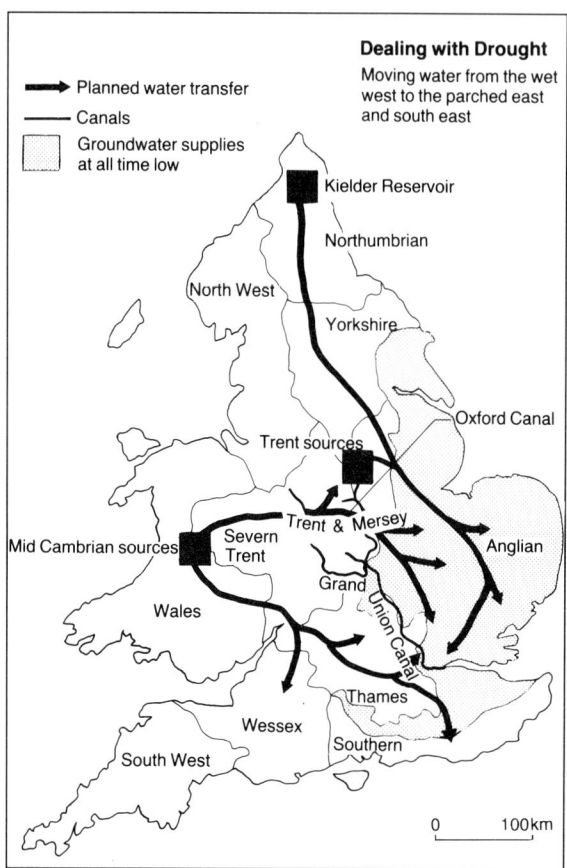

Dealing with Drought
Moving water from the wet west to the parched east and south east

Legend:
- ➤ Planned water transfer
- — Canals
- Groundwater supplies at all time low

Regions/locations labelled: Kielder Reservoir, Northumbrian, North West, Yorkshire, Oxford Canal, Trent sources, Trent & Mersey, Mid Cambrian sources, Severn Trent, Anglian, Wales, Grand Union Canal, Thames, Wessex, South West, Southern

Fig. 4.5.9

NRA urges water transfer by canal to ease drought

By Susan Watts
Technology Correspondent

BRITAIN'S network of canals could rescue the drought-stricken east and south-east of Britain by transporting water from the wetter west, the National Rivers Authority said yesterday.

Better management of water is becoming increasingly inportant as the drier regions of the UK face a fourth year of drought and groundwater levels sit at an alltime low, the authority said.

Jerry Sherriff, head of water resources at the NRA, said that some form of major water transfer between regions looks almost inevitable. Such schemes would require reservoirs, dams and pumping stations are at least 10 years away. Even so they could prove four or five times cheaper than a national water grid.

Canals could play a vital role. "The canals are already in place. With some pumping and engineering to get rid of bottlenecks we could move reasonable amounts of water to areas of greatest need," Mr Sherriff said.

British Waterways, which manages Britain's canals, welcomes the idea. "This would benefit parts of the country requiring water, and also our regular customers, who would enjoy the increased flow of water," a spokesman said. "Canals are also cheaper than pipelines, because you are just pumping water around our lock system, rather than fighting against the friction of a pipe."

One option is to extract water in the Severn Trent region, transfer it to the Trent and Mersey canal near Stafford, then send it on its way, perhaps into the Grand Union canal and Oxford canal to serve the Thames region, then on into the River Ouse and Nene in the Anglian Water region.

Water-transfer schemes already exist, but could be linked together, the NRA said. The Trent-Witham Link and the Ely Ouse-Essex scheme could be joined to bring about 600 million litres a day from the Severn Trent region into the Anglian and Thames regions.

The Kielder reservoir in Northumbria, completed in 1982, could send about 900 million litres a day of surplus water into the River Ouse then on to Lincolnshire, Suffolk and Essex.

Schemes for moving water from the mid-Cambrian region could help the South and South-east. They might include reviving a 1960s plan to transfer about 1200 million litres a day by pumping from the rivers Wye and Severn to the River Thames and Wessex region. Many of these plans would need large storage reservoirs to help in monitoring the quality of the water being sent around.

The NRA wants comments and suggestions on its proposals, and admits that it has made no attempt to suggest who should pay for these projects.

The outlook for water suppplies this summer remains grim. The Essex, Cambridge and the Three Valleys water companies have been worst hit by the continuing drought, largely because they have little or no surface water reserves to turn to when boreholes dry up.

Groundwater levels are at an all-time low in the chalks of south-east Yorkshire, the Anglian region, the extreme east of the Thames region, and the northern part of the Southern region. Records in these regions go back more than a 100 years.

"Even if it rained every day all through the summer we would not be able to replenish groundwater to satisfactory levels," Mr Sherriff said. "In three to four weeks time any rainfall which occurred would just evaporate away."

The past 43 months have seen only 80 per cent of normal rainfall in parts of eastern England. But it would take only one "average or wettish" winter to get Britain's groundwater back into a reasonably healthy shape, he added.

The NRA favours wider use of metering, and is urging water companies to reduce leakage from pipes before water reaches houses. The authority also says that the UK could re-use more of its effluent, probably indirectly, with effluent flowing into a river then extracted again and passed through treatment plants.

Fig. 4.5.8
Independent 12 March 1992

RECREATION IN THE UPLANDS

Fig. 4.6.1 Climber in the Brecon Beacons in mid-winter

It is estimated that visitors spend approaching 20 million days in the Lake District each year. The planning Board will have failed in its aim if by providing facilities for large numbers of people it reduces the quality of the enjoyment. Essentially the National Park should be enjoyed for what it is – an area of relatively wild and beautiful country suitable above all for walking and quiet recreation.

(Lake District: National Park Plan 1978)

The capacity of the National Park to accept recreational use is limited by the duty to conserve the landscape. It is not possible to quantify ultimate capacity, nor to predict when it may be reached. Increased levels of management may postpone the day; changing leisure patterns could mean that it never arrives. However, the potential problem is there, and the protection of the nationally significant resource demands that pressures upon it shall be minimised.

(Dartmoor National Park Plan, 1977)

I am delighted when I read about the phenomenal popularity of hill-walking, for our countryside is there to be enjoyed. Provided we take sensible decisions on conservation issues our environment will stand increased pressure.

(Richard Gilbert, Wild Walks 1988)

About 100 million visitor days are spent in the National Parks each year. Evidence suggests that after rapid growth in the 1950s and 1960s, and levelling off in the late 1970s and early 1980s, a further rise occurred in the late 1980s. Although sight-seeing and walking are the main activities, twice as many people as in 1980 now participate in such active pursuits as horse-riding, trail-biking, caving, canoeing, sailing, climbing and hang-gliding. These activities, together with walking, are frequently focused on particular routes and small areas: such intensive use can lead to erosion and other problems.

(Report of the National Park Review Panel 1991)

1 What is the main issue raised in these quotations?

2 How does the comment of 1991 differ from those in the 1970s?

3 Broad policy on recreation is stated in the first two quotes. Does the 1991 Report suggest a change in policy is necessary?

SNOWDON'S FOOTPATHS AND SUMMIT

The map (Fig. 4.6.2) shows the main footpaths leading to the summit of Snowdon, and other tourist facilities. More than half a million people walk on Snowdon each year, and the issue of Snowdon News (Fig. 4.6.3) indicates the management problem that now faces the National Park Authority. Snowdon has the attraction of being the highest summit (1085 m) in England and Wales, but its slopes and corries provide habitats for a range of rare and sensitive arctic-alpine communities. The diagram (Fig. 4.6.4) shows some aspects of the vulnerability of the mountain mass to heavy visitor pressure.

Fig. 4.6.2

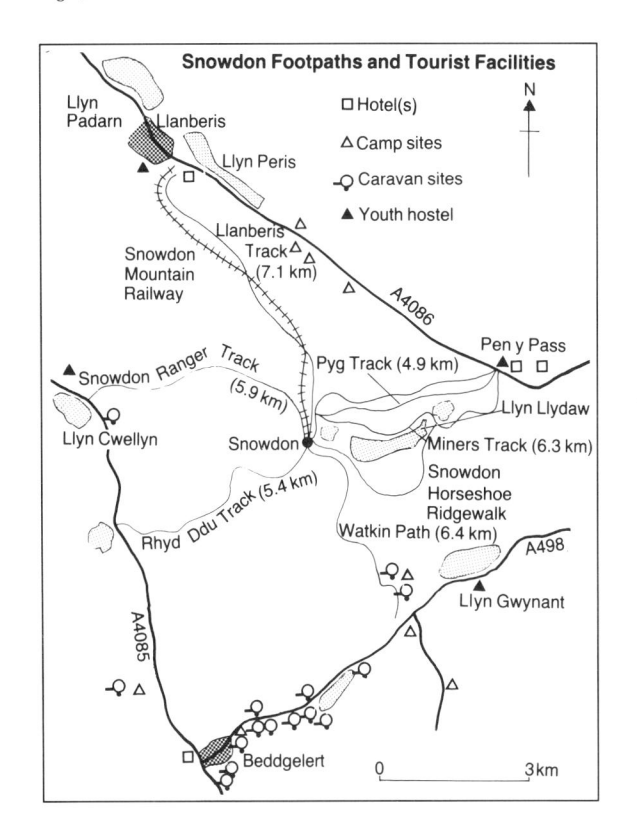

Fig. 4.6.3

SNOWDON NEWS

ISSUED BY SNOWDONIA NATIONAL PARK   FIRST ISSUE 1981

WOULD YOU BELIEVE YOU CAN WEAR OUT A 3,560 FOOT MOUNTAIN ?

More than half a million people walk on Snowdon every year. This is causing serious footpath erosion and increasing conflicts with farming and conservation interests.

SNOWDON MANAGEMENT SCHEME

It is the aim of the Snowdon Management Scheme to restore the eroded footpaths and to resolve the conflicts between recreation, agriculture and conservation. This is a five year programme supported by the Countryside Commission and run by the Snowdonia National Park.

BUT WE NEED YOUR HELP...

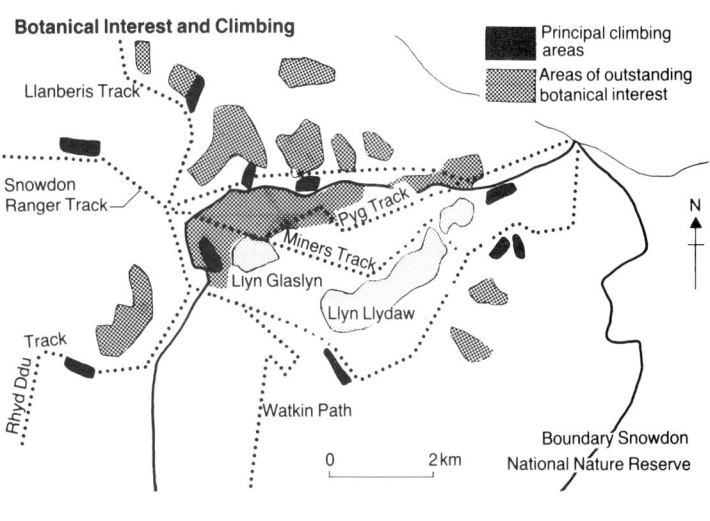

Fig. 4.6.4

The footpaths

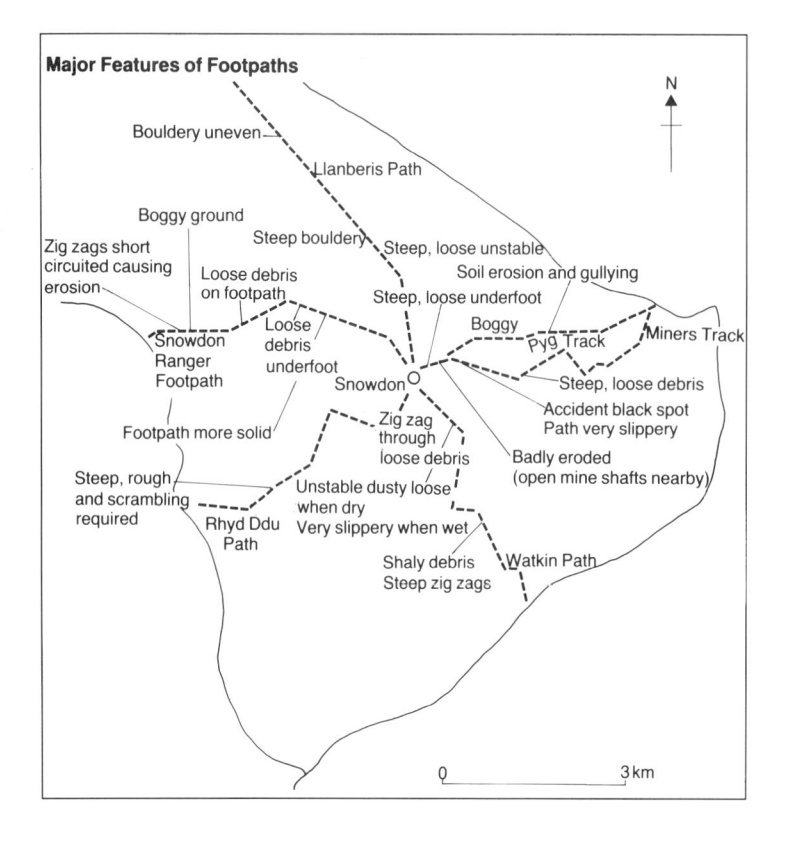

Fig. 4.6.5a

Major Features of Footpaths

(map showing footpaths around Snowdon with labels: Bouldery uneven; Llanberis Path; Boggy ground; Steep bouldery; Steep, loose unstable; Soil erosion and gullying; Zig zags short circuited causing erosion; Loose debris on footpath; Steep, loose underfoot; Snowdon Ranger Footpath; Loose debris underfoot; Boggy; Pyg Track; Miners Track; Steep, loose debris; Accident black spot, Path very slippery; Footpath more solid; Zig zag through loose debris; Badly eroded (open mine shafts nearby); Steep, rough and scrambling required; Unstable dusty loose when dry, Very slippery when wet; Rhyd Ddu Path; Shaly debris Steep zig zags; Watkin Path. Scale 0–3 km)

Six footpaths ascend Snowdon from the surrounding valleys: their location and major features are shown in Fig. 4.6.5a. Levels of use vary on the different footpaths: the following figures represent use of the paths (Fig. 4.6.5b).

Fig. 4.6.5b Snowdon: use of footpaths by walkers

	PEAK SUMMER MONTHS (JULY – AUGUST)	ANNUAL
Llanberis	17 781	32 527
Miners' Track	14 885	67 000
Pyg Track	12 619	41 429
Watkin Path	14 519	44 000
Rhyd Dhu Path	7 001	19 200
Snowdon Ranger	5 126	16 043
Total	71 921	220 199

The figures are the result of automatic counter readings and may not reflect the total usage, but their relative values are reliable.

1 Using the Figs 4.6.2 and 4.6.5a discuss the **reasons why some paths are used more than others**.

2 Using Figs 4.6.5a and b and the statistics, draw a diagram to show where erosion is likely to be heaviest on the various paths.

3 Before footpath restoration can take place a strategy for acceptable future use of the paths has to be decided. Using your answers to questions 1 and 2, suggest a suitable strategy: levels of future high, medium and light use should be indicated.

Fig. 4.6.6a
The Watkin Path, here resurfaced with flagstones (see Map Fig. 4.6.2)

Fig. 4.6.6b
The Pyg Track up Snowdon (see Map Fig. 4.6.2)

Footpath management

Different methods of footpath management are required for different situations. Four main types of erosion may be recognised:

RECREATION IN THE UPLANDS

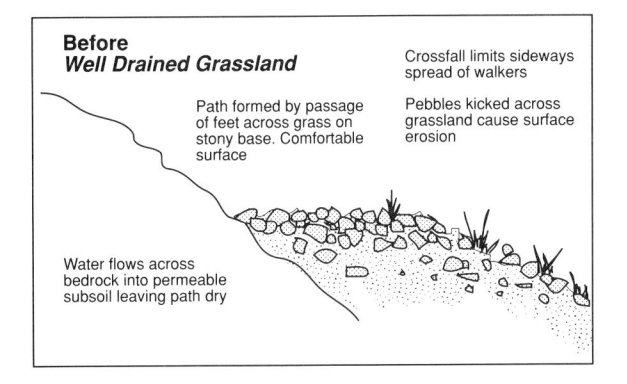

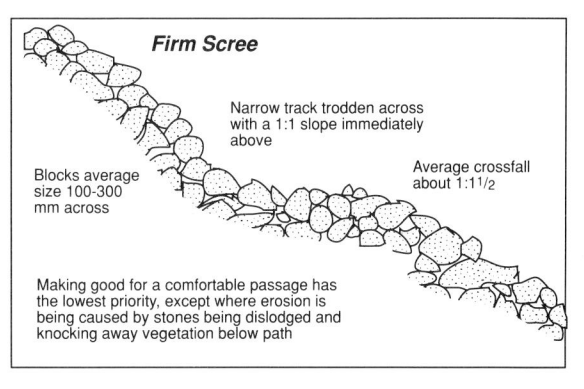

Before Well Drained Grassland
- Path formed by passage of feet across grass on stony base. Comfortable surface
- Crossfall limits sideways spread of walkers
- Pebbles kicked across grassland cause surface erosion
- Water flows across bedrock into permeable subsoil leaving path dry

Firm Scree
- Narrow track trodden across with a 1:1 slope immediately above
- Blocks average size 100-300 mm across
- Average crossfall about 1:1½
- Making good for a comfortable passage has the lowest priority, except where erosion is being caused by stones being dislodged and knocking away vegetation below path

Fig. 4.6.7

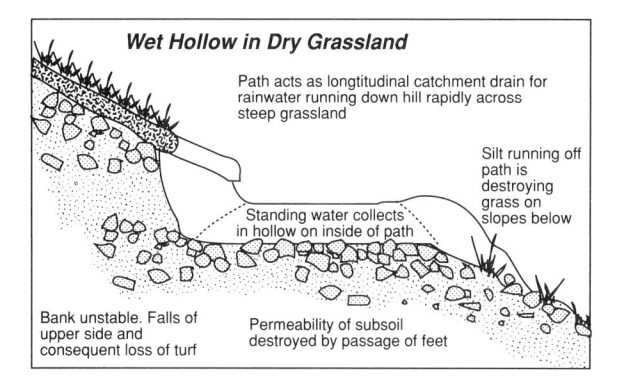

Wet Hollow in Dry Grassland
- Path acts as longtitudinal catchment drain for rainwater running down hill rapidly across steep grassland
- Silt running off path is destroying grass on slopes below
- Standing water collects in hollow on inside of path
- Bank unstable. Falls of upper side and consequent loss of turf
- Permeability of subsoil destroyed by passage of feet

Boulder Scree
- Rock 4000mm high
- Very uncomfortable going and dangerous in bad light owing to large hollows and protusions
- Rock 1500 x 800mm
- Erosion impossible over such gargantuan boulders
- Plant life survives close to path

1 erosion on well-drained grassland
2 erosion in wet hollow in dry grassland
3 erosion on firm scree
4 erosion on boulder scree.

The technique for repairing the erosion on well-drained ground is shown in Fig. 4.6.8.

The summit

The summit building on Snowdon has been referred to as 'the highest slum in Wales'. Built in 1934, it has to withstand more severe weather conditions than any other building, and to cope with a level of visitor activity well above that in the 1930s. On a fine summer's day up to 1000 people may be on the

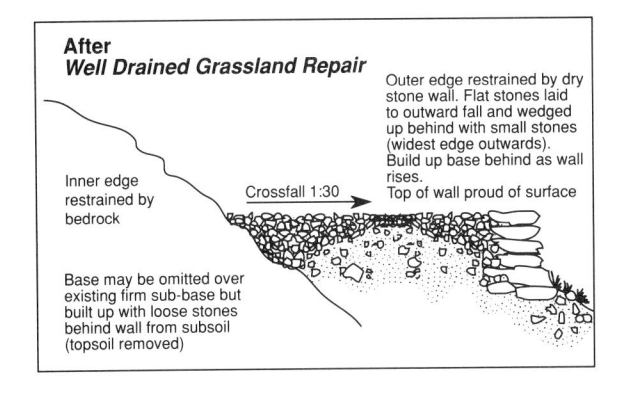

After Well Drained Grassland Repair
- Inner edge restrained by bedrock
- Crossfall 1:30
- Outer edge restrained by dry stone wall. Flat stones laid to outward fall and wedged up behind with small stones (widest edge outwards). Build up base behind as wall rises. Top of wall proud of surface
- Base may be omitted over existing firm sub-base but built up with loose stones behind wall from subsoil (topsoil removed)

Fig. 4.6.8
Path on well-drained ground after repair

1 Draw diagrams to show suitable repair techniques for the other three types of erosion.

2 Draw a map showing where the different types of repair would best be used on the six footpaths.

Fig. 4.6.9
Snowdon Summit

205

summit at any one time between 1.30 p.m. and 3 p.m. In such a relatively confined space, the problems of litter, pollution and erosion are potentially severe (Fig. 4.6.10). The summit building has a shop, a self-service cafe, and toilets and can cater for up to 250 people at any one time. Drinking water is brought up by the train, which serves the summit, and there is a small generator to supply power.

Fig. 4.6.10

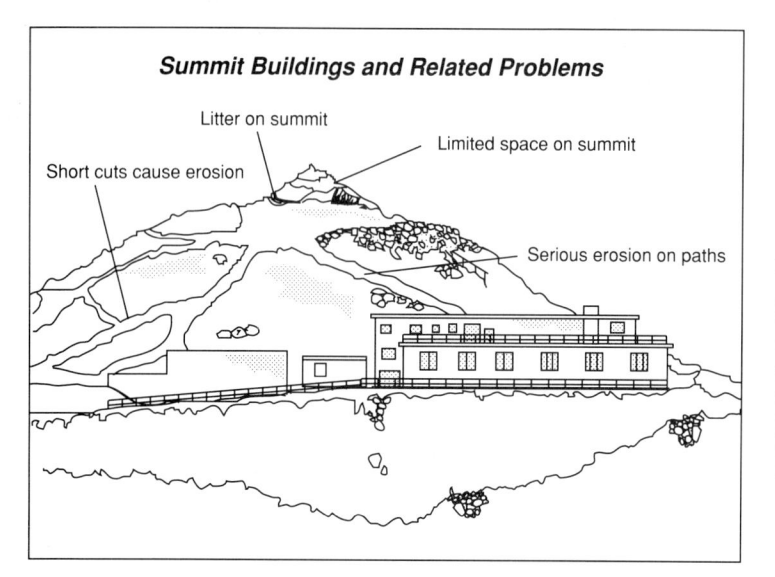

Summit Buildings and Related Problems
- Litter on summit
- Short cuts cause erosion
- Limited space on summit
- Serious erosion on paths

A firm of consultants made a study of the problems of the Snowdon summit, and proposed five different options for improving conditions and facilities at the summit.

The five different options are listed below (Fig. 4.6.11).

Fig. 4.6.11

1 Remove all provisions at the summit.

Options for the summit One
Removal of all provisions

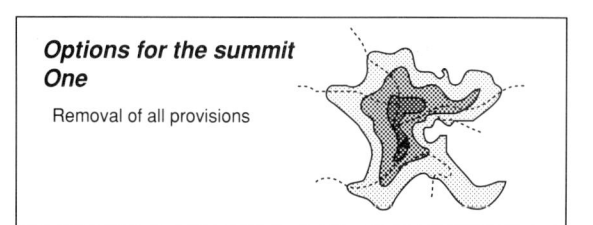

2 Removal of the railway, with minimum provisions at the summit.

Options for the summit Two
Removal of railway, minimum services at Summit.

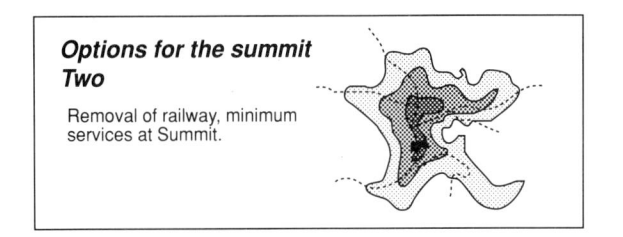

3 Railway to the summit in the season, terminating at Clogwyn (the next station below) out of season: main building there, with minimum services at the summit.

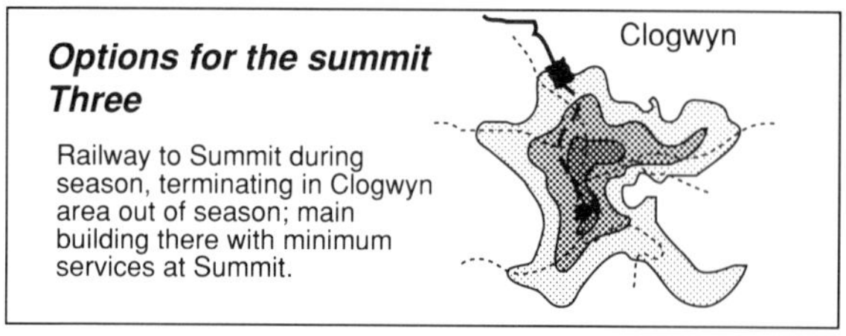

Options for the summit Three
Railway to Summit during season, terminating in Clogwyn area out of season; main building there with minimum services at Summit.

4 Railway to summit during the season, terminating at Clogwyn out of season. Main building at summit with optional minimal services at Clogwyn.

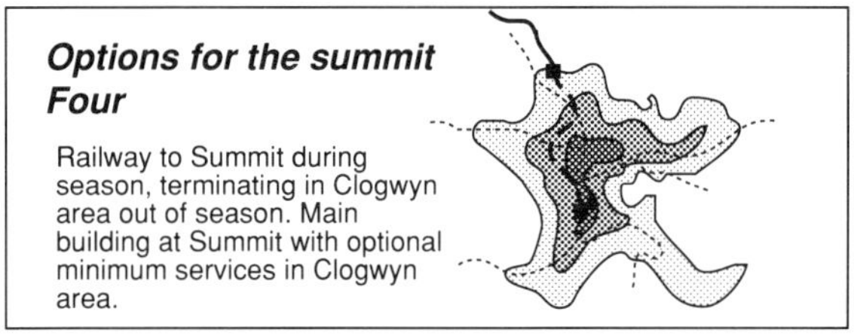

Options for the summit Four
Railway to Summit during season, terminating in Clogwyn area out of season. Main building at Summit with optional minimum services in Clogwyn area.

5 Railway to summit all the year.

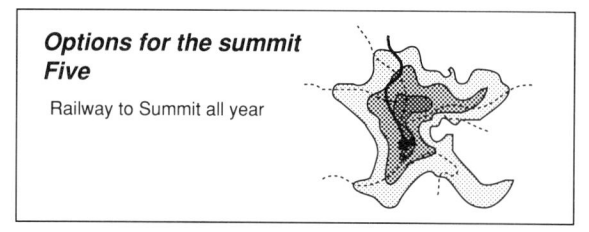

Options for the summit Five
Railway to Summit all year

Snowdon summit building was bought by the National Park Authority in 1984. A three-year programme was then planned for the refurbishment of the building.

1 Discuss the merits and disadvantages of all five of the proposals.

2 Consider the reasons for the decision of the National Park Authority.

3 In the light of the comments above concerning the conditions at the summit suggest what other improvements could be made to the summit environment. The improvements could be shown on a diagram of the summit.

SKIING IN THE CAIRNGORMS

Fig. 4.6.12
(a) The summit plateau of the Cairngorms and the cliffs of the northern corries

(b) The Cairngorms: the northern corries from Loch Morlich

Conservation and recreation

The Cairngorms are the most extensive area of high ground in Britain (Fig. 4.6.12a and b). The boulder-strewn granite plateau, rising to over 1200 m, is cut into by some of the finest and most distinctive glacial landforms, including magnificent Northern Corries and the troughs such as the Lairig Ghru that thrust deep into the mountain mass. This environment possesses equally distinctive ecosystems, from the arctic-alpine summit moorlands to the ancient pine forests in the lower glens. Both its wildlife and plant communities are unique in Britain. The bird life of the Cairngorms is rich and varied: species such as the snow-bunting and the dotterel breed here at the southernmost extent of their range; grouse and ptarmigan are common and golden eagle is prominent amongst the predators. Animal life is equally rich, and includes red deer and reindeer (re-introduced in 1952).

The Cairngorms are also important as a major recreational resource. The height of the plateau has enabled the area to establish itself as the main skiing centre in Britain, as well as offering a whole range of opportunities for other mountain pursuits such as rock and ice-climbing, and hill walking. At lower levels the lakes such as Loch Morlich, and the woodlands present a range of other recreational options.

The Cairngorms are one of the designated National Scenic areas of Scotland, and considerable tracts are afforded the protection of a National Nature Reserve, another area has SSSI status, and the RSPB have an important bird reserve, (see Fig. 4.6.13). It is thus not difficult to see why there is a conflict between recreational and conservation interests in the Cairngorms. The greatest controversy in recent years has been over proposals to extend the skiing area in Coire na Ciste and Coire Cas westwards into the Lurcher's Gully area.

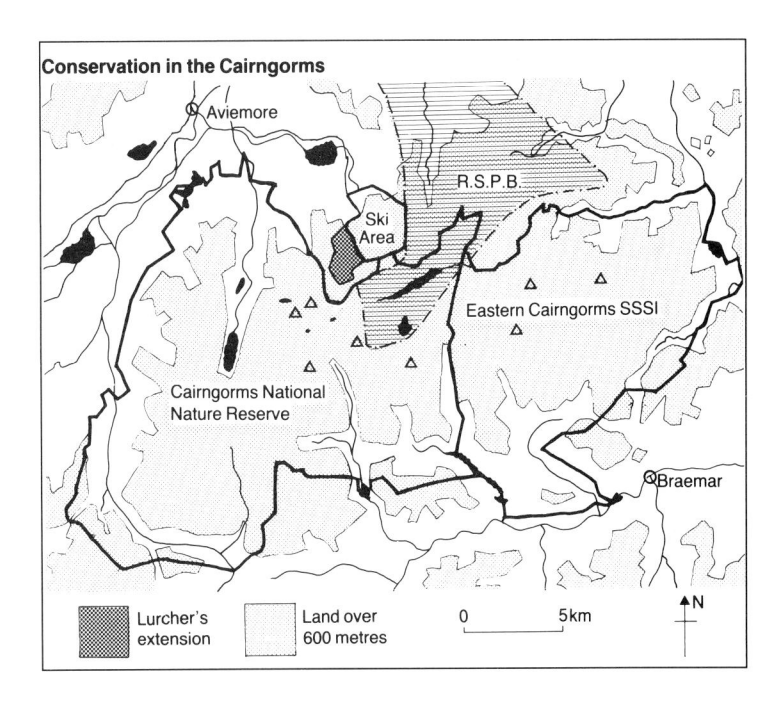

Fig. 4.6.13

The Lurcher's Gully controversy

Fig. 4.6.15

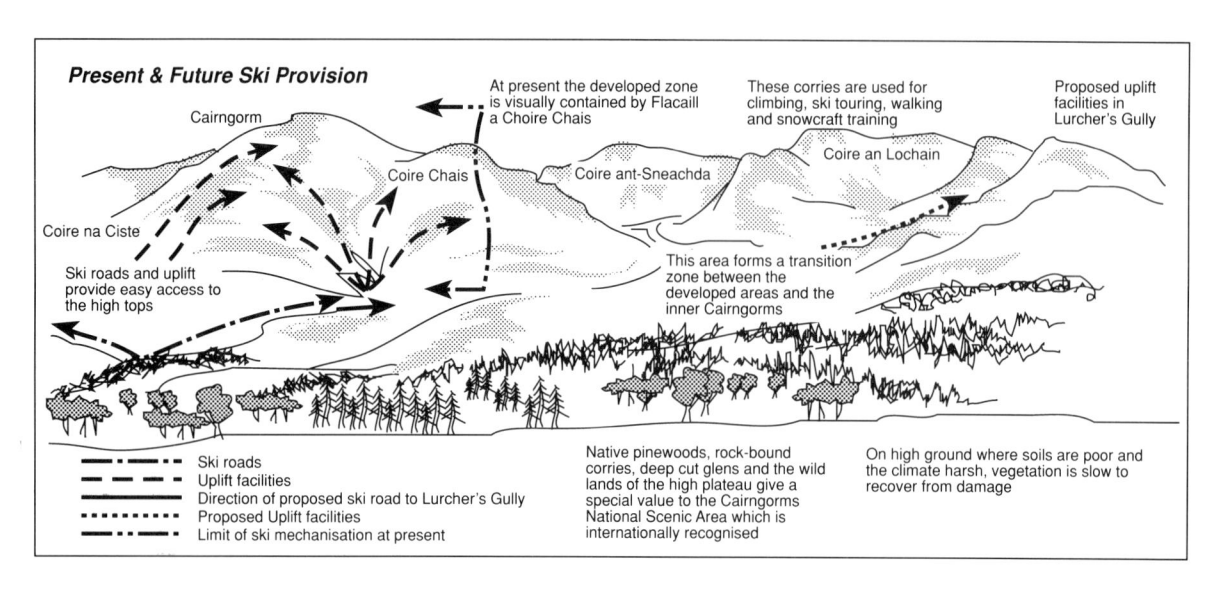

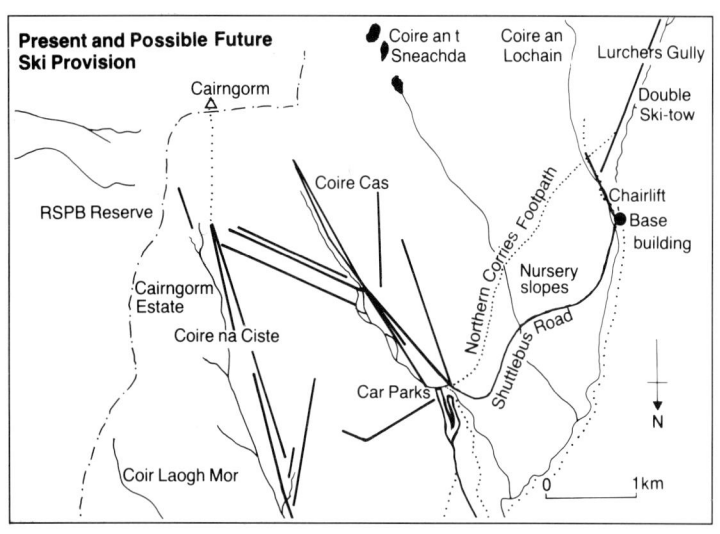

Fig. 4.6.14

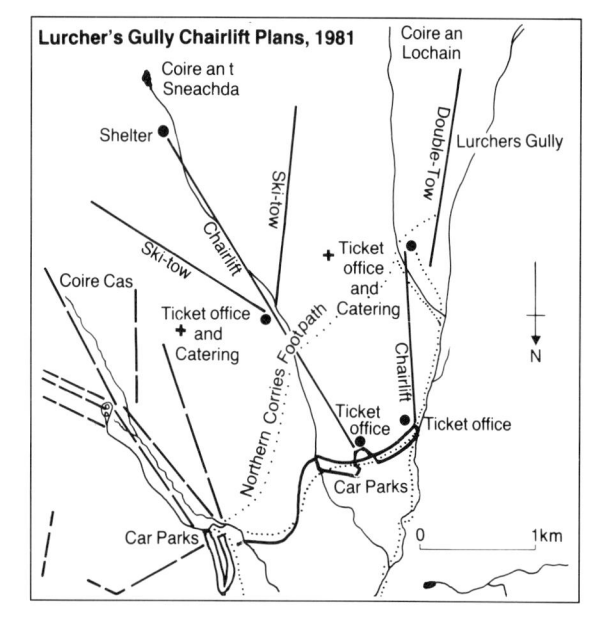

Fig. 4.6.16

The present and future ski facilities in the Cairngorms are shown in the map (Fig. 4.6.14) and the sketch (Fig. 4.6.15). The new road from Glenmore to Coire Gas was built in 1960 and this was followed by the construction of the first chair lift in 1961. Four areas were designated for development: Coire Cas, the White lady, Coire na Ciste and Lurcher's Gully all of which were within the Highland and Islands Development Board's Estate that was purchased from the Forestry Commission in 1971.

Development in the first three areas continued until 1979, when the Cairngorm Chairlift Company produced its plans for Lurcher's Gully (Fig. 4.6.16). These plans went to a Public Enquiry in 1981. The Secretary of State for Scotland refused permission, but suggested that a more limited scheme might be an acceptable compromise in the future.

The second proposal

Modified proposals for Lurcher's Gully were put forward in 1989. The map (Fig. 4.6.17) shows the main features of the new scheme:

- 400 m chair lift from floor of Lurcher's Gully to a 1300 m double ski-tow running up to a point below the top of the gully

- snow fences to be erected, but excellent snow holding properties of the area mean that this would be kept to a minimum

- base building at the foot of the Gully to house catering, toilets, and maintenance facilities
- single track access road 2 km in length from Coire Cas Car park.

Justification by Cairngorm Chair lift Company

- Many of the present lifts cannot be used in strong winds when snow is scarce. This results in overcrowding and long queues.
- Nursery slopes are at a premium in the present ski area: the extra shelter in Lurcher's Gully will provide intermediate skiing for 1200 skiers (a 20 per cent increase over present capacity).
- No increase in car park capacity; skiers would be spread out over a wider area.
- Effective capacity in poor conditions will be increased to give 100 000 extra skier-days season (20 per cent increase).
- Skiing industry is worth £12 million to local economy: extra jobs (number not specified) will be created.

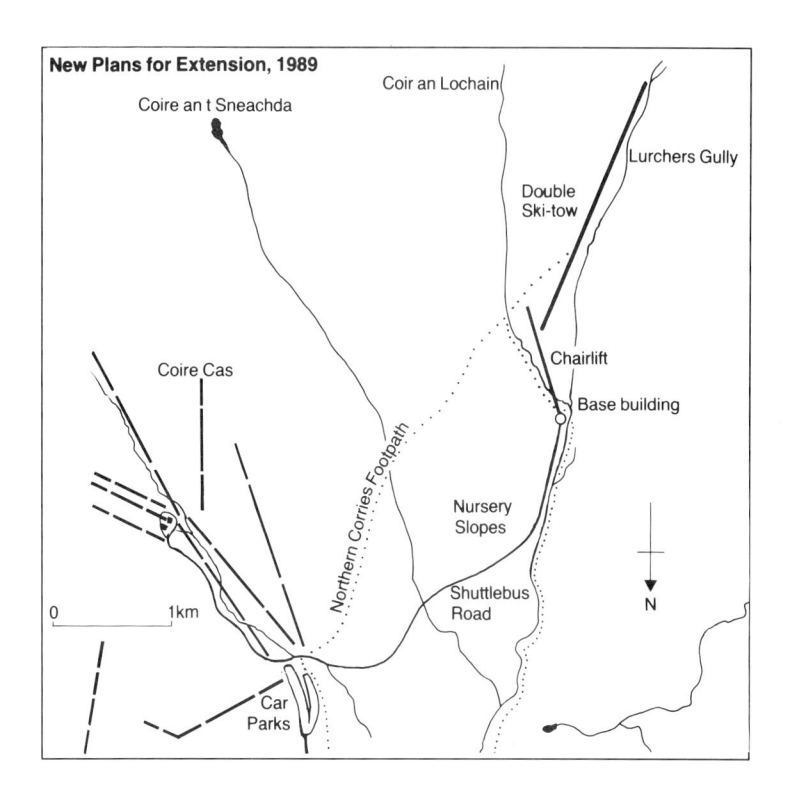

Fig. 4.6.17

The case against the extension

The case against the extension has been brought principally by The Countryside Commission for Scotland.

- Cairngorms has been identified as a National Scenic area (see Fig. 4.6.12). The cumulative effect of the change has impaired the natural qualities of the Cairngorm scene.
- Not withstanding careful design and implementation, the extension would alter the physical environment of the Lurcher's Gully area irrevocably.
- The proposal does not carry development into the inner Cairngorms, but carries development close to this wilder area and removes the buffer zone between the developed area and the wild areas.
- The impact of trampling has led to widespread damage that has not been managed in the past.
- Easy access to the high ground as a result of the new road and the chairlifts has led to the Northern Corries and their hinterland becoming a mountain recreational resource for a wide range of outdoor activities. The western extension would impair enjoyment of these activities by destroying the wilderness and the solitude that walkers and others seek to enjoy. A balance needs to be struck between the different outdoor interests.

You are required to write an objective report on the proposal to extend ski facilities from Coire Cas to Lurcher's Gully. You are provided with the proposals from the Cairngorm Chairlift Company, their justification for the extension, and a summary of the case put for the Countryside Commission opposing the proposal. The report should follow the guidelines below:

1 an evaluation of the need for the extension

2 a critical review of the proposal (with a comparison with the proposals that were put forward in 1981 (see Fig. 4.6.16))

3 an assessment of the case against the extension

4 a recommendation suitable for presentation to officials at the Scottish Office.

In February 1990, the Countryside Commission for Scotland lodged its formal objection to the Lurcher's Gully proposals.

In November 1990 the Scottish Office announced its proposal for the designation of the Cairngorms as a World Heritage Site.

COASTAL BRITAIN

THE CHANGING COASTLINE

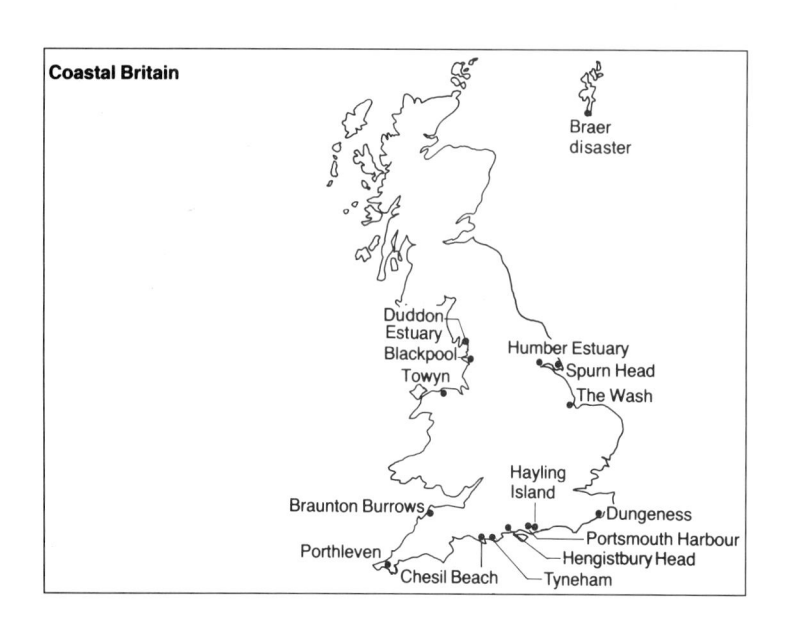

Coastal Britain

[Map showing locations: Braer disaster, Duddon Estuary, Blackpool, Towyn, Humber Estuary, Spurn Head, The Wash, Hayling Island, Braunton Burrows, Dungeness, Porthleven, Portsmouth Harbour, Hengistbury Head, Chesil Beach, Tyneham]

No one in Britain lives more than 120 km from the coast. Its cliffs (Fig. 5.1.1) and estuaries, its resorts and villages, and less happily, its industries (Fig. 5.1.2) and their pollution, are part of the experience of most of Britain's population. The coast, extending to 4400 km in England and Wales alone, possesses an infinite variety that owes much to the frequent changes in geology. Few other coasts can match the contrast between the granite cliffs of Penwith in Cornwall and the boulder clay-capped chalk cliffs of Flamborough, or the differing attractions of the dark sea lochs of north-west Scotland, and the salt marsh fringed estuaries of the Essex and Suffolk coasts.

Sand dunes and shingle spits, cliffs and salt marshes provide habitats for some of the most sensitive and fragile ecosystems in Britain.

Coastal Britain provides one of the most fundamental of challenges to people. It is no static landscape, and constant change and renewal within the coastal system test our ingenuity to manage natural processes, which, even now, we do not understand completely. Even if Sir Archibald Geikie, the famous Scottish geologist could write '... before the sea, advancing at a rate of ten feet [3 m] a century, could pare off more than a mere

No one can walk the clifftops and shorelines of Britain and remain unmoved by the fickle sea and the natural sights and sounds which convey so many moods and impressions that cannot be detected further inland. In terms of formal ownership the coast has a complex history, but exploitation over the last two centuries, as well as the legitimate development of areas for commerce and industry, have made it a cause for shame rather than pleasure.

(Sir Archibald Foster, in the introduction to 'In Search of Neptune: A Celebration of the National Trust's Coastline.')

Fig. 5.1.2
British Steel: coastal plant at Redcar

Fig. 5.1.1
The coast of North Devon to the east of Ilfracombe

marginal strip of land between seventy and eighty miles in breadth, the whole land might be washed into the sea by atmospheric denudation', coastal erosion and its management is a matter of serious concern along many parts of the coastline. His 'ten feet a century' would probably be regarded as a mere guess now, for we have ample evidence that at worst, the coast is receding at 2 m a year! We now have an additional threat to the coastline, for, if the worst scenarios are to be believed, then low-lying coastlines will be inundated by the rising sea level consequent on global warming. Coastal flooding already makes headline news, and presents coastal management teams with a new challenge to solve old problems.

People may be responsible indirectly for some of the problems of the management of coastal processes, but pollution of coastal waters is a more explicit consequence of tidewater concentrations of population and industry. Nowhere is this more clearly seen than in estuaries, whose shores are increasingly lined with processing industries and whose ecosystems are increasingly under threat. Water quality has become as important a theme along the coast as it is in inland waters. In particular it can be a threat to leisure activities, and the enjoyment of the coastline's rich recreational resources. The sheer volume of people that seek recreation on the coastline is in itself a threat to the relaxation and relief that they seek. Traditional resorts may still attract many, but it is only the most remote parts of the British coastline that do not attract the urban visitor now (Fig. 5.1.3). The burgeoning caravan parks of Lincolnshire and North Wales, and the new generation of theme parks and holiday complexes, present problems of coastal management that few could have foreseen in the early part of the century.

Fig. 5.1.3
Caravan Park at Freshwater near Burton Bradstock, West Dorset

SPURN HEAD: THE UNSTABLE COAST

Few places on the coast of Britain represent the dynamics of the coastal system better than Spurn Head (Fig. 5.1.4). It is a 5 km stretch of sand and shingle extending out from Holderness in east Yorkshire into the Humber Estuary. The management of the Spurn peninsula has exercised people's inventiveness and imagination for a very long time. It illustrates well the basic problem in coastal management: unless we understand the processes at work along the coastline, we cannot seek to control them.

Fig. 5.1.4
Spurn Head from the air

Read the extracts from the two newspaper articles and study Fig. 5.1.5c.

1   How does the basic instability of the Spurn Peninsula affect the lives of local people?

2   What form of management is likely to be most effective?

3   Summarise your answer in a sketch diagram.

At the mercy of time and the tide

Military defences have ensured Spurn's survival since Napoleonic times. Now the sea is destroying it. Mike Prestage reports

AT TIMES, as you drive along the narrow road that runs virtually to the tip of Spurn, the milk-chocolate coloured waves of the North Sea seem too close for comfort and the piles of hastily dumped concrete rubble an inadequate deterrent to the pounding water. It is not just the road, which is of Second World War vintage, that is threatened (it has already had to be diverted twice as the sea has claimed part of the original route). Now, the whole sand and shingle spit faces destruction. Bitterly cold winds whip across the flat, three-and-a-half-mile stretch of land that reaches half-way across the mouth of the Humber.

On one side the mud flats of the estuary enjoy a gentle calm. A steady procession of ships makes its way past to the thriving port of Hull, or vessels anchor off the point and await a pilot to guide them through hazardous waters.

But where the spit faces the full brunt of the sea, a century and a half of sea defences and the fragile sand and shingle are being rapidly stripped away. The small stretch of land, only 50 yards wide at its narrowest point, is littered with a military presence dating back to Napoleonic times. Other buildings have been demolished to provide a rubble defence against the sea: the old coastguard tower, lifeboat cottages, the pub and some other wartime buildings have gone.

Spurn is one of the most valuable sites of its kind in the country and has been designated a Site of Special Scientific Interest. Since 1960 it has been a nature reserve managed by the Yorkshire Wildlife Trust. The dunes, held together by marram grass, provide a rare habitat for many plants. Sea holly and spring beauty are found there. The spit is also an extremely valuable site for migrating birds. More than 320 species have been recorded and the mud flats are a feeding ground for wading birds such as dunlin, knot, redshank and curlew.

At Spurn Point, a cluster of modern detached houses provides homes for the country's only full-time lifeboat crew. Last year the lifeboat averaged one emergency call a week. A tall control tower houses the pilots whose boats are moored at the end of a long grey jetty. They make 27 000 ship boardings and landings during a year, but the lighthouse is no longer in use.

Yet time is running out for Spurn and those based on it. Military defences ensured the spit's survival while the rest of the coastline succumbed to tides and storms, leaving the peninsula vulnerable and out of alignment with the coast. Once the Army abandoned its maintenance in 1953, the sea quickly started to make up for lost time. A combination of spring tides and storms from the north-west could see the spit breached this year.

After the initial breakthrough, the powerful waters will quickly enlarge the gap, sweeping away tons of material and beginning the process of rebuilding a new spit to the west. One theory estimates this could be the sixth Spurn in 1400 years. Previous incarnations have had a port on their tip, and provided the embarkation point for the remains of the Scandinavian army defeated by Harold.

Barry Spence, bird warden on Spurn for 26 years, watches the encroachment of the sea with a philosophical air. His bungalow, a former First World War army building, is constantly in danger of being flooded.

A bank, built with clay from excavations for the building of the nearby natural gas terminal, was breached in a 1983 storm. The neck of the peninsula was almost swamped and brought home the fact that the spit's finite life had come close to running its course.

Spurn is formed from the sand and shingle washed from the eroding cliffs of Holderness and deposited in more sheltered waters inside the mouth of the Humber. Waves pile the material into an embankment and the wind heaps sand from the beaches into dunes.

While the deposits are increasing, the coast to the north of the base of the peninsula is being eroded so that, eventually, the shoulder of the spit becomes threated by storms and tides. When the sea breaks through, the end of the spit is left as an island and a new Spurn begins to form.

"From an environmental point of view, nature is takings its course," says Mr Spence. "The birds will move to the island that survives and the process starts again: there is some interest in seeing just what happens."

Turning the peninsula into an island would, however, hit the tourist trade – 60 000 visitors were attracted to the reserve last summer. A pair of Wellington boots and a low tide might bring the new island within reach of trippers, but it would not be the same.

Many believe the sea should have its way. There was a fear that the sand and silt would be swept into the Humber shipping lanes, but studies by the port authority have largely discounted this theory. David Kilpatrick, the coastal protection officer for Humberside County Council, says the rest of the coast retreated at two metres a year while Spurn was protected by well-maintained defences.

"Everbody is frightened to spend any money to save it. Wading birds are not worth the cost involved and nature is moving Spurn to where it should be," he says.

At the moment, small-scale defence measures are being made where the sea is at its most threatening. But they are unlikely to prevent this most unusual feature, at least for the time being, from disappearing beneath the waves.

Fig. 5.1.5a
Independent 23 February 1991

COASTAL BRITAIN

Sea defences for Spurn Head

... fears that a natural cycle, believed to have washed away and then rebuilt the peninsula five times in the past by the tidal movement of sand, was approaching its destructive climax.

But the Institute of Estuarine and Coastal Studies at Hull University challenges the cycle theory, and the 150 years of shoring-up Spurn which it has inspired.

Dr John Pethick of the Institute concludes that the peninsula constantly changed shape but was never destroyed, regenerating itself naturally until man started trying to keep it in one place.

"Sea defences have prevented natural movement," says the report. "Without man's obstructions, sand would be picked up from the eastern side and dropped on the west, either by being carried round the point by the tide, or by being blown or washed over the top."

The height of the peninsula's artificial defences has upset the pattern, says the report, preventing sand reaching the west shore to compensate for the erosion on the east.

The report warns that renewing artificial defences would be extremely costly and ineffective without a barrier continuing up the coast of Holderness, the fastest-eroding seaboard in Europe. A Commons report recently rejected a Holderness sea wall as uneconomic.

The Yorkshire Wildlife Trust is likely to approve the "natural" option – constructing wash-over points along the peninsula to encourage the interrupted regeneration cycle nad boosting the western beaches with gravel and rubble.

Fig. 5.1.5b
Guardian 24 January 1992

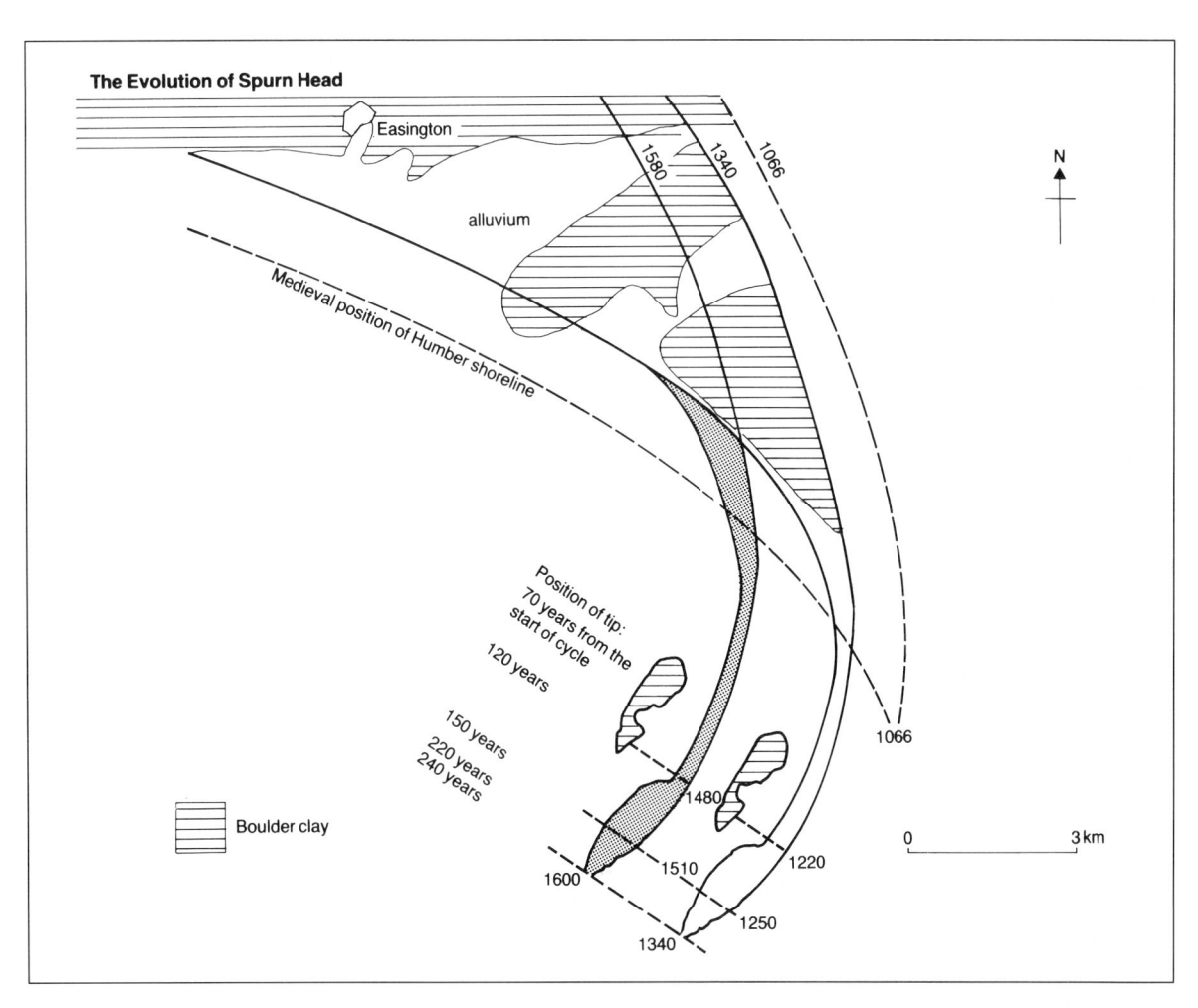

Fig. 5.1.5c

THE MANAGEMENT OF COASTAL PROCESSES AND LANDFORMS

Fig. 5.2.1a Hengistbury Head from the air

Spurn Head, used in the intoductory section illustrates just how difficult it is to manage coastal landforms. Two writers, Joliffe and Patman have suggested that there are five essential requirements for successful coastal management:

- the need to integrate natural and human systems
- the need for a sound environmental data base
- the need to identify the respective roles of the public and private sectors
- the need to match management organisation with the scale of the problems
- the need to co-ordinate the coastal policies of different agencies.

Success depends on effective planning, and a programme of well considered and effectively designed coastal engineering. According to writers Inman and Brush the latter depends on:

- identification of the important processes operating in an environment
- the understanding of their relative importance and their mutual interactions
- the correct analysis of their interaction with the contemplated design.

Several case studies follow to illustrate different modes of coastal management designed to meet different problems.

Hengistbury Head, Dorset: a combination of coastal defence measures.

Porthleven, Cornwall: construction of a new sea-wall.

Hayling Island, Hampshire: beach nourishment.

THE MANAGEMENT OF COASTAL PROCESSES AND LANDFORMS

HENGISTBURY HEAD: DORSET

The aerial photograph of Hengistbury Head, just to the east of Bournemouth (Fig. 5.2.1a), was taken in the 1970s, before any of the current coastal defence works began. The built-up outskirts of Bournemouth can be seen in the far distance, just within the frame of the photograph; Christchurch Harbour lies to the right of the picture. It should be studied in conjunction with the accompanying sketch and map (Figs 5.2.1b and 5.2.2).

1 Make a copy of the sketch, then mark in on the copy:

- areas of active cliff, where erosion appears to be occurring
- areas of dead cliff, where erosion is slow, or appears to have ceased and there is a measure of stability in the cliff profile
- areas of rocky foreshore
- areas of wide sandy beach, backed by sand dunes
- areas of salt marsh
- the main groynes that have been used for coastal defence work.

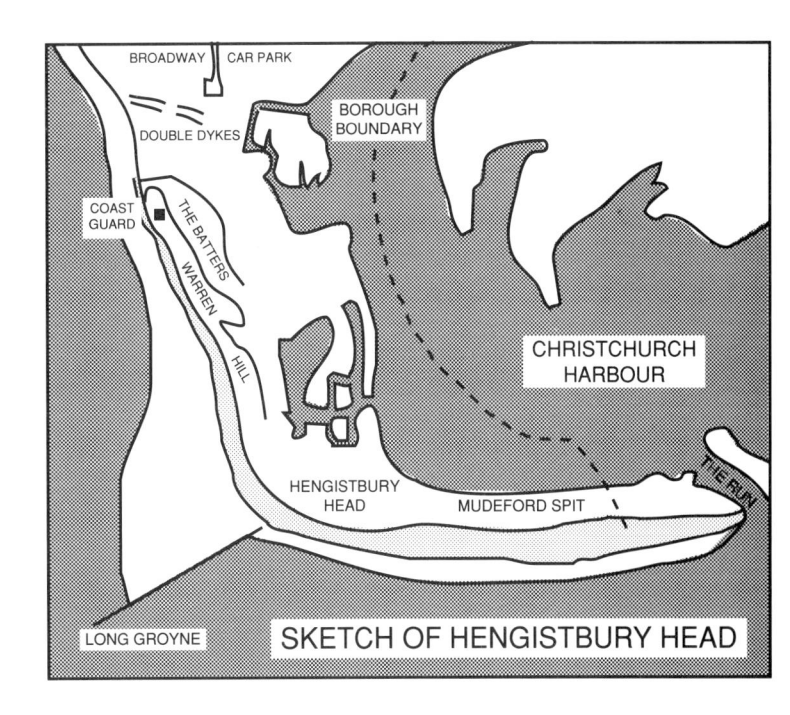

SKETCH OF HENGISTBURY HEAD

Fig. 5.2.1b

2 What would you consider to be the main aim of the large groyne in the foreground?

3 How effective do you think it has been?

4 What is the main type of land use on Hengistbury Head?

5 Will this type of land use be of any consequence when considering coastal management policies in the area?

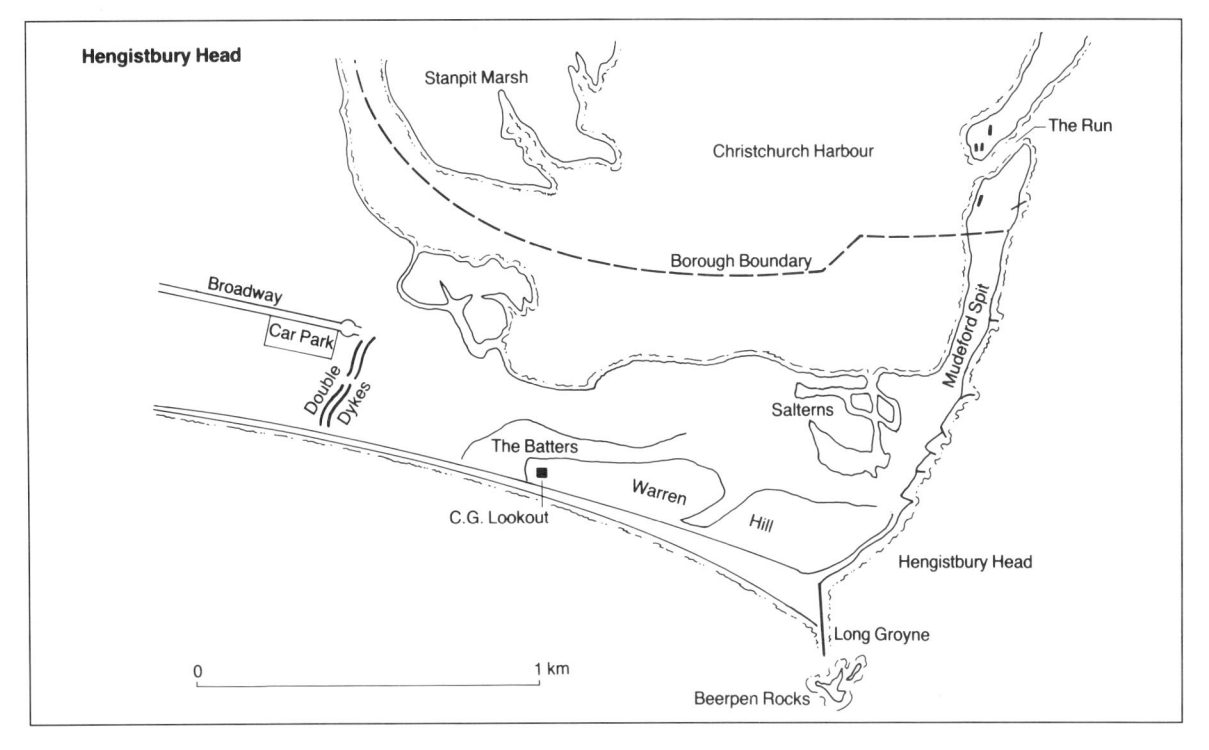

Fig. 5.2.2

215

Coastal management has been a problem in the Hengistbury Head area for many years. Reference to the map showing loss of land (Fig. 5.2.3) gives some idea of the amount of coastal retreat since 1785. it also gives some indication of the instability of Mudeford spit, which is inextricably bound up with the problems of the Head itself.

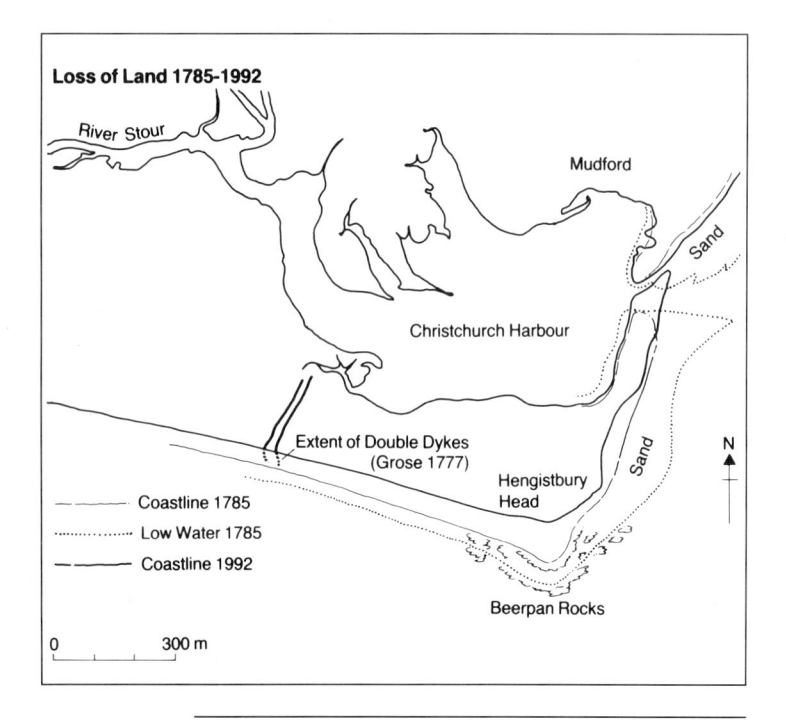

Fig. 5.2.3

Use the scale of the map to work out the annual average rate of retreat of the coastline over 200 years. Compare this with one of the fastest rates of retreat in Britain, in Holderness, on the east coast of Yorkshire at 1.88 metre per year.

Fig. 5.2.5
Low cliffs to the west of Double Dykes, Hengistbury Head: note the capping of blown sand, supporting marram grass

The geological section (Fig. 5.2.4) shows that the Head is made up of relatively soft and poorly consolidated rocks of Tertiary Age (an exception being the layers of ironstone or doggers as they are locally known), with a capping of Quaternary gravels and blown sand. In the section of low cliffs between Hengistbury Head and the outskirts of Bournemouth the coastal section is made up almost entirely of Quaternary gravels and blown sand (Fig. 5.2.5). The nature of the geology is important, not only for its control over the rate of cliff-foot erosion by the sea, but also for the way in which it controls the movement of surface water down the cliff-face, and within the rock itself. There are three main formations that make up the bulk of the cliff (see Fig. 5.2.6): the Boscombe Beds, mainly sands but with some bands of clay; the Hengistbury Beds,

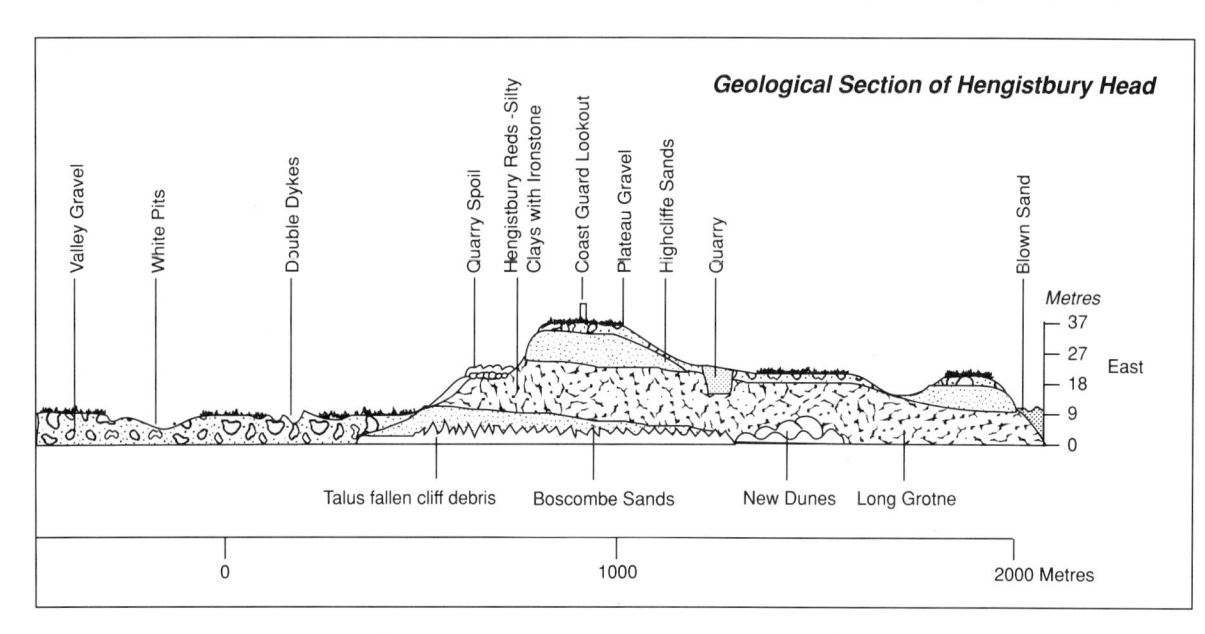

Fig. 5.2.4

mostly sands or sandy clays, with the important bands of tough ironstone, and the Highcliffe Sands that are only preserved in the higher parts of the cliff on Warren Hill (where the cliff reaches heights of some 40 m). The significance of the bands of clay will be seen when discussing the feature of sub-aerial erosion on the cliff.

Coastal management at Hengistbury Head is complicated by the fact that different local authorities are responsible for adjacent parts of the coastline: this is shown on the map (Fig. 5.2.2). Policies pursued by one authority may not always operate in harmony with those of the neighbouring one.

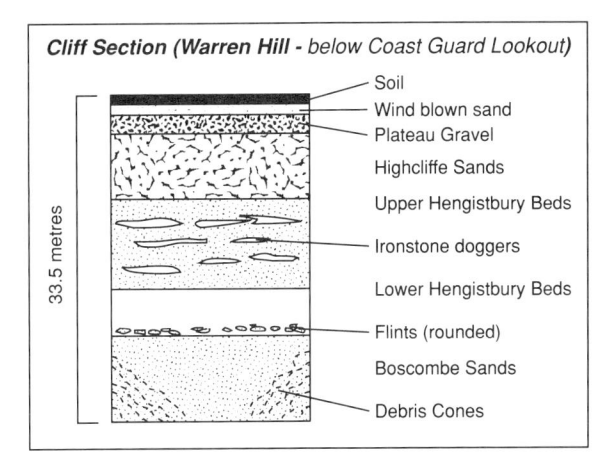

Cliff Section (Warren Hill - below Coast Guard Lookout)

33.5 metres
- Soil
- Wind blown sand
- Plateau Gravel
- Highcliffe Sands
- Upper Hengistbury Beds
- Ironstone doggers
- Lower Hengistbury Beds
- Flints (rounded)
- Boscombe Sands
- Debris Cones

Fig. 5.2.6

Hengistbury Head carries no population now, apart from the summer visitors that live in the beach huts and chalets along the Mudeford sandspit. It is however a major archaeological site, having been occupied from Stone Age to Roman times. Probably the most important feature is the defensive rampart of Double Dykes that spans the narrow isthmus that joins Hengistbury Head to Bournemouth (Fig. 5.2.7).

Fig. 5.2.7
Double Dykes: ancient earthwork at Hengistbury Head

Erosion processes along the coastline

The map of Poole Bay (Fig. 5.2.8a) shows the general orientation of the coastline in the area, and the rose diagram (Fig. 5.2.8b) shows wind frequency and direction.

Fig. 5.2.8a

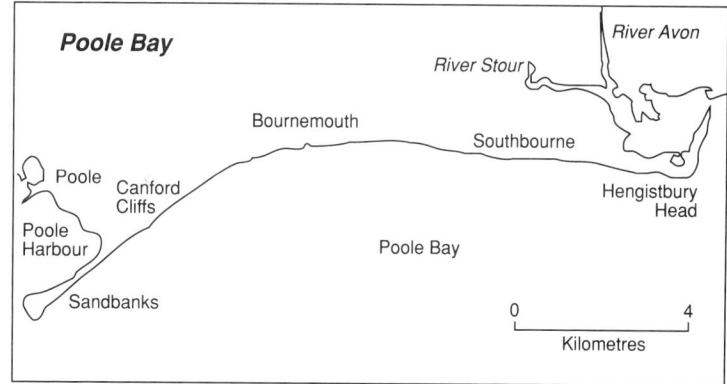

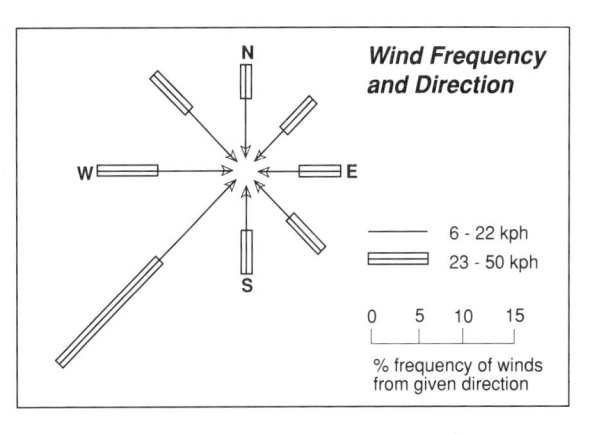

Poole Bay — showing River Avon, River Stour, Bournemouth, Southbourne, Hengistbury Head, Poole, Canford Cliffs, Poole Harbour, Sandbanks, Poole Bay. Scale 0–4 Kilometres.

Wind Frequency and Direction
- 6 - 22 kph
- 23 - 50 kph
- 0, 5, 10, 15 % frequency of winds from given direction

Fig. 5.2.8b

1. Use the map and the original aerial photograph, to determine the main direction of longshire drift (ignore the spit at Sandbanks, which is due to special local features of the coastal movement of beach material).

2. Use the rose diagram to determine the direction from which the wind most frequently blows. Wave direction will generally reflect wind direction and therefore this can also be ascertained.

Two types of waves which relate to direction may be distinguished:

a those waves that are experienced most frequently are known as prevalent waves, and their direction can be determined as above

b those waves that develop when the wind is blowing over the maximum distance of open sea (known as dominant waves), and are powerful in terms of both size and effectiveness (in terms of erosional capability).

Prevalent waves are responsible for the direction of longshore drift, and dominant waves are responsible for maximum concentration of wave-attack on the shoreline.

1 Use a good atlas map of the south coast of England to determine this maximum distance (the distance of greatest fetch) for the coast at Hengistbury.

2 Which parts of the coastline of Poole Bay are most susceptible to powerful wave-attack?

The area of Hengistbury Head beyond the Long Groyne (built in 1938) is also susceptible to wave-attack, either because of the occasional storm waves from the south-east, or because waves from the south-west are refracted by the groyne and the submerged Beerpan Rocks offshore (Fig. 5.2.9b).

Fig. 5.2.9(b)
Cliffs undergoing erosion (1973) to the east of the Long Groyne

Processes affecting the cliff-foot at Hengistbury Head

All of the cliff-foot from Solent Road, on the eastern outskirts of Bournemouth, to the end of Hengistbury Head is potentially vulnerable to wave-attack. The cliff-foot is protected by an apron of sand and shingle beach but if this is removed for any reason, then the soft, poorly consolidated rocks will yield readily to wave-attack. The photograph of the cliffs to the east of Double Dykes at the Batters (Fig. 5.2.9a) shows a notch cut by recent erosion, and the subsequent collapse of overlying cliff material.

Fig. 5.2.9(a)
Wave cut notch, collapsed cliff at the Batters, east of Double Dykes

Processes affecting the cliff-face at Hengistbury Head

There appear to be a number of processes operative on the cliff-face.

a Gullying: water flowing over the surface of the cliff is capable of cutting into its face, and in many places the face is seamed with a series of parallel gullies. Sometimes these gullies commence at the top of the cliff: in other places they seem to be related to seepages where water is leaking out because of the frequent impermeable bands in the sands that make up most of the cliff (Fig. 5.2.12).

b Splashback: where the water from the gullies drips or falls on to soft material below, the splashes erode a hollow that increases in size until the cliff above it is undermined, and collapses (Fig. 5.2.12).

c Rock falls: these are usually the result of undermining, as indicated in **b**.

d Flows: where material becomes saturated (particularly in the sandy clays and clays where it is liable to flow down the cliff).

Debris from all of these cliff-face processes tends to collect in large talus cones at the foot of the cliff, and these do, of course, offer some protection from wave-attack.

THE MANAGEMENT OF COASTAL PROCESSES AND LANDFORMS

Fig. 5.2.10

Fig. 5.2.11
Cliff at Warren Hill: Hengistbury Head

Make a copy of the sketch (Fig. 5.2.11) of the photograph (Fig. 5.2.12) which represents the cliff face at the highest point of Warren Hill: mark in the main geological horizons (refer back to Fig. 5.2.6) and then identify and label as many features mentioned in the previous section as possible.

The coastal management problem

The present problem of coastal management needs to be studied in the light of the first major defence work undertaken, the building of the Long Groyne by Bournemouth Corporation in 1938. The aim of the building of this groyne is fairly clear from the aerial photograph (Fig. 5.2.1a) at the beginning of the section. Although it has been successful in causing a build-up of beach material to the west, and thus protecting the cliff-foot from marine attack, this has resulted in what has been described as the terminal groyne problem. The groyne, while

Fig. 5.2.12
Splash-back erosion and gullying: Hengistbury Head

allowing accumulation of sand and shingle to the west, has been responsible for the stopping of longshore drift of material, and thus causing beach starvation to the east and north-east. Serious cliff-foot erosion has occurred in this area as a result, and this has had a knock-on effect on the sandspit. Ironically enough the problem had been transferred to the very doorstep of the area that Christchurch Borough Council had obtained on a 98 year lease from Bournemouth in 1931 – an unwitting case of coastal beggar-my-neighbour!

Two major problems exist today.

1 The above problem referred to beyond the Long Groyne. Serious coastal erosion was causing concern in this area, and this was accompanied by very active processes on the cliff-face, including the widening and deepening of a very prominent gully. Starved of sediment, the Mudeford sandspit was also clearly at risk. Beyond the sandspit lies the unstable exit of the rivers Stour and Avon from Christchurch Harbour. Although not unconnected with coastal management at Hengistbury Head, this was an entirely different problem that Christchurch Borough Council was called upon to address, which has involved them in expensive defence works.

2 The problem in the area from Solent Road to just beyond Double Dykes. Return to the aerial photograph of Hengistbury Head (Fig. 5.2.1a). Notice the narrow neck of land that connects Hengistbury Head to the main area of Bournemouth.

1 With such rapid coastal erosion along this coast, what could eventually happen here to isolate the Head?

2 Are there any catastrophic circumstances in which it could happen more quickly?

Clearly this section is very vulnerable, and the need for protection is heightened by the presence of the Iron Age earthwork at Double Dykes, and by the important recreational facilities on the area of land between the coast and Christchurch Harbour.

Selecting two problems is not to ignore the remainder of the coast at Hengistbury Head. The cliffs at Warren Hill are still under some threat, but with limited local resources priorities have to be decided. The map (Fig. 5.2.13) shows the areas most vulnerable to further erosion at Hengistbury Head.

Fig. 5.2.13

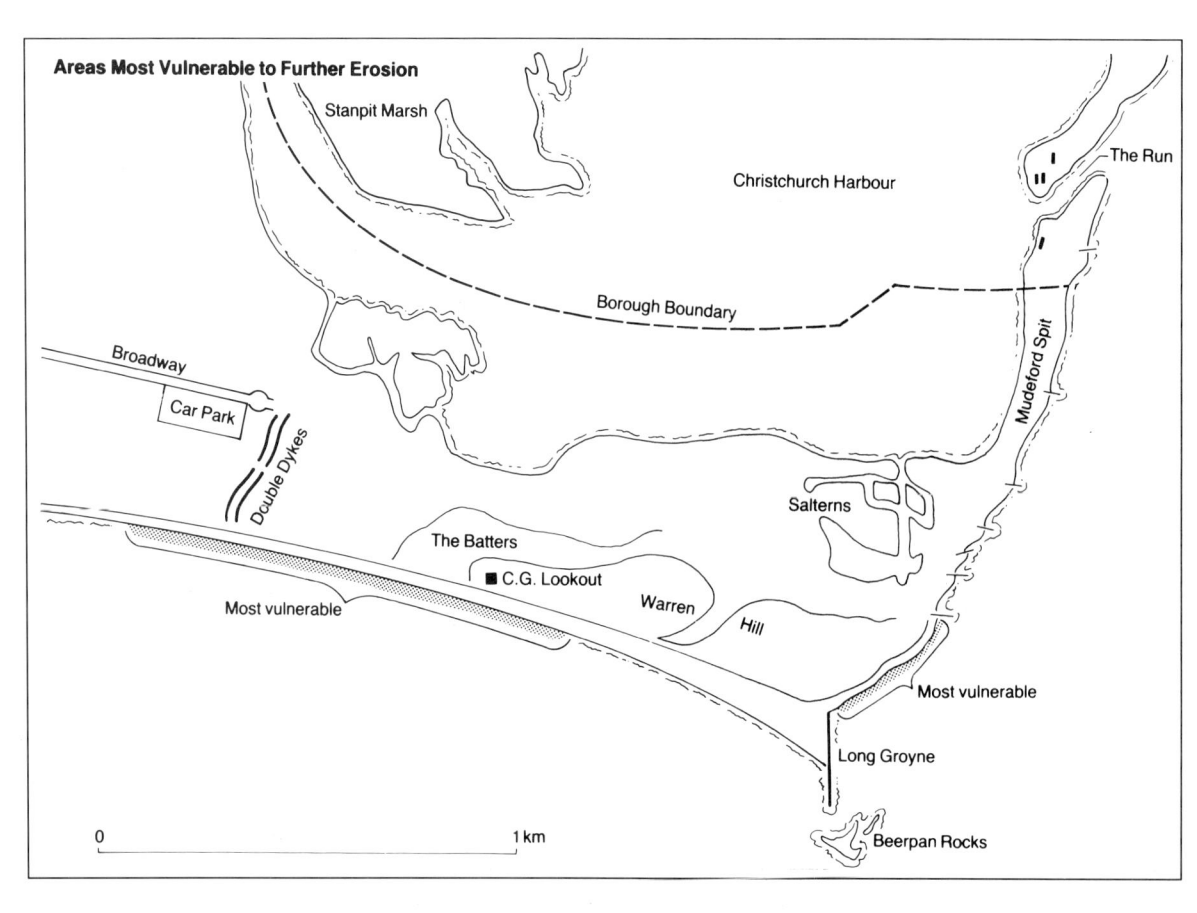

Dealing with the problems

With coastal defence work, there are a number of possible solutions that are available to the engineers dealing with the erosion problem. These may be broadly divided into two: those that are used at the cliff-foot and those that are used on the cliff-face, and clearly there is a degree of interdependence between the two.

Cliff-foot solutions include:

- the building of a seawall
- the building of a series of groynes, using either stone or timber
- using gabion armouring along selected sections of the coast that most need protection
- the use of revetments (pilings that are buttressed by a backfill of boulders to dissipate wave energy)
- the use of rip-rap protection (large fragments of broken rock tipped at high water mark).

Cliff-face solutions include:

- the use of brushwood to fill in small gullies
- the use of gabion cages (metal cages filled with rock fragments) in stepped formation in gullies, or at the base of cliffs to check earthflow
- draining off groundwater behind the cliff-face
- regrading of the cliff to produce a more even slope.

The map (Fig. 5.2.14) shows the methods of coastal protection currently used at Hengistbury Head.

1   Compare this map with the one showing the areas most vulnerable to further erosion at Hengistbury Head (Fig. 5.2.13). Comment on the likely effectiveness of the methods being used.

2   Draw up a table to show the advantages and disadvantages of the different methods of coastal protection. Make a separate table for cliff-face methods and cliff-foot methods.

Fig. 5.2.14

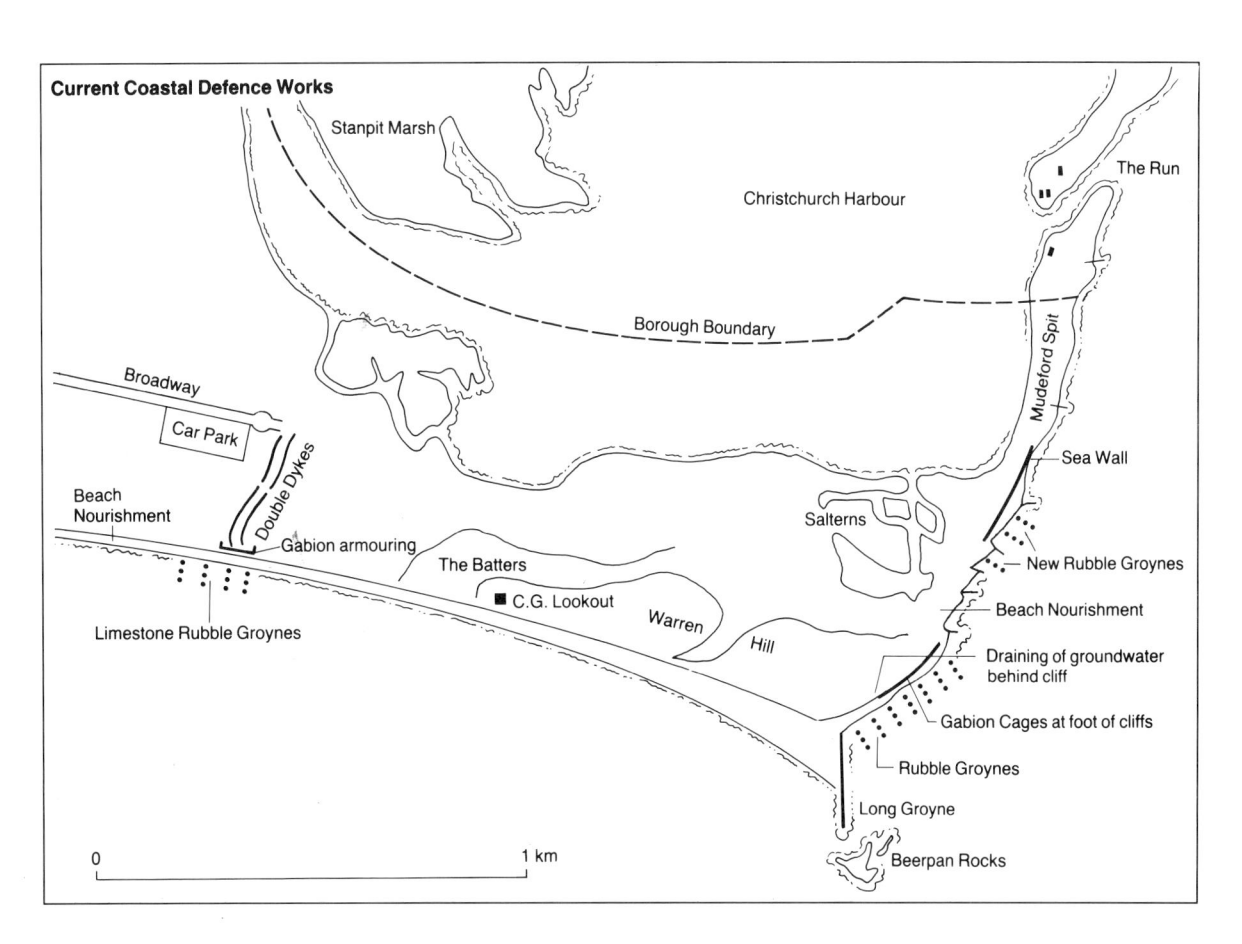

A NEW VIEW OF BRITAIN

Evaluating coastal defence measures

Study the map (Fig. 5.2.15) which shows all of the coastal defence schemes undertaken by Bournemouth and Christchurch Borough Councils in the period between 1937 and 1984.

1 Which areas seem to have been the site of most of the defence work that has been undertaken during this period?

2 Is there any evidence of a clear coastal defence policy as shown by the works completed during this period?

3 What would you consider to be the main shortcoming of such a policy?

Fig. 5.2.15

Coastal Defence Work 1937–1984

BBC Bournemouth Borough Council
CBC Christchurch Borough Council

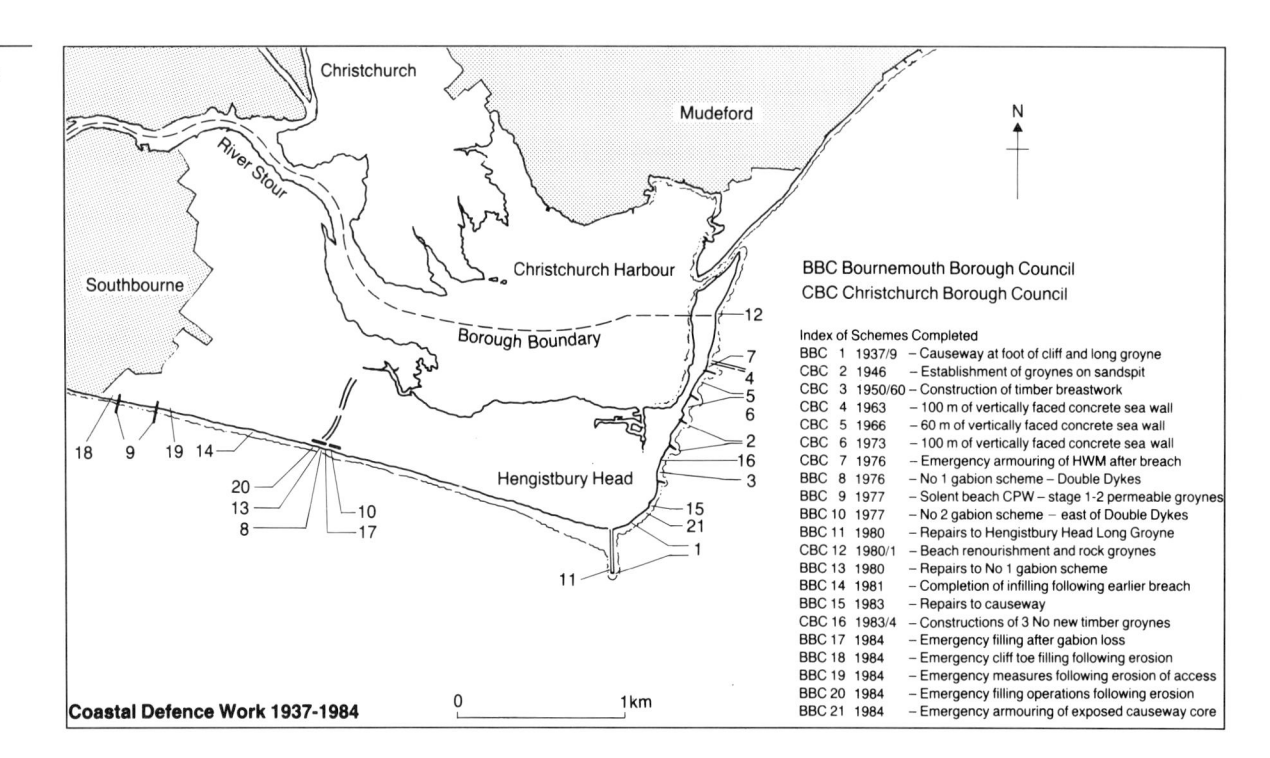

Index of Schemes Completed
BBC 1 1937/9 – Causeway at foot of cliff and long groyne
CBC 2 1946   – Establishment of groynes on sandspit
CBC 3 1950/60 – Construction of timber breastwork
CBC 4 1963   – 100 m of vertically faced concrete sea wall
CBC 5 1966   – 60 m of vertically faced concrete sea wall
CBC 6 1973   – 100 m of vertically faced concrete sea wall
CBC 7 1976   – Emergency armouring of HWM after breach
BBC 8 1976   – No 1 gabion scheme – Double Dykes
BBC 9 1977   – Solent beach CPW – stage 1-2 permeable groynes
BBC 10 1977   – No 2 gabion scheme – east of Double Dykes
BBC 11 1980   – Repairs to Hengistbury Head Long Groyne
CBC 12 1980/1 – Beach renourishment and rock groynes
BBC 13 1980   – Repairs to No 1 gabion scheme
BBC 14 1981   – Completion of infilling following earlier breach
BBC 15 1983   – Repairs to causeway
CBC 16 1983/4 – Constructions of 3 No new timber groynes
BBC 17 1984   – Emergency filling after gabion loss
BBC 18 1984   – Emergency cliff toe filling following erosion
BBC 19 1984   – Emergency measures following erosion of access
BBC 20 1984   – Emergency filling operations following erosion
BBC 21 1984   – Emergency armouring of exposed causeway core

Fig. 5.2.16

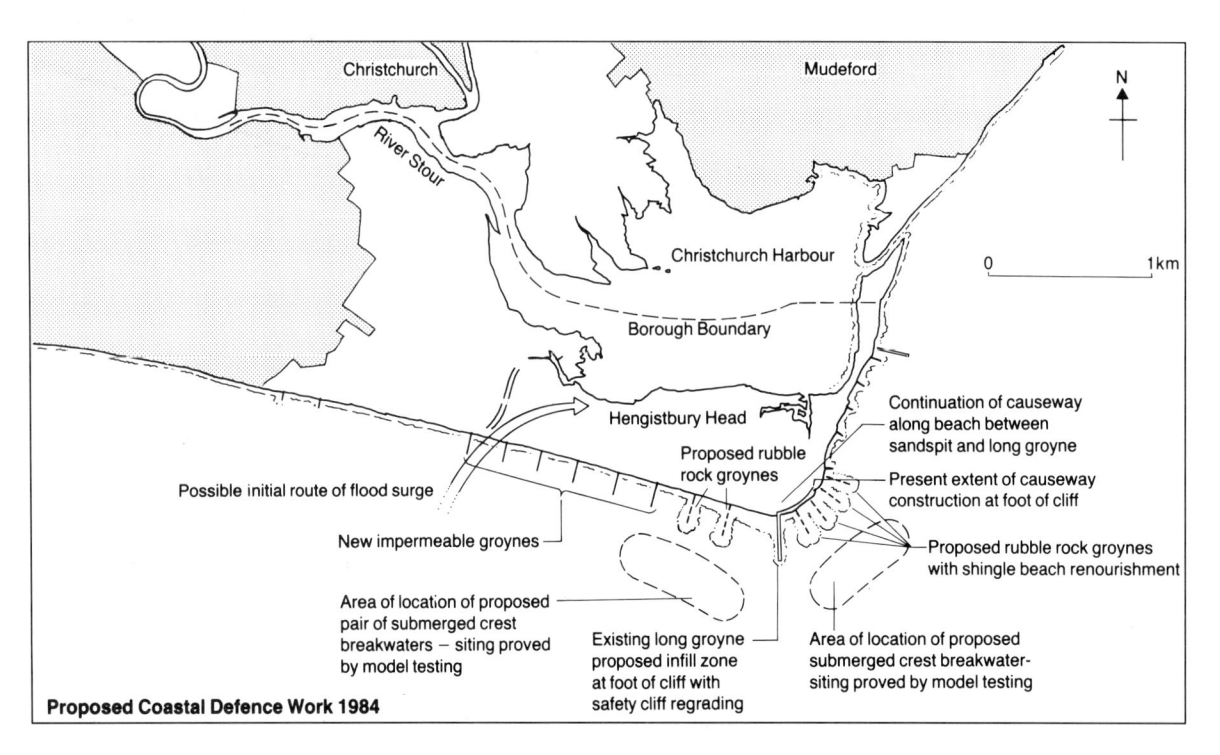

Proposed Coastal Defence Work 1984

THE MANAGEMENT OF COASTAL PROCESSES AND LANDFORMS

No confidence in erosion scheme

Nearly £1 000 000 may be spent by Bournemouth Borough Council on solving the problems of erosion at Double Dykes and Hengistbury Head.

The Council have produced a scheme to build nine breakwaters between Solent Road and Warren Hill. They believe this will halt the problem of wind, rain and sea battering down the cliffs, threatening property, and possibly eventually turning Hengistbury Head into an island.

A total of £970 000 will be needed to complete the proposed breakwaters plus further work to the east of the causeway.

A letter has been sent to the Chief Executive of Bournemouth Council, Mr. Keith Lomas by Independent Southbourne councillor Ken Thresher. Mr. Thresher is Chairman of 'Determined Action to Save Hengistbury Head' (D.A.S.H.), and his letter is highly critical of the action so far by Bournemouth Council.

One point he raises says, 'the July '84 report (by Bournemouth Council) indicates £60 000 to be spent in 1986–87 for HRS model testing. We are concerned on several points on this particular issue. We have no confidence in the Council, and their past handling of this complex and difficult problem gives considerable cause for concern. We think it downright stupid to spend nearly £1 million before seeking the advice and guidance of HRS."

In another section of the letter he accuses the Council of making a 'crass mistake' in removing 'doggers' from the foot of the cliff"... which led directly to the recent massive cliff fall."

As to the cost of the desperately needed reclamation project, Mr. Thresher says, "We do not accept the arguments about lack of funds, moratoriums, etc., as this Council has demonstrated quite clearly that it can raise large sums of money to fund projects which it desires; witness the £18 million raised without recourse to the Ratepayers' pockets for the BIC, a development which, it is still argued, will assist the basic economy of the town. We believe Hengistbury Head and all it stands for is of similar, if not greater, importance.

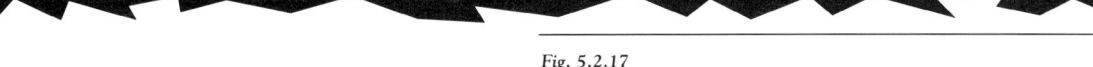

Fig. 5.2.17
Evening Echo, Bournemouth 24 January 1985

The second map produced by Bournemouth Borough Council in 1984 (Fig. 5.2.16) shows work to be undertaken in the period from 1984 onwards.

1 How does this scheme differ from those that were operating in the period from 1937–84?

2 Why do you think that Bournemouth Borough has opted for such a scheme?

3 Read the short article from the local newspaper (Fig. 5.2.17). Summarise the points being made in the article (HRS = Hydraulic Research Services, BIC = Bournemouth International Centre, major conference centre in Bournemouth). What are the essential values that underly these views? How far do you agree with them?

Photographs of two locations are shown here (Figs. 5.2.18 and 5.2.19): at Double Dykes and at the coastal zone to the east of the Long Groyne. Both of these photos were taken in 1992, seven years after the works were completed.

1 Examine the Double Dykes photograph (Fig. 5.2.18). How effective do you think the gabion armouring will be in protecting this earthwork? What is the main disadvantage of such a strengthening of this small stretch of coast?

Fig. 5.2.18
Gabion Armouring, Double Dykes

Fig. 5.2.19
Groynes to East of Long Groyne. The Long Groyne is the last groyne to the right

A NEW VIEW OF BRITAIN

2 Examine the photograph of the coastline (Fig. 5.2.19) to the east of the Long Groyne and compare it with the view of this piece of coastline in the aerial photograph (Fig. 5.2.1a) which shows what it was like before. What are the main differences?

Cost-benefit analysis

Almost all schemes such as the one under consideration here are subjected to what is known as cost-benefit analysis. In such an analysis the known cost of the scheme (in this case the initial estimate was of the order of £5 million) is measured against the benefits accruing from the construction of the coastal defence works. Known costs are fairly easy to calculate (although 1985 prices will be very different in the 1990s!). Non-quantifiable costs also have to be considered. Once the costs are known it is then necessary to work out the benefits that the scheme will bring.

Working out the detailed figures for a cost-benefit analysis is beyond the scope of the following simple exercise. You are required to construct a table that shows, in non-quantified terms, the main costs, and the main benefits (these could be divided into tangible and non-tangible) of the current schemes at Hengistbury Head.

What are the main difficulties in setting up such a cost-benefit analysis? To what statistics would you need access in order to complete such an exercise?

A NEW SEA WALL FOR PORTHLEVEN

The Map (Fig. 5.2.20) shows the position of the town of Porthleven on the western side of the Lizard Peninsula in Cornwall. It occupies a very exposed position, open to the strong waves generated by winds blowing over an almost unlimited fetch that extends across the North Atlantic Ocean.

Fig. 5.2.20

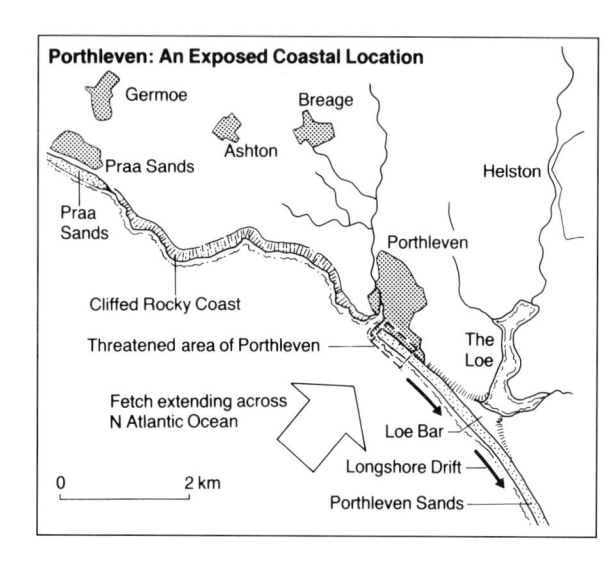

Fig. 5.2.21

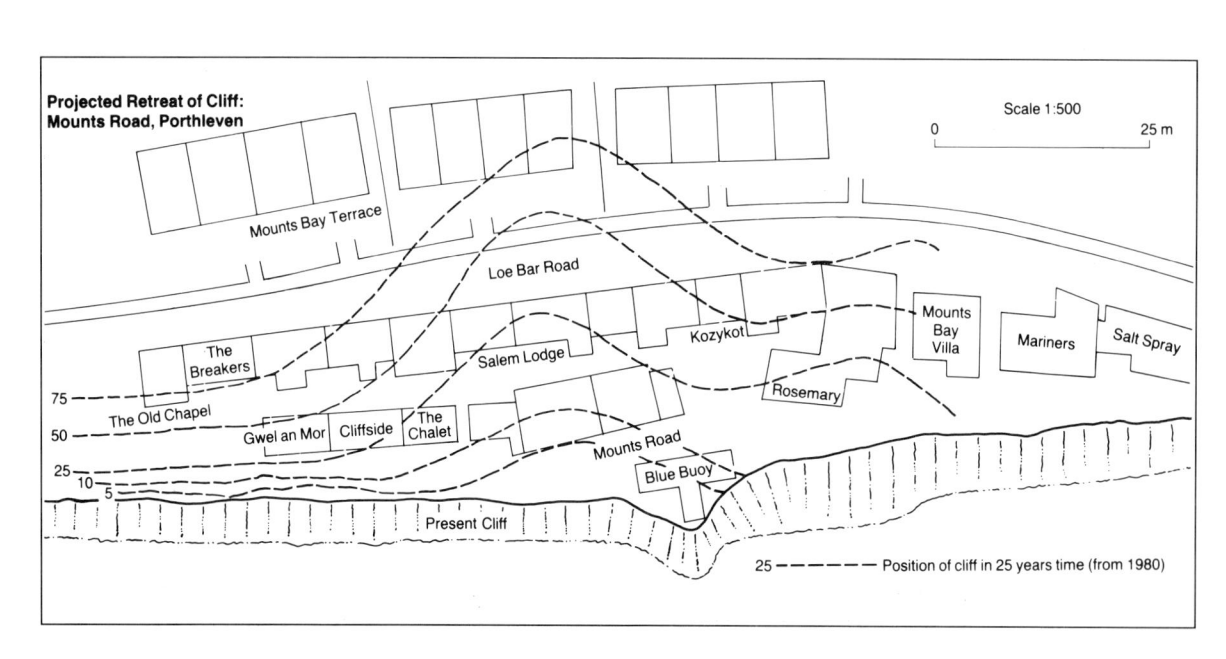

THE MANAGEMENT OF COASTAL PROCESSES AND LANDFORMS

Porthleven itself is built around the small sheltered harbour, but the settlement extends eastwards around the coast along the cliff-top. It is this section that has become increasingly vulnerable owing to its exposure to the south-west (see Fig. 5.2.20). The cliffs here are of the Devonian Mylor slates and greywackes, relatively tough and resistant rocks, very different to the weaker rocks at Hengistbury Head that have proved so susceptible to erosion.

The large-scale map (Fig. 5.2.21) shows the coastal road to Loe Bar and the properties on either side together with the estimated extent of erosion that would have occurred up to the year 2055. The road carries a number of important public utilities, such as sewerage, gas and electricity.

1 Using the map, estimate the total number of properties that are likely to suffer damage, or total destruction over the 65 year period to 2055.

2 Identify the other principal losses to the community of Porthleven if the erosion were to proceed over the next 65 years.

The erosion problem was recognised in a Department of Environment Coast Protection Survey in 1980, and Kerrier District Council approved the scheme shown on the map (Fig. 5.2.22) in the mid 1980s. It was to be carried out in a number of phases that are shown on the map.

Fig. 5.2.23 Coastal Defences, Porthleven

The main protection was to be a sea wall built in a number of stages. It is a mass concrete sea wall with a curved wave deflector, backed at the top of the cliff with granite stone walling acting as protection to mass concrete behind. Above this section there is slope stabilisation with gabions and geotextile materials (see Fig. 5.2.23). Protection is thus afforded to the toe of the cliff, and to the upper part of the cliff, which was badly damaged in some areas in the storms of late 1989 and 1990.

The protection scheme

Fig. 5.2.22

Coastal Defence Scheme: Porthleven

Cost-benefit analysis

Initial ideas concerning cost-benefit analysis were developed in the previous section dealing with Hengistbury Head. Here they are developed more fully with reference to the coastal defence works at Porthleven. Cost-benefit analysis was carried out on the proposed scheme as shown below.

Total benefits accruing from protecting:

- 80 metres of serviced highway
- main trunk sewer
- 60 m of footway
- 285 m of serviced highway
- properties that would be damaged or lost.

Total benefits accruing over 75 years = £1 176 814

Total cost of scheme = £1 125 111

1   Calculate the cost-benefit ratio.

2   In 1985 Cornwall County Council spent £40 000 on works trying to prevent the existing highway from slipping into the sea. If the works were not carried out then it was likely that a similar amount would have to be spent every ten years over the 75 year period that the scheme was planned to last.

Calculate a new cost-benefit ratio taking this additional spending into account (if the scheme did not go ahead) i.e. cost to Cornwall County Council of this additional work for the next 75 years = $^{75}/_{10}$ × 40 000. Add this amount to the existing benefits (£1 176 814) and then divide the new amount by the cost of the existing scheme (£1 125 111).

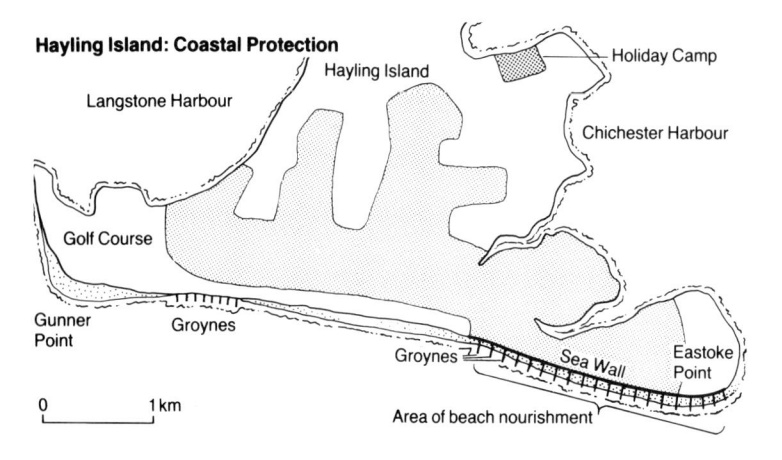

a   How has this additional consideration altered the cost-benefit ratio?

b   What factors have not been taken into account? How would they affect the cost-benefit ratio?

3   Coastal management schemes often have a 'knock-on effect' along the coast. Study the map (Fig. 5.2.20) and suggest what such an effect might be in the immediate area of Porthleven.

BEACH NOURISHMENT ON HAYLING ISLAND, HAMPSHIRE

Hayling Island beach extends for some 6.5 km from Gunner Point at the entrance to Langstone Harbour to Eastoke Point, at the entrance to Chichester Harbour (Fig. 5.2.24). The beach is one of the most important amenity features in East Hampshire. It consists of two main facets, the shingle upper beach, and the lower foreshore of sand, exposed mainly at low tide (Fig. 5.2.25). The soft rock geology of the area (Eocene sands) means that left to itself, the coastline would adjust to the waves and tidal forces acting upon it. However, Hayling Island is now quite densely populated, not only serving as a resort, but also as a major commuting centre in the Portsmouth conurbation. The frontage of Hayling Island, because of its exposure to storm surges in the English Channel, is protected over a length of 2 km in the east by a sea wall or timber revetment, together with a series of groynes along the beach (Fig. 5.2.26). The beach to the west requires less in the way of protection, with only 0.5 km requiring protection (Fig. 5.2.24).

The broad features of sediment circulation in the coastal area around Hayling Island are shown in the map (Fig. 5.2.27). The eastern part of Hayling Island Beach has suffered over the years from loss of sediment, particularly shingle. This has led to overtopping of sea walls and extensive repair and maintenance bills on existing structures.

Fig. 5.2.24

Hayling Island: Coastal Protection

1   Describe the main features of the circulation of sediment in the Hayling Island area.

2   What would appear to be the principal reason for reduced supply of shingle to the eastern part of Hayling Island in recent years? (Another case of coastal beggar-my-neighbour?)

THE MANAGEMENT OF COASTAL PROCESSES AND LANDFORMS

Fig. 5.2.25
Beach at Hayling Island: site of the replenishment scheme

Fig. 5.2.26
Aerial photograph of Hayling Island showing groynes

Fig. 5.2.27

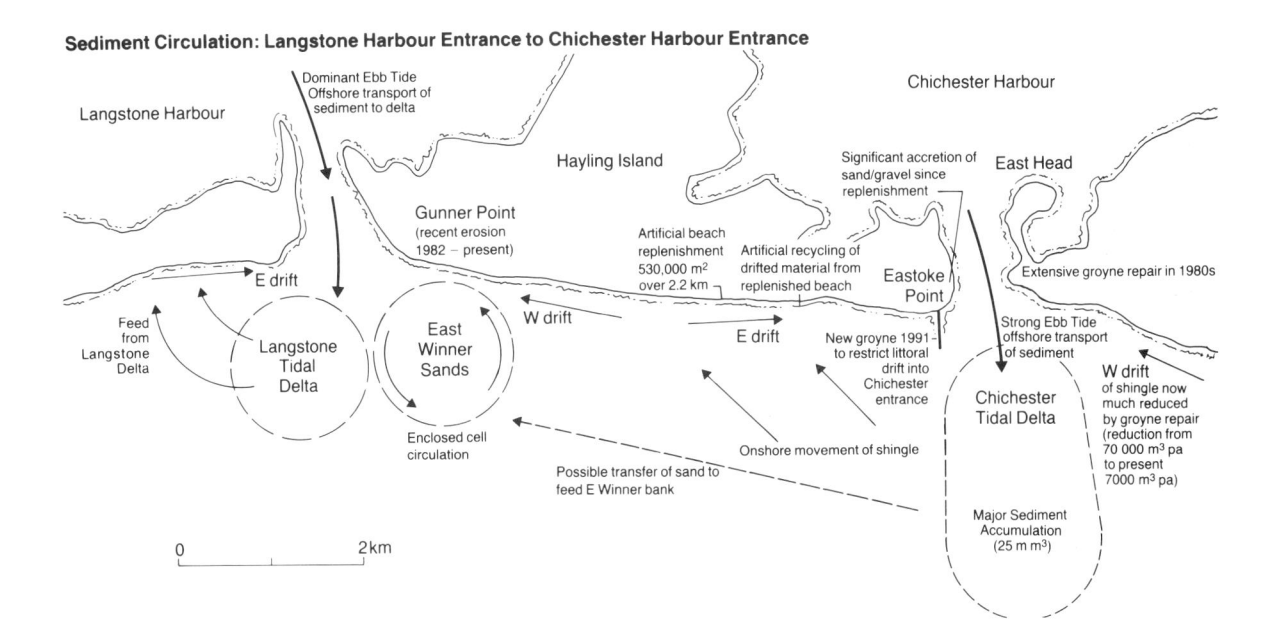

Sediment Circulation: Langstone Harbour Entrance to Chichester Harbour Entrance

The beach replenishment scheme

Beach replenishment by sand is a fairly common form of coastal management, but the concept of using shingle is a relatively new one. In 1985 500 000 m³ of shingle, dredged from an offshore location was supplied to the beach from shallow draft barges, and then bulldozed on to the eastern beach for a frontage of 2 km. The crest of the beach was raised to a height of 6 m above O.D. The full benefits of the replacement scheme were felt in the following winters. Overtopping no longer took place when anticipated, particularly in the very severe storm surge of March 1987.

The effects of the replenishment on coastal dynamics

(Reference to the map of sediment circulation (Fig. 5.2.27) will help to answer some of the questions below).

1   Suggest what effects the beach replenishment scheme may have had on

- existing groynes in this section of the beach
- the beach profile in the replenished area
- longshore drift along the eastern part of the beach
- shingle accumulation at the entrance to Chichester Harbour.

2   What new measures have been adopted to manage changes in shingle accumulation at Eastoke Point and the entrance to Chichester Harbour?

SUMMARY: THE OPTIONS FOR COASTAL MANAGEMENT

The table below (Fig. 5.2.28) summarises the main options for coastal management. For each of the case studies in this section, discuss the way in which it might be said to fit one of the options identified in the summary. Some case studies might represent a mix of more than one option.

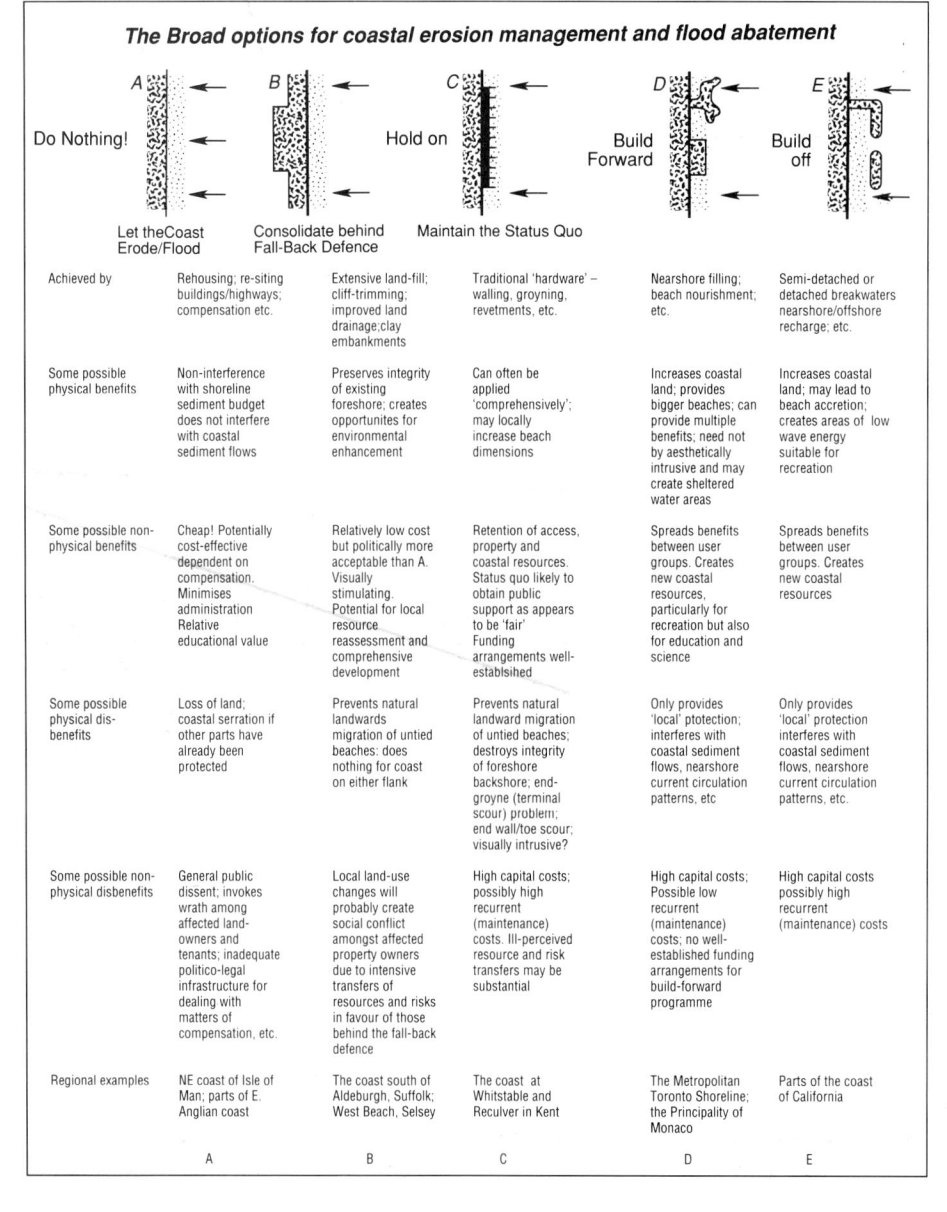

The Broad options for coastal erosion management and flood abatement — Fig. 5.2.28

	A — Do Nothing! Let the Coast Erode/Flood	B — Consolidate behind Fall-Back Defence	C — Hold on Maintain the Status Quo	D — Build Forward	E — Build off
Achieved by	Rehousing; re-siting buildings/highways; compensation etc.	Extensive land-fill; cliff-trimming; improved land drainage; clay embankments	Traditional 'hardware' – walling, groyning, revetments, etc.	Nearshore filling; beach nourishment; etc.	Semi-detached or detached breakwaters nearshore/offshore recharge; etc.
Some possible physical benefits	Non-interference with shoreline sediment budget does not interfere with coastal sediment flows	Preserves integrity of existing foreshore; creates opportunites for environmental enhancement	Can often be applied 'comprehensively'; may locally increase beach dimensions	Increases coastal land; provides bigger beaches; can provide multiple benefits; need not by aesthetically intrusive and may create sheltered water areas	Increases coastal land; may lead to beach accretion; creates areas of low wave energy suitable for recreation
Some possible non-physical benefits	Cheap! Potentially cost-effective dependent on compensation. Minimises administration Relative educational value	Relatively low cost but politically more acceptable than A. Visually stimulating. Potential for local resource reassessment and comprehensive development	Retention of access, property and coastal resources. Status quo likely to obtain public support as appears to be 'fair'. Funding arrangements well-established	Spreads benefits between user groups. Creates new coastal resources, particularly for recreation but also for education and science	Spreads benefits between user groups. Creates new coastal resources
Some possible physical dis-benefits	Loss of land; coastal serration if other parts have already been protected	Prevents natural landwards migration of untied beaches; does nothing for coast on either flank	Prevents natural landward migration of untied beaches; destroys integrity of foreshore backshore; end-groyne (terminal scour) problem; end wall/toe scour; visually intrusive?	Only provides 'local' protection; interferes with coastal sediment flows, nearshore current circulation patterns, etc	Only provides 'local' protection interferes with coastal sediment flows, nearshore current circulation patterns, etc.
Some possible non-physical disbenefits	General public dissent; invokes wrath among affected land-owners and tenants; inadequate politico-legal infrastructure for dealing with matters of compensation, etc.	Local land-use changes will probably create social conflict amongst affected property owners due to intensive transfers of resources and risks in favour of those behind the fall-back defence	High capital costs; possibly high recurrent (maintenance) costs. Ill-perceived resource and risk transfers may be substantial	High capital costs; Possible low recurrent (maintenance) costs; no well-established funding arrangements for build-forward programme	High capital costs possibly high recurrent (maintenance) costs
Regional examples	NE coast of Isle of Man; parts of E. Anglian coast	The coast south of Aldeburgh, Suffolk; West Beach, Selsey	The coast at Whitstable and Reculver in Kent	The Metropolitan Toronto Shoreline; the Principality of Monaco	Parts of the coast of California

COASTAL FLOODING

The Towyn disaster has provided many lessons on the risks of planning in hazardous environments around the British coast. How the planning system responds to flooding and the possibility of higher risks in the future due to the greenhouse effect needs to be addressed. The knowledge gained from these floods must be quickly implemented if the number of future Towyn disasters is to be reduced.

(Robert Kay and Angela Wilkinson, 'Lessons from the Towyn Flooding', The Planner, 17 August 1990)

Flooding represents one of the biggest challenges to management on the British coastline. Overtopping of seawalls and the consequent risk to properties behind prompted the action at Hayling Island. Since the major disaster caused by the East Coast floods in January 1953, threatened coastal communities around the British coast have had a heightened awareness of the potential dangers. Although defences against flooding do exist in many locations this inadequacy has been only too clearly illustrated in the disaster that struck the coastal settlements of North Wales in February 1990. Global warming, and the consequent rise in sea level pose another sinister threat to many coastal settlements and clearly raise some fundamental questions concerning the suitability of the design of existing structures.

IN THE WAKE OF THE TOWYN DISASTER: PLANNING FOR THE FUTURE

Towyn lies on a low-lying part of the coast of North Wales (see Fig. 5.3.1). Most of the area around the town is below normal spring tide levels, and is protected from the sea by an embankment owned by British Rail, which runs from Pensarn to Kinmel Bay. The embankment was originally built in 1800, by the Rhuddlan Marsh Commissioners, to protect newly reclaimed land from incursions by the sea. The railway was built behind the sea embankment, which caused flooding of the railway, and the low-lying land behind. The Railway Company took over control of the embankment in 1880. Residential development of the area behind the embankment began in the 1920s (Fig. 5.3.2). Much of this development proved attractive to retired people, since the land was flat and undemanding, so inevitably, bungalows were predominant amongst the housing.

In the last week of Februry 1990, a large low pressure area moved east from Greenland into Scandinavia, accompanied by the movement further south of a series of depressions that moved east across Britain into the North Sea. The most severe of these storms crossed southern Scotland on 26 February, with the

Fig. 5.3.1
The Planner 17 August 1990

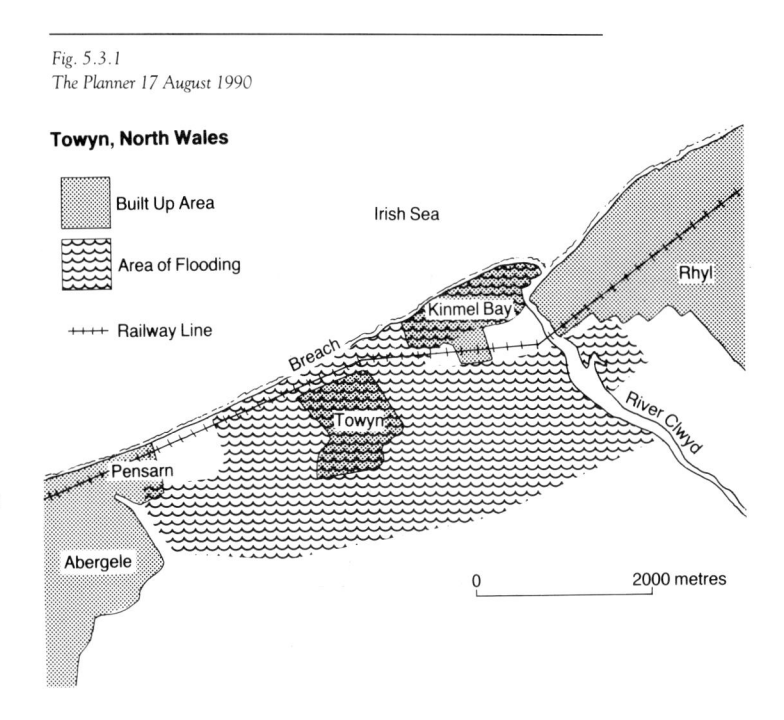

Fig. 5.3.2
Seawall and residential development, Towyn, North Wales

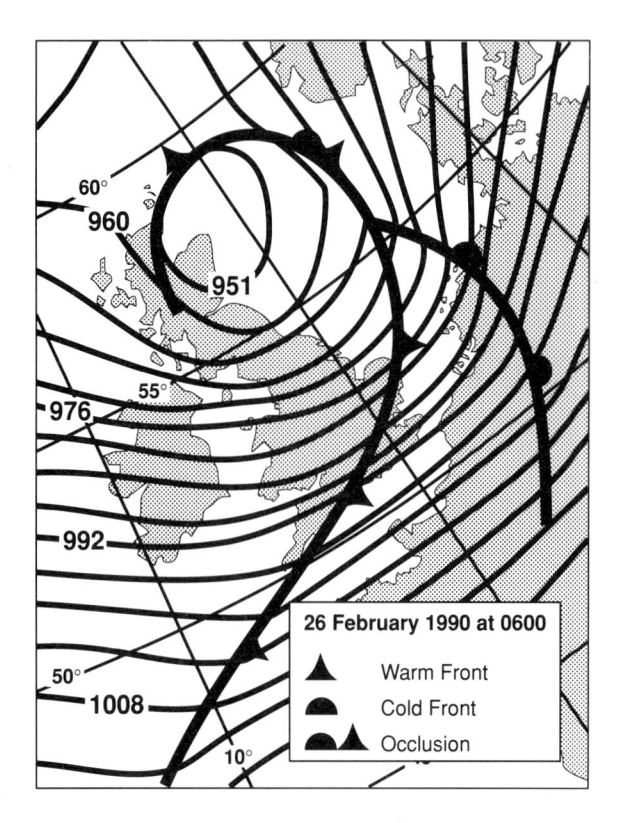

Fig. 5.3.3
Meteorological Chart 26 February 1990

Fig. 5.3.4
Breach in sea wall

Fig. 5.3.5
Flooding in Towyn

pressure reduced to 951 mb at 6 a.m. (Fig. 5.3.3). Recorded wind speeds in the area gusted to 83 knots, and continued until late afternoon, but strengthened again the following morning when renewed speeds of 70 Knots were recorded. To give some idea of the extreme nature of the conditions, it was later calculated that wave heights combined with the surge heights at almost the exact time of the highest tides of the year gave a combination return period of between 500 and a 1000 years! As the official report put it – 'We had hit the jackpot!'

At 10.30 a.m. on the 26 February, an hour before high tide, the area between the embankment and the railway was completely inundated by sea water, caused by overtopping. At 11 a.m. the sea wall could no longer withstand the force of the huge surge, and was breached, over a period of three hours for a distance of 467 m (Fig. 5.3.4). By midday on the 27 February, a tide level of 6.3 m was being experienced, with flooding extending up to depths of 1.8 m over 10 sq km in Towyn and Kinmel Bay (Fig. 5.3.5). Approximately 2800 properties were flooded, about 12 per cent of the housing stock of Colwyn Borough. Since the majority of the buildings were bungalows the level of damage to house contents was particularly high, and over 5000 people had to be evacuated from their homes. Within the area 31 per cent of the inhabitants were elderly, and later investigations showed that nearly 40 per cent of those affected had no insurance on their house contents, and 6 per cent had no property insurance. Serious flooding of this magnitude affects electricity supplies and this resulted in the failure of pumps at the many sewage stations in the area, resulting in contamination of floodwater with sewage. Similarly the main pumping station that drains surface water from the area was put out of action resulting in a delay in ridding the area of the floodwater.

1 What were the essential physical and human ingredients of the Towyn disaster?

2 With hindsight, what precautions might have been taken to lessen the likelihood of this disaster occurring? Consider the following points.

a The distribution of shingle along this section of the coast: it is concentrated in the Pensarn, Kinmel Bay and Rhyl areas, but is largely absent in front of the sea wall at Towyn.

b The age and levels of maintenance of the embankment in front of Towyn.

c Planning policies that allowed the unrestricted growth of residential development in the low-lying area landward of the railway.

Aftermath: repair, renewal and future planning

Immediate repairs to the breach were put into operation within hours of the rupture occurring. British Rail accepted full financial responsibility for emergency repairs to the wall. Fortunately North Wales has ready supplies of rock suitable for armouring an emergency repair. Armour stone was placed in a series of layers along the breach, and voids between the armour were filled with ready mixed concrete. The breach section was built up until its crest level was at least equal to the parapet level of the undamaged wall. Later a rock breakwater was built parallel with the sea wall to dissipate the force of the waves (Fig. 5.3.6). Further offshore breakwaters and groynes are due to be completed in 1992 sufficient to prevent a further breach, but the possibility of overtopping remains.

Fig. 5.3.6

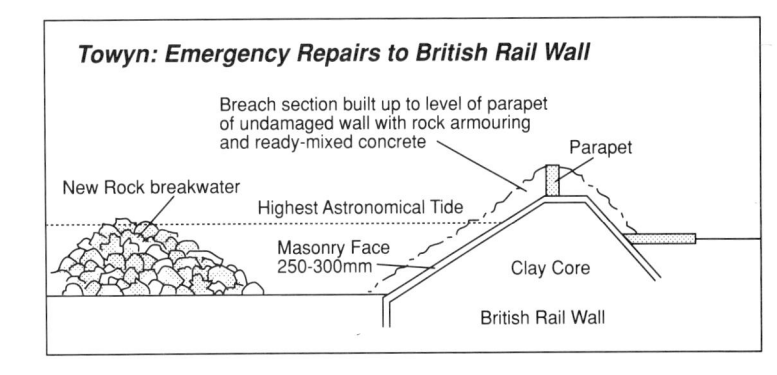

Towyn: Emergency Repairs to British Rail Wall

The map (Fig. 5.3.7) shows the current planning proposals for the Pensarn – Towyn – Kinmel Bay Area. The floods of February 1990 resulted in a moratorium on planning permission for any new residential development being proposed by the National Rivers Authority (NRA), and accepted by Colwyn Borough Council. The Local Plan allowed for the building of 1780 new dwellings in the Towyn – Kinmel Bay areas in the period 1986–96. By the end of 1989 some 244 dwellings had been built, and

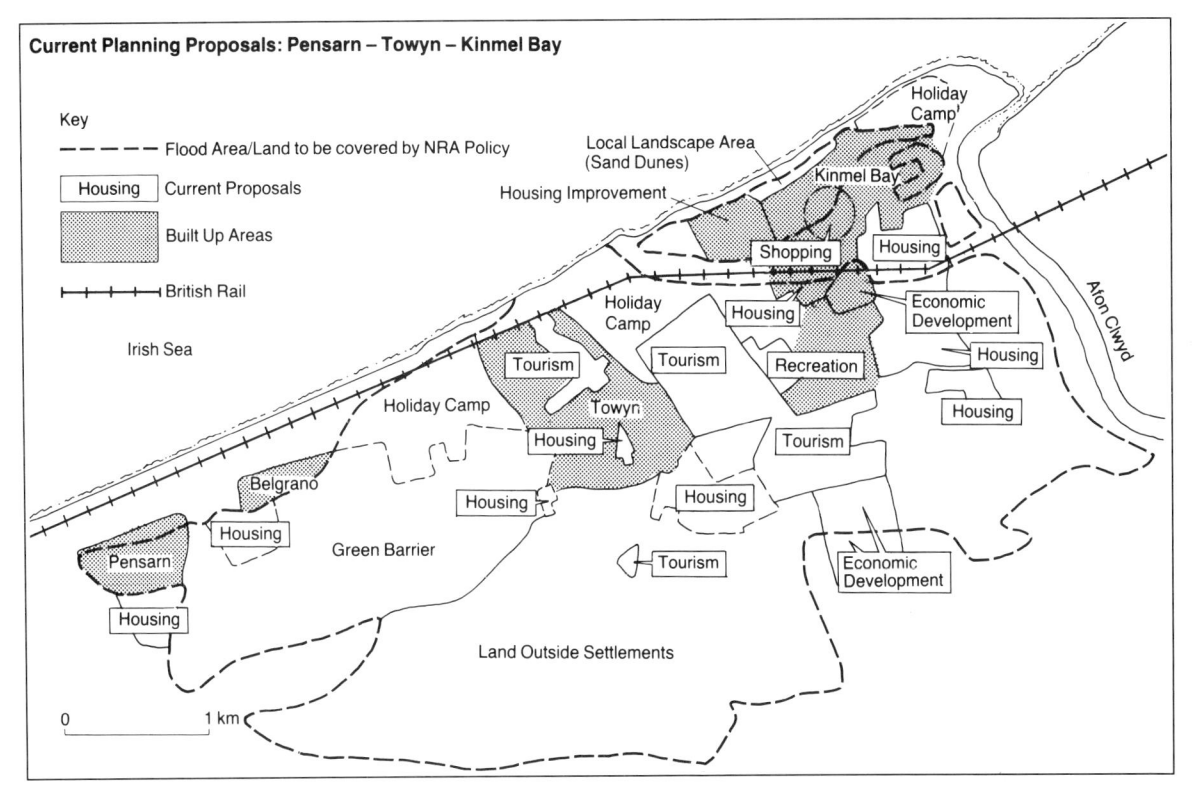

Fig. 5.3.7

Current Planning Proposals: Pensarn – Towyn – Kinmel Bay

with the slow-down in late 1989, only 12 new dwellings had been completed in the three month period before the floods. Although the NRA wished to extend the moratorium to commercial and industrial development, Colwyn Borough Council was unwilling to extend the moratorium to these categories.

Options for the future

In considering the future development of the area, Colwyn Borough Council put forward a number of options for the Pensarn – Towyn – Kinmel Bay area, bearing in mind the possible influences of the greenhouse effect i.e. potential rise in sea level, and increased likelihood of storms.

- Revocation of planning permissions.

- Refusal of future planning permissions.

- The creation of a contour map to show pockets of low-lying land on which development would be undesirable – contours to be mapped at 0.5 m intervals – cost £20 000).

- Release of higher land in the Borough for housing development: the only area of such available land is south of the A55 from Abergele eastwards, on high quality land in a special landscape area.

- Building design: bungalows were seen to be at a grave disadvantage in the flooded area. Planning permission could be given only for houses or dormer bungalows with a staircase.

- A note could be attached to all future permissions granted that the land had been flooded, thus putting the onus on the developer.

DECISION-MAKING EXERCISE

1 Discuss the main reasons for the placing of a temporary moratorium on all residential development.

2 Evaluate the options, considering both their advantages and disadvantages.

3 Select and justify which option, or combination of options would be the most suitable.

THE EFFECTS OF CLIMATIC CHANGE

It can be seen from the previous discussion of the aftermath of the Towyn floods that local authorities have already begun to take note of the implications of global warming in their deliberations. Forecasts of the rise in mean sea level over the next century, consequent on global warming, vary from 0.5 m to 3.5 m. A middle estimate of 0.8 m and a maximum of 1.65 m is now something of a consensus view. The effect on coastal areas in Britain is complicated by the fact that in the south-east the land is sinking relative to the sea already (as a result of isostatic adjustment) at an estimated rate of 3 mm a year.

The map (Fig. 5.3.8) shows the area of Britain most vulnerable to a rise in sea level.

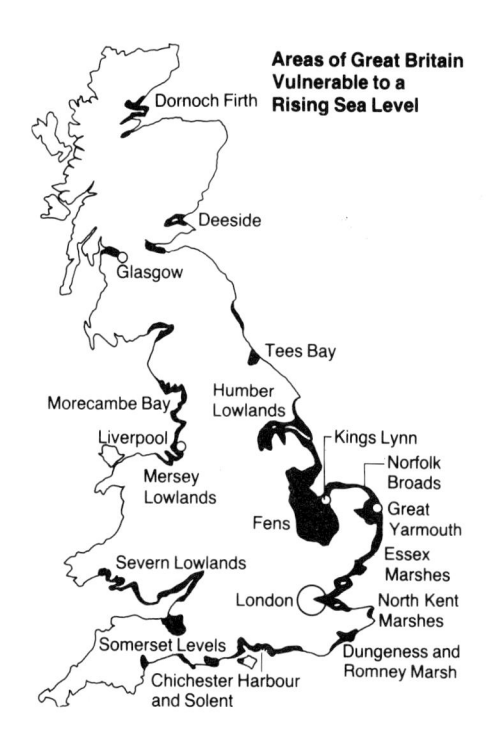

Fig. 5.3.8

1 Comment on the distribution of the areas under threat.

2 Using a good atlas, classify these areas into groups according to their densities of population and land use.

3 Which parts of the British economy would be hardest hit if this sea level rise were to become reality?

Consequences of a sea level rise on coastal systems

Sea level rises of the order indicated above would have a dramatic effect on coastal systems.

- Increased erosion along coastlines with cliffs of weakly resistant rocks.
- Increased supplies of sediments from this erosion to salt marsh and sand dune areas further down the coast.
- Sandy beaches and shingle banks would suffer increased erosion. On unprotected shores these features would probably reform inland, but on shores protected by a sea wall, beaches would be lowered and the sea wall threatened.
- Sand dune sites would suffer some initial erosion, but the sand released would be available for renewed building of dunes on the landward side. A possible beneficial effect would be a rise in the fresh water table and the regeneration of dune slack communities.
- Salt marshes on open coasts would become cliffed and creeks within the marshes would become rejuvenated.
- Salt marshes in narrow protected estuaries would be drowned in their lower parts, but this might be partially offset by increased rates of sedimentation.
- The invertebrate fauna of intertidal flats would become poorer and less diverse.
- The number of birds that roost, feed or breed in the estuaries would be much reduced.

All of the above consequences assume little or no reaction of people to the rising sea level. Where sea walls are raised or barriers constructed in estuaries the effects are likely to be much more complex.

THE FUTURE OF CHESIL BEACH

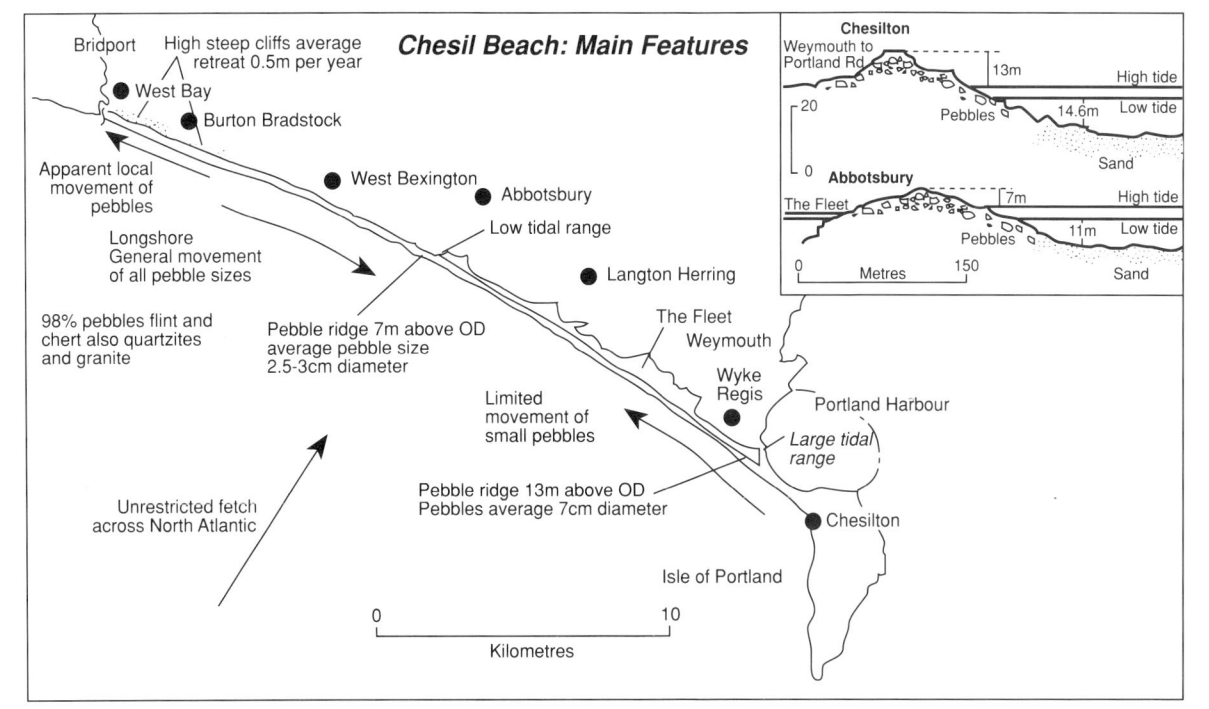

Fig. 5.3.9 Chesil Beach, seen from Verne Yeates, Isle of Portland

Fig. 5.3.10

Under their front, at periods of a quarter of a minute, there arose a deep hollow stroke, like the single beat of a drum, the intervals being filled with a long-drawn rattling, as of bones between huge canine jaws. It came from the vast concave of Deadman's Bay, rising and falling against the pebble dyke.

(Thomas Hardy, The Well-Beloved)

Hardy's description of the shingle movement on Chesil Beach is a dramatic introduction to one of the most remarkable features of the British coastline (Fig. 5.3.9). The beach extends from the Isle of Portland westwards to West Bay for some 28 km. There is much dispute about the precise western limit, others preferring its western end to be at Abbotsbury or Burton Bradstock (see Fig. 5.3.10). For 13 km it is backed by the shallow tidal waters of the Fleet. The beach was probably formed from debris carried by periglacial streams into the area now occupied by Lyme Bay, which was exposed as a dry area in the low sea levels of Pleistocene times. The rising sea levels at the end of the Pleistocene gradually pushed this material shorewards until it occupied its present position. Rapid cliff erosion at Burton Bradstock may be part of a process that is slewing the beach round to occupy a position at right angles to the direction of maximum fetch.

1 Read the article from *The Independent* (Fig. 5.3.11).

2 Draw an annotated map to show the likely consequences of a rise in sea level on Chesil Beach and the Fleet.

3 Explain the likely consequences of a rise in sea level for:

a the salt marsh areas within the Fleet

b the fish population of the Fleet

c bird populations of the Fleet.

4 Why are Chesilton and West Bay likely to be at risk from a rising sea level? (Both have recently had their sea defences strengthened.)

Fig. 5.3.11 Independent

Rising sea puts villages under threat

Global warming is destroying Chesil Beach. **David Nicholson-Lord** reports

CHESIL BEACH, the raised shingle bank on the Dorset coast which is one of the best know landscape features of the British Isles, will be a casualty of global warming. The rising sea level is slowly destroying it, threatening to flood towns and villages and to cut off the Isle of Portland from the mainland.

According to the National Rivers Authority and English Nature, the renamed English section of the Nature Conservancy Council, bigger waves and more frequent storms caused by global warming will push the 18-mile beach back and split it up.

The beach is not only an important geological and wildlife site but a vital natural flood defence. Dr Alan Brampton, a maritime engineering specialist from Hydraulics Research, which has studied the formation of the beach, said there was a "very serious danger" that the A354 road between Weymouth and the Isle of Portland could be cut off early in the next century.

Settlements at risk include West Bay and Chesil, where coastal defences costing more than £1m were built in the 1980s. The NRA says there are between 30 and 40 houses at Chesil and "there is a limit to how much money you can spend".

Chesil Beach is already retreating because of the subsidence of the south coast since the Ice Age. Global warming could triple the rate of sea rise. The NRA predicts a 6mm annual rise.

"Chesil Beach is being trapped between rising seas and the land behind it," Dr Brampton said. "Bits of shingle are picked up by high tides or a big wave and thrown to the back of the beach. Once they are there they can't get forward again. Slowly the whole thing marches gently back."

Chesil Beach's most notable feature is the sorting of the pebbles by longshore drift: at the northern end they are pea sized but at the southern tip they are two inches across. Commercial gravel extraction was halted in the 1980s after studies showed that the beach was not, as thought, gaining new shingle.

There is also a threat to the Fleet, the largest tidal lagoon in Britain, and the Abbotsbury Swannery, a unique colony of mute swans. The Fleet lies between the beach and the land and is home to up to 1200 swans as well as several rare plants and birds, including little terns and Cetti's warbler. The lagoon is part of the Chesil Beach Site of Special Scientific Interest.

Geologists regard Chesil as a "transient" feature: even without global warming, it might erode eventually. Biologists and botanists want to preserve it because of its wildlife role.

Dr Brampton said; "If you were living in a situation where the only consequence of not interfering with transient features like Chesil was that they would disappear, there would be no problem. Unfortunately they have another role: they are coastal defences.

"As a person, rather than as a scientist, I would be horrified to see Chesil broken up into pieces."

MANAGING ESTUARIES

Estuaries and their wildlife are a key part of our natural heritage. Our challenge is to work together to maintain and enhance this common estuarine heritage.... We have treated estuaries in the past as wastelands. Let us treat them as treasures in the future.

(Sir William Wilkinson, Chairman: English Nature, in the introduction to – Nature Conservation and Estuaries in Britain)

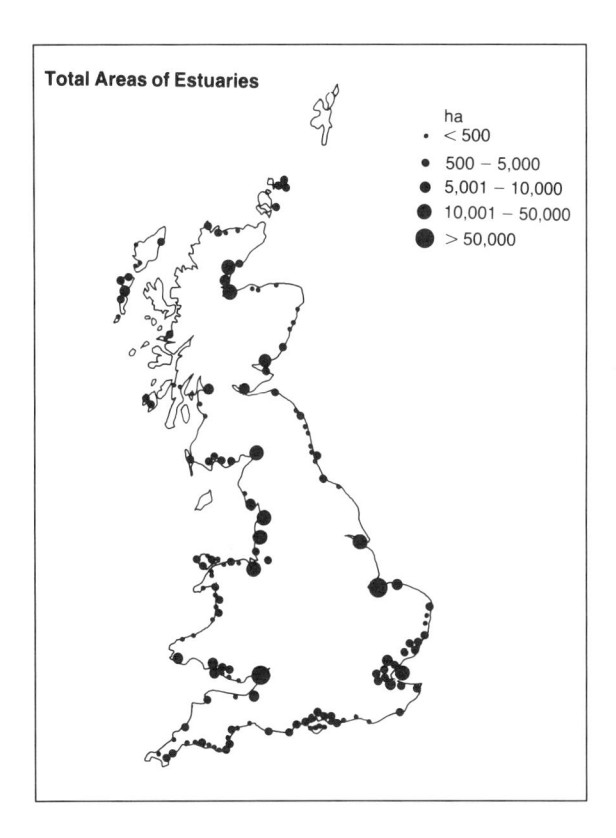

Fig. 5.4.1

Fig. 5.4.2
The Mawddach Estuary, one of Britain's most beautiful estuaries

Fig. 5.4.3
The Humber Estuary

Britain's estuaries possess some of the finest scenery, and shelter some of the richest wildlife around the coast. Our estuaries form a quarter of the whole estuarine resource of Europe (see Fig. 5.4.1). Great variety exists, from the mountain-backed sands and creeks of the Mawddach in North Wales (Fig. 5.4.2) and the Duddon in Cumbria, to the industry-scarred sweep of the Humber (Fig. 5.4.3) and the vast emptiness of the Wash. Estuaries are where sea and river meet and mix. Sediments and nutrients are brought down by rivers and in from the sea. They create the conditions for a rich abundance of flora and fauna, well adapted to the constantly changing environment of tide, sediment and salinity. People have found estuaries a rich resource too. Originally they were seen as sources of food and grazing for animals, functions that they still fulfil, but they have increasingly become centres of urban and industrial growth, resulting in inevitable pollution and threat to wildlife.

ESTUARIES UNDER THREAT

Fig. 5.4.4

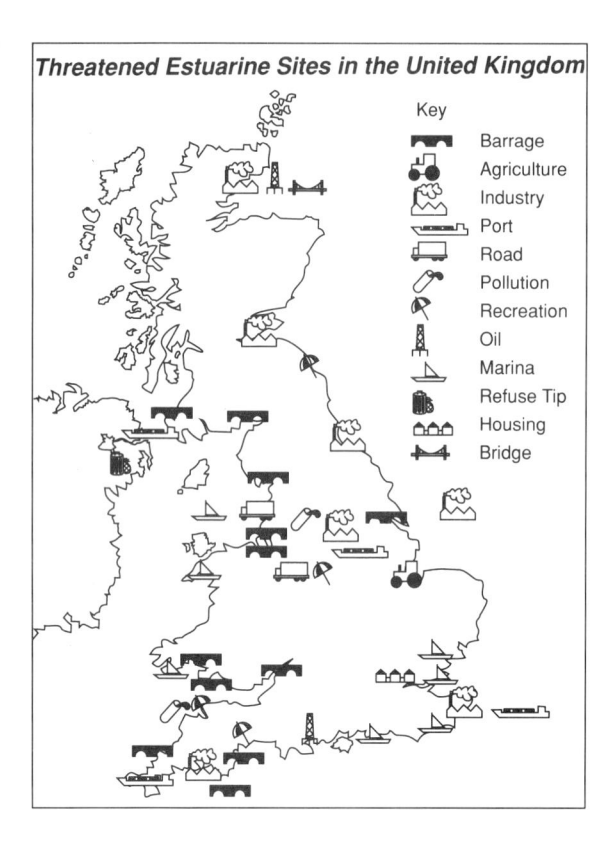

Threatened Estuarine Sites in the United Kingdom

Key:
- Barrage
- Agriculture
- Industry
- Port
- Road
- Pollution
- Recreation
- Oil
- Marina
- Refuse Tip
- Housing
- Bridge

The map (Fig. 5.4.4) shows the range of threats to estuaries in Britian.

1 Comment on this method of displaying information on a map.

2 What appear to be the most serious threats to estuaries shown on the map? Why might this be misleading to the uninformed viewer of the map?

The diagrams (Fig. 5.4.5) show current and proposed land-claims involving estuarine habitat loss.

Fig. 5.4.5

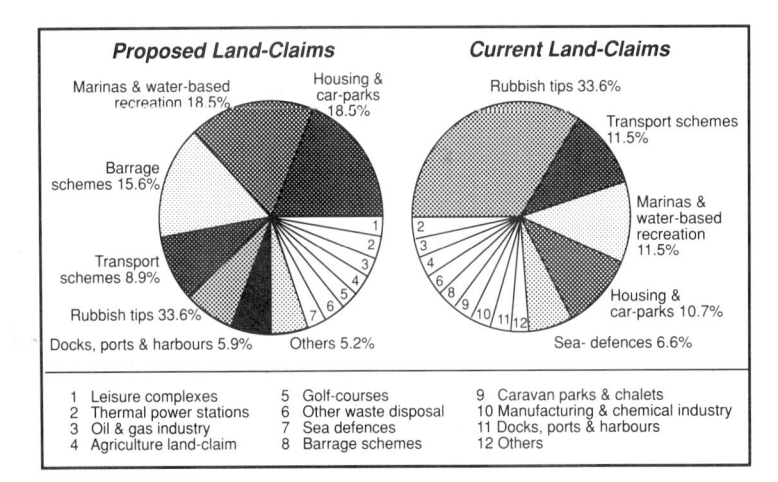

Proposed Land-Claims
- Marinas & water-based recreation 18.5%
- Housing & car-parks 18.5%
- Barrage schemes 15.6%
- Transport schemes 8.9%
- Rubbish tips 33.6%
- Docks, ports & harbours 5.9%
- Others 5.2%

Current Land-Claims
- Rubbish tips 33.6%
- Transport schemes 11.5%
- Marinas & water-based recreation 11.5%
- Housing & car-parks 10.7%
- Sea-defences 6.6%

1 Leisure complexes
2 Thermal power stations
3 Oil & gas industry
4 Agriculture land-claim
5 Golf-courses
6 Other waste disposal
7 Sea defences
8 Barrage schemes
9 Caravan parks & chalets
10 Manufacturing & chemical industry
11 Docks, ports & harbours
12 Others

3 Compare this method of showing estuarine areas under threat with the map (Fig. 5.4.4). What are its relative merits and disadvantages?

4 Comment on the range of threats that are indicated in the diagram. How do they compare with the range shown in the map (Fig. 5.4.4)?

PORTSMOUTH HARBOUR: AN 'INDUSTRIAL' ESTUARY

Portsmouth Harbour may be regarded as typical of estuaries on 'soft' coasts that are under a variety of threats from development. The harbour (Fig. 5.4.6) is a tidal basin of some 1537 hectares and possesses a shoreline of 57 km. At low tide nearly 60 per cent of the harbour is exposed as mud flats. Land reclaim has been piecemeal over many centuries, with the early claims being for the expansion of the naval dockyards and other port facilities. The main losses have occurred from the nineteenth century onwards. Since 1540 nearly a quarter of the tidal basin has gone, and since 1860, 20 per cent has disappeared in reclamation projects (Fig. 5.4.7). Most of the tidal mudflats that are in the upper reaches of the harbour have been designated a SSSI, which reflects its international importance as a particularly rich and diverse ecological resource (Fig. 5.4.8).
Undeveloped foreshore and the intertidal flats act as ecologically interdependent areas, the land acting as important feeding and roosting sites. In particular open grasslands around the Harbour such as Bedenham, Cams Hall and Wicor, are important wader roosts.

The maps (Figs. 5.4.7 to 5.4.9) show:

- past reclamations and present proposals for reclamation in Portsmouth Harbour
- landscape character zones
- boat moorings
- Maritime Heritage attractions.

The main issues concerning the future use of the harbour have been identified as:

- the changing role of the Royal Navy resulting in the possible release of land adjoining the Harbour
- the growth of commercial activities at Mile End, particularly the Continental Ferry Port, and the demand for more back up land and quay space

MANAGING THE ESTUARIES

Fig. 5.4.6
Portsmouth Harbour from the air

Fig. 5.4.7

Portsmouth Harbour Plan: Past Reclamations

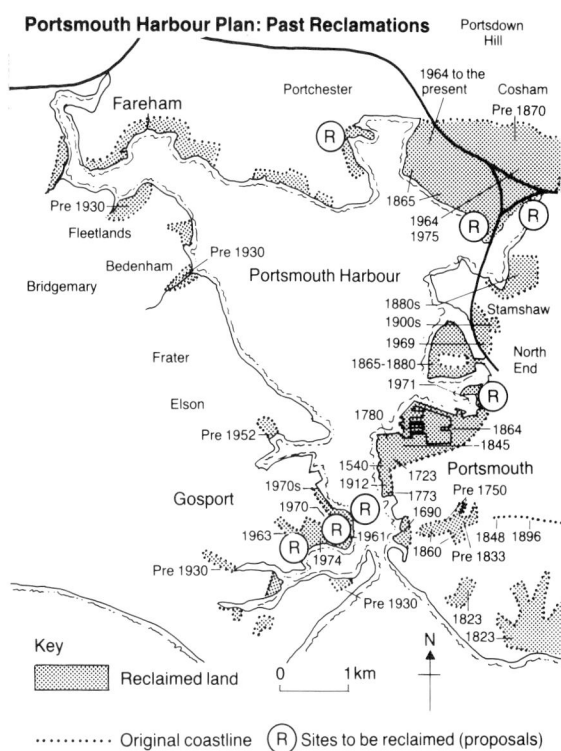

Key
▓ Reclaimed land
·········· Original coastline
(R) Sites to be reclaimed (proposals)

Fig. 5.4.8

Portsmouth Harbour Plan: Landscape Character Zones

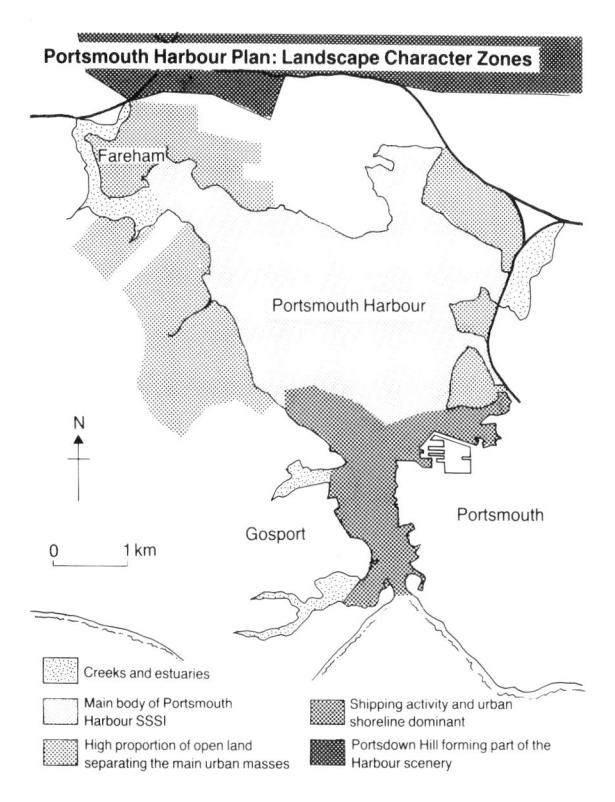

- Creeks and estuaries
- Main body of Portsmouth Harbour SSSI
- High proportion of open land separating the main urban masses
- Shipping activity and urban shoreline dominant
- Portsdown Hill forming part of the Harbour scenery

Fig. 5.4.9

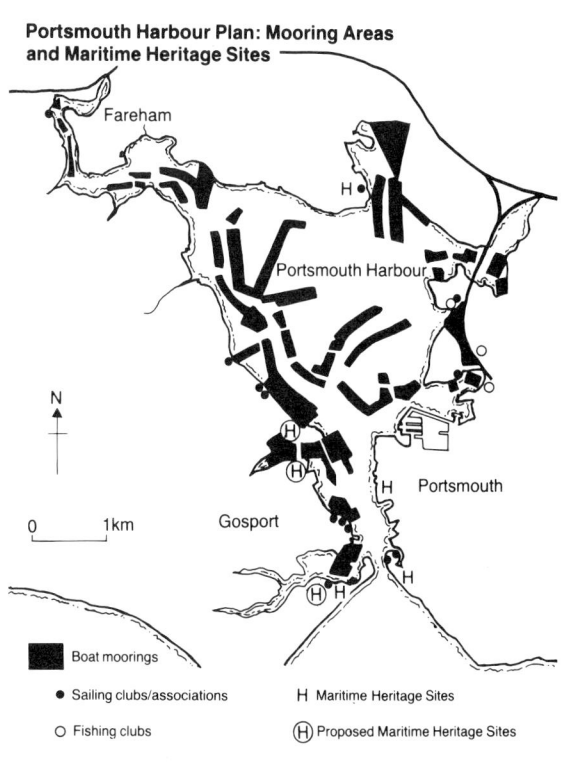

Portsmouth Harbour Plan: Mooring Areas and Maritime Heritage Sites

- ■ Boat moorings
- ● Sailing clubs/associations
- ○ Fishing clubs
- H Maritime Heritage Sites
- (H) Proposed Maritime Heritage Sites

237

- the need to safeguard the Harbour's landscape setting and its importance for nature conservation

- the desirability of providing greater public access to the Harbour and foreshore, particularly for leisure pursuits

- the development of the Harbour's maritime heritage attractions for tourism purposes

- pressure to increase the number of recreational boat berths in the Harbour and the further provision of slipways, boat compound space and car parking

- the need to find space for a new marine aggregate quay.

1 Consider each of the issues in turn, and discuss its implications for nature conservation and the harbour ecosystems (there is no need to discuss the third issue, since that deals directly with conservation).

2 Produce a compatibility matrix (after Fig. 5.4.10) to show the extent to which these different uses are reconcilable.

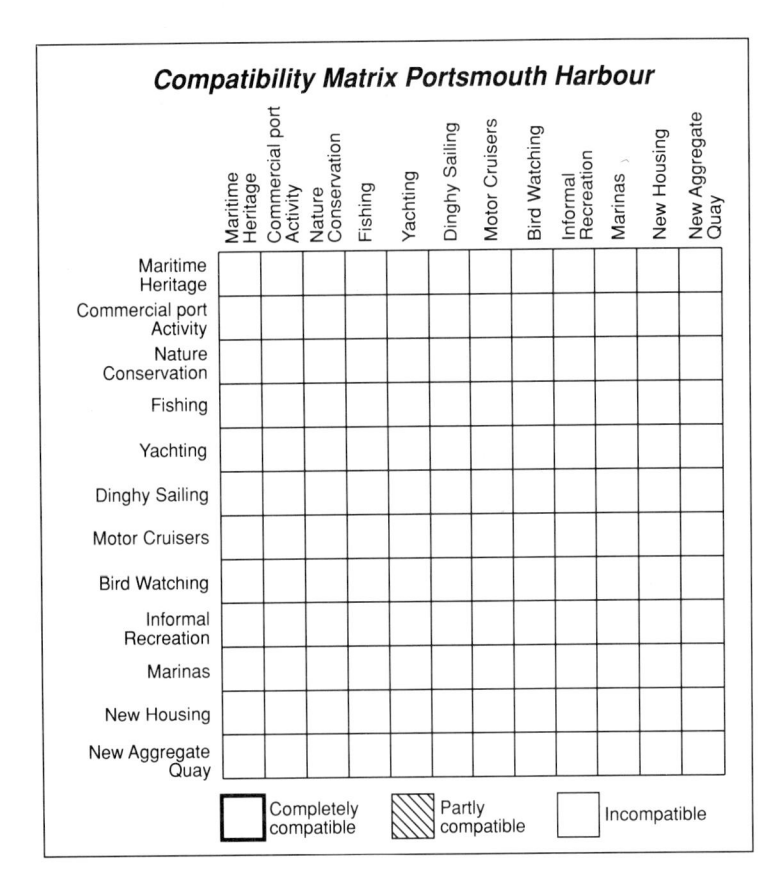

Fig. 5.4.10

THE WASH: LAND RECLAMATION FOR FARMING

Fig. 5.4.11
The Wash from the air

Compared with Portsmouth Harbour, an average sized estuary, the Wash is the largest British estuary (see Fig. 5.4.1 for comparative sizes), occupying some 66 600 hectares. Unlike Portsmouth Harbour it does not have a large urban population on its shores, possesses little industrial development, and has only small ports such as King's Lynn compared to a major naval base and a growing continental ferry port. Pressures on the ecological resource exist, but they are clearly different.

The Wash is an internationally important site for wildfowl and waders, and contains one of the largest areas of saltmarsh in Britain. Its importance lies in the interrelationship between the intertidal habitats, saltmarsh (their breeding birds and invertebrates) and the mudflats with their invertebrate animals and plant life which provide food for large numbers of wintering birds.

(Dr. P. Doody, English Nature, Peterborough: Conference on the Wash and its Environment)

The Wash seems to have originated as a depression cut in Jurassic Clays by both fluvial and glacial action. During the glacial period alternate transgressions and recessions of the sea led to the deposition of a variety of sediments. As the sea finally rose in the post-glacial Flandrian transgression the lowland forests of the area were destroyed and the present area of the Wash became part of a lagoon – marshland complex protected by a sand barrier. In Roman times as the sea continued to rise, the barrier moved inland to form the coastal sand barrier of the coasts of north Norfolk and Lincolnshire. Coastal accretion began to replace coastal retreat within the sheltered embayment of the Wash. Around the edge of the Wash there is a sequence of marshes, mudflats and sandflats. Behind the marshes there is an embankment that protects reclaimed land from tidal incursions (see Fig. 5.4.12). Sediment is supplied by both rivers draining into the Wash and also from the North Sea. The diagram (Fig. 5.4.13) shows the sediment sources and transport paths in the Wash and neighbouring areas.

Reclamation and its impact

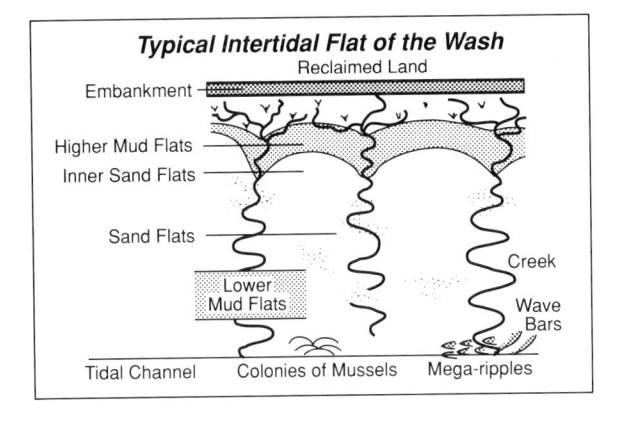

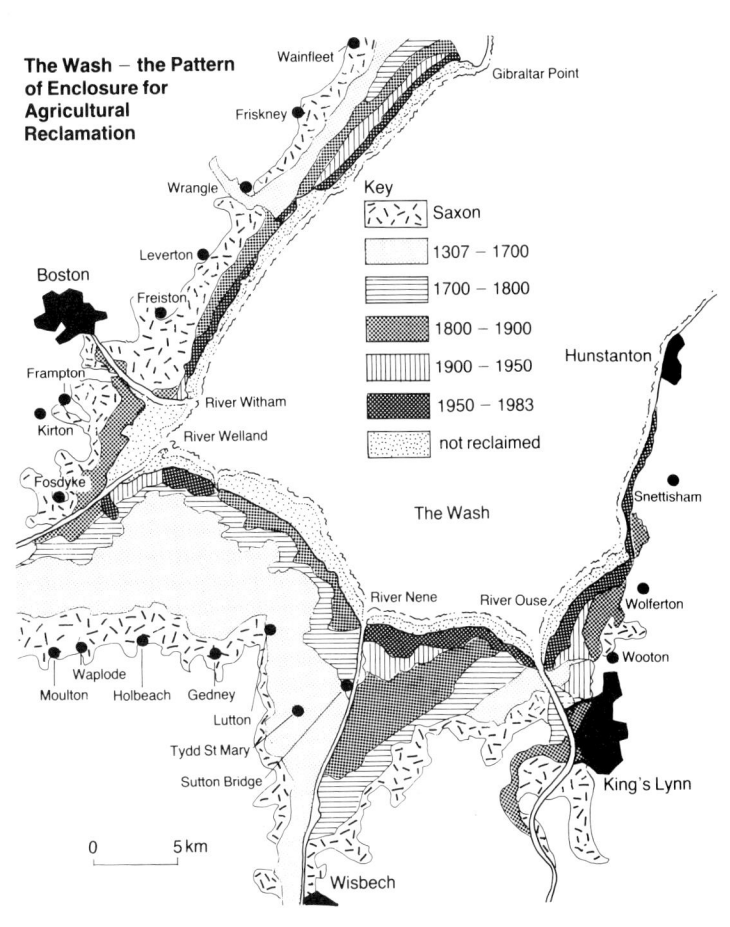

Fig. 5.4.12

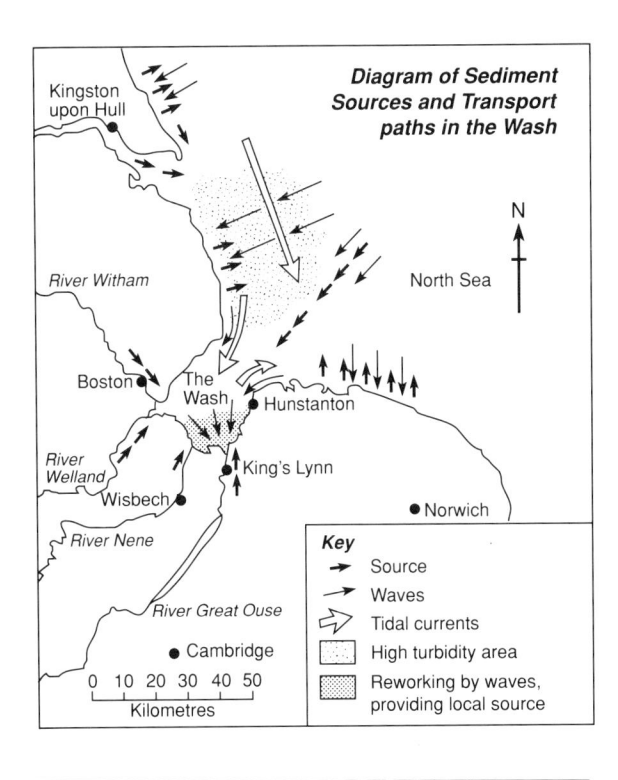

Fig. 5.4.13

Fig. 5.4.14

The map (Fig. 5.4.14) shows successive stages of reclamation of the Wash. It is a process that has been continuing since the Middle Ages. A total of some 32 000 hectares have been reclaimed (a sizeable figure when the present area of the Wash (66 600 hectares) is considered). The land reclaimed will normally be of relatively high quality agricultural value (Grades 1 and 2 of the five grade agricultural land classification). Delay in reclaiming may lead to increased clay content, reducing the range of crops that can be grown. Once the land has been reclaimed and improved there is virtually no limitation to its agricultural use. The importance of this reclaimed land can be gauged from the fact that the total amount of Grade 1 agricultural land in England and Wales is some 360 000 hectares: so 9 per cent of all Grade 1 land is that reclaimed from the Wash! Very high outputs of the order of £1600 per hectare are obtained from this land when it is given over to cash roots and field vegetables. High yields can be obtained from relatively low inputs, hence unit costs are low.

The likely effect of reclamation on the natural environment of the Wash will include:

- loss of species – rich upper saltmarsh communities on enclosure, drainage and leaching

- loss of all invertebrates and breeding birds that depend upon the upper salt marsh for their survival

- renewed growth of salt marsh on the seaward side of the enclosing embankment

- more frequent tidal inundation of the unreclaimed lower marsh and any newly accreted marsh, tending to flood nests and destroy young birds

- loss of some of the intertidal flats, which is detrimental to the population of wader birds (see Fig. 5.4.15). Waders tend to be concentrated where the feeding conditions are best, and this depends on the abundance of food (see Fig. 5.4.16). As new salt marsh accretes, the inner mud zone accretes towards the mud zone near the tidal channel, squeezing out the vital sand-flat zone where wading populations feed.

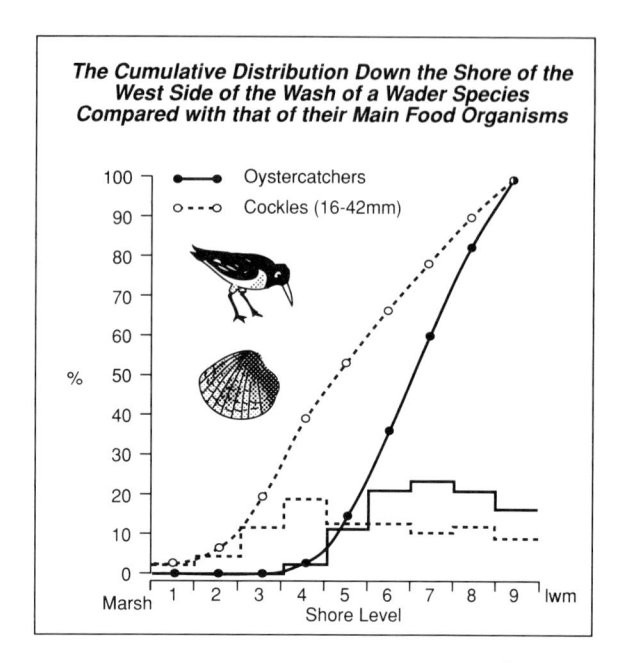

Fig. 5.4.15

1 Draw an annotated diagram based on Fig. 5.4.12 to show the effects of reclamation on invertebrate and bird populations.

2 What are the merits and disadvantages of reclamation for farmland? On balance, can the reclamation of such land be justified?

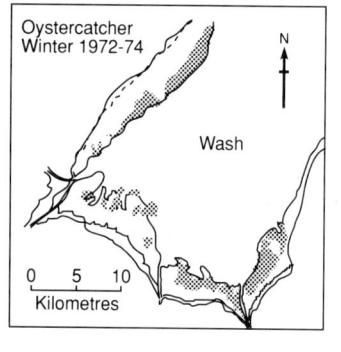

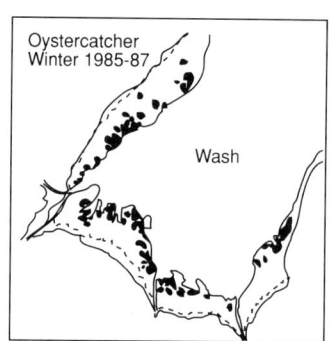

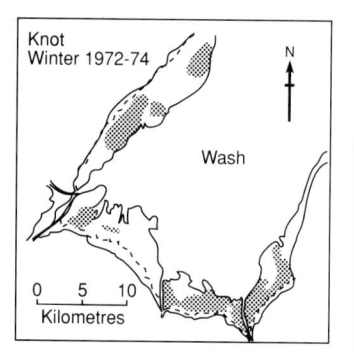

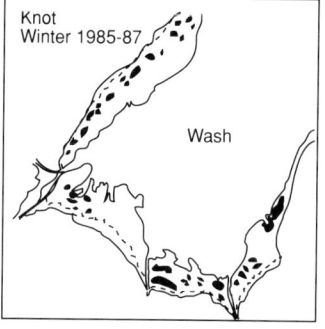

The Cumulative Distribution Down the Shore of the West Side of the Wash of a Wader Species Compared with that of their Main Food Organisms

Fig. 5.4.16

The Feeding Areas used by Two Wader Species: Oystercatcher and Knot

Other proposals for the Wash

Several proposals have been put forward for the development of the Wash. Three possible enclosures are shown on the map (Fig. 5.4.17). None of these went beyond the planning stage. Various water storage proposals were put forward in 1970 by the Water Resources Board, including the erection of three or four bunded reservoirs offshore to be pump-filled with water from the Ouse and the Nene. Detailed studies of the likely impacts of these water storage schemes were carried out. The studies concluded that the schemes would all cause reduction in areas of vegetation, faunal habitats and birds' feeding grounds, with some loss of habitat. Final conclusions were that the adverse environmental effects would be small, but the exercise became purely academic because of the fall in demand forecasts for water at the time. Such a scheme could well be reviewed in the future.

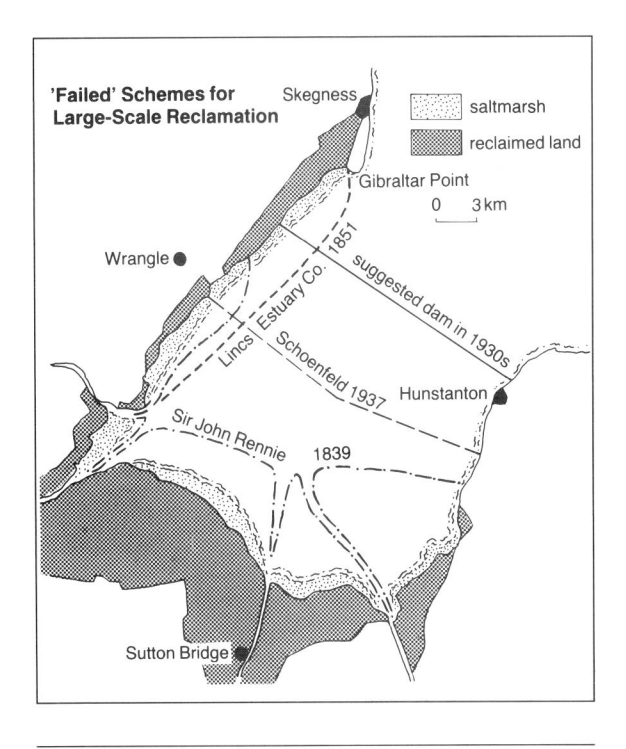

Fig. 5.4.17

The tidal barrage issue

Barrage schemes have been proposed for a number of the estuaries of Britain. In the 1970s as well as the water storage schemes proposed for the Wash above, other water storage schemes were proposed for the Severn, Cheshire/Welsh Dee, Morecambe Bay, and Solway estuaries. All of the schemes were very ambitious and highly expensive, requiring technology not readily available in Britain. With the drop in levels of water demand consequent on lower population predictions, these schemes are likely to remain in abeyance, at least for the time being.

Barrage construction for the generation of tidal power is currently at the stage of feasibility studies for a number of British estuaries. The Severn and the Mersey are major contenders, together with seven other major estuaries, as well as ten smaller ones.

Majestic Duddon, over smooth flat sands,
Gliding in silence with unfettered sweep!

(William Wordsworth)

Fig. 5.4.18
The Duddon Estuary

Tidal barrage scheme poses green dilemma

An electricity-generating project that should delight conservationists may threaten a wildlife sanctuary. Michael Morris reports

CONSERVATIONISTS who favour alternative energy are in a dilemma over plans to build a £126 million barrage across Wordsworth's beautiful Duddon estuary in west Cumbria, famous for its birds and other wildlife.

The Department of Energy is to consider a proposal for a three-mile tidal barrage between Millom and Askam in Furness that would generate 99–135 megawatts of electricity, enough to power 60 000–80 000 homes.

The development companies Sir Robert McAlpine and Balfour Beatty Projects are applying for a government contribution to the £180 000 cost of a nine-month feasibility study covering the environmental effects of the project and its financing.

Mike Houston, secretary of the Friends of the Lake District, summed up the opposition line: "We favour alternative technology, but not at the expense of damaging the environment and beauty of the estuary."

The line-up against the barrage includes the Nature Conservancy Council, the Royal Society for the Protection of Birds, Cumbria Wildlife Trust and the National Trust. It also includes local people, like fisherman Joe Stevenson, who called out "We want no barrage!" when he mistook a surveyor for an NCC officer spreading a map on his car.

"It would ruin the place," said Mr Stevenson, who earns much of his living from fishing. "Most Askham people think it will spoil the look of he estuary, and there will be no birds."

He and his brother, Eric, believe that recreational use of a lake formed by a barrage would be for the rich man's yacht rather than small boats moored on the salt marshes.

The barrage would be flung across the estuary sands on one of two possible alignments, both from Millom. It is unlikely to pay for itself without a road on top to encourage tourism.

Standing on Askam pier, David Shaw, an NCC scientific officer, said as he gazed upriver at mountains rising beyond Ulpha: "Putting a barrage, almost certainly concrete, across here would damage what we can see."

A barrage would alter the tidal pattern, leaving less sand for wading birds to feed on invertebrates such as worms, he said. Birds feed on the wet sand left by the tide and on dry sand as it dampens afresh.

A blizzard of dunlins swept low on the sands as Mr Shaw pointed to the home of a large population of natterjack toads.

"We have nationally and internationally important numbers of waders and wildfowl that occur on the Duddon as a result of this mosaic of habitats of saltmarsh, sand dunes, and inter-tidal sand and mud."

The scheme's opponents will point out that Duddon has five sites of special scientific interest. It qualifies as a special protection area under an EC directive.

The RSPB has told the development companies of its 'grave concern' over the proposals. Its consevation officer, Tim Cleeves, likened the Duddon to a motorway service station for "thousands and thousands of birds that come from Siberia, Iceland, Scandinavia and Canada."

They depended on an area such as the Duddon for winter survival or as a stopover en route south.

A barrage could change tidal velocity, sedimentation, and salinity, he said. Pollutants might not be flushed out.

"Birds are international travellers. So how would we like it if Africans destroyed our swallow population?

"It is wrong to tamper with an estuary that has been here for millions of years, a dramatic and wild landscape that is one of Britain's last true wildernesses."

Fig. 5.4.19
Guardian 19 January 1991

1   Read the article from *The Guardian* by Michael Morris (Fig. 5.4.19).

2   What is the nature of the dilemma facing the conservationists?

3   What are the main environmental changes that are likely to result from the possible construction of a Duddon Barrage?

COASTAL ECOSYSTEMS

Way out beyond the mudflats, across the patches of salty samphire and seablite, the low tide line is audible from its silty gurglings, but is invisible as the horizon.

(Hammond Innes, East Anglia and the Wash, in Coastline; Britain's Threatened Inheritance, Greenpeace)

Some of Britain's most valued ecosystems are found around the coast. Sand dunes and salt marshes, shingle spits and cliff faces all provide habitats for a wide variety of plants and wild life. Coastal dunes are under increasing threat from their increased use for recreation, and also from their value as military training grounds. Estuarine salt marshes are vulnerable to new developments taking place along the shores of estuaries. Both may well be at risk if global warming leads to a rising sea level.

Fig. 5.5.1 The Distribution of Major Sand-Dunes. Nationally Important Dune Systems are Named

SAND DUNES AND THEIR ECOSYSTEMS (PSAMMOSERES)

The map (Fig. 5.5.1) shows the major concentrations of sand dunes around the coast of Britain. Of the locations marked 31 lie on west-facing shores, and 18 on east-facing shores. The main source of sand is the shore. The foreshore, between low water and high tide, although much of it remains wet, will supply up to 20 per cent of the sand for dune growth from its higher levels that do dry out sufficiently. The backshore, from the high tide line up to the dunes supplies some 80 per cent of the requirements. The accretion and stabilisation of sand dunes depends on the fixing and stabilising influence of vegetation. Sand couch grass, and sea lyme-grass, both of which can tolerate moderate levels of salinity, are important colonisers. Marram grass (Fig. 5.5.2) with a much lower tolerance, is responsible for the colonisation of the higher parts of the dunes and can survive in areas where the accretion rate of sand is as high as 1 m per year.

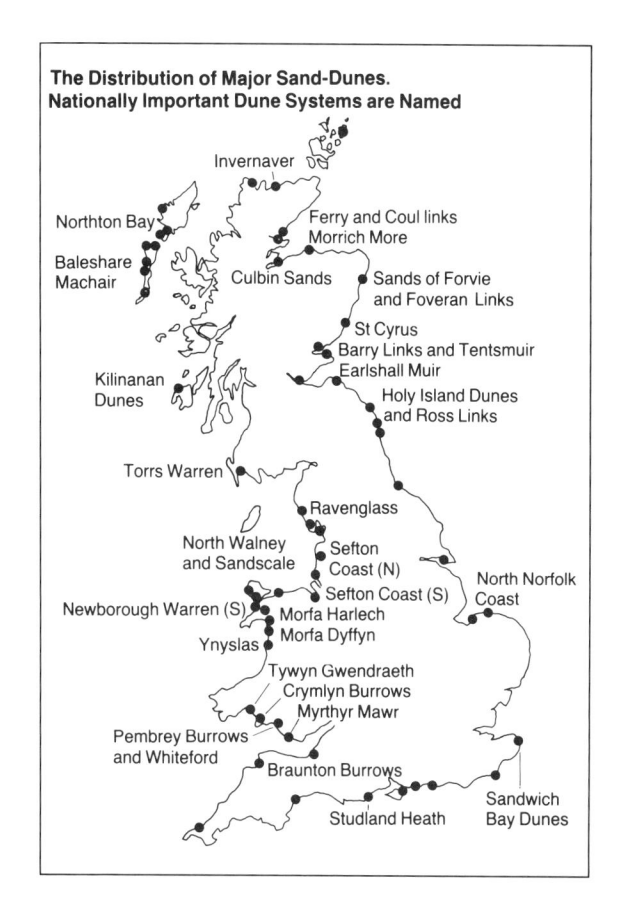

Fig. 5.5.2
Marram Grass on sand dunes at Studland, Dorset: Old Harry Rocks in the background

As the dunes grow, they are invaded by other species, and a more or less continuous sward may develop. New, embryo dunes (Fig. 5.5.3) will appear on the seaward side, and a dune complex will develop (Fig. 5.5.4), with a range of habitats and species.

BRAUNTON BURROWS

The map (Fig. 5.5.5) shows the location of Braunton Burrows (970 hectares) on the North Devon Coast. To the seaward lies the exposed stretch of Saunton Sands (Fig. 5.5.6), which is the principal source of supply of sand, derived from the huge sediment trap at the head of Bideford Bay. Sand blown from the dried out foreshore, and the backshore is trapped by obstacles such as drift wood and vegetation, and begins to form the embryo dunes, which are colonised by sea couch grass. This species cannot tolerate an accretion rate of more than 30 cm a year, and the building of the high dunes behind is left to marram grass. Marram grass thrives under these conditions of high rates of accretion. The tangle of stems, rhizomes and roots holds the sand and stabilises the dunes (Fig. 5.5.7). Inland there is a series of three ridges parallel to the embryo dunes, shown in the diagram (Fig. 5.5.8). These represent successive stages of growth of the complex, although mobile dunes may move inland at about 7 m a year. Separating the dune ridges there is a series of hollows which are usually intermittent, rather than in continuous lines, known as slacks. These may originally have contained areas of salt marsh, but

Fig. 5.5.3
Embryo sand dunes, backed by the main dune system, Braunton Burrows, North Devon

Fig. 5.5.4

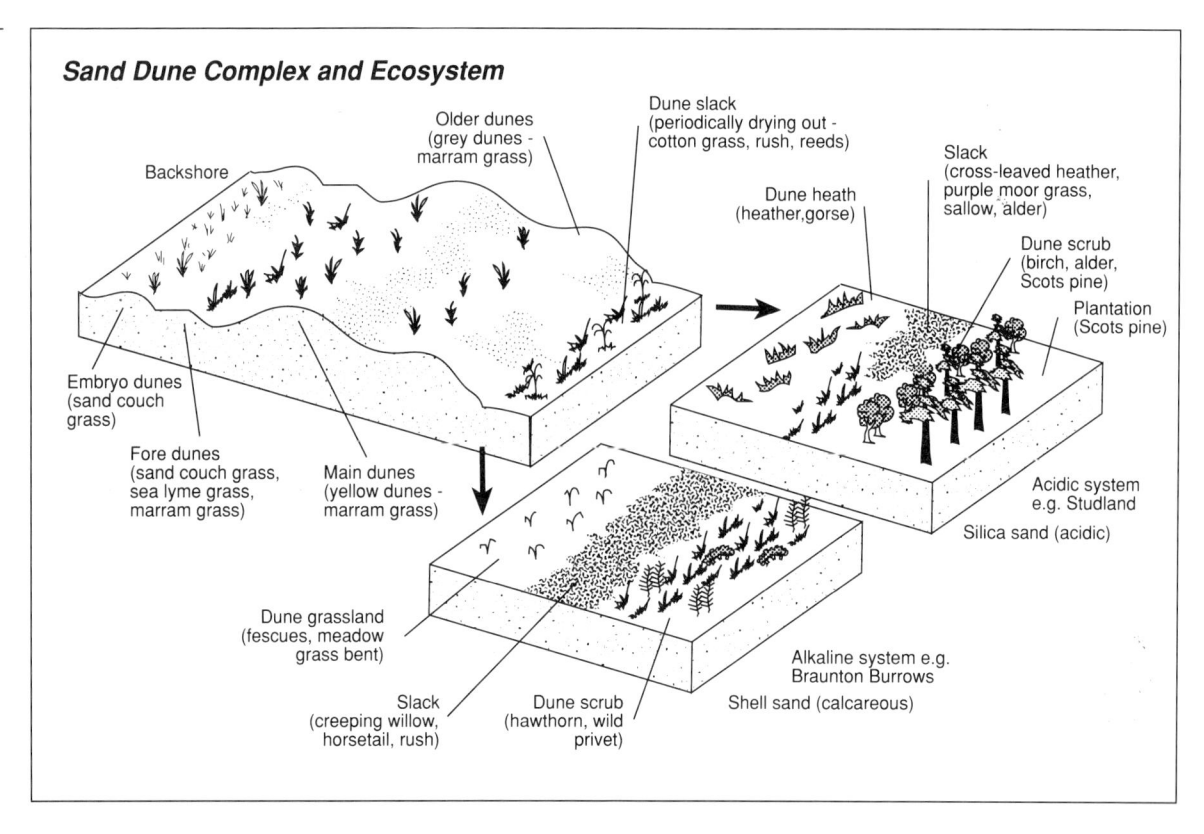

Sand Dune Complex and Ecosystem

COASTAL ECOSYSTEMS

Fig. 5.5.6
Saunton Sands, North Devon, backed by the sand dune complex of Braunton Burrows

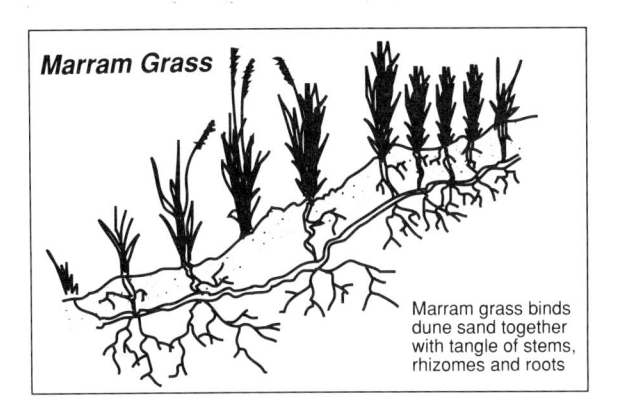

Marram Grass

Marram grass binds dune sand together with tangle of stems, rhizomes and roots

Fig. 5.5.7

Fig. 5.5.5

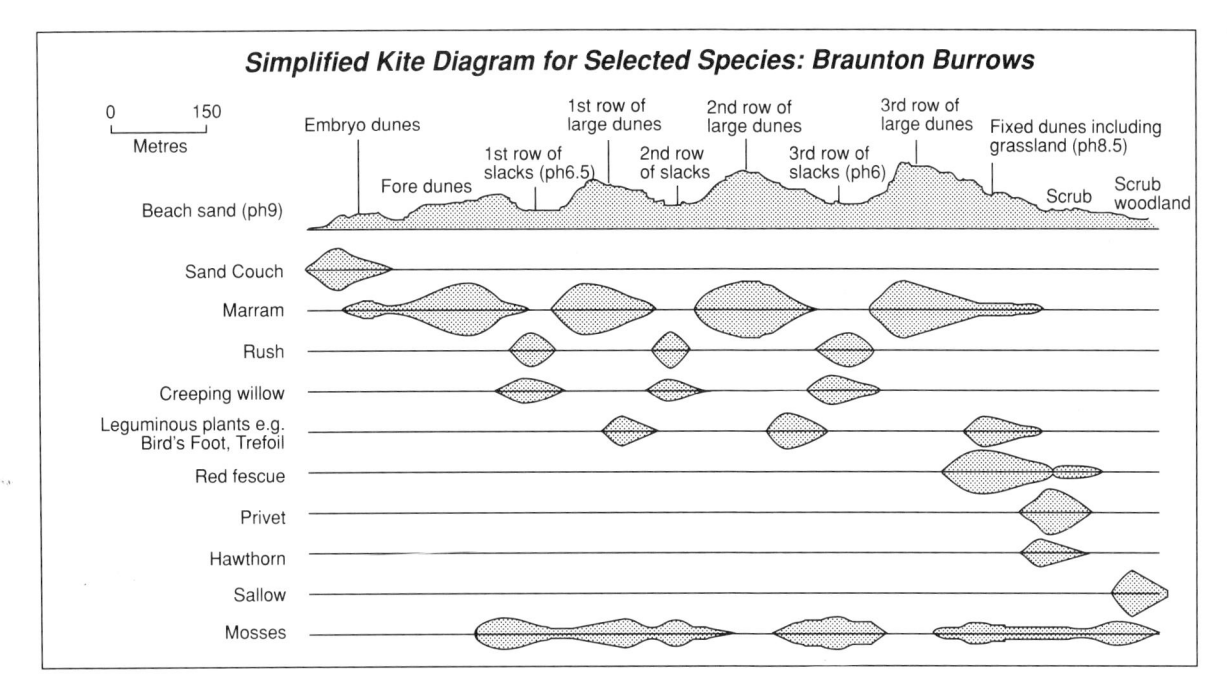

Simplified Kite Diagram for Selected Species: Braunton Burrows

Fig. 5.5.8

Embryo dunes — Fore dunes — 1st row of large dunes — 1st row of slacks (ph6.5) — 2nd row of large dunes — 2nd row of slacks — 3rd row of large dunes — 3rd row of slacks (ph6) — Fixed dunes including grassland (ph8.5) — Scrub — Scrub woodland

Beach sand (ph9)

- Sand Couch
- Marram
- Rush
- Creeping willow
- Leguminous plants e.g. Bird's Foot, Trefoil
- Red fescue
- Privet
- Hawthorn
- Sallow
- Mosses

245

Fig. 5.5.9
Horsebreaker's Slack, Braunton Burrows, a blowout feature

Evolution of Horsebreaker's Slack

1. Primary slack
   Ridge of mobile dune — Dune crest

2. Blowout develops in crest of dune when vegetation is disturbed

   Direction of prevailing wind

3. Blowout deepens until moist sand is exposed at **X**

4. Primary slack
   Parabolic dune forms and open west end fills in with small dunes: **W**
   Secondary slack
   Trailing arms form E-W ridge

now freshwater marsh has replaced it. With current falling water tables, standing water is found only after heavy rain. Some hollows such as Horsebreaker's Slack (Fig. 5.5.9) may have developed as a result of a blowout (see Fig. 5.5.10). Beyond the innermost of the three ridges there is a zone of fixed dunes which gives way to scrub and scrubby woodland.

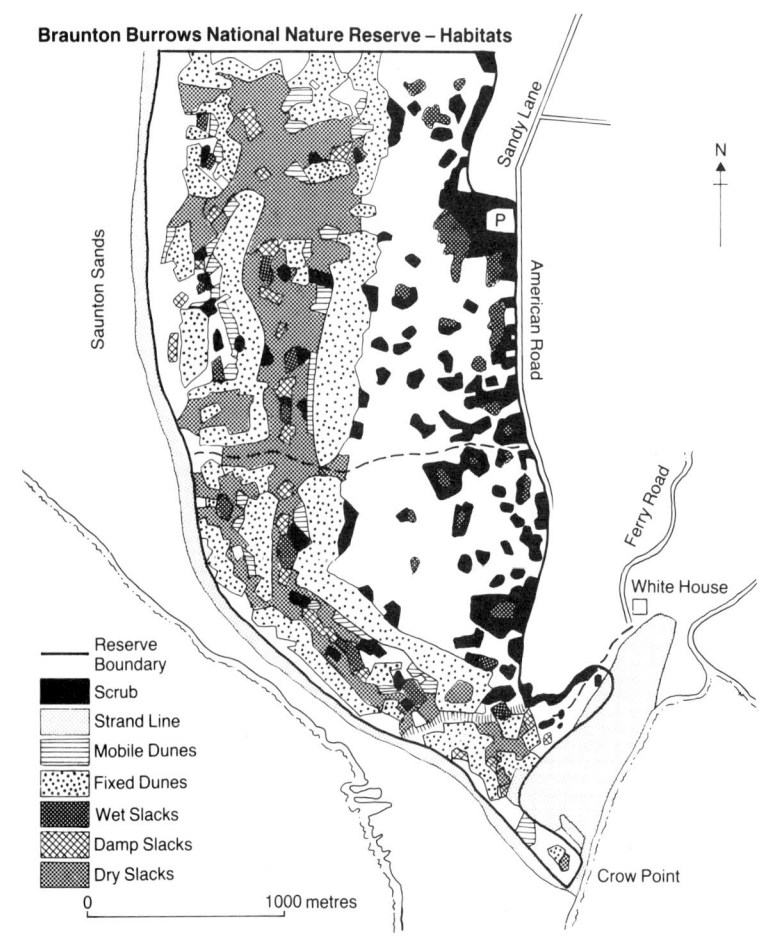

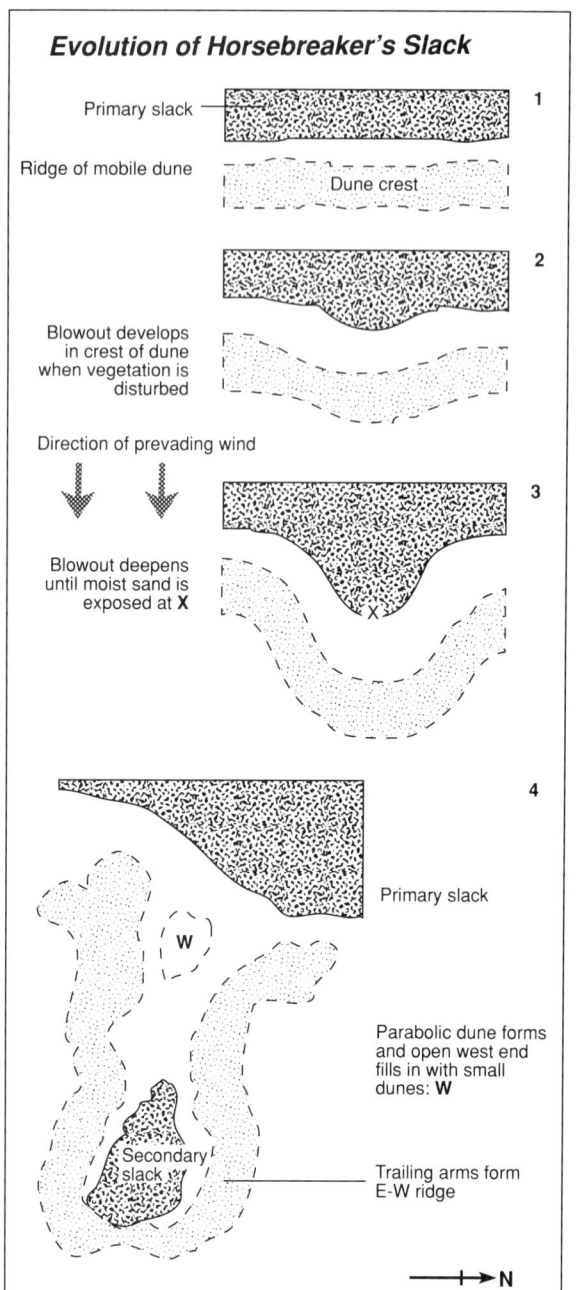

Braunton Burrows National Nature Reserve – Habitats

Legend:
- Reserve Boundary
- Scrub
- Strand Line
- Mobile Dunes
- Fixed Dunes
- Wet Slacks
- Damp Slacks
- Dry Slacks

Fig. 5.5.10

Seven different habitats are recognised at Braunton Burrows, and they are shown on the map (Fig. 5.5.11). They represent successive stages of colonisation from the strand line to the scrub. The major variations in plant species are shown in the diagram, Fig. 5.5.8.

Fig. 5.5.11

1   Explain the sequence of habitats that have developed from the strandline to the eastern boundary of the dune complex.

2   How might this sequence be affected by human intervention?

Management of Braunton Burrows

Of the 970 hectares of Braunton Burrows, 603 are leased to the Ministry of Defence for military training. The training takes place intermittently, and is confined to certain areas (see Fig. 5.5.5). Since 1964 English Nature have sub-leased the 603 hectares and manages the area to conserve the dune system and its wildlife. The northern part of Braunton Burrows supports a golf course (where water extraction for spraying the greens may be lowering the water table and affecting moisture-loving species in the slacks). Other important uses of the dune complex apart from military training are for recreation and sand extraction. Four main forms of management are practised at present.

- **Control of sea-buckthorn.** This was originally planted at the northern end of the complex in 1937 in order to induce some stability. It spread at 2 m a year to form an impenetrable thicket. It also encouraged plants such as nettles to colonise the dunes by virtue of the nitrogen-fixing bacteria in its roots enriching the soil. Water drawn up through the extensive root system of the sea-buckthorn has led to the drying out of some of the slacks and species loss. Sea-buckthorn is controlled by chemical spraying.

- **Control of the grassland and scrub on the landward side of the complex.** Left to itself this would degenerate into thicker scrub. Mowing of certain areas is carried out once a year, and treatment by burning and herbicides is also used. The main scrub in this section is of hawthorn, wild privet, and sallow. Recently a flock of Soay sheep (Fig. 5.5.12) has been introduced in order to improve the quality of the grassland. Rabbits could perform the same function, but their numbers have never recovered from the outbreaks of myxomatosis in the 1950s. The ultimate aim in this section is to produce a herb-rich turf and in the future cattle may be introduced to give more balance to the grazing.

Fig. 5.5.12
Soay Sheep on Braunton Burrows

- **Monitoring of and provision for recreation.** Two main car parks are provided, shown on the map (Fig. 5.5.5). Visitor levels are relatively low, compared to other sand dune complexes in the south of England, such as Studland in Dorset. A boardwalk has been provided to give better access to Crow Beach from the southern car park, and to prevent further erosion to the dune complex through trampling. Motorbike scrambling is strictly forbidden on the dunes, and this has to be carefully monitored.

- **Monitoring of the mobile dunes.** Normal practice is to allow natural evolution in the area of the mobile dunes and old blowouts. However, if it appears that a sensitive plant community is likely to be threatened (in damp slacks for instance) then appropriate action is taken, including the erection of fencing and planting to renew stability.

1   The population of North Devon has increased by 8 per cent in the 1981–91 period, and visitor numbers have increased by a similar amount during the same period. What are the main implications of this increase in indigenous and visitor population for a future management strategy for Braunton Burrows?

2   What priorities for such a strategy could be established?

3   What active management proposals can you make for achieving these priorities? Draw a sketch map of the area to illustrate your proposals.

SALT-MARSHES AND THEIR ECOSYSTEMS

The map (Fig. 5.5.13) shows the distribution of the major salt-marsh communities around the coasts of Britain (Fig. 5.5.14).

Fig. 5.5.13

The Distribution of Salt-Marsh Habitat

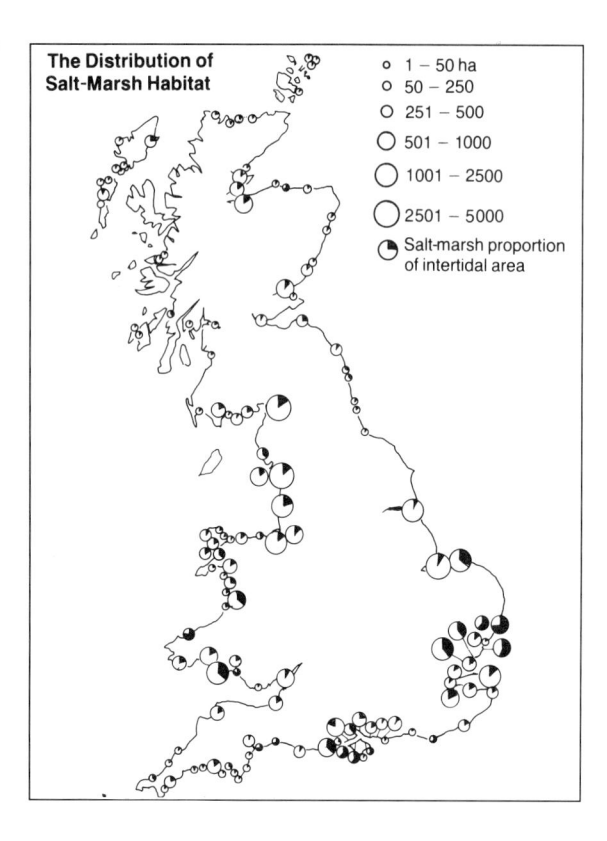

- 1 – 50 ha
- 50 – 250
- 251 – 500
- 501 – 1000
- 1001 – 2500
- 2501 – 5000
- Salt-marsh proportion of intertidal area

Fig. 5.5.14
Salt-marsh: Arne Bay, Poole harbour, the salt-marsh now occupies almost all of the bay

1 Compare the distribution of salt-marsh with the distribution of sand dunes (Fig. 5.5.1). What are the main similarities, and what are the main differences?

2 How might these similarities and differences be explained?

Salt-marsh accumulation occurs wherever tidal waters are sufficiently quiet for fine sediments to be deposited. Initial stabilisation of the sediments begins within the binding action of algae. Mud-dwelling invertebrates aerate the sediments and the first plants may well establish themselves by attachment to loose invertebrate shells. At higher levels progressively more terrestrial types of vegetation appear. Lower levels of the marsh are colonised by plants tolerant of high levels of salinity, and those capable of withstanding water movement, since immersion will occur with nearly every tide. At higher levels plants such as seablite, annual glasswort and perennial cord grass appear. As newly established plants assist with sedimentation so the marsh is built up to provide a wider range of environments. The classic succession in salt-marshes is shown in the diagram (Fig. 5.5.15), but it is unlikely to be found in its entirety in many of the larger estuaries.

The diagram (Fig. 5.5.16) shows the different trophic levels in the salt-marsh ecosystem.

1 Comment on the relative simplicity of the structure.

2 Comment on the susceptibility of the detritus food chain to contamination, and the likely affects of this contamination.

Enclosure, grazing and the management of salt-marshes

The significance of enclosure of salt-marshes for farming, and in particular, crop-growing, has been mentioned in the section on the Wash in 'Managing Estuaries'. Enclosure often means that the transitional and upper marsh communities, which are the most species rich, are removed from the tidal regime. Where enclosure has been extensive there is a dominance of the low marsh and pioneer communities which are generally species poor. This is illustrated well in the diagram (Fig. 5.5.17) which compares the species diversity of the ten largest salt-marsh areas in Britain. Where extensive enclosure has taken place, the salt-marsh grass and red fescue

COASTAL ECOSYSTEMS

A Salt Marsh Profile Showing a Generalised Scheme of 'natural' Saltmarsh Succession

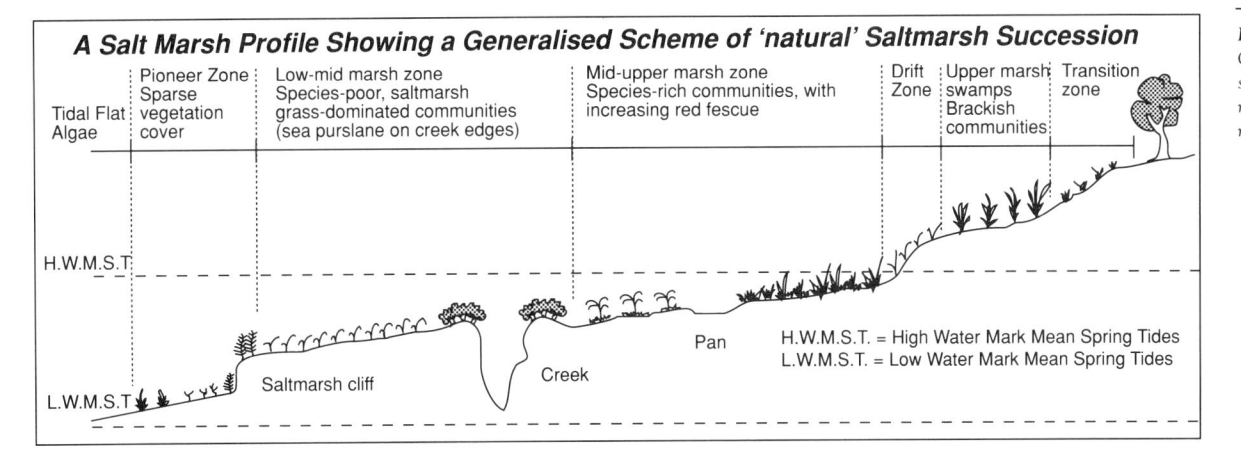

Fig. 5.5.15 *Generalised scheme of natural salt-marsh succession*

General Scheme for the Tropic Structure of a Salt-marsh

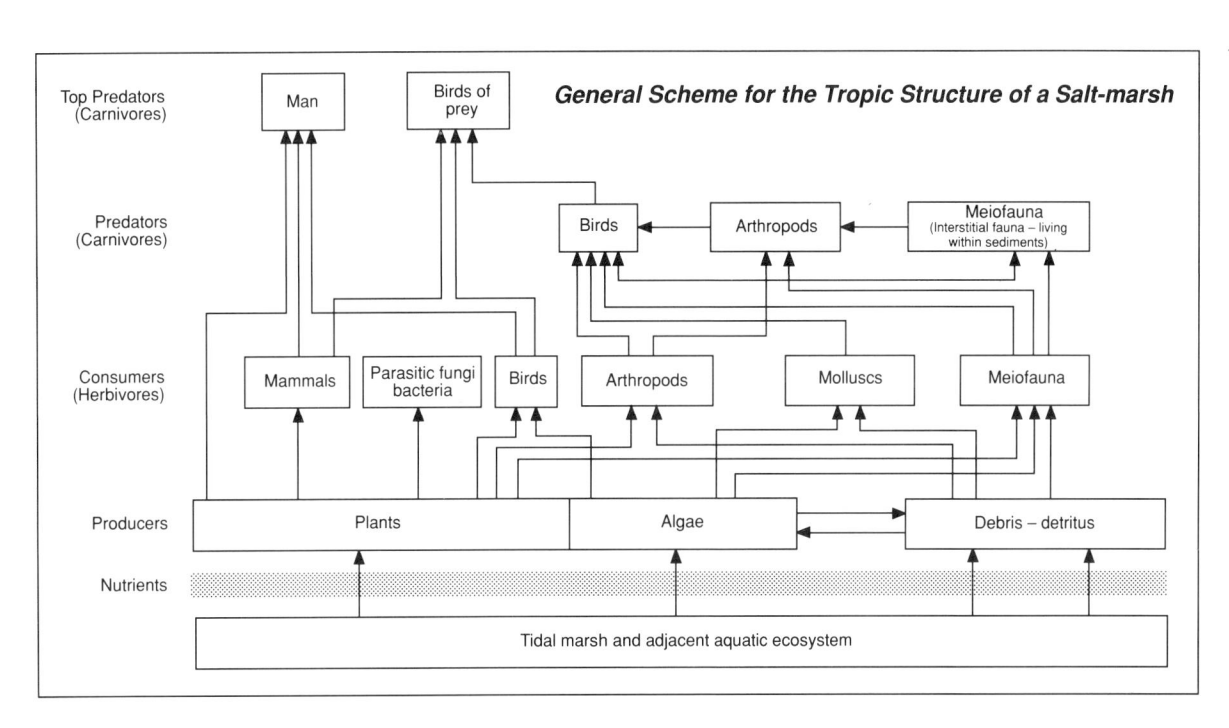

Fig. 5.5.16 *Trophic structure of a salt-marsh*

The Proportional Distribution of the 14 Main Salt-marsh Community Types in the 10 Largest Salt-marsh Areas

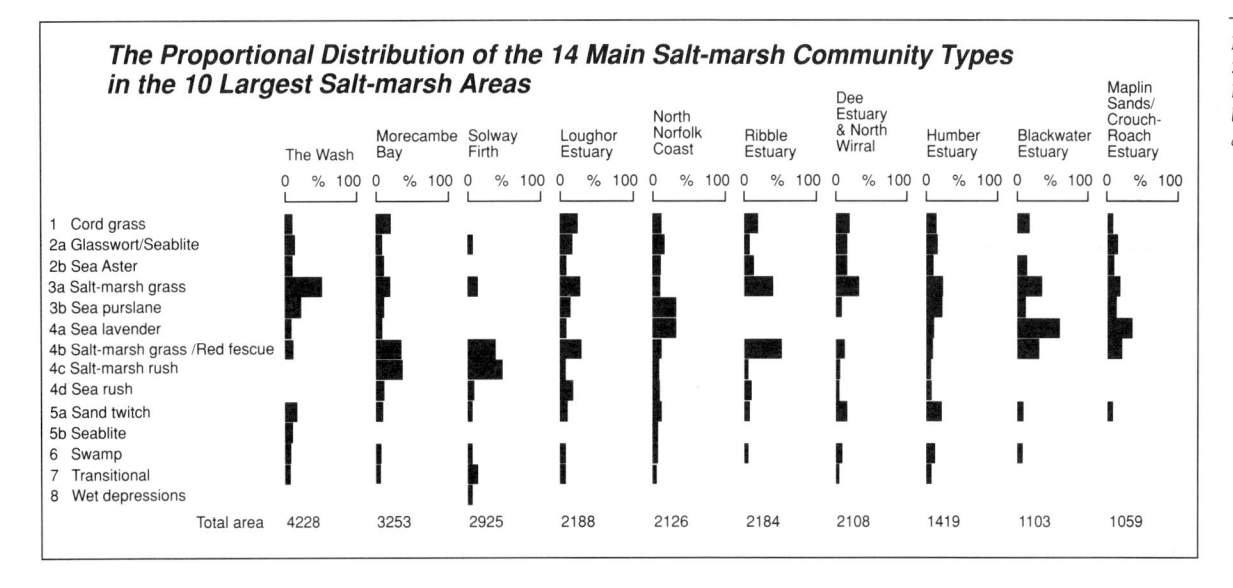

Fig. 5.5.17 *Species Diversity: 10 largest salt-marsh areas*

Communities:
1. Cord grass
2a. Glasswort/Seablite
2b. Sea Aster
3a. Salt-marsh grass
3b. Sea purslane
4a. Sea lavender
4b. Salt-marsh grass / Red fescue
4c. Salt-marsh rush
4d. Sea rush
5a. Sand twitch
5b. Seablite
6. Swamp
7. Transitional
8. Wet depressions

Area	Total area
The Wash	4228
Morecambe Bay	3253
Solway Firth	2925
Loughor Estuary	2188
North Norfolk Coast	2126
Ribble Estuary	2184
Dee Estuary & North Wirral	2108
Humber Estuary	1419
Blackwater Estuary	1103
Maplin Sands/Crouch-Roach Estuary	1059

grass of the upper marshes is missing e.g. in the Wash and in the Dee Estuary. Where these marshes have not been reclaimed they are clearly represented by their dominant species e.g. in Morecambe Bay and the Solway Firth.

1 Comment on the species diversity in the ten salt-marsh communities.

2 What other factors, apart from enclosure, may have led to these variations in species diversity?

Grazing management has important implications for the quality of the salt-marsh ecosystem. With little or no grazing the salt-marsh will:

- be rich in a wide variety of salt-marsh plants

- have a good structural diversity which will provide food for terrestrial invertebrates and habitats for breeding birds

- provide winter feeding grounds for small populations of grazing ducks and geese.

1 What effects will increased grazing have?

2 Discuss the possible effects of other forms of human influence on salt-marsh ecology.

DUNGENESS: A SHINGLE STRUCTURE UNDER THREAT

Fig. 5.5.18
Dungeness

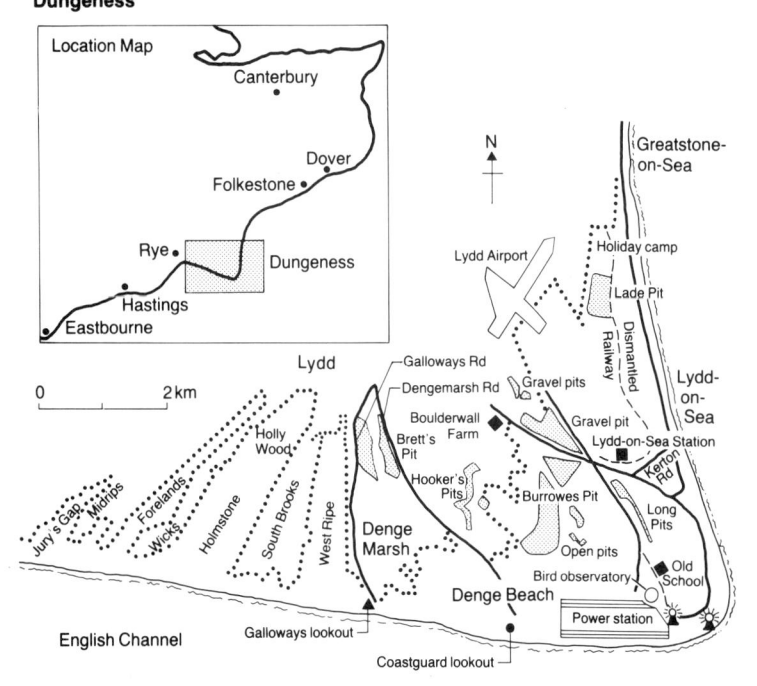

Fig. 5.5.19
Dungeness from the air

Dungeness (Fig. 5.5.18 and 5.5.19) is the largest cuspate shingle foreland in Britain. It seems to have evolved in a series of stages over the last two thousand years (Fig. 5.5.20). It is also of considerable ecological importance, the different habitats include shingle beaches, salt-marsh, freshwater marsh, some open water and grassland (Fig. 5.5.21). Flora on the shingle beaches include a number of species of limited distribution in Britain (including an ancient fragment of holly scrub) and a number of lichen communities of national importance. Both grassland and salt-marsh are of ecological value.

Human intervention on this shingle structure has included:

- military training using a range leased from the RSPB

- extraction of sand and gravel (up to 10 000 tonnes per hectare)

- water abstraction – the water is of exceptional purity and only needs chlorination

- two nuclear power stations occupy an area of 108 hectares (now excluded from the SSSI)

- sea defences; beach feeding is necessary at Denge Beach in order to protect the nuclear power stations – up to 35 000 m^3 of shingle is moved each year from the east of the Ness to the west

- recreation; a number of threats to ecological survival have been noted, including increased pressure from recreational fishermen and bait diggers, new wind-surfing proposals, extensions to the Water Sports Centre at Lydd and a new holiday village at Romney Sands.

COASTAL ECOSYSTEMS

Fig. 5.5.20

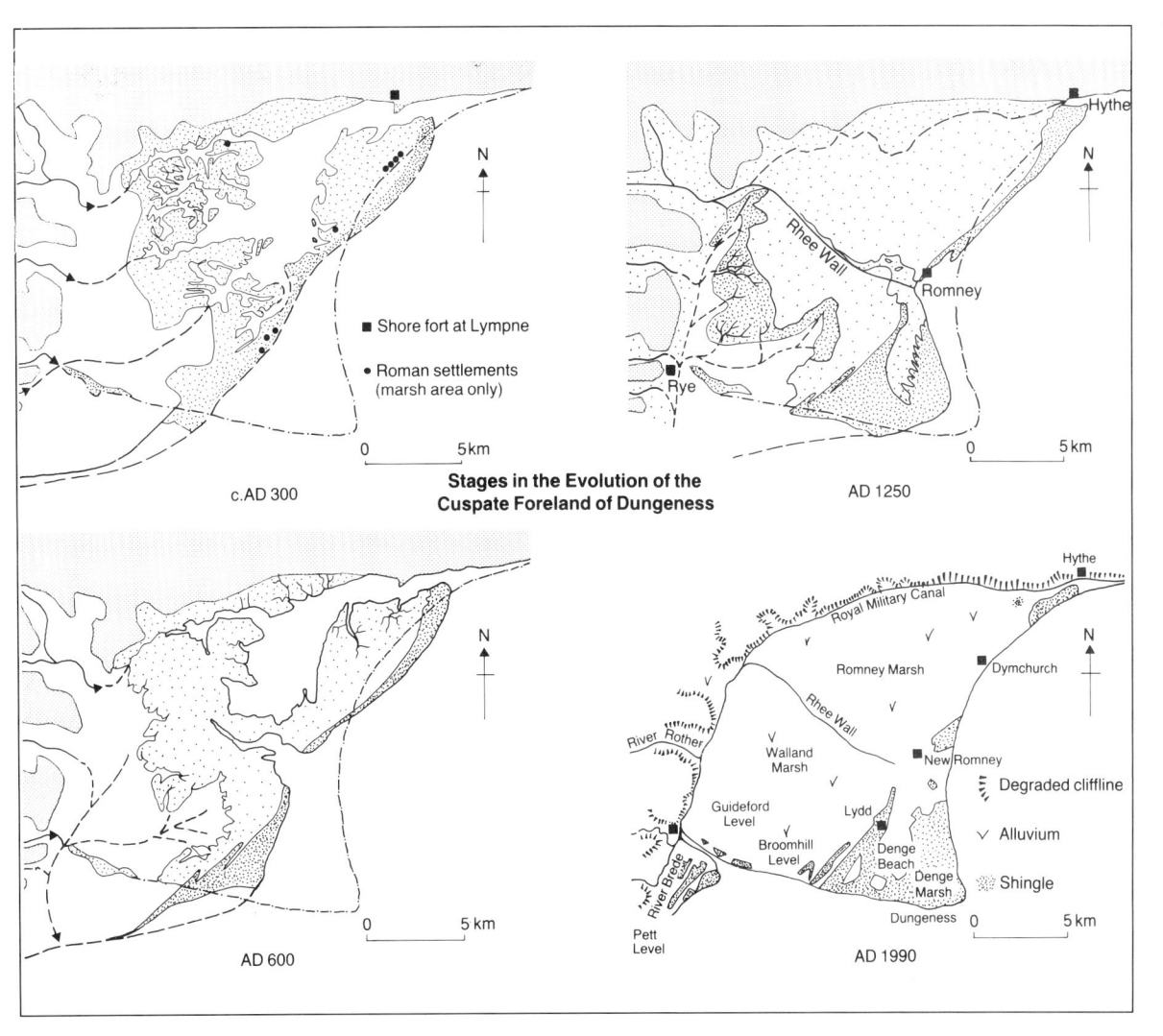

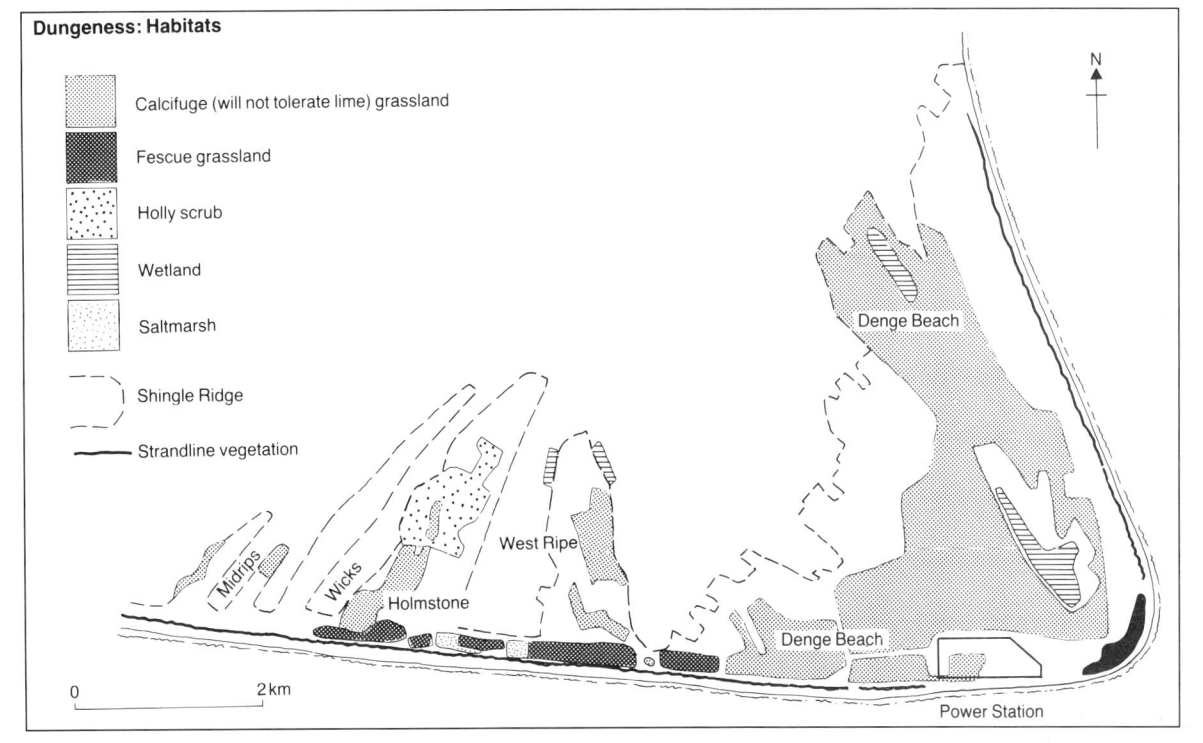

Stages in the Evolution of the Cuspate Foreland of Dungeness

c.AD 300
- ■ Shore fort at Lympne
- • Roman settlements (marsh area only)

AD 1250

AD 600

AD 1990

- Degraded cliffline
- v Alluvium
- Shingle

Fig. 5.5.21

Dungeness: Habitats
- Calcifuge (will not tolerate lime) grassland
- Fescue grassland
- Holly scrub
- Wetland
- Saltmarsh
- Shingle Ridge
- Strandline vegetation

POLLUTION AROUND THE COAST

The sea has long been regarded as a vital resource, ready to offer up its riches for human benefit, and as an enormous dumping ground, ready to accept some of the huge quantities of waste generated by civilised society. Britain's seas and coastal waters are now threatened by the demands placed on them by this heavily industrialised and populated island. In part some of these pressures are unavoidable and constitute the inevitable compromise that accompanies human interaction with the natural environment. However, it can be argued that many of the practices that can threaten coastal waters stem predominantly from economic considerations. These ignore the wishes and needs of all who share in the environment to the advantage of the few who have a direct commercial interest in it.

(Paul Johnston, Greenpeace Scientist, Coastline: Britain's Threatened Heritage)

BATHING WATER QUALITY AROUND BRITAIN'S COASTS

The sketch diagram (Fig. 5.6.1) shows a number of possible pollution sources likely to adversely affect the quality of bathing water.

1 Identify each of the pollution sources.

2 Show how it can contribute to the contamination of bathing water.

Fig. 5.6.1

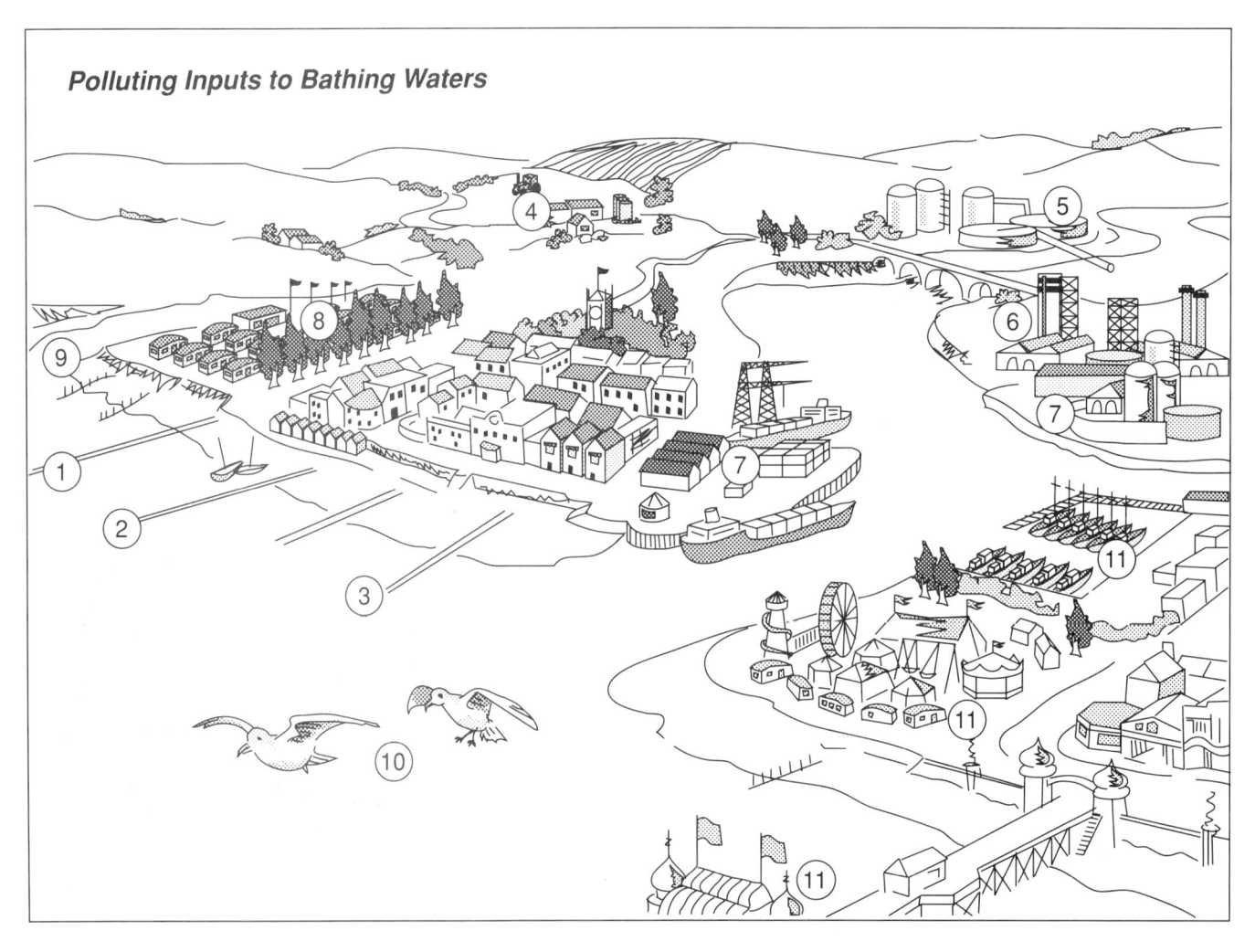

Polluting Inputs to Bathing Waters

POLLUTION AROUND THE COAST

Most sewage from coastal areas is discharged directly into the sea with the only treatment being screening for removal of solids. It is discharged into the sea through one of three main routes:

- short sea outfalls
- long sea outfalls
- stormwater overflows.

Of the first two types there are over 1000 outfalls around the coast of England and Wales, 50 per cent of which were built before 1959, and 25 per cent were constructed before 1925. There are between 10 000 and 20 000 storm water overflows responsible for about 5 per cent of the sewage discharge. Many of the short sea outfalls discharge at or above low water mark, and storm water overflows often discharge directly across a beach (Fig. 5.6.2). These conditions of discharge have meant beaches do not pass the standards laid down by the EC Bathing Water Directive. Replacing these outdated structures with long sea outfalls appears to be the main way in which standards can be improved. The Government now requires all sewage disposal agencies to complete works to ensure that all bathing waters are brought into compliance with EC standards by 1995. A capital investment programme of £31 400 million should bring 95 per cent into compliance by the mid 1990s.

EC Directives on Bathing Water Quality require water to be tested on a number of parameters, which will indicate the extent to which the water is contaminated by sewage-derived materials. In the UK 446 beaches are tested each year to see whether they reach the required standards. The diagram (Fig. 5.6.3) shows the level of compliance of beaches in England and Wales in the period between 1980 and 1990. Two awards are now available for beaches which satisfy stringent standards of cleanliness and competent beach management.

- The Blue Flag Award
- The Golden Starfish Award

For a beach to be awarded the Blue Flag status it has to satisfy 27 different criteria, under headings of:

- Water Quality
- Intertidal sediment and coastal quality
- Environmental Education and Information
- Beach Management and Safety.

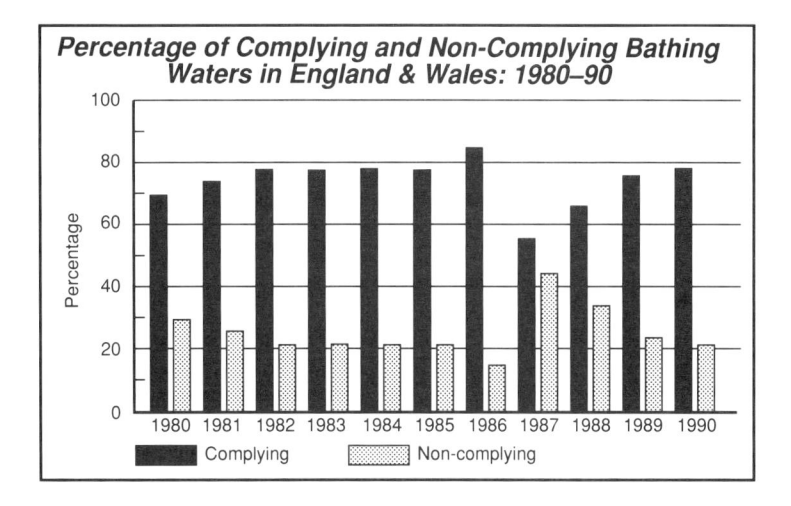

Fig. 5.6.3

Fig. 5.6.2
Sewage outfall across beach

Fig. 5.6.4
Blue Flag Award Beach (1991) Swanage, Dorset

A NEW VIEW OF BRITAIN

Blue Flag Beaches 1991

1 Cullercoats	19 Jacobs Ladder (Sidmouth)
2 Filey	20 Budleigh Salterton
3 Bridlington North	21 Teignmouth
4 Bridlington South	22 Oddicombe
5 Hunstanton	23 Redgate (Anstey's Cove)
6 Lowestoft	24 Meadfoot
7 Southwold	25 Paignton (Paignton Sands)
8 Clacton	26 Crinnis
9 Sheerness	27 Sennen Cove
10 Camber	28 Porthmeor
11 Bexhill	29 Porthminster
12 Eastbourne	30 Woolacombe
13 Christchurch (Friar's Cliff)	31 Weston-Super-Mare
	32 Caswell
14 Bournemouth	33 Pembrey
15 Poole (Sandbanks)	34 Tenby
16 Swanage	35 Magilligan
17 Weymouth	
18 Seaton	

Fig. 5.6.5 Blue Flag Beaches 1991

The Blue Flag Awards for 1991 are shown on the map (Fig. 5.6.5).

Golden Starfish awards are awarded to beaches that cannot satisfy the Blue Flag Criteria because of their relative isolation and the fact that they are little used. Criteria for Golden Starfish Awards are designed to acknowledge the unspoilt and undeveloped nature of these beaches and to encourage considerate use by visitors. Golden Starfish Award Beaches for 1991 are shown on the map (Fig. 5.6.6).

Golden Starfish Beaches 1991

1. Sandend
2. Inverboyndie
3. Bamburgh
4. Beadnell Bay
5. Kessingland
6. Bournemouth Hengistbury
7. Harlyn Bay
8. Polzeath
9. Sandymouth
10. Trebarwith Strand
11. Treyarnon Bay
12. Crackington Haven
13. Constantine Bay

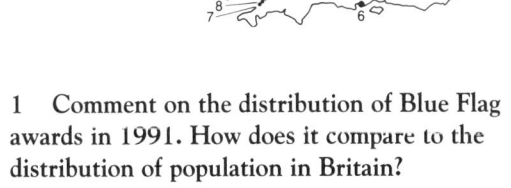

Fig. 5.6.6 Golden Starfish Award Beaches (1991)

1 Comment on the distribution of Blue Flag awards in 1991. How does it compare to the distribution of population in Britain?

2 Suggest what detailed criteria might be used under the main headings given above to determine the award of Blue Flag status.

3 Comment on the distribution of Golden Starfish awards in 1991. Why is it unlikely to reflect distribution of population in main resorts?

4 Suggest a range of criteria for rural isolated beaches suitable for the Golden Starfish Award.

POLLUTION IN ESTUARIES: THE CASE OF THE HUMBER

The map (Fig. 5.6.7) shows some of the major sources of pollution which discharge into the Humber Estuary. The main rivers that empty into the Humber (the Trent and the Yorkshire Ouse) drain some 250 000 sq km of England's land area. The Humber receives the effluent from nearly 11 million people in addition to the liquid refuse from industrial areas in West and South Yorkshire, the East and West Midlands and Humberside itself.

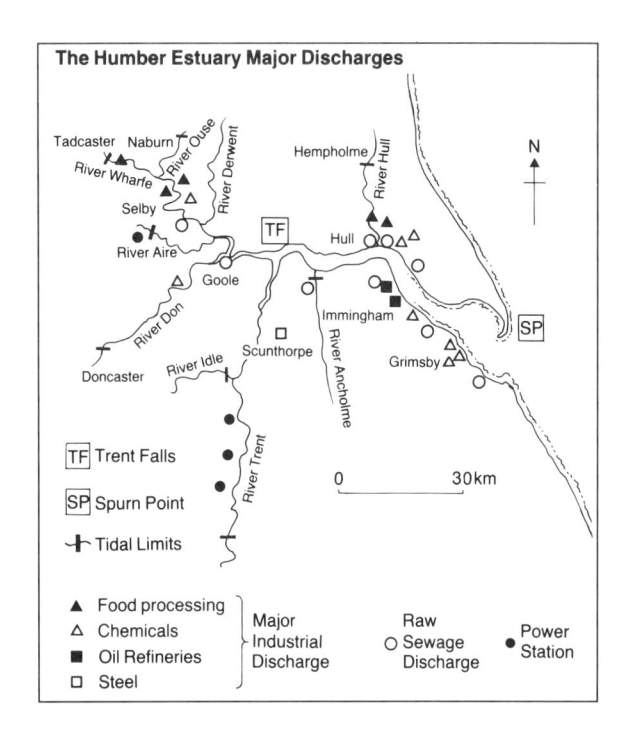

Fig. 5.6.7 Geography, Volume 76 (part 1) 1991

Between Immingham and Grimsby, a collection of chemical works backs on to the estuary near the grimly named village of Killingholme, their skeletal frames visible for many miles over the bleak landscape. Here, near the mouth of the estuary, within sight of the nature reserve at Spurn Point, SCM, Courtaulds, Ciba-Geigy and Tioxide all dump tonnes of unwanted substances – unwanted by them and unwanted by the sea.... SCM has extended its outfall pipe so that waste products are carried further from the shore and so diluted in a larger volume of water which carries them away faster into the North Sea.

(Roger Cowe, Chemical producers battle to turn the tide on Britain's rivers of death, *The Guardian* 2 May 1991)

POLLUTION AROUND THE COAST

The Humber: Estuarine Dynamics

A Ebb → Flood — 6hrs | 6hrs
Tidal curve is symmetrical here. Flood and ebb equal duration and velocity

B Ebb → Flood — 9hrs | 3hrs
Tidal curve asymmetrical here. Flood tide is much shorter than ebb and thus much faster velocity. Sediments are thus moved into the estuary on the flood and are not removed on the ebb

Water quality poorer on south bank

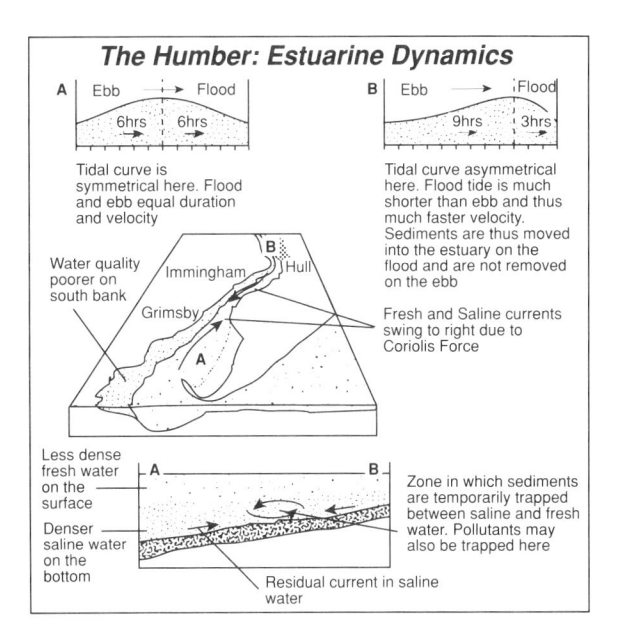

Fresh and Saline currents swing to right due to Coriolis Force

Less dense fresh water on the surface
Denser saline water on the bottom
Residual current in saline water
Zone in which sediments are temporarily trapped between saline and fresh water. Pollutants may also be trapped here

Fig. 5.6.8
Environment Now

Freshwater discharge from the Humber is 13 million m3 daily, but 160 million m3 of seawater flood into the estuary on a spring tide. Estuarine dynamics of the Humber mean that pollutants delivered to the estuary follow a complex path (Fig. 5.6.9).

Water quality standards in the Humber

Water samples are regularly monitored in the Humber and are assessed against Environmental Quality Standards (EQS). The following tables (Fig. 5.6.9) show water quality standards in 1987 together with the EQS for comparison. The monitoring stations are shown on the map (Fig. 5.5.7).

1 Study the tables, and comment on the spatial variations indicated.

2 At which locations was the EQS not attained? Is there any pattern revealed in the locations that failed to attain the EQS?

3 What would appear to be the most serious problem? What could have caused this problem, and how might these levels be improved?

4 How does the overall pattern in these figures compare with popular imagery of towns 'pouring their filth into the rivers that wend their way to the Humber'?

Fig. 5.6.9
Humber Estuary: Chemical Survey 1987

1987 HUMBER ROUTINE CHEMICAL SURVEY DATA COMPARISON WITH ENVIRONMENTAL QUALITY STANDARD

STATION	TEMP °C	DISSOLVED OXYGEN (%)	PH RANGE	AMMONIA (mg/1 N)
Tidal River				
Ouse				
Cawood	19	77	7.4–7.9	0.010
Selby	19	60	7.3–7.9	0.007
Drax	18	40	7.1–7.7	0.008
Boothferry	21	32+	7.2–7.7	0.009
Blacktoft	19	32+	7.3–8.0	0.009
Aire				
Snaith	19	59	7.1–7.6	0.019
Don				
Kirk Bramwith	19	61	7.5–7.8	0.071
Rawcliffe	21	38	7.4–7.7	0.040
Trent				
Dunham	23	82	7.7–8.4	0.010
Gainsborough	20	80	7.7–8.0	0.013
Keadby	18	62	7.7–8.1	0.010
EQS	25	40	5.5–9.0	0.021
Estuary				
Humber				
Brough	14	63	7.7–8.1	0.010
New Holland	18	49	7.2–7.9	0.002
Albert Dock	18	54	7.1–7.8	0.003
Saltend	18	51	6.8–7.8	0.002
Killinghome	18	43	7.2–7.7	0.003
Spurn	16	83	7.3–7.9	0.002
EQS	25	55	6.0–8.5	0.021

STATION	ANNUAL AVERAGE (ug/l)								
	Cd T	Cr D	Cu D	Ni D	Pb D	Zn T	Hg T	As D	Fe D	HCH T
Tidal River										
Ouse										
Cawood	0.2	1	4	3	2	37	0.2	1	140	0.007
Selby	0.2	1	4	3	2	56	0.2	4	164	0.008
Drax	0.2	2	5	7	2	43	0.2	3	152	0.021
Boothferry	0.2	3	6	8	1	51	0.9	4	139	0.024
Blacktoft	0.2	7	20	8	7	115	0.2	15	37	0.032
Aire										
Snaith	0.2	5	8	9	1	40	0.2	4	167	0.075
Don										
Kirk Bramwith	0.2	10	7	32	3	41	0.1	–	138	0.024
Rawcliffe	0.4	2	6	13	1	67	0.2	4	100	0.024
Trent										
Dunham	0.7	3	11	21	3	118	0.1	3	60	0.027
Gainsborough	1.0	2	17	14	2	86	0.1	4	70	0.024
Keadby	0.5	2	18	12	2	85	0.2	10	110	0.020
EQS	5	250	28	200	250	500	1.0	50	1000	0.100
Estuary										
Humber										
Brough	0.1	1.0	13.2	6	4	15	<0.1	27	27	0.022
New Holland	0.3	0.3	4.9	7	1	14	1	5	12	0.077
Albert Dock	0.3	0.8	9.9	6	5	34	0.1	5	103	0.025
Saltend	1.5	1.1	13.4	5	6	36	0.1	4	81	0.027
Killinghome	0.3	0.2	3.1	5	2	14	0.1	1	471	0.007
Spurn	0.2	1.1	5.5	5	6	32	0.1	1	199	0.014
EQS	5D	15	5	30	25	40D	0.5D	25	1000	0.02

Notes

T total
D dissolved
HCH Hexachlorocylohexone

Metals and HCH data based on high water samples

THE RISK OF OIL SPILLAGE ON THE COAST

Fig. 5.6.10 The Braer wrecked on Shetland's coast

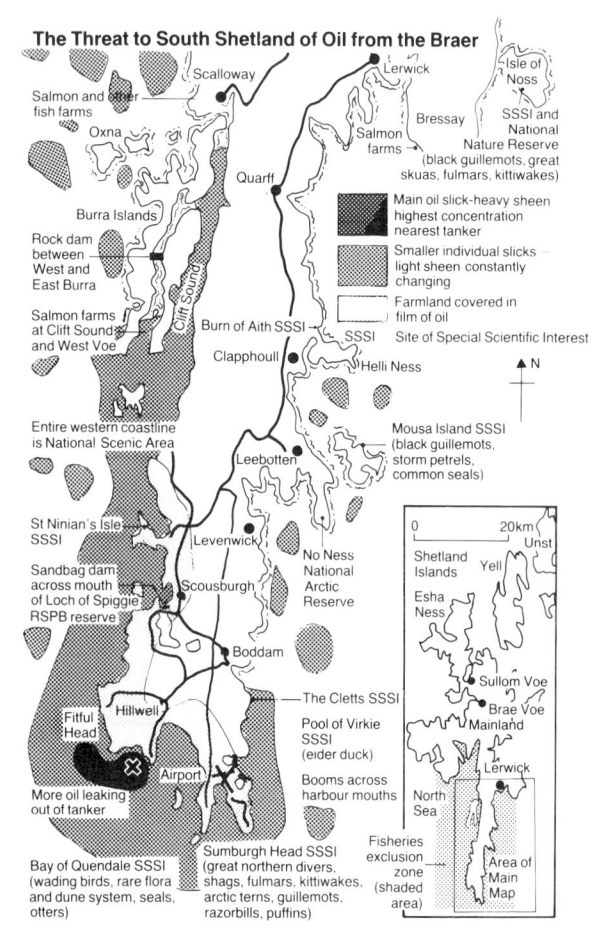

Fig. 5.6.11

Risk became stark, desperate reality on 5 January 1993 when the Liberian-registered oil tanker *Braer* (Fig. 5.6.10) lost power as it entered the stretch of water between the Shetland Islands and Fair Isle. Force ten winds and high seas forced the vessel inexorably northwards and six hours later it struck rocks on the Shetland Coast just to the west of Quendale Bay. Immediately crude oil began to leak from ruptured tanks into the sea. The tanker was carrying a load of 84 000 tonnes of crude oil from the Norwegian terminal at Mongstad to an oil refinery near Quebec in Canada. After a week of severe storms, the tanker broke up, and virtually all of its cargo of crude oil had been spilled into the coastal waters of southern Shetland.

The map (Fig. 5.6.11) shows the situation in the southern part of mainland Shetland four days after the *Braer* went aground. Some comments from people involved:

Willie Mainland, owner of Noss Farm: *'Right up by the cliffs, we lie right on the west coast and overnight it (oil-contaminated spray) spread right over our land. We've had eight acres of vegetables condemned by the health people today'*.

Jim Conroy, Institute of Terrestrial Ecology: *'The spill could not have come at a worst time for the common otter. Shetland otters' food is lowest at this time of year. They feed on the bottom and dive in close to the shore. Oil will affect the fish they feed on; if they ingest poisoned fish they are liable to contract stomach ulcers'*.

Sandra Hogben, manager of the SSPCA centre at Inverkeithing, Fife: *'The survivors (sea-birds) are in a pretty bad state. This particular type of oil has burned the skin and caused feather loss around the eyes and there is obvious internal damage as well.'*

Paul Johnston, research fellow at the University of Exeter: *'It is an extremely unusual situation for which there are no precedents, but the health of the population, the animals and the food plants will have to be monitored for some time. My advice would be to stop breathing in the oil – but presumably the oil is everywhere, so people cannot take my advice'*.

Alistair Grains, salmon farmer: *'God knows the outcome of it. There's 15 per cent loss – or maybe 20 per cent – before they (insurance companies) step in. And the value of the fish now is not what it will be in a year's time. Nothing like this has ever happened to them (the insurers), you see'*.

Allison Duncan, sheep farmer: *'All my land is contaminated. There's a thin film everywhere. Here we have the best, healthiest sheep in Britain, and we face a disaster that could ruin everything'*.

Jim Wallace, M.P. for Orkney and Shetland (who wrote to the Government in 1991 about the inadequacy of contingency plans for dealing with an oil spill): *'The Government has largely ignored the dangers of a major spill off the UK coast – now it looks as if the Shetland islanders will have to pay over many years for the Government's complacency.'*

POLLUTION AROUND THE COAST

1 Use the map (Fig. 5.6.11) and the comments above to prepare a short summary of the main immediate effects of the oil spill from the *Braer* on the people and environment of Shetland.

2 Use the extract from *The Independent* (Fig. 5.6.13) and the map from *The Guardian* (Fig. 5.6.12) as a basis for discussion.

a How effective are methods of dispersal of oil slicks? (Spraying of the slick from the *Braer* was stopped after Shetlanders protested about the toxic effects of the dispersant where high winds carried it on to land.)

b Should much stricter controls be introduced on tanker routes around Britain?

By late 1993 it was apparent that the effects of the disaster were much less serious than feared.

Oil Tanker Movements and Routes around the Coasts of Britain

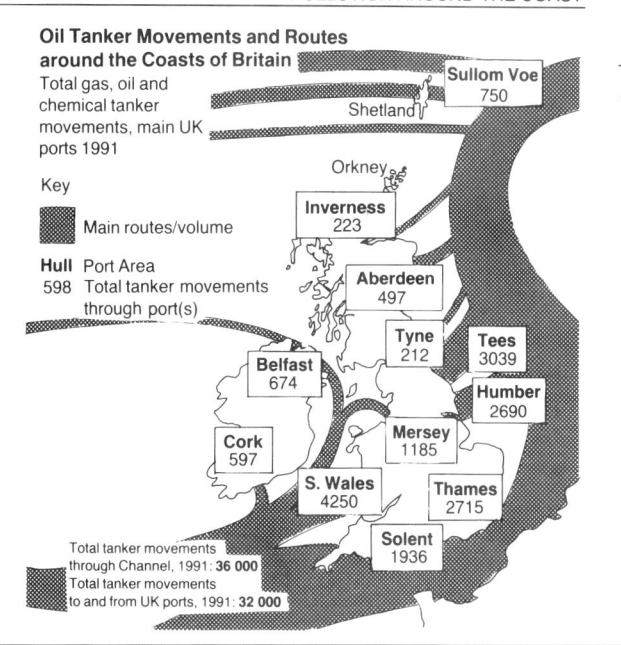

Total gas, oil and chemical tanker movements, main UK ports 1991

Key
- Main routes/volume
- Hull Port Area
- 598 Total tanker movements through port(s)

- Sullom Voe 750 (Shetland)
- Orkney
- Inverness 223
- Aberdeen 497
- Tyne 212
- Tees 3039
- Belfast 674
- Humber 2690
- Mersey 1185
- Cork 597
- S. Wales 4250
- Thames 2715
- Solent 1936

Total tanker movements through Channel, 1991: **36 000**
Total tanker movements to and from UK ports, 1991: **32 000**

Fig. 5.6.12

Nature knows best how to overcome oil slick damage

By Tom Wilkie
Science Editor

FOR ALL practical purposes, no one will be able to tell in five to ten years time that the *Braer* ran aground and polluted the Shetland coast, so powerful are natural processes of regeneration.

The best way to clean up the oil spill is to let Mother Nature do the work, according to marine biologists. But nature can be helped. By spraying dispersants from aeroplanes, the oil can be broken up into droplets, sinking to deeper water where naturally occurring oil-loving bacteria can degrade and digest it.

Dr Peter Donkin, from the Plymouth Marine Laboratory, said a technique used successfully with the *Exxon Valdez* spill was to spray a mixture of microbes and fertiliser on to the oil to kick-start the natural process of degradation. The fertiliser supplies extra nutrients to accelerate bacteria growth and reproduction.

Some of the *Braer's* cargo of light crude will evaporate and blow away with the winds, while some of the remainder will be water-soluble, so a substantial proportion will disappear without any human intervention.

According to Dr Donkin, the normal procedure in Britain is to attack slicks out at sea to sink and disperse oil, to let waves clean contaminated rocks on shore and to skim the top layer of sand off affected beaches.

The priority is to spray dispersants on oil to remove it from the surface and to prevent wind and waves fluffing it up into "chocolate mousse" – an emulsified mixture of about 70 per cent water, and oil with some air.

Dane Neilson, of the Oil Spill Services Centre at Southampton said the "chocolate mousse" phase greatly increases the volumes which have to be sucked up by skimmers and other pumping equipment and makes life more difficult for oil-removing teams. However, Mr Neilson said oil from the *Braer* appears to have fewer asphaltines in its composition, suggesting that it would be less liable to form such emulsions than some varieties.

If the oil reaches rocky shores, British practice is to leave it alone. American authorities have been criticised for steam cleaning rocks contaminated when oil from the *Exxon Valdez* came ashore in Alaska: that is alleged to have parboiled organisms which would have survived the oil.

Life on the shoreline and in rock pools, such as limpets or whelks, that gets covered by oil will die anyway, but detergents could increase the toll in the short term because they are toxic.

"There will be mortality onshore," Dr Donkin said, but affected areas "will re-recruit from adjacent shores over five years or so". If only pockets are damaged, they will re-recruit next spring.

For small marine life, oil is not quite the instant killer portrayed. It acts first as a narcotic: limpets can be pulled off rocks "because they are half-asleep," he added. if the oil is washed away on the next tide, the organisms will recover.

Fig. 5.6.13
Independent 7 January 1993

RECREATION AT THE COAST

Switched off after summer, the litter cleared ... the cafes and bed-and-breakfasts closed, these towns have to earn a living between the crowded August breaks and the next Spring Bank Holiday. Winter's compensation is that this lovely coast can be enjoyed at its most natural – both by its people and its wildlife.

(John Fowles, Coastline: Britain's Threatened Heritage)

The British make 300 million trips to the coast every year. As a recreational resource the coast offers unending variety, satisfying the widely differing needs and aspirations of those who visit it. Once excursion trains funnelled thousands of visitors into the resorts that lay at the end of a railway line: today almost universal car mobility has meant that millions of visitors are not restricted to the resorts. Isolated beach and remote headland, if they lie within walking distance of a road or track will have their visitors, albeit fewer in number than those who still flock gregariously to crowded promenades, full hotels and sprawling caravan parks. So conservation becomes an issue at the coast, as indeed it is everywhere in Britain at the end of the twentieth century. To say that the coast is worth protecting is to state the obvious, but to put such good intentions into practice is far from easy.

BLACKPOOL: THE RESORT IN THE 1990s

In Blackpool it is practically a condition of entry that you do not take life there too seriously. Consider Blackpool's most famous visitors: the family Ramsbottom. 'There's a famous place called Blackpool that's noted for fresh air and fun', begins the famous Stanley Holloway monologue, which used to be the staple of BBC Radio's Children's Favourites with Uncle Mac. The

Fig. 5.7.1 Blackpool Beach

Ramsbottoms were driven to visit the lions in Blackpool due to a lack of maritime excitement: 'They didn't think much to the ocean: the waves, they were fiddlin' and small. There were no wrecks and nobody drowned. Fact nothing to laugh at, at all.'

Last summer the Daily Mail ran a particularly vicious article by A.N. Wilson. 'The whole town is an ugly, smelly, down at heel dump' he wrote unequivocally....

While people may argue about whether Blackpool is the best of our resorts, there is no doubt that it is the biggest. In its heyday in the Thirties it had enough hotel and guesthouse rooms to accommodate half a million visitors every night. Before the war, family holidays were limited to annual factory and mill closures, when for a fortnight entire communities would decamp to the seaside. An idea of this extraordinary migration is given by the fact that during August in the Thirties, 20 million more people would use the railways than in October. Holiday habits have changed. The rise of the cheap Mediterranean package holiday hit Blackpool hard. But the decline in the manufacturing industry has hammered the resort just as badly. But Blackpool has survived – it continues to survive. It can no longer offer accommodation for half a million, but with 120 000 holiday beds it comes way ahead of Brighton – its nearest rival – which has accommodation for just 15 000. There are more hotel rooms in Blackpool than in the whole of Portugal: but it is a question of never mind the quality, feel the width....

'Visitors to Blackpool tend not to stay for the traditional fortnight – nor even for a week. These days the resort is in the business of the short break booked at the last minute. Short-stay holidaymakers make the visitor numbers look good, but they play hell with the economy. There were visible casualties on the waterfront: boarded up hotels and restaurants that smelled of damp and reeked of despair. Blackpool's main problem is that its most ardent fans have been the wrong side of sixty.... The town's tourist department has attempted to rejuvenate its clientele with a marketing campaign that accentuates bright lights – but not those of the illuminations. With 36 night clubs and discos, Blackpool is being sold as a sort of Manchester club scene by the sea.... But aside from straightforward tourism, Blackpool has another valuable string to its bow: the conference business. With a major party conference every other year.... Blackpool has itself a nice little earner, and an attractive walk on part in political history.

(Frank Barrett, To Blackpool Just for a Laugh, the Independent 27 July 1991)

1 What does this article tell us about the changing seaside holiday patterns in Britain?

2 How have resorts responded to these changes?

3 How are resorts such as Blackpool likely to change in the future?

THE TYNEHAM CONTROVERSY IN DORSET

The hardest fate of all has befallen Tyneham which, with its exquisite village and Worbarrow Bay – surely the loveliest in England – was taken over during the War for training the British and American Armies for their great mission on D-Day... but then... by a lamentable breach of faith retained as part of a Gunnery Range.

(Arthur Bryant, historian, in the Introduction to Tyneham by L.M.G. Bond)

The village of Tyneham, and Worbarrow Bay (Figs. 5.7.2 and 5.7.3) both lie within the Purbeck Area of Outstanding Natural Beauty, and the Bay is part of the Purbeck Heritage Coast. Practice firing for tanks began on the Lulworth Ranges in 1916 and the area was commandeered by the Government in 1943. Approximately 1200 hectares were expropriated, which included four farms and Tyneham House, the home of the Bond family. Consequently 250 people scattered in farms, hamlets and Tyneham village were evacuated from the area. Although the land has never reverted from its military use, much improved access to the Ranges, in the form of the Lulworth Range Walks (see Isle of Purbeck Map, inside front cover), was introduced in 1974, as a result of the Nugent Report in 1974. Access is normally possible at weekends, during the main school holidays at Easter, Whitsun, throughout August and at Christmas.

The area, apart from its intrinsic natural beauty has a wide appeal to naturalists. It is an important area for a variety of reasons, including:

- chalk and limestone plants occurring on the Purbeck Hills and the Portland and Purbeck Limestones along the southern fringe
- the Dorset heathland habitat to the north of the chalk ridge
- fragments of copse and woodland, particularly the woodland in the 'gwyle' that is drained by the Tyneham stream to Worbarrow Bay

A NEW VIEW OF BRITAIN

- colonies of sea-birds, including gulls, kittiwakes, cormorants, guillemots and razorbills which nest and breed along Gad Cliff and the chalk cliffs to the west

- groups of rare butterflies, such as the Small Skipper, the Dark Green Fritillary and the Marbled White which flourish on the chalkland.

Fig. 5.7.2
Tyneham, Dorset

Fig. 5.7.3
Worbarrow Bay, Dorset

Role play exercise: Should Tyneham be released by the Ministry of Defence?

Despite the increased access since 1974, there is still some controversy over the continued presence of the Army on the Lulworth Ranges. The value issues surrounding this debate are explored in a role play exercise. Eight roles are assumed in the exercise.

1   Allot each of the roles to various members of the group.

2   Each member should speak to the brief for approximately two minutes.

3   When the speakers have finished, draw up a list of the values that have become apparent in the course of the debate.

4   Examine each of the value positions and see, by questioning, how well it can be sustained.

5   Draw up a table to show the strengths of the various positions held, and then attempt to come to a decision on the controversy.

The Roles, and their Briefs:

Lieutenant Colonel, Commanding Officer, Lulworth Camp

You are a Career Army Officer, educated at Haileybury, and Sandhurst. Commissioned in 1974, you have spent most of your career with armoured units in Britain and Germany, and took part in the brief ground campaign in the Gulf War against Iraq in 1991. Although the threat of the Cold War has now virtually disappeared, your experience in Iraq confirmed your belief in the maintenance of highly-trained quick response armoured groups in the British Army. Off the record, you quite like living in this beautiful part of Dorset, have many local friends, and would be sorry to see this posting disappear.

Ministry of Defence Civil Servant

You are an official who has worked for the last three years in the M.O.D. As a civil servant you have been trained to view issues such as this impartially. Your Minister is convinced of the need for retention of the Lulworth Ranges and you find it difficult to see any other point of view. However you have a second home in a village just to the east of Dorchester, and it does annoy you that you cannot always have access to this section of coastline as much as you would like.

Planner, Dorset County Council

As a planner it is required of you that you keep a balanced view of development in the County. You are well aware of the important source of employment that Lulworth and Bovington Camps provide for local civilian labour. It is equally your concern that much of the coast around Lulworth is suffering from its attraction as a honeypot site, with footpath erosion quite severe in places. Access to Worbarrow Bay is by a narrow winding road that simply cannot carry heavy traffic. On the other hand the Lulworth Ranges are part of the Heritage Coast, and people should be able to visit the area. You have to argue that the loss of the Lulworth Ranges would be something of a mixed blessing.

Farmer at Steeple Leaze Farm

Your family have been tenant farmers at Steeple Leaze, since they had to abandon Lutton Farm in 1943. You have built up a thriving sheep business, and rent 400 hectares from the M.O.D. between the farm and the sea at Worbarrow. You are given due warning of when firing will take place across the Tyneham Valley, and find the land that you rent from the M.O.D. quite invaluable. The farm offers bed-and-breakfast accommodation for tourists who visit Purbeck in the summer. You have mixed feelings about the release of Tyneham.

Representative of English Nature

You are attached to the local English Nature office at Slepe Farm in the north of Purbeck. You are well aware of the important ecological value of the Tyneham valley. Although the idea of gunfire damaging the flora is not an appealing one, it is easy to see that, with the present restricted access, the ecosystems are not suffering damage that might occur with unlimited opening of this area.

Representative of Purbeck District Council

You were elected as a member of the District Council ten years ago. In each of your successful defences of your seat, you have vigorously opposed the retention of the Lulworth Ranges. The people that you represent are constantly telling you that enough is enough, and that the Ranges should be released. Local Government elections are approaching, and considerable pressure is being brought to bear on you that the time for return of the Ranges is long overdue. However you are also aware that unemployment is also a local issue, so that the loss of jobs at Lulworth Camp is something that you have to consider.

Retired University Lecturer, living at Church Knowle

You have retired to the village of Church Knowle, a few miles to the east of the Lulworth Ranges. You care passionately for the coastal landscapes of Dorset, and have spent many summer holidays in your present house, which was your second home until you moved here permanently. You rate open access to the Dorset Coast as one of your top priorities. Other locations exist for tank training, such as Castlemartin (in Dyfed), and Salisbury Plain. You will argue that alternative locations such as these should be used.

Chairman, Dorset Branch of the Ramblers' Association

You have been a member of the Dorset Branch for 30 years and have been campaigning against the retention of the Ranges for as long as you can remember. Although you appreciate that a partial victory was won in 1974, with the granting of greater access to the Ranges, you feel it is now time for the complete release of the Ranges. The paths are hemmed in with fences, so there is no real sense of freedom when you walk the Coast path in this section. Furthermore, in your approaching retirement you would like to be able to stroll along the coast whenever you wished, not just on weekends and in the major school holidays, when it does tend to get rather overused.

PROTECTED BRITAIN

SAFEGUARDING BRITAIN'S HERITAGE

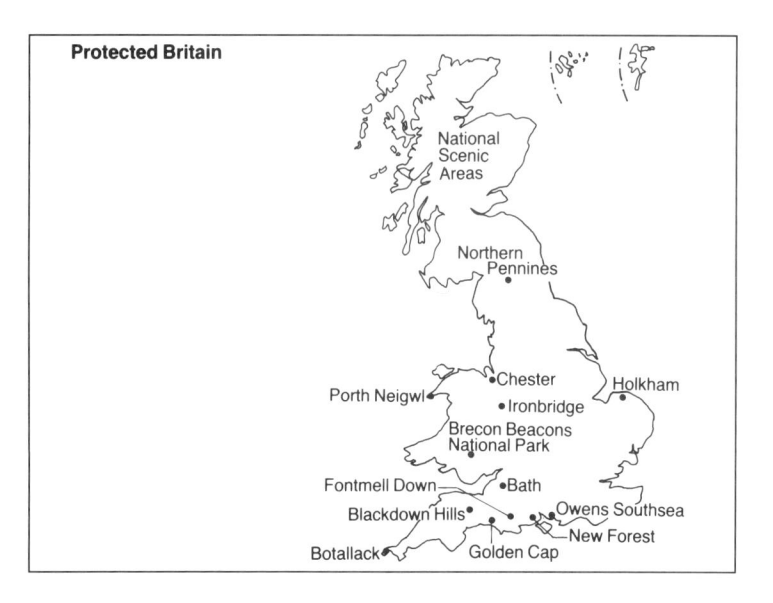

Protected Britain

(map showing National Scenic Areas, Northern Pennines, Porth Neigwl, Chester, Ironbridge, Holkham, Brecon Beacons National Park, Fontmell Down, Bath, Blackdown Hills, Owens Southsea, Botallack, Golden Cap, New Forest)

Many parts of the English landscape remain just as our forefathers left them a long time ago. It is to these quiet solitudes... that we can still gratefully turn for refreshment and sanctuary from noise and meaningless movement.

(W.G. Hoskins, in the Introduction to *The Making of the English Landscape*)

Nearly 20 years on, Hoskins words are still apposite. Much of the best in Britain's landscape is now in need of protection, whether it be the finest examples of our urban tradition (Fig. 6.1.1), the frail intimacies of the countryside (Fig. 6.1.2), the primeval power of our early industrial heritage (Fig. 6.1.3), or the most spectacular of our mountain and coastal scenery (Fig. 6.1.4). Since the 1950s, the conservation movement has gathered pace, and the growing consciousness of our national heritage is increasingly evident in both town and country. A whole range of bodies, both public and private, local and national, are engaged in promoting a new culture of environmental awareness in Britain. Its aspirations are many, but a better appreciation of the infinite variety of Britain's landscapes, a clearer understanding of the fragile balance in so much of our ecology, the protection of our best building and townscapes, and the preservation of the industrial past are all worthy and achievable goals.

The idea of conservation and protection is introduced in two small contrasted case studies.

- Conservation of chalk downland.
- Urban Conservation; the case of Owen's Southsea.

Fig. 6.1.1
High Street Fareham, Hampshire – an area of urban conservation

Fig. 6.1.2
Stourhead Gardens, Wiltshire, managed by the National Trust

Fig. 6.1.3
Our industrial heritage: The Big Pit, Blaenavon, South Wales – a working pit museum

Fig. 6.1.5
Fontmell Down, Dorset, National Trust property, protected Chalkland

Fig. 6.1.4
The Heritage Coast of Gower

Fig. 6.1.6
Cultivated Chalkland, West Dorset, with the now familiar bales of straw in the fields after harvest

CONSERVATION OF CHALK DOWNLAND IN DORSET

Two views on the chalk downland 1793 and 1986:

The most striking feature of the county (Dorset) is the open and unenclosed parts, covered by numerous flocks of sheep scattered over the downs, which are in general of a delightful verdure and smoothness affording a scene beautifully picturesque.

(J. Claridge, 1793)

Chalk downland is Dorset's main scenic feature, but little now remains of its characteristic grassland. Some remnants that have escaped the modern plough are enshrined in reserves like Fontmell Down (see Fig. 6.1.5) but survival of chalk plant life and its invertebrates often depends on steepness of the hill slope and the legal protection afforded to the county's rich legacy of archaeological sites.

(**Macmillan guide to Britain's Nature Reserves, 1986**)

Extracts from two letters to the Western Gazette 1971:

Recently the downland turf which covers the hills on the side of the Plush valley has been removed in the name of good husbandry.... No more will there be a wealth of wild flowers growing on the turf in spring. Instead there is barbed wire and ploughed earth... (Fig. 6.1.6).

For how long do the general public wish or expect the farmer to bear the burden of waste acres of unprofitable land. For they may rest assured that if it was economically sound the farmer himself would like nothing more than to leave these expanses in their state of natural beauty.

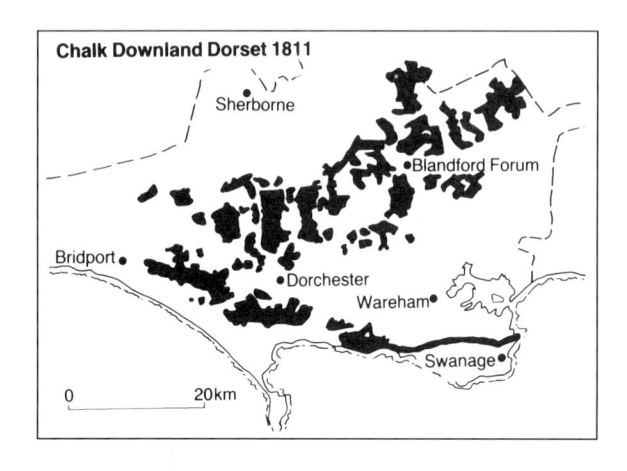

Fig. 6.1.7

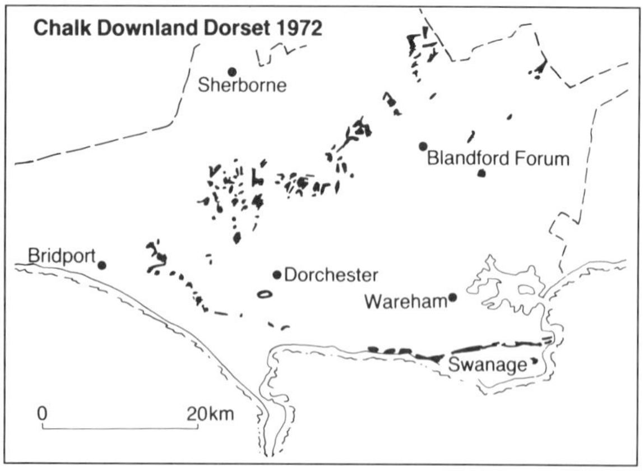

Fig. 6.1.8

1 Study the two maps of the distribution of chalk downland in Dorset (1811 and 1972) (Figs. 6.1.7 and 6.1.8). Comment on the changes that have taken place over 160 years. How would you expect the map to look in the year 2000?

2 Contrast the views expressed by the two writers in 1793 and 1986.

3 Consider the values expressed in the extracts from the two letters. What is your view on conservation of chalk downland?

Fig. 6.1.9

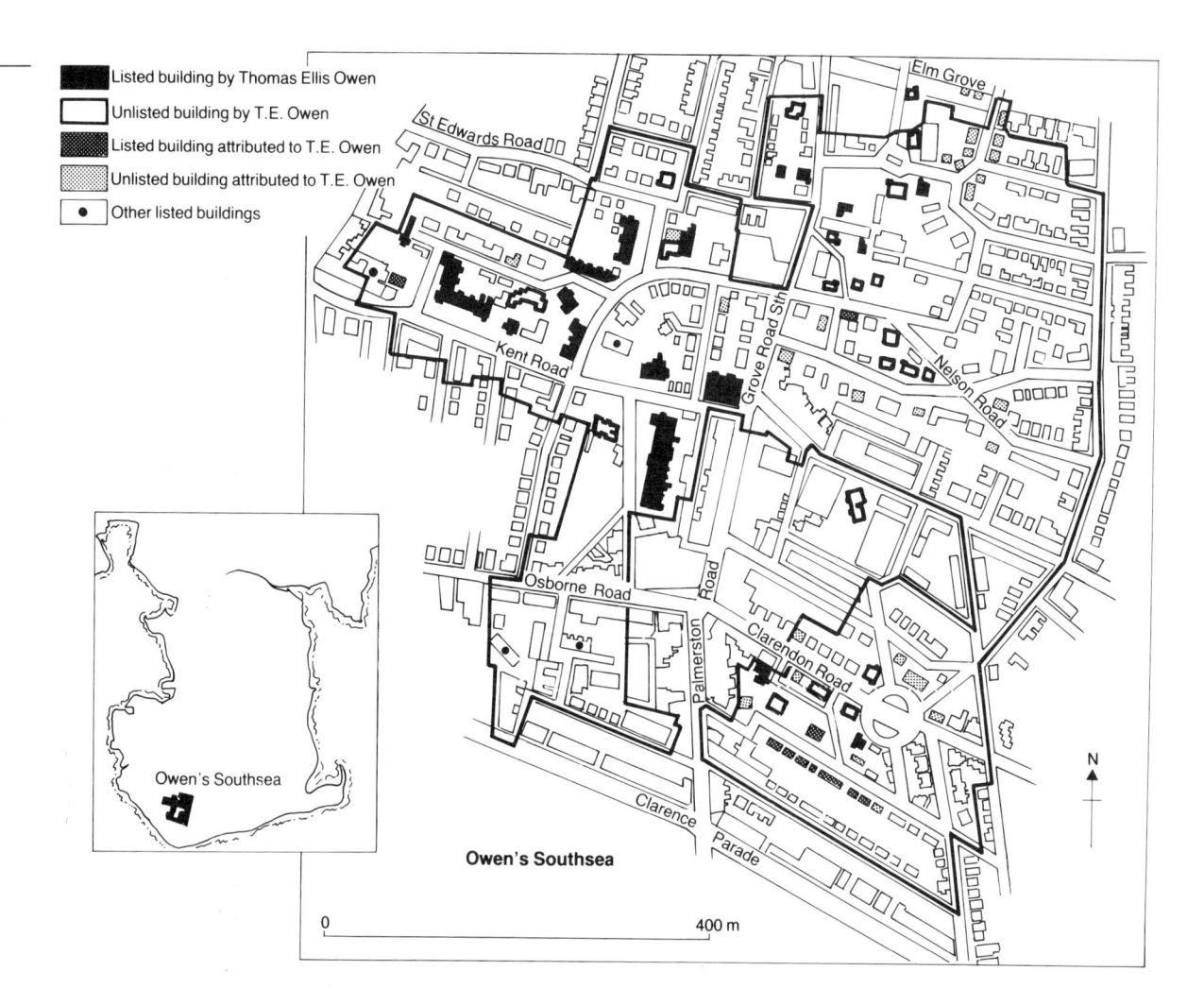

Owen's Southsea

URBAN CONSERVATION: OWEN'S SOUTHSEA

The map (Fig. 6.1.9) shows the limits of that part of Portsmouth known as Owen's Southsea. Thomas Ellis Owen was a local businessman, twice mayor of Portsmouth, whose activities as an architect and builder have created a distinctive part of Southsea. He built most of the villas and terraces in the mid-nineteenth century, and created in Southsea, homes for the emerging middle classes. The road pattern of Owen's Southsea is distinctive: Owen knew how to use a curve in the road to maximum effect, to show off his villas at their best. He used trees and shrubs to create a semi-rural atmosphere in which his buildings could be set. There was little uniformity in his houses: some stood in their own grounds, others fused together in elegant terraces.

Owen's Southsea is now officially recognised as an Outstanding Conservation Area. Once awarded this status, an area can be allocated funds from the Department of the Environment which can be used for restoration work and enhancement. Many of the villas and terraces are listed buildings because of their particular architectural merit. Numerous trees in the area, which contribute so much to its distinctive character are subject to tree preservation orders.

During the twentieth century many changes have taken place in Owen's Southsea: new buildings have been erected; new uses, such as offices and clubs have appeared. Many of the substantial buildings that Owen built, now too large for late twentieth century families have been converted into flats and bed-sitter accommodation.

1 Study the four photos (Fig. 6.1.10) that show examples of Owen's architecture and street layout. What do you think is their distinctive appeal?

2 What are your views on the conservation of such an area?

3 It is the view of some that urban conservation is not a salient issue in towns. It is 'just a nice thing to do'. How far do you agree with this view?

Here two widely differing conservation issues have been briefly examined.

4 Although one is very much an issue of the natural environment, and the other of the built environment, do they have any features in common?

5 Which, in your opinion, is the more worthwhile project: conservation of the remaining areas of chalk downland or creating an Outstanding Conservation Area in a town?

Fig. 6.1.10
Owen's Southsea
(a) Sussex Terrace 1854–55
(b) Portland Terrace 1846
(c) Dovercourt House 1848
(d) Annesley House 1844

CONSERVATION IN TOWNS

If a healthy complexion and good looks in a person are normally signs of a sound physical constitution and a healthy way of life, much the same may be said of a town.

(Gerald Burke, Townscapes)

Conservation of a town's buildings preserves a sense of history, and recognises architectural merit. It can apply to individual buildings, but often seeks to maintain the style and character of a whole area, such as Owen's Southsea, or the Railway Village in the centre of Swindon, or the stern, magnificent buidings that make up Liverpool's Business District. In the second half of the twentieth century, when so many of Britain's town and city centres have been refurbished, and redeveloped, the importance of conserving worthy reminders of the municipal past has necessarily been an important part of urban policy.

BATH: CONSERVATION OF A CLASSICAL LANDSCAPE

From the romantic aspect of visual beauty Bath is incomparably more lovely than its theoretically more perfect rivals, and in the same way the unorthodoxy of its individual buildings gives them a liveliness and interest lacking in the more correct designs.

(Walter Ison, The Georgian Buildings of Bath)

Fig. 6.2.1
Georgian Bath: Queen's Square

In 1987 the city of Bath became a World Heritage site, recognised as being of outstanding universal value from the historical, aesthetic, ethnological or anthropological point of view. Although Bath is best known for its imposing Georgian terraces, crescents and squares (see Fig. 6.2.1) it has other important features, such as the Roman baths, the old medieval street pattern within the walled city and its Victorian villas. Open space has also created a feeling of spaciousness that has contributed towards the city's unique character (see Fig. 6.2.2).

Fig. 6.2.2
Open space, Bath

The map (Fig. 6.2.3) shows the main periods of construction of the buildings in the inner areas of Bath. Four main areas (see Fig. 6.2.4) may be recognised, each related to the age of building:

- the altered and comparatively irregular central area

- the well-proportioned squares and terraces from Queen's Square to King's Circus

- the more open and curved building in the north west of the town

- the more massive and regular forms dominated by Great Pulteney Street.

1 **Discuss the ways in which the built environment varies in these different areas.**

2 **Use a bi-polar semantic scale to assess the environmental quality in the four different areas.**

CONSERVATION IN TOWNS

Main Periods of Construction of Buildings in Inner Bath

Fig. 6.2.3

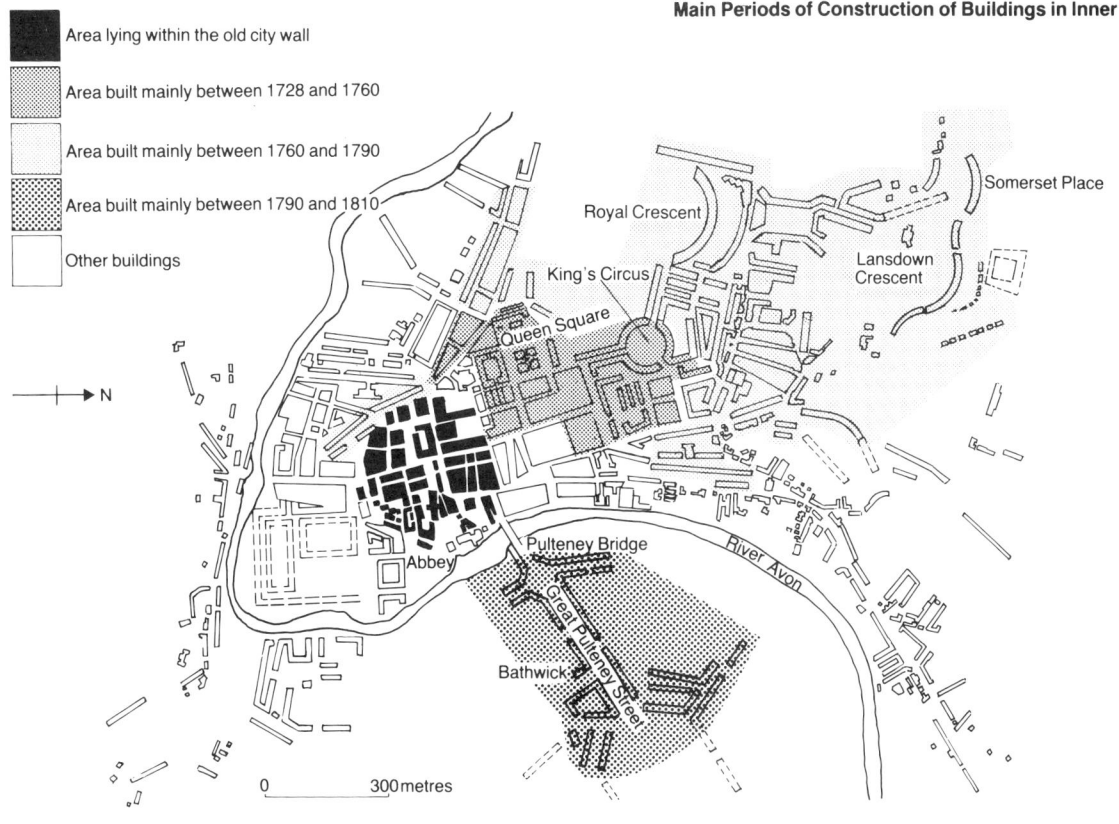

- ■ Area lying within the old city wall
- ▨ Area built mainly between 1728 and 1760
- ░ Area built mainly between 1760 and 1790
- ▦ Area built mainly between 1790 and 1810
- ☐ Other buildings

Fig. 6.2.4
The urban landscape of Bath
(a) Central Bath: Church Yard, pedestrianised central area – Baths and Pump Room on left of photograph
(b) King's Circus 1754–1758
(c) Royal Crescent 1767–1774
(d) Great Pulteney Street 1788–1793

267

Although the City of Bath has been recognised as a World Heritage Site by UNESCO, there is, as yet, no funding available to underpin the designation or its objectives. At the moment conservation is sustained to a certain extent by grants through the Bath Town Scheme, which is a partnership between English Heritage and Bath City Council. This scheme has run since 1955. Normally English Heritage will reduce schemes after 10 years and terminate them after 20. In recognition of the importance of Bath as a historic town and a World Heritage site a further renewal of the grant was made in 1990.

It is interesting to compare surveys carried out in 1975 and 1989 of an area just to the north-east of King's Circus known as Lower Lansdown. It is an area that includes a complete cross-section of Georgian buildings from terraces to elegant town houses. It has a wide variety of ownership patterns and types of land use. Four categories of condition of building were recognised (Fig. 6.2.4a):

1 Very bad – structural failure or likely anticipation

2 Poor – clearly deteriorating

3 Fair – structurally sound but in need of minor repairs

4 Good – structurally sound, no need of repairs.

Fig. 6.2.4a Condition of buildings: Lower Lansdown Bath

RESULTS FOR 1975

CONDITION	NUMBER	PER CENT
1	55	8
2	304	47
3 } 4 }	291	45

RESULTS FOR 1989

CONDITION	NUMBER	PER CENT	CHANGE
1	7	1	–48
2	185	28	–119
3	316	49	
4	142	22	+167

Cost of repairs to a condition 2 building in 1975 = £3000 + fees + VAT
Cost of repairs to a condition 2 building in 1989 = £12 500 + fees + VAT

1 Comment on the changes in the condition of buildings surveyed at the two dates.

2 Explain why grants towards maintenance are essential (calculate the increase in cost for repairs to a condition 2 building from 1975 to 1989).

A policy for the future

Once World Heritage status had been awarded Bath was faced with something of a dilemma. The city is under pressure for growth and change. Bath has to provide additional and improved housing, new employment and recreational facilities for its population and it has to respond to increasing pressures from tourism, and to maintain its position as a regional shopping centre for a catchment area of some 300 000 people. All of this growing pressure has to be balanced against maintaining its position as a World Heritage Site.

Bath City Council produced a manifesto which addressed the above dilemma. Suggest a series of policy initiatives that might appear in the manifesto.

CHESTER: TWENTY-FIVE YEARS OF CONSERVATION

Conservation is not about living in the past; it is the creation of an environment within which our architectural heritage can survive for future generations.

(Donald Insall, *Conservation Review Study: Chester*)

Fig. 6.2.5 Half-timbered buildings, central Chester

CONSERVATION IN TOWNS

Chester presents some of the classic problems that confront every moderate-sized town with a historic core (Fig. 6.2.5). In the 1960s the walled centre was in a state of some decay. Residential growth in the suburbs had led to the abandonment of many substantial houses in the centre. Traffic congestion led to the choking of the main streets in the central area. Many of the historic buildings were badly in need of repair, and demolition and redevelopment seemed a real possibility. In 1967, the Minister for Housing and Local Government commissioned four studies to examine how conservation policies might work within the new Civic Amenities Act, that was in the process of preparation at the time. The cities of Bath, Chester, Chichester and York agreed to commission these studies. Chester instituted its conservation Programme as a result of this study.

The Rows: The historic centre of Chester

The Rows gives Chester its distinctive character as a historic town (Figs. 6.2.6 and 6.2.7). A series of continuous covered galleries at first-floor level runs above the shops at street level. The buildings themselves belong to a number of different ages and represent a number of different architectural styles. Most of Chester's shopping centre lies within the Rows. Commercial pressures for change are inevitable within such an area, and it has to be able to respond to these tensions that develop. Problems of multiple ownership and tenancy make repair and renovation difficult in such an area. Furthermore it is often the case that the upper floors and the rear extensions were in disuse and, in some cases, in a state of some dereliction. In 1968 a report identified 12 Rows buildings with serious defects, 57 that were deteriorating and only 20 in a satisfactory condition.

This report recommended:

- better maintenance of all the properties in the Rows
- bringing the upper floors into use
- the construction of a series of pedestrian bridges at Row level across side streets.

A variety of grants are available for the renovation and refurbishment of buildings such as those in the Rows. In some cases, such as that of the famous Dutch Houses (Fig. 6.2.8) in Bridge Street the City Council was able to purchase part of the building so that repairs could proceed. By the late 1980s the Council could report that the condition of the houses in the Rows had improved considerably, but the economic vitality of parts of the area still needed improving.

Fig. 6.2.6
The Rows, central Chester

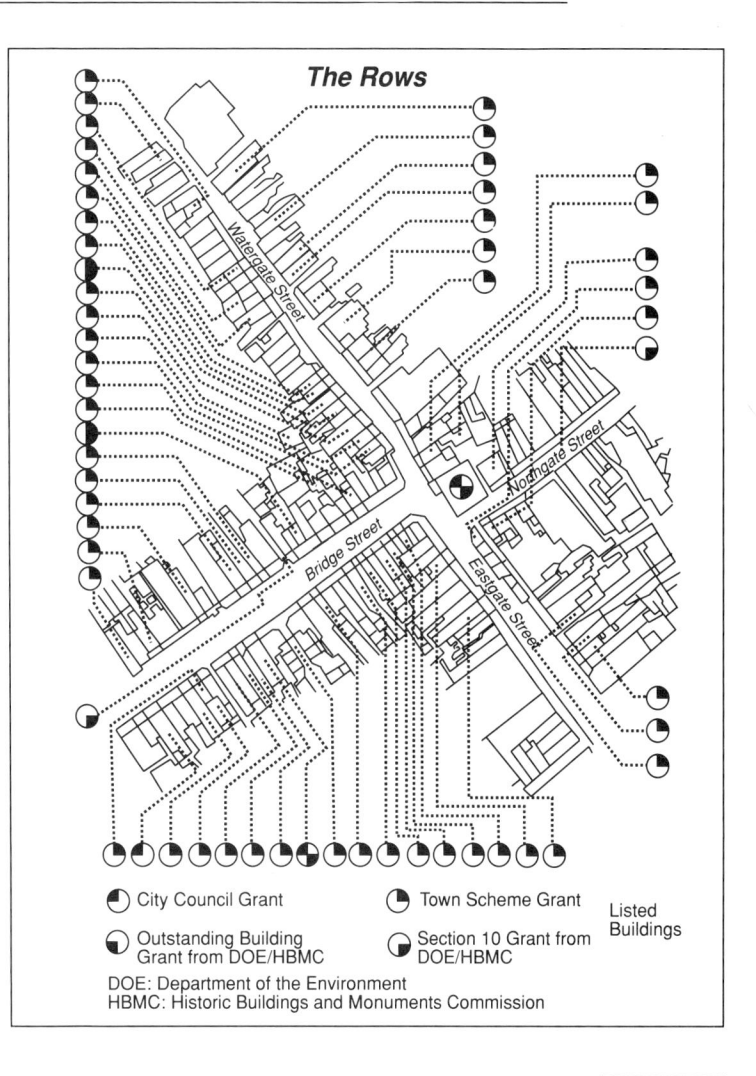

Fig. 6.2.7

A NEW VIEW OF BRITAIN

Fig. 6.2.8
The Dutch Houses, Chester

Fig. 6.2.9

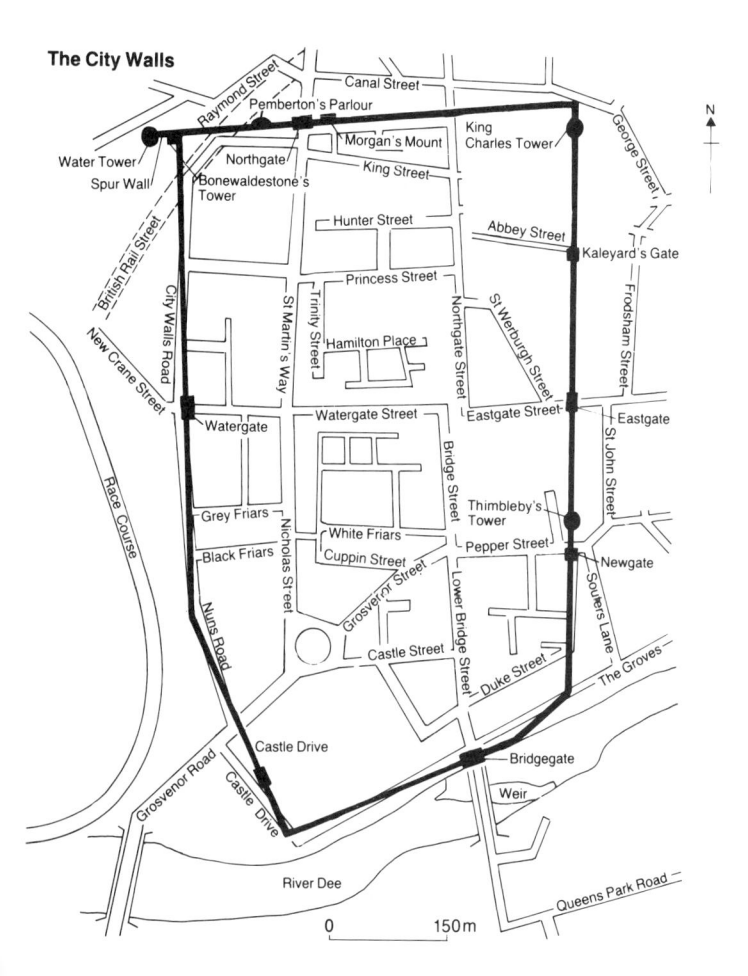

The City Walls

1 Study the photograph of the buildings in the Rows: why are buildings such as these worth preserving?

2 What are the commercial pressures to which an area such as the Rows has to respond in the 1990s?

3 What measures can be taken to reconcile the importance of the conservation of these buildings and the pressure for commercial response in central areas of cities such as Chester?

The City Walls

Fig. 6.2.10
Chester City Walls

The City Walls (Figs. 6.2.9 and 6.2.10) lend a further sense of history to Chester. Although they were originally built by the Romans in the second century they have been substantially modified at a number of later stages. Repair and maintenance has been a constant concern throughout the city's history, but the 1968 Report identified a number of opportunities for improving and conserving the Walls. In a city with such a high tourist potential, access to the Walls and uninterrupted views of their extent are of prime importance. Constant

270

CONSERVATION IN TOWNS

Fig. 6.2.11

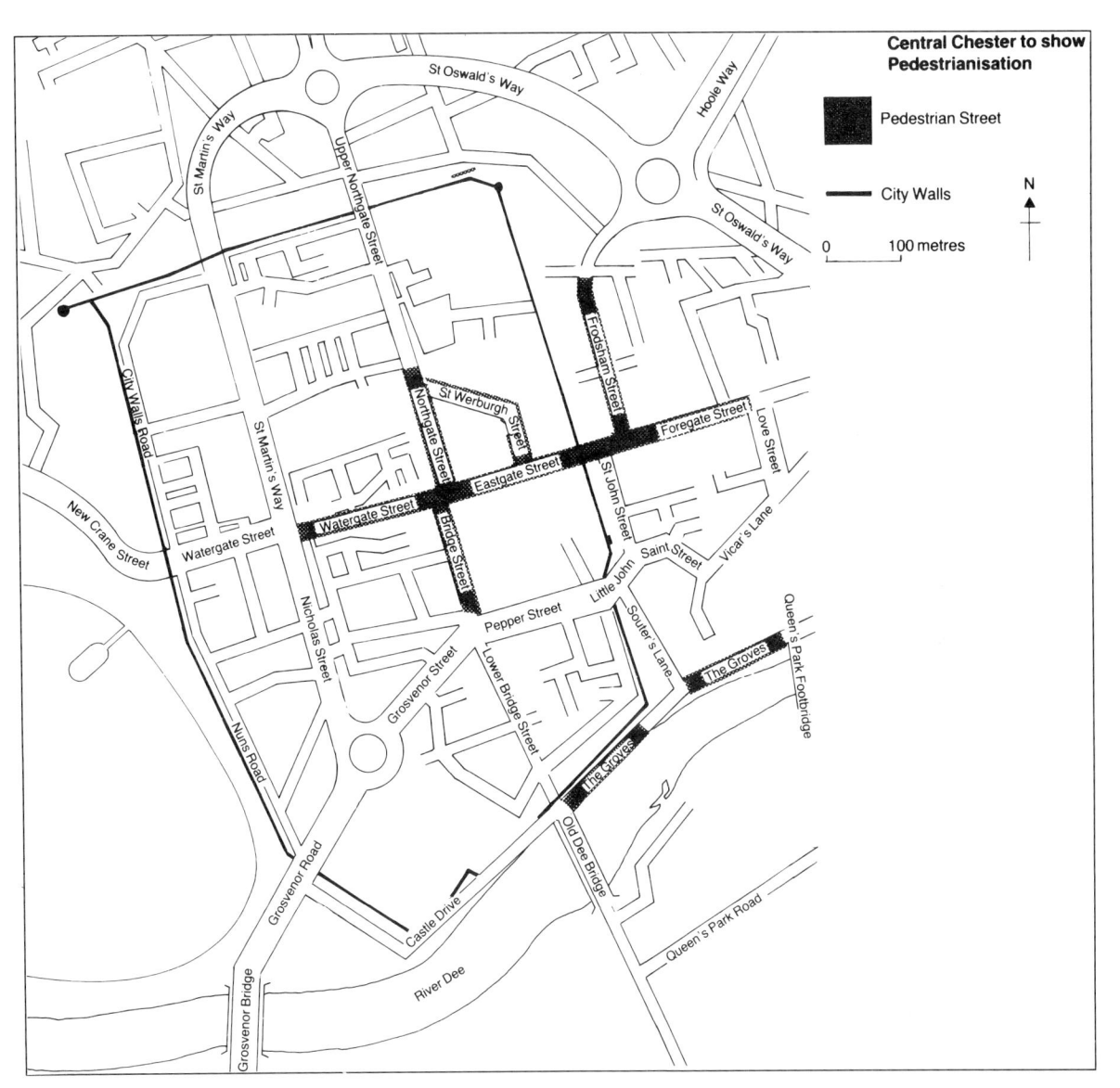

Central Chester to show Pedestrianisation
- Pedestrian Street
- City Walls
- 0 — 100 metres

monitoring of the structural stability of the Walls is necessary in order to check for subsidence and movement of material within the body of the structures. Support for the City's work on the conservation of the Walls is available from the Historic Buildings and Monuments Commission of the Department of the Environment.

The future for Chester

Chester's prosperity will depend ultimately on its ability to attract the region's custom, and this it can do not only by the variety of its goods but by making the best use of its environmental advantages.

(Insall Report 1968)

Chester has a significant role to play in the urban hierarchy of its region. Despite new developments in Liverpool, and the creation of New Towns in Runcorn, Widnes and Telford, it has remained an important service centre and 75 per cent of all jobs in the city are now in this sector. As a shopping centre it was ranked by one study in 1985 as the top non-metropolitan centre in the country and obviously benefits from the specialised pedestrianised shopping atmosphere provided by the Rows. Within the city 4000 jobs are directly related to tourism, and its high profile conservation image (with its Heritage Centre and Visitor Centre) enhances its attraction. It is also becoming increasingly important as a residential centre for people who work outside of the City. Chester has been awarded two important European Awards for Conservation and this has done much to enhance its international reputation in this field.

BRITAIN'S INDUSTRIAL HERITAGE

Urban conservation is now an important theme in many of Britain's towns and cities, and the recognition of the need to promote Britain's industrial heritage is growing. De-industrialisation has left many areas scarred with derelict factories and workshops, disused canals and railways and unsightly ruins and spoil heaps. Yet it is this very tradition of industry which is now giving birth to a new enterprise – the heritage industry. Although heritage awareness is not confined to industry alone, some of its most important sites are inevitably industrial ones. Industrial archaeology has, to some extent, been the preserve of the specialist, but the heritage business seeks to give Britain's industrial past a more popular appeal. Abandoned coal-mines, empty textile mills and deserted iron-making furnaces are all seeing a new industry rise, Phoenix-like, from the ashes of the past. From small beginnings, the heritage industry may well be an important provider of employment as well as promoting a new tourism in areas that had little to offer in this field of attraction in the past.

IRONBRIDGE: BIRTHPLACE OF THE INDUSTRIAL REVOLUTION AND WORLD HERITAGE SITE

In 1837 the valley was called the most extraordinary district in the world.

There aren't any working people left in Ironbridge ... because there aren't any factories left for them to work in!
(Eustace Rogers)

I thought it was interesting that it was going to be saved.

Why was the Ironbridge District called the 'most extraordinary place in the world' in 1837 (Fig. 6.3.3)? Nearly a quarter of all the iron produced in Britain was being smelted in and around the Ironbridge Gorge. Coal had first been exploited in the Coalbrookdale Coalfield in Elizabethan times and this eventually gave rise to a whole series of industries on the banks of the Severn. Although there were several furnaces and forges in the area by 1700, the crucial innovation in iron making was made in 1709 by Abraham Darby in his blast furnace at Coalbrookdale, where he successfully smelted iron ores using coke instead of the orthodox charcoal. Rapid expansion of iron-making occurred throughout the eighteenth century, using local coal, iron ore and limestone. Settlements grew to accommodate the growing numbers of miners, iron workers and quarrymen, and the famous Iron Bridge was constructed in 1779 (Fig. 6.3.2).

Fig. 6.3.1
Ironbridge: The Museums

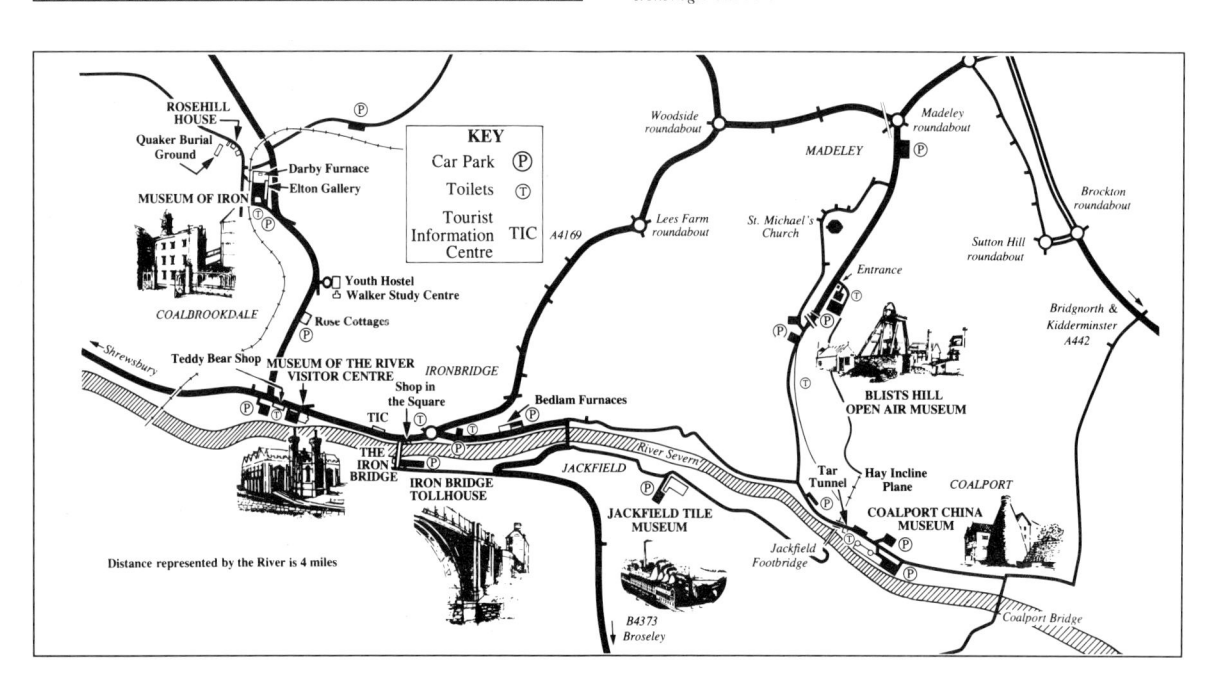

Fig. 6.3.2
The Ironbridge over the Severn at Ironbridge

Fig. 6.3.3
Coalbrookdale by night

In the mid-nineteenth century new ironworks opened, engineering became an important new industry, and the manufacture of fine pottery and porcelain continued at Coalport. By the late nineteenth century raw materials were beginning to be exhausted, and after 1870, furnaces, forges and mines began to close in increasing numbers, although the industries based on clay still prospered. Miners and ironworkers began to migrate from the area as it no longer could provide jobs, and new construction of houses virtually ceased. De-industrialisation came early to Coalbrookdale and Ironbridge, which was described in 1912 as 'tiers of dirty cottages above riverside tips – a dull and squalid little town'. Virtually all the blast furnaces had been closed by the end of the First World War, the Coalport China Works ceased production in 1926 and tile manufacture ended in the 1950s. By 1963 the area was described as, *'an extraordinary landscape (as it was in 1837, but for very different reasons) ... a dense pattern of settlements and industry, set amidst tumbled mounds, quarries and claypits, with old buildings, long-disused canals and railway tracks, with occasional woods and fields of smallholdings and the last surviving farms'*.

The rise of Industrial Heritage at Ironbridge

Few people before the 1950s had any interest in the history of industry, but for those who had, the Gorge was a storehouse of riches.

In the early 1960s the Ironbridge area was one seriously affected by de-industrialisation, with major landscape and dereliction problems. The opportunities to revive the area were first stimulated by the announcement in 1963 of a New Town at Dawley, which included all of the Ironbridge Gorge (which was to be protected 'as a notable beauty spot and a place of recreation for the New Town'). In 1968 the area to be administered by the New Town was much enlarged, and it was renamed Telford.

The awareness of a site worthy of preservation as part of Britain's industrial heritage began to gather pace in the post-war period. The original furnace where Abraham Darby made his initial breakthrough with the use of coke for ironsmelting was saved in the 1950s, and a small museum was set up by Allied Ironfounders, the successors to the Coalbrookdale Company. In 1967, the Ironbridge Gorge Museum Trust was formed with the encouragement of the Telford Development Corporation. In the early 1970s the Department of the Environment gave listed building status to every significant building in the area built before 1900. The Trust now occupies 35 different sites, looks after 5 ancient monuments and numerous listed buidings (see Fig. 6.3.2). World Heritage Status was conferred on Ironbridge in 1986: it was the first site in Britain to receive its official plaque, and shares this status alongside places such as Stonehenge and the City of Bath.

The main museum sites

Museum Visitor Centre: the old warehouse which was used for storage of goods brought down by plateway (tram line) from factories to the Severn before shipment by barge.

Coalbrookdale Furnace and Museum of Iron: the site of the furnace in which Abraham Darby first used coke to smelt iron. The adjacent Long Warehouse houses the Elton Gallery, the Museum Library and the Institute of Industrial Archaeology.

Fig. 6.3.4 William Reynolds, 1758–1803, one of Ironbridge's foremost Industrialists

The Ironbridge and Tollhouse Information Centre: the first iron bridge in the world, erected in 1779 and opened in 1781.

Fig. 6.3.5

Bedlam Furnaces: coke blast furnaces built from 1757–8 and only ones of the 1755–9 period which remain.

Fig. 6.3.6

Blists Hill Open Air Museum: a series of industrial monuments that have been preserved in situ, including a stretch of the Shropshire Canal, the Hay Inclined Plane, three blast furnaces and the remains of a brick-and-tile works.

Fig. 6.3.7

Coalport China Museum: one of the most celebrated porcelain factories in the world closed in 1926, reopened as a museum in 1976.

The Tar Tunnel: a site giving access to natural sources of bitumen, discovered when constructing a tunnel for a canal to link coal pits to the River Severn.

Jackfield Tile Museum: displays tile manufacture, which was very important in the latter nineteenth century, when the iron industry was in decline.

1 A World Heritage Site must 'be authentic and have exerted great architectural influence or bear unique witness, or be associated with ideas or beliefs of universal significance, or it may be an outstanding example of a traditional way of life that represents a certain culture'.

Which of the above criteria best fits Ironbridge? Produce a list of more detailed criteria which you think would have qualified Ironbridge for World Heritage status.

2 The Museum attracts nearly 400 000 visitors a year. The majority of these visitors arrive by car, and then drive round the nine main museum sites (see Fig. 6.3.1). A Park and Ride Scheme would obviously be beneficial to avoid traffic congestion in a site that is relatively confined and has narrow, tortuous roads. Using the map, select a suitable site for the Park and Ride Car Park and then devise a route that would enable visitors to see the major attractions of Ironbridge.

3 For the future, a Management Plan is clearly necessary for Ironbridge. Suggest a set of guidelines that would be necessary for such a Management Plan.

PROTECTING BRITAIN'S LANDSCAPES

Fig. 6.4.1
The Malvern Hills, the British Camp; Area of Outstanding Natural Beauty

... almost all of our landscapes bear the indelible impression of man's continuous habitation, farming and industrial activities. But the landscapes created often have great beauty and charm and the British have a strong emotional relationship with their countryside. One consequence has been a consistent national desire "to prevent undesirable and destructive development in the countryside and to conserve the best of it for the nation to enjoy."

Protected landscapes: The United Kingdom Experience,
Countryside Commission

Protection of Britain's landscapes has been a Government concern for over 40 years. England and Wales now have twelve National Parks. In Scotland 40 National Scenic Areas are recognised, and there is substantial pressure for the creation of National Parks within Scotland. National Parks, by definition, are extensive areas of beautiful and relatively wild country, but there are many areas within Britain that, although they lack extensive tracts of open country, possess high quality landscapes that require some measure of protection. Areas of Outstanding National Beauty (AONB) (Fig. 6.4.1), of which there are 38 in England and Wales, and 9 in Northern Ireland receive this degree of protection. More recently the designation of Environmentally Sensitive Areas has extended the area where landscape conservation is supported by the Government (see pages 117 et seq.). On a smaller scale National Nature Reserves and Sites of Special Scientific Interest are offically recognised. The National Trust is an important non-Government Body that is also involved in the preservation of Britain's landscapes, with 228 000 hectares in its care.

The National Parks

The National Parks were set up under The National Parks and Access to the Countryside Act in 1949. (see Fig. 4.4.3). Two main aims were in mind when the idea of National Parks was conceived in the immediate post-war period.

1 To provide protection for the countryside while also taking care of the distinctive ways of life found within them.

2 To provide opportunities for relaxation and outdoor recreation.

One of the main tasks of National Park Authorities is to ensure that these two principal aims are not mutually exclusive and that each may be reconciled with the other. Ten National Parks were created in the 1950s, and two have been designated more recently, the Broads and the New Forest. Four major reports on the National Parks have been published and the latest, *Fit for the Future*, 1991, made some important recommendations, concerning the two principal aims above:

- Higher profiles should be given to nature conservation, archaeological and architectural conservation.

- Only outdoor recreational pursuits that involve the quiet enjoyment of the Parks should be encouraged.

- Higher levels of maintenance and signposting of rights of way should be achieved.

- Access arrangements should be agreed over all open land by the year 2000.

- Clear policies directed towards tourism and traffic management should be introduced.

Fig. 6.4.2

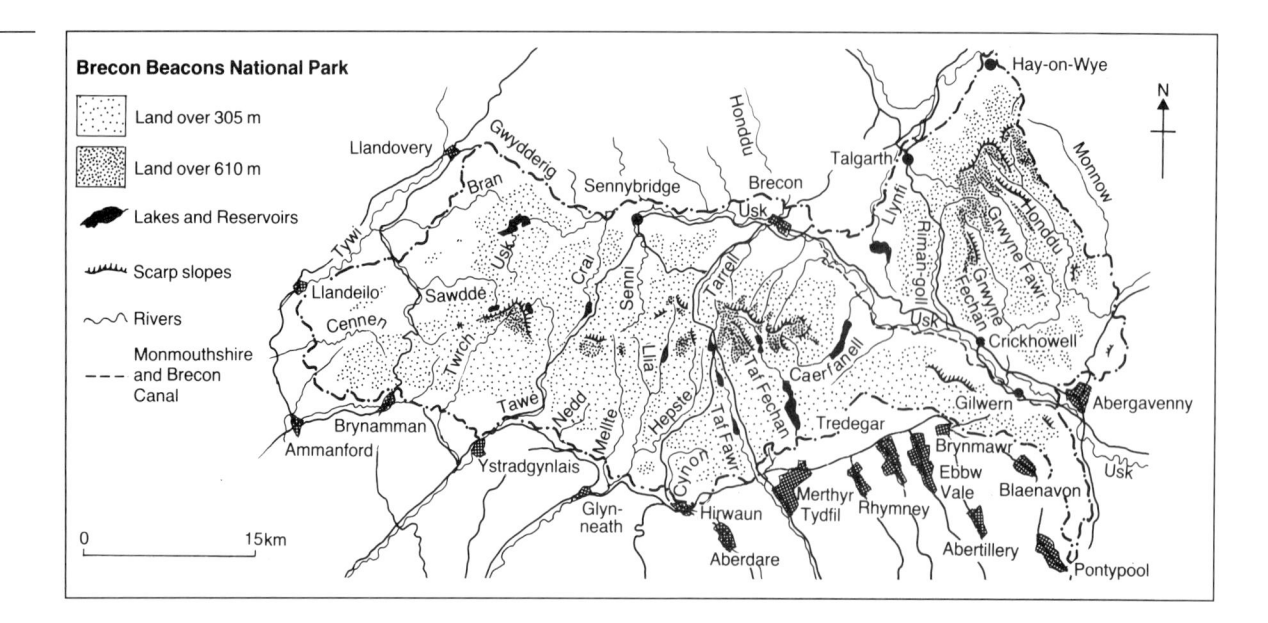

Brecon Beacons National Park

[Map legend:
- Land over 305 m
- Land over 610 m
- Lakes and Reservoirs
- Scarp slopes
- Rivers
- Monmouthshire and Brecon Canal]

Some important new recommendations were also made concerning various forms of land use in the Parks.

- A new farm support system should be introduced in the Less Favoured Areas of the Parks.

- Coniferous afforestation in the Parks should be actively discouraged at Government level. Broadleaved woodland management should be encouraged.

- High priority should be given to schemes that will help in job creation.

- Local needs housing should be made available.

- All planning applications for major developments should be accompanied by an Environmental Assessment.

- A review of military training needs, with a view to reducing the areas used at the present time.

Consider all of these recommendations, both for the refining of the present aims of National parks, and for future land use.

1 How are they likely to alter the way in which the National Parks are managed?

2 What changes in land use could result from the implementation of these proposals?

3 What objections could be raised against such recommendations?

A CASE STUDY: THE BRECON BEACONS NATIONAL PARK

Mountains are the beginning and end of all scenery.

(John Ruskin)

The Brecon Beacons National Park, with an area of 1350 sq km, fifth in size, was the last of the original ten to be designated in 1957 (Fig. 6.4.2). The Park lies astride four counties, and embraces a wide variety of scenery. The upland core of the Park extends from the Black Mountain in the west, through the sweeping moorlands of Forest Fawr, and the dominating summits of the Beacons themselves in the centre (see Fig. 6.4.3) to the sloping tableland of the Black Mountains in the east. The valleys of the Usk and Wye form the northern boundary, whilst in the south the Park reaches to the edge of the South Wales Coalfield. A glance at a road map shows that the Brecon Beacons National Park is within easy reach of a considerable urban

Fig. 6.4.3
The summits of the Brecon Beacons looking west: Cribyn, Pen y Fan and Corn Du in the distance

PROTECTING BRITAIN'S LANDSCAPES

Land use in the park

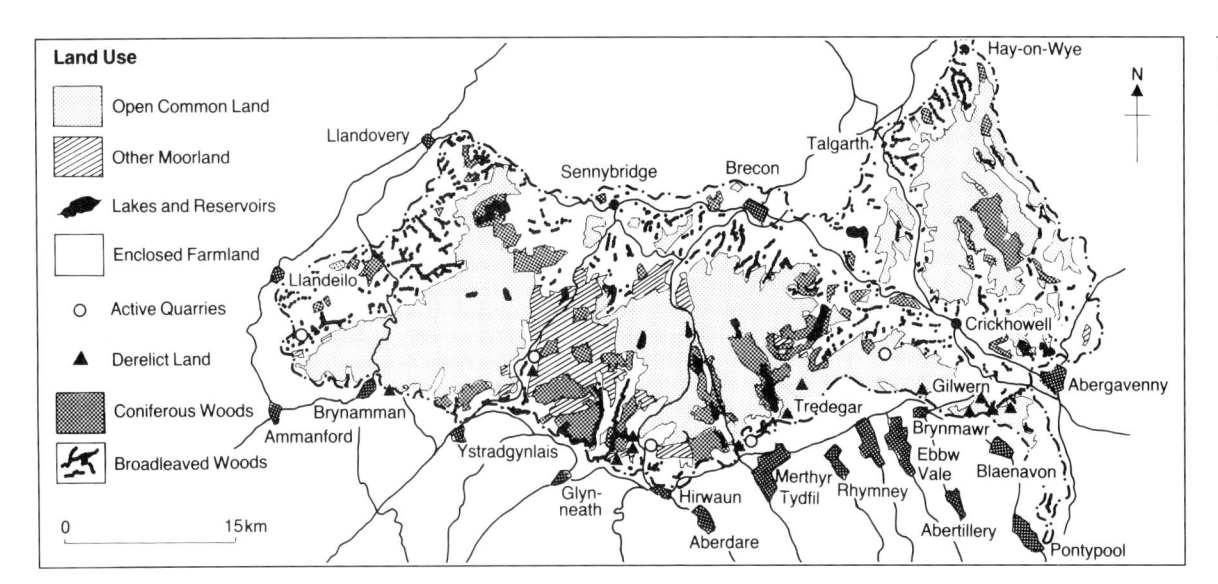

Fig. 6.4.4 Beacons National Park

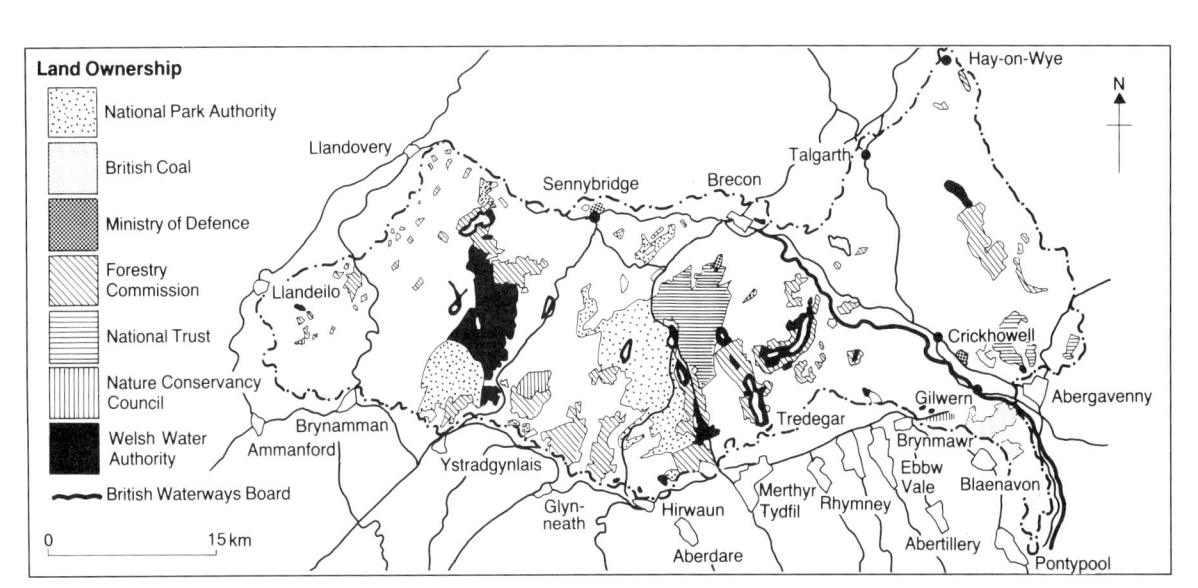

Land Use
- Open Common Land
- Other Moorland
- Lakes and Reservoirs
- Enclosed Farmland
- Active Quarries
- Derelict Land
- Coniferous Woods
- Broadleaved Woods

Fig. 6.4.5

Land Ownership
- National Park Authority
- British Coal
- Ministry of Defence
- Forestry Commission
- National Trust
- Nature Conservancy Council
- Welsh Water Authority
- British Waterways Board

population. Two million people in industrial South Wales are only one hour's drive away, those in Bristol a little more, and those in the West Midlands two hours and even those in London only three.

The map (Fig. 6.4.4) shows the main types of land use within the Park. Agriculture occupies 85 per cent of the area, with 50 per cent in agricultural holdings and 35 per cent registered as common land. Almost all of the farms lie within the Less Favoured Areas, which qualifies them for the higher rate of subsidy. Almost all are involved in livestock production, with a trend towards larger farms with more improved grassland and less rough grazing. The map of land ownership (Fig. 6.4.5) shows that two major owners of land have important effects on land use. Most of the major supply reservoirs (19) for industrial South Wales are located within the valleys of the Park. Much of the Welsh Water Authority's original land was leased to the Forestry Commission in order to reduce erosion, but it was soon realised that planting could reduce run-off and affect water quality. Planting began in the 1920s and 1930s, and reached its maximum during the 1940s and 1950s. New planting has now fallen away. Mining and quarrying only affects the southern fringe of the Park, mainly along the outcrop of the Carboniferous Limestone and the restricted outcrop of the coal measures; 400 hectares of derelict land result from abandoned quarries and spoil heaps. Parts of the open moorland, particularly around Sennybridge are used from time to time for military training.

Strategies for conservation and recreation

The strategies for conservation in the Park are listed as:

- the natural beauty of the landscape will be conserved and enhanced
- wildlife and scientific interest will be conserved and the variety of habitats conserved
- qualities of remoteness will be protected
- archaeological, historical and traditional features, buildings, landscapes and townscapes will be conserved and enhanced
- public awareness of the Park will be developed.

The strategies for enjoyment are listed as

- public access, enjoyment and appreciation of the Park will be promoted
- unsuitable or conflicting recreation activities will be discouraged
- the life of local communities will be supported
- the Welsh language will be supported
- sustainable economic activities will be encouraged
- damaging development will be discouraged.

1 Are there any elements within the two strategies that appear to be incompatible?

2 If there is any incompatibility between the two strategies, which should take preference?

The map (Fig. 6.4.7) shows some of the pressure and vulnerable areas within the Brecon Beacons National Park. Pressure areas are defined as areas where an increase in recreation use is likely to make worse any existing problems. Vulnerable areas are those where wildlife habitats or qualities of remoteness are especially sensitive to local increases in visitor numbers.

Fig. 6.4.6
(a) Pressure zone in the Brecon Beacons National park: heavily used footpaths on the approach to Pen y Fan
(b) Remote area in the west of the Brecon Beacons National Park: Llyn y Fan Fawr. Areas such as this are seen as particularly vulnerable to more widespread recreational activity in the Park.

Fig. 6.4.7 Brecon Beacons National Park: Pressure and vulnerable areas

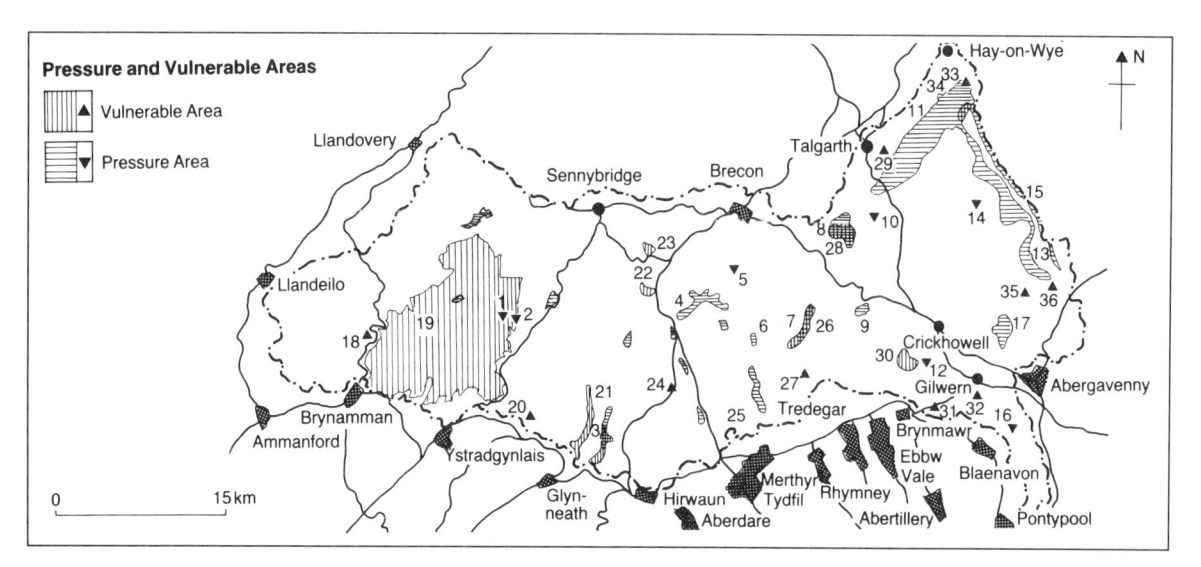

Pressure areas

Definition: areas where an increase in recreation use is likely to make existing conditions a good deal worse, eg physical erosion, traffic congestion, trespass or disturbance or inconvenience to local communities.

1. **Bwlch y Giedd** Unsightly and dangerous erosion on steep path.
2. **Cerrig Duon area** Roadside parking can cause obstruction on hot weekends.
3. **Cwm Porth/Clun Gwyn/Sgwd yr Eira** Parking problems, river bank erosion, dangerous paths and falls, path erosion and proliferation on shaly substrate. Pressure on Porth yr Ogof by large numbers of caving groups.
4. **Pont ar Daf – Pen y Fan path/Cor Du/Pen y Fan/Cribin/Gap** Well-used paths are heavily eroded and very unsightly.
5. **Cwm Cynwyn** Small roadside parking area, served by narrow minor roads. An important access to the Beacons, becomes very congested at times.
6. **Torpantau** Path up to Craig y Fan-ddu has become very eroded in recent years.
7. **Talybont** Many visitors to: 4 outdoor centres, 4 pubs/hotels, canal including aqueduct, 2 camping sites. Casual visitors owing to position near reservoir, river Usk and through routes. The character of the village could be threatened by further expansion of recreation facilities or visitor accommodation.
8. **Llangorse Lake, common and village** Many day and resident visitors to: several camp/caravan sites, outdoor activity centre, 2 pubs, lake and common. Lake is well used by several water sports and at summer weekends its access points are also heavily used in relation to their capacity. Existing and potential conflicts between various interests are not yet reconciled within a mangement plan. The character of the village could be threatened by further expansion of recreation facilities or visitor accommodation.
9. **Llangynidr** Parking problems at: river bank, canal, lock; many visitors to the above and pubs. Erosion on river bank.
10. **Cockit Hill** Serious erosion problem on the steep access to Llangorse Hill, owing to heavy use by pony trekkers, motor cyclists (illegal) and walkers.
11. **Black Mountains north scarp/Hay Common** (including access roads) Heavy use causing path erosion by pony trekkers, also walkers and hang gliders. Cars and motor cycles driven and parked on common without authority. (Heavy non-recreational pressure from unauthorised itinerant campers at some times of year.)
12. **Pant-y-rhiw** Access point for well-used caving and climbing area also popular with some casual visitors. Lack of parking space and narrow approach roads lead to congestion and inconvenience to residents. New parking area planned; toilets under consideration.
13. **LLANTHONY VALLEY** The valley road and some public paths are heavily used by recreation traffic: motor vehicles, pony trekkers and walkers. The popularity of the area leads to considerable traffic congestion during holiday periods and summer weekends and causes much inconvenience to local users and friction between the different kinds of traffic on some routes.
14. **Yr Eithen** A picnic site at the end of the metalled public road in the Grwyne Fawr valley. Its popularity at certain times during holiday periods leads to congestion of people and cars in a limited area with few facilities.
15. **Offa's Dyke Path** Erosion of the peaty surface of the Long Distance path is becoming worse and is difficult to control.
16. **Keepers Pond** Pond and viewpoint are very popular with local people and visitors. Lack of car park prior to August 1986 led to damage and unsightly parking on sensitive common land.
17. **Sugar Loaf** Path erosion on approaches to summit is becoming increasingly damaging and unsightly.

Vulnerable areas

Definition: areas where qualities of remoteness or wildlife habitats are especially sensitive to local increases in visitor numbers or recreation activity. Their vulnerability arises from the difficulty of close management and control of visitors, and the fact that they tend to be attractive and/or more readily accessible to the general public.

18. **Clogau Mawr/Bach** (SSSI) Rare plants around headwater of stream and in old quarries. Rich moss and lichen communities (with rare species) in old quarrying areas. Common land.
19. **Carmarthen Black Mountain remote area** (part SSSI) 'Remote area' where 'wilderness qualities' should be retained. Common land.
20. **Craig Llech woodlands** Several rare plant species and luxuriant moss communities with some locally scarce species. Public path. Part owned by the National Trust.
21. **Blaenau Nedd and Mellte woodlands** (SSSI) Rich moss flora with rare species. Also ferns and flowering plants of ecological interest and rarity. Much is owned by the FC and NPA and has paths which are well used by the public.
22. **Craig Cerig-Gleisiad** (National Nature Reserve) Rare plant species on crags and in gullies. Close to main road. Permit access system often disregarded.
23. **Traeth Mawr/Bach** (SSSI, Nature Reserve) Uncommon invertebrates, several rare plants. Waders and waterfowl: breeding and passage birds of interest. Common land owned by NPA, with public access.
24. **Spring-fed pool by A4059** Rare Odonata (dragonflies), near unfenced main road over common land owned by NPA.

25 **Ogof y Ci/Nant Glais** (SSSI, Nature Reserve) Cave features and fauna, woodland and stream habitats with tufa deposits and unusual flora and fauna. Downstream of Rhyd Sych is especially delicate. Public paths present but not properly maintained.

26 **Talybont Reservoir** (SSSI, Local Nature Reserve) Important refuge, feeding or breeding place for many species of wildfowl and waders. Susceptible to disturbance by general public or organised water sports (including angling). Permitted access is restricted but there is a degree of trespass and public rights of way nearby.

27 **Blaen Dyffryn Crawnon** (part Nature Reserve) Lower plant communities on limestone are species-rich with a few rarities. Rare species of higher plants on cliffs. FC land with public paths adjacent.

28 **Llangorse Lake** (SSSI) Important site for breeding and passage birds, and aquatic and wetland fauna. Diverse community of aquatic and wetland plants with some rare species. Public access for water sports etc.

29 **Pwll y Wrach Wood** (SSSI, Nature Reserve) Important woodland habitat with very diverse ground flora containing several rare species. Well-used public path with "de facto" use of other paths causing some erosion in places.

30 **Craig y Cilau/Pwll Gwy-rhoc** (National Nature Reserve, SSSI) Pool is of ornithological importance with breeding teal, waders and gulls, and is surrounded by an important area of blanket mire. Rare plant species in vicinity. Rock outcrops, cliffs and scree have several rare plant species. Common land; some public paths.

31 **Cwm Clydach** (National Nature Reserve) Rare plant species. Public paths through wood.

32 **Gilwern Hill Quarries** (SSSI) Cave fauna, together with locally scarce moss and vascular plant species within and around quarries. 'De facto' public access.

33 **Hay Bluff/Ffynnon y Parc** (Part SSSI) Heather and bilberry, cotton grass blanket mire, and crowberry. An important habitat already disturbed and to some extent eroded by recreation activity. Common land.

34 **Hay Common** Pools and grassland contain some rare plant species. Common land with public access.

35 **Cwm Coed-y-cerrig** (SSSI, Local Nature Reserve) Wet woodland and mire habitat with scarce plant species. Particular species very susceptible to damage by trampling. Natural change in vegetation being monitored. Unfenced from public rights of way.

36 **Strawberry Cottage Wood** (SSSI, Nature Reserve) Some scarce plants near to public bridleway through wood.

1 Summarise the problems that occur in the pressure areas.

2 Summarise the problems that occur in the vulnerable areas.

3 Suggest how the strategies might be put into operation in order to lessen these problems.

BRITAIN'S NEWEST NATIONAL PARK: THE NEW FOREST

Within equal limits perhaps few parts of England afford a greater variety of beautiful landscape. Its woody scenes, its extended lawns, and vast sweeps of wild country, unlimited by artificial boundaries, together with river views and distant coasts; all are in a great degree magnificent.

(William Gilpin, Vicar of Boldre 1791)

Gilpin's description of the New Forest is as accurate today as it was two hundred years ago. Much of the New Forest is at present administered by the Forestry Commission, whose responsibilities not only encompass the management of the timber resources of the Forest, but also extend to the control and supervision of recreation within its limits. Proposals are now well advanced for the establishment of a new governing body with National Park status, similar to the one responsible for the Norfolk Broads. It is felt by many that the New Forest, lying between the expanding conurbations of Bournemouth and Southampton, will benefit from the enhanced status that designation as a National Park will bring.

A mosaic of landscape

Fig. 6.4.8
New Forest: Open landscape looking across Holmsley Bog to Turf and Shappen Hills

New Forest Mosaic of Landscape Types

Map legend:
- Ancient and Ornimental Woodlands
- Timber Enclosures
- Heathland
- Major Areas of Grass Lawn
- Enclosed and Settled landscape
- River and Coastal Landscape
- – – – Limit of New Forest

Places labelled: Frogham, Linwood, Fritham, Branshaw, Minstead, Lyndhurst, Burley, Brockenhurst, Beaulieu, Southampton Water. Insets marked Fig. 6.4.10a and Fig. 6.4.10b.

Scale: 0–5 Kilometres

Fig. 6.4.9

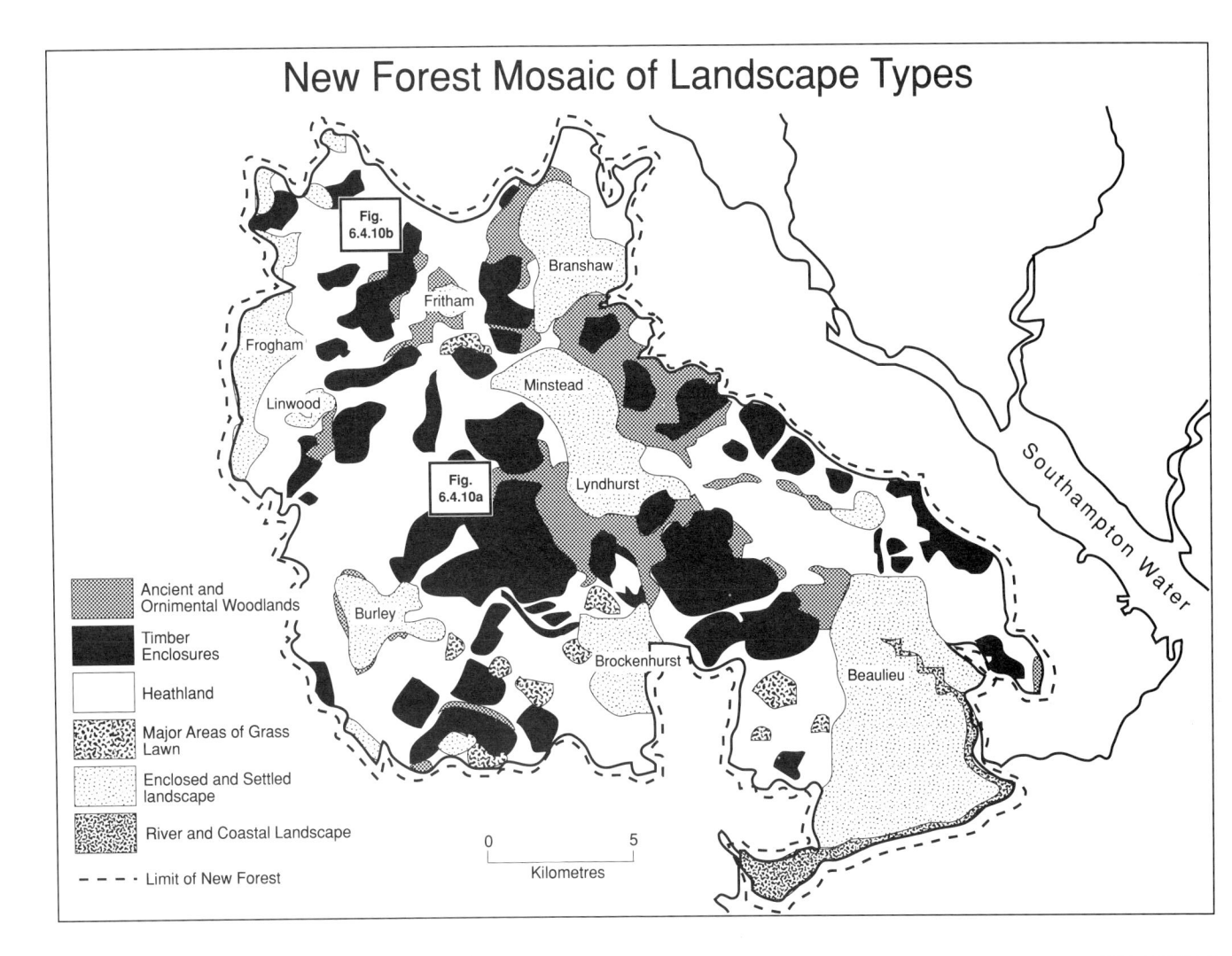

The variety in the New Forest landscape is also shown in the map (Fig. 6.4.9), which shows the different landscape elements that make up the mosaic. Such a mosaic exists at two different scales; it exists not only on the scale shown on the map of the whole Forest, but it also exists on a more intimate scale. Two areas are identified on the map of the New Forest as a whole, and then shown on the smaller scale (Fig. 6.4.10).

1 Discuss the broad pattern of the mosaic that is displayed on the map of the whole of the New Forest (Fig. 6.4.9).

Woodland Landscapes

Legend:
- Ancient and Ornamental beech, oak and birch woods
- Alder, carr, woodland, some coppiced
- Self-sown pine
- Plantation oak or beech
- Mixed oak or beech plantations
- Mixed plantations of conifers with oak and beech
- Conifer plantations
- —— Division between Ancient and Ornamental woodland and timber inclosures
- —— Forestry Commission compartment boundary

Places labelled: Scrub, Smoky Hole, Pound hill, Mark Ash Wood, Blackensford Lawn, North Oakley Inclosure, Church Moor, Beech Bed Inclosure, Hart Hill, Burley Rails Cottage, Anderwood Inclosure.

Scale: 0–500 Metres

Uncoloured areas are not part of the woodland landscapes

Open Landscapes

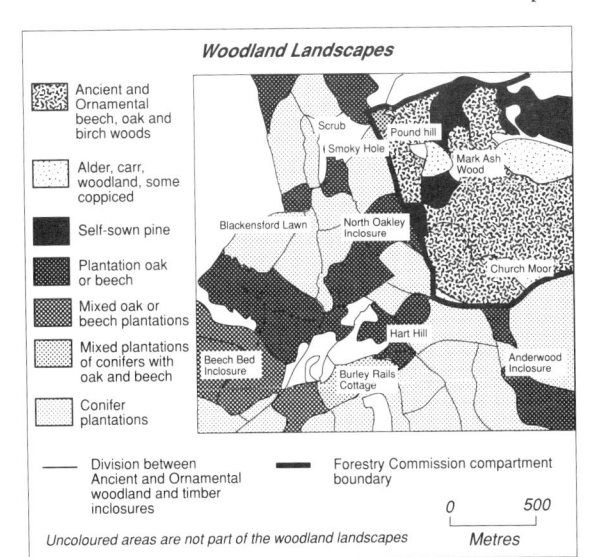

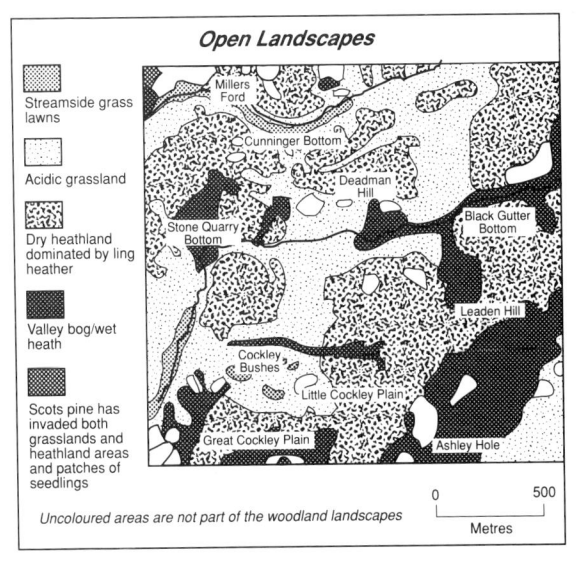

Legend:
- Streamside grass lawns
- Acidic grassland
- Dry heathland dominated by ling heather
- Valley bog/wet heath
- Scots pine has invaded both grasslands and heathland areas and patches of seedlings

Places labelled: Millers Ford, Cunninger Bottom, Deadman Hill, Stone Quarry Bottom, Black Gutter Bottom, Leaden Hill, Cockley Bushes, Little Cockley Plain, Great Cockley Plain, Ashley Hole.

Scale: 0–500 Metres

Uncoloured areas are not part of the woodland landscapes

Fig. 6.4.10

A NEW VIEW OF BRITAIN

(a)

(b)

(c)

(d)

Fig. 6.4.11
Landscape elements in the New Forest
(a) Ancient and Ornamental Woodland
(b) Timber Inclosure
(c) Heathland
(d) Open lawns (grazed grasslands)

MAIN TYPES OF VEGETATION IN THE NEW FOREST

TYPE OF VEGETATION	HECTARES
Ancient and ornamental Broadleaved woodland	3381
Self-sown pine woodland	211
Calluna heathland	5913
Valley bogs and wet heath	2835
Coarse grassland with bracken and gorse	5712
Forest lawns	324

2 Describe the small scale pattern that is revealed in Fig. 6.4.10. What factors would be responsible for the variations shown in the small scale patterns?

3 Discuss the contributions made to the scenic attraction of the New Forest by both large and small-scale mosaics. Use the photographs (Fig. 6.4.11) to help in this evaluation of scenery.

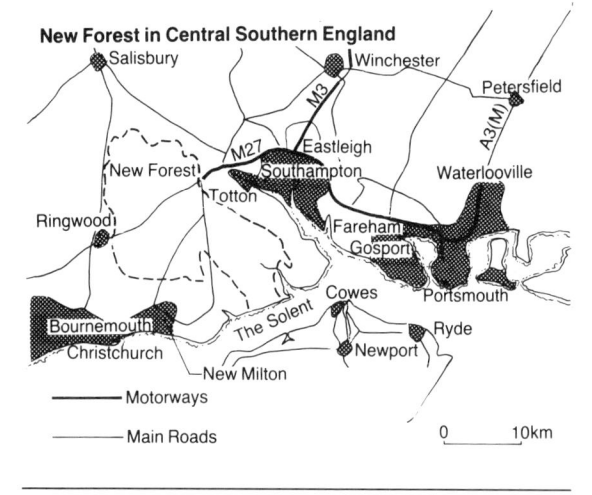

Fig. 6.4.12

The map (Fig. 6.4.12) shows the New Forest in its setting in central southern England.

1 Annotate a copy of this map to show the development and recreation pressures from which the New Forest is likely to suffer.

2 Refer back to the recommendations of the Edwards Report, *Fit for the Future* (page 275) and the strategies proposed in the Brecon Beacons National Park (page 278), and then suggest a twofold set of strategies, one for conservation, and one for recreation in the New Forest for the first decade of the next century.

PROTECTING BRITAIN'S LANDSCAPES

National Parks for Scotland

Scotland has no National Parks, since National Park legislation only extends to England and Wales. Areas in Scotland that were proposed as National Parks by the Ramsay Committee were named National Park Direction Areas, and in 1978 they became known as National Scenic Areas (see Fig. 6.4.13). Much of the land is privately owned and therefore protection is largely exercised through special development control procedures.

Fig. 6.4.13

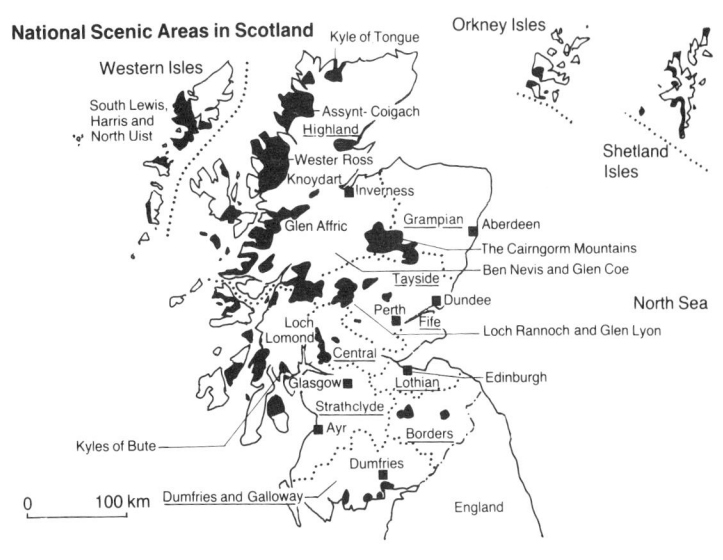

Read the article from *The Guardian* (Fig. 6.4.15), Scots warm to national parks.

1. Discuss some of the main objections to the establishment of National Parks in Scotland.

2. How far do you think that they are justified?

Fig. 6.4.15
Guardian 22 February 1991

Fig. 6.4.14
National Scenic Area, Scotland: Loch Torridon and Beinn Alligin, Wester Ross

Scots warm to national parks
Parks plans are on the move. Alastair Hetherington reports

HOPES for national parks in Scotland are looking up. During the last week's second reading of the Natural Heritage (Scotland) Bill, the Secretary of State, Ian Lang, said there were "many valuable ideas" contained in the report by the Countryside Commission for Scotland (CCS), though he found "no enthusiasm" among the people of the Highland areas.

Only around the proposed Loch Lomond-Trossachs park was there apparent support, he said, and he was not yet ready to come to a conclusion.

The Shadow Secretary, Donald Dewar, gave a warm welcome to CCS proposals. His main worry was whether there would be enough money for the work. Liberal Democrats Robert Maclennan and Ray Michie concerned themselves with the "bad footing" of the Nature Conservancy Council (NCC) rather than with the mountain areas. They took no stand for or against national parks.

The Bill is now in committee, with the hope of finishing its work by Easter. Meanwhile, Mr Lang awaits the final CCS report due to be published next week. For Mr Lang, with a general election not far away, it is important not to lose any votes, though it would be unjust to think he would make his decisions on a political basis.

He has instructed Magnus Magnusson – the upcoming chairman of the merged CCS and NCC in Scotland – that develoments such as national parks must be based on "voluntary" action. If landowners object to a national park, then there will be no park. Mr Magnusson has nevertheless said we must 'think positively' about parks.

However hostile to parks some wealthy landowners may be, the Secretary of State will look wider. The Crofters Union is another opponent but not many of its members are likely to vote for the Conservatives.

Nor will many members of Highland Regional Council – probably the strongest opponent to the parks, since it believes it could manage them itself if it had the money.

The Scottish Office is said to have put pressure on the CCS to modify its proposals. It is virtually certain that Roger Carr, CCS Chairman, will stand firmly by the CCS recommendations of last autumnn. That report, The Mountain Areas Of Scotland, was the result of many months of study, and its recommendations were endorsed by 12 of its 15 members.

Proposals for the four parks – the Cairngorms, Loch Lomond/Trossachs, Nevis/Glen Coe/Black Mount, and parts of Wester Ross – were recommended as the primary need. Later, if money became available, these might be followed by extending the Wester Ross park to include Assynt and Coigach. That Mr Lang found "no enthusiasm" is open to debate: there is plenty of enthusiasm nationally for parks. And it is no accident that, three weeks ago, the World Conservation Union gave its full support for the parks proposed by the CCS.

AREAS OF OUTSTANDING NATIONAL BEAUTY

The map (Fig. 6.4.16) shows the distribution of Areas of Outstanding Natural Beauty (AONB), and the National Parks.

Areas of Outstanding Natural Beauty and National Parks in England and Wales

Areas of Outstanding Natural Beauty
1. Llyn
2. Gower
3. Wye Valley
4. Shropshire Hills
5. Clwydian Range
6. Anglesey
7. Cotswolds
8. Cannock Chase
9. Malvern Hills
10. Chilterns
11. North Devon
12. South Devon
13. Quantock Hills
14. Isles of Scilly
15. East Devon
16. Cornwall
17. Mendip Hills
18. Chichester Harbour
19. East Hampshire
20. North Wessex Downs
21. Isle of Wight
22. Dorset
23. South Hampshire
24. Cranborne Chase and West Wiltshire Downs
25. Northumberland Coast
26. North Pennines
27. Solway Coast
28. Howardian Hills
29. Forest off Bowland
30. Arnside and Silverdale
31. Suffolk Coast and Heaths
32. Norfolk Coast
33. Dedham Vale
34. Lincolnshire Wolds
35. Sussex Downs
36. High Weald
37. Surrey Hills
38. Kent Downs
39. Blackdown Hills

National Parks
1. Lake District
2. Northumberland
3. Yorkshire Dales
4. North York Moors
5. Peak District
6. Snowdonia
7. Brecon Beacons
8. Pembrokeshire Coast
9. Exmoor
10. Dartmoor
11. The Broads
12. The New Forest

Fig. 6.4.16

1. Comment on the respective distributions of the National Parks and the AONBs

2. What does the distribution indicate about their differing landscape qualities?

The main aims of AONBs are similar to those for National Parks:

- to conserve and enhance the natural beauty of the landscape
- to meet the need for quiet enjoyment of the countryside
- to have regard for the interests of those that live and work there.

It should be remembered, however, that in many of the AONBs the land, unlike that in the National Parks, is quite heavily farmed. Therefore the last of the three aims takes on an added importance. AONBs are protected by planning laws and all development proposals receive rigorous examination at a variety of levels. Practical help is available to farmers and other landowners for positive countryside management. Visitor awareness and education is another important aim of the work of AONBs; many of the AONBs receive large numbers of visitors each year, and successful management of this pressure is important.

Two landscapes compared

AONB: The North Pennines

AONB: The Blackdown Hills

The North Pennines AONB (Fig. 6.4.16) was confirmed in June 1988, after some protracted discussion about its extent (finally about 15 per cent less than the original proposals). The Blackdown Hills (Fig. 6.4.16) were originally included in a wider area in 1947 which was recommended for AONB status, but were eventually omitted from the East Devon area that was designated. Final confirmation of the status of the Blackdown Hills as an AONB was given on 26 June 1991.

THE NORTH PENNINES AONB: ENGLAND'S LAST WILDERNESS

The landscape of the North Pennines AONB is largely the creation of hill farming, and to a lesser extent of lead mining. Changes in hill farming have led to abandonment and amalgamations, the decay of vernacular farm buildings and the pattern of stone walls (Fig. 6.4.17). Improvement of hay meadows

Fig. 6.4.17
Northern Pennines: Area of Outstanding Natural Beauty, isolated barn on summit plateau near Hartside Cross

and overgrazing of moorland and woodland have also taken their toll. Although this area has suffered a continuing decline in population, there is reason to believe that, with better farm management and a greater awareness of countryside issues, this loss can be stemmed, and a modest growth of population will occur.

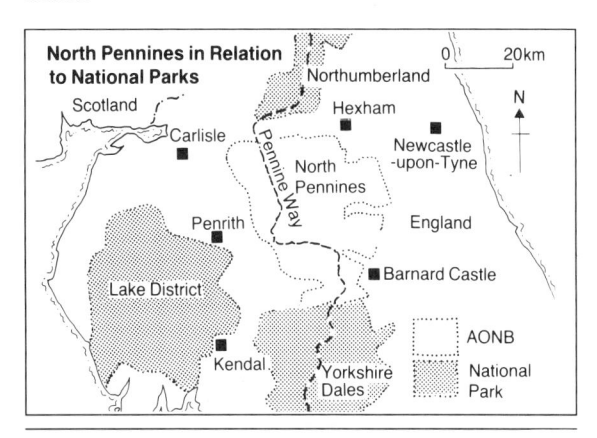

Fig. 6.4.18

The map (Fig. 6.4.18) shows the relative positions of the North Pennines AONB and the three adjacent National Parks – the Lake District, the Northumberland National Park, and the Yorkshire Dales. The different landscape elements in the North Pennines are shown in Fig. 6.4.19, with brief comments on their physical and human character.

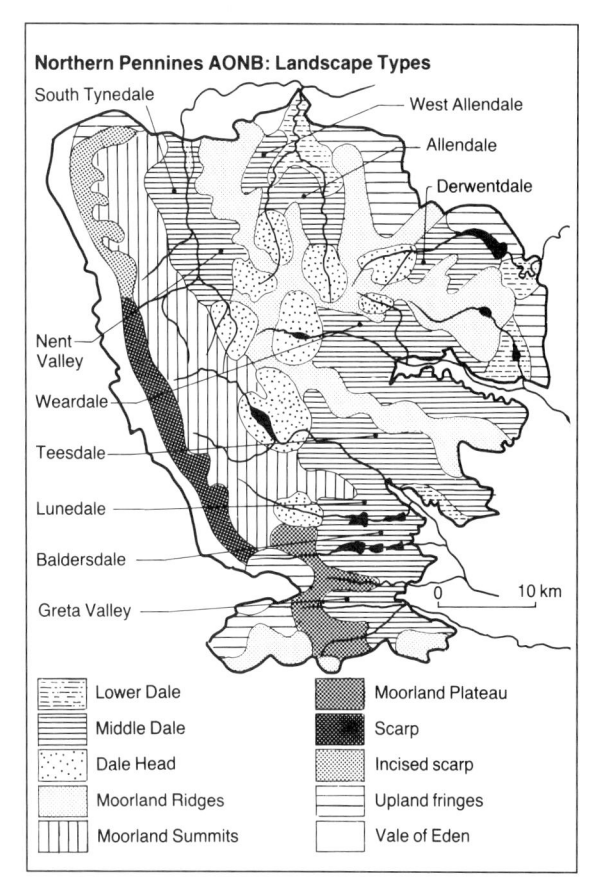

Fig. 6.4.19

Summary of landscape features

Lower dale
Enclosed, initmate character: deciduous woodlands on valley sides; strong field patterns; dense tree cover with hedgerows, hedgerow trees, copses and woodlands.

Middle dale
Complexity and diversity; rich pattern of textures and colours; sense of historical continuity; flower-rich valley bottom meadows and pastures; strong pattern of stone walls; scattered farmsteads with clumps; field trees along stone walls; vernacular style of stone buildings and settlements; tree-lined river landscapes; dale-side ghylls and woodland; contrasting colours of moorland fringe; evidence of mining history including high level farmsteads: stepped landform.

Dale head
Shallow, even gradient of valley sides; deeply incised becks; uninterrupted encircling moorland skyline; marginal character; cultivation, enclosure and settlement at their limits; reverting, rush-infested pastures; derelict walls and buildings; mining remains; forestry plantations; bleak character.

Moorland ridges
Apparent naturalness and lack of human influence; gently rolling ridges and summits, uniform dry heath vegetation; colours and patterns of heather; prominent burning patterns; uninterrupted extent; views over sequences of dales and moorland.

Moorland summits
Wild, remote character, severe climate; extent and uniformity of blanket bog vegetation; openness and apparent naturalness; relative lack of man-made structures and human influence; landform of sweeping interlocking ridges; prominent and distinctive millstone grit caps on high summits; dramatic distant views; patterns on the land; peat haggs; hidden valleys of becks.

Moorland plateau
Relatively flat and featureless topography; apparent naturalness; lack of human influence; continuous blanket peat bog; bleak, wilderness character; lack of landmarks; upstanding; rocky band of Shackleborough.

Scarp
Dramatic landforms: outlying conical hills; unbroken sweep of unimproved rough grazing; colour contrasts between different types of grassland; exposed rock features; lack of enclosure except on lower slopes; bare, treeless character; long views out to Vale of Eden and beyond; prominence in views from the west; dramatic lighting effects of low sun.

Incised scarp
As above but more gentle slopes; complex topography with incised valleys, outlying hills and a secondary escarpment.

Upland fringes
Gently rolling or terraced landform; small valleys; enclosed pasture; some arable land; mixed field boundaries of fences, hedges and walls; prominent farmsteads; large farm buildings; coniferous shelterbelts; sycamore clumps. Limestone fringe in the south west is distinct.

Vale of Eden
Rich, diverse character; complex, rolling drumlin topography; rich, red soils; pasture and arable fields; red sandstone villages often with village green; dense network of tree-lined lanes, small, wooded valleys and hedgerows with trees.

A NEW VIEW OF BRITAIN

Fig. 6.4.20
Four landscapes in North Pennines

1 Fit a landscape type to each of the four photographs (Fig. 6.4.20) shown.

2 Analyse the landscape types to show the marked contrasts that exist between the moorlands, ridges and dales.

3 Summarise in your own words the reasons why this area of the North Pennines should be designated an AONB.

THE BLACKDOWN HILLS

… a thoroughly English piece of countryside…quite unspoit by any mass admiration and exploitation and quite unconscious of being picturesque.

(Donald Maxwell, Unknown Somerset 1927)

The Blackdown Hills are situated astride the border of Devon and Somerset (Fig. 6.4.21). The different landscape types are shown on the map (Fig. 6.4.22), and are illustrated in the photographs (Fig. 6.4.23). Donald Maxwell's comment sums up the attraction of the Blackdown Hills. There could be no greater

Fig. 6.4.21

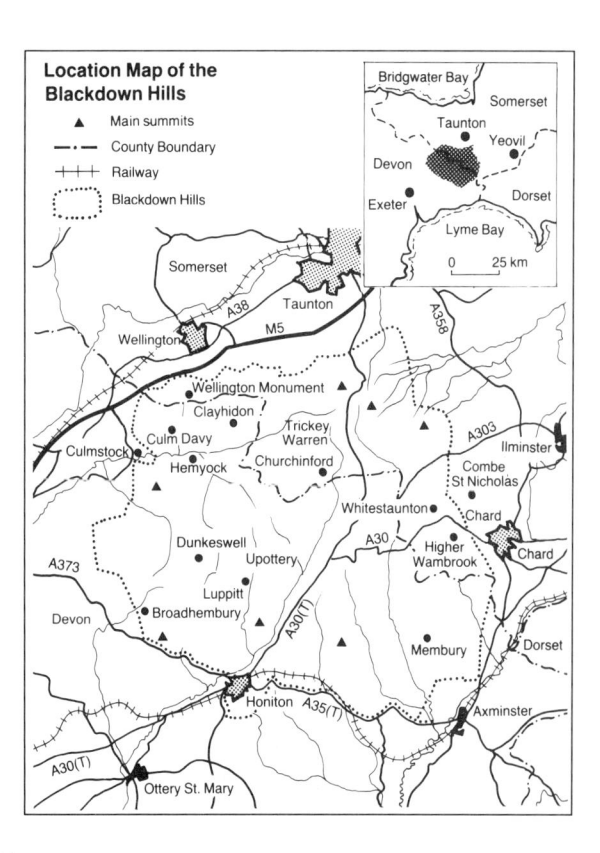

contrast of landscapes than the one between the North Pennines and the Blackdown Hills. The broad desolate sweep of the high moorlands of the Northern Pennines bear witness to a totally different landscape evolution to the patchwork of woodland and field, hedgerow and country lane that characterises the Blackdown Hills. Yet to the discerning eye, the two landscapes have features in common. Both rise from a fringing lowland (the Vale of Taunton Deane to the north of the Blackdown Hills, and the Vale of Eden to the west of the Northern Pennines). Both have the same contrast between high ridges and plateaux, as well as the enclosed dales of the Northern Pennines and the combes of the Blackdown Hills. Latitude, aspect, ecology, and human occupance have all contributed to the differences.

Change is evident in the Blackdown Hills just as it is in the Northern Pennines. More heathland existed on the high Blackdown ridge in the past and much has now been reclaimed for pasture. Few grassland areas are now of conservation value and coniferous woodlands have replaced broadleaved areas. Hedgerows and hedgerow trees have inevitably been lost with farm amalgamations and consolidation. Pressure for change in the future does not seem to be a great threat. The Blackdown Hills will continue to be an attractive place to live in for commuters to Exeter and Taunton, and tourism may develop also.

Using the Northern Pennines and the Blackdown Hills as examples, examine why the designation of AONBs is important in landscape protection.

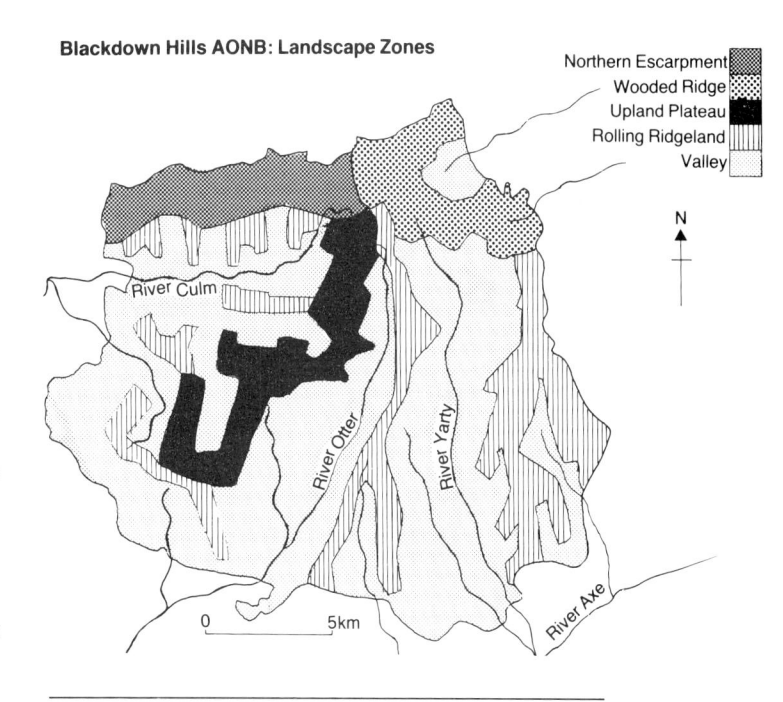

Fig. 6.4.22

Fig. 6.4.23
Landscape elements: Blackdown Hills
(a) Otter Valley
(b) Upland Plateau
(c) North Escarpment

PROTECTING BRITAIN'S COASTLINE

The dragon – green, luminous, the dark, the serpent haunted sea....

James Elroy Flecker

Many of Britain's finest landscapes are coastal ones and need protection in the same way as inland areas of countryside and moorland, of downland and mountain. Some National Parks have a coastal fringe, such as the North Yorkshire Moors, as do some AONBs such as Dorset, and east Devon, but there are considerable stretches of coastline that lack this protection. The concept of Heritage Coasts was developed in the late 1970s and was accepted by the Government in 1972. Since that time 43 stretches of coastline, in total some 1460 km or one third of the coastline of England and Wales, have been declared Heritage Coasts. Nearly one third of Heritage Coasts are owned by the national Trust, whose own Operation Neptune was established in 1965. Its principal aim was to acquire up to 1400 km of coastline in order to protect it from development. Although this aim has not been achieved yet, much of our best coastline now has some measure of protection. Two further areas of Heritage Coast, Exmoor and the stretch around Hartland Point in North Devon and Lundy Island have yet to be defined in detail.

The stretches of coastline that have been designated as Heritage Coasts are shown in Fig. 6.5.1.

Comment on the distribution of the Heritage Coasts.

Fig. 6.5.1

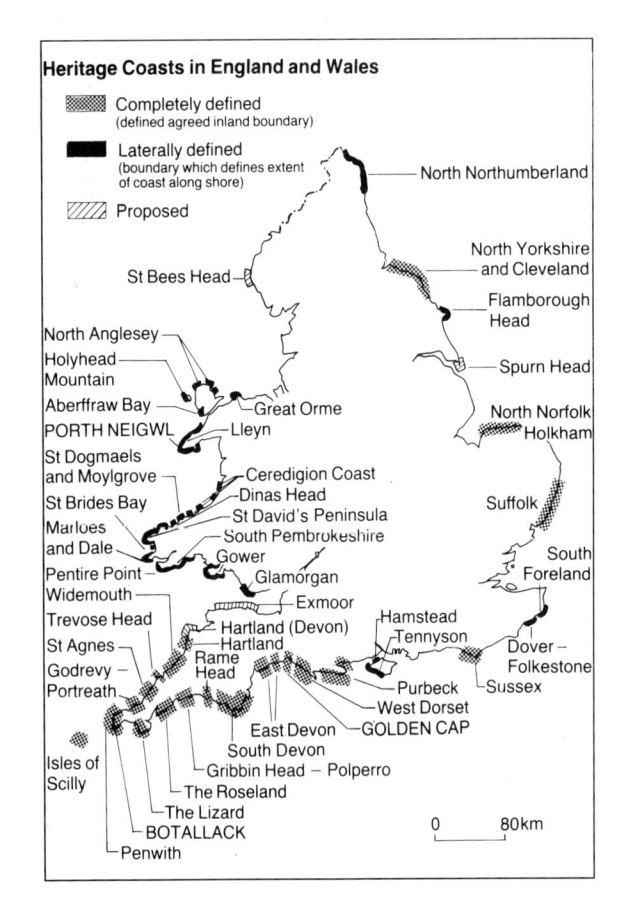

Heritage Coasts in England and Wales
- Completely defined (defined agreed inland boundary)
- Laterally defined (boundary which defines extent of coast along shore)
- Proposed

Labels on map: North Northumberland, North Yorkshire and Cleveland, Flamborough Head, St Bees Head, Spurn Head, North Anglesey, Holyhead Mountain, Aberffraw Bay, PORTH NEIGWL, Great Orme, Lleyn, North Norfolk, Holkham, St Dogmaels and Moylgrove, Ceredigion Coast, Dinas Head, Suffolk, St Brides Bay, St David's Peninsula, Marloes and Dale, South Pembrokeshire, Gower, South Foreland, Pentire Point, Glamorgan, Widemouth, Exmoor, Hamstead, Trevose Head, Hartland (Devon), Tennyson, St Agnes, Hartland, Rame Head, Dover-Folkestone, Godrevy, Purbeck, Sussex, Portreath, West Dorset, East Devon, GOLDEN CAP, South Devon, Isles of Scilly, Gribbin Head – Polperro, The Roseland, The Lizard, BOTALLACK, Penwith

0 — 80km

HERITAGE COAST POLICY

The original objectives of Heritage Coast policy, as stated in 1970 were:

- to identify the finest stretches of undeveloped coast

- to conserve and manage them comprehensively

- to facilitate and enhance their enjoyment by the public through the promotion and encouragement of recreational activities consistent with the conservation of their fine natural scenery and heritage features.

Twenty years later in 1991, the Countryside Commission felt able to redefine and broaden the objectives.

- To conserve protect and enhance the natural beauty of the coasts, including their terrestrial, littoral and marine flora and fauna, and their heritage features of architectural, historical and archaeological interest.

- To facilitate and enhance their enjoyment, understanding and appreciation by the public by improving and extending opportunities for recreational, educational, sporting and tourist activities that draw on, and are consistent with, the conservation of their natural beauty and the protection of their heritage features.

- To maintain and improve, (where necessary) the environmental health of inshore waters affecting Heritage Coasts and their beaches through appropriate works and management measures.
- To take account of the needs of agriculture, forestry and fishing, and of the economic and social needs of the small communities on these coasts, through promoting sustainable forms of social and economic development, which, in themselves conserve and enhance natural beauty and heritage features.

1 Examine carefully the old and the new sets of objectives for the Heritage Coasts.

a What are the main ways in which the objectives have been refined and extended?

b What do you consider have been the main forces behind this revision of the objectives?

2 What are the principal management guidelines that will have to be observed in order to achieve these objectives?

Four locations with Heritage Coast areas are shown in the photographs (Fig. 6.5.2) and maps (Fig. 6.5.3). The maps show current recreational activities and possible developments in those areas.

- The Penwith coast near Botallack.
- The Lleyn coast at Porth Neigwl.
- The North Norfolk coast near Holkham.
- The West Dorset coast near Golden Cap.

Fig. 6.5.2
Four protected coastlines
(a) Botallack, West Penwith, Cornwall; (b) Porth Neigwl (Lleyn Peninsula, North Wales); (c) Holkham Bay (North Norfolk); (d) Golden Cap (West Dorset)

A NEW VIEW OF BRITAIN

1 For each of the photographs suggest the qualities of landscape that have been considered worthy of Heritage status.

2 Discuss the nature of the possible threats to each stretch of coastline.

3 Suggest how such threats could be managed.

4 Rank these four stretches of coastline

a in terms of landscape quality

b in terms of their degree of risk from development.

Explain the reasons behind your ranking.

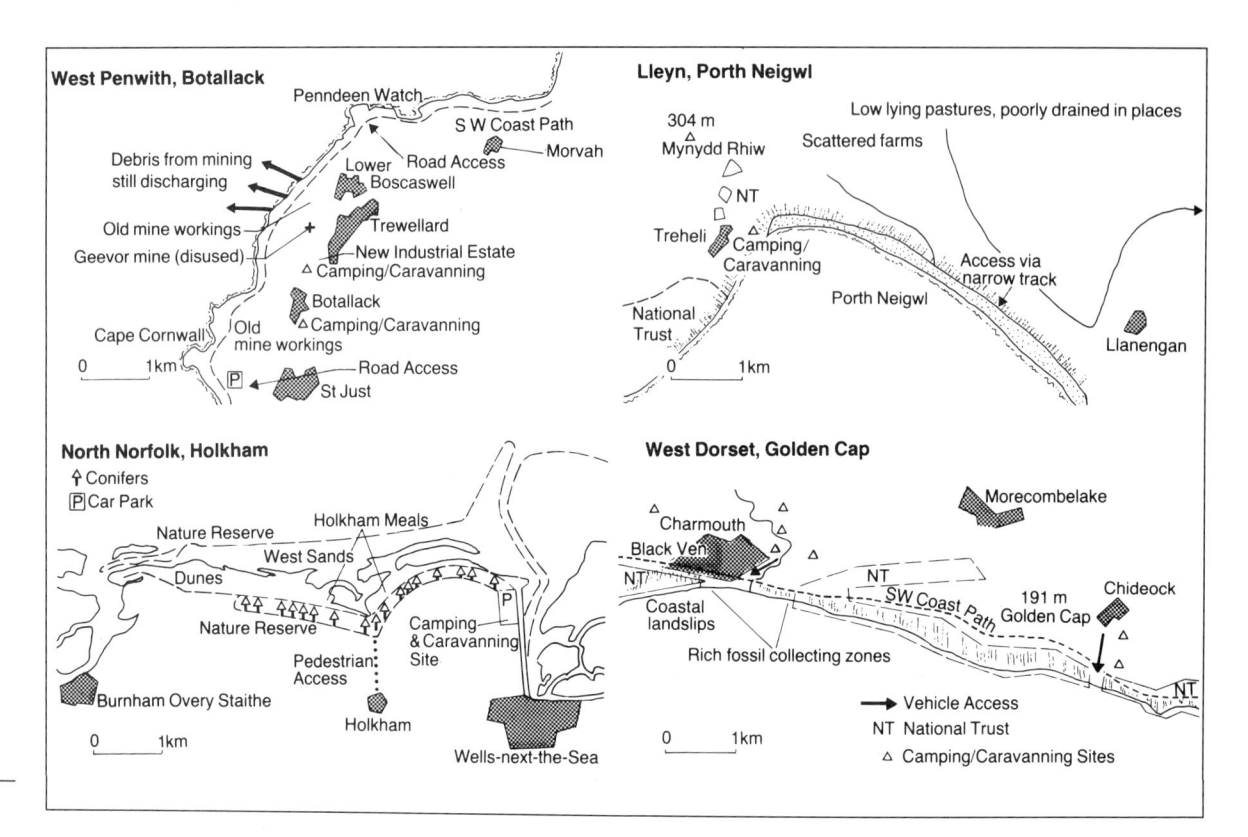

West Penwith, Botallack
- Penndeen Watch
- S W Coast Path
- Morvah
- Debris from mining still discharging
- Lower Boscaswell
- Road Access
- Trewellard
- Old mine workings
- New Industrial Estate
- Geevor mine (disused)
- Camping/Caravanning
- Botallack
- Camping/Caravanning
- Cape Cornwall
- Old mine workings
- Road Access
- St Just
- 0 1km

Lleyn, Porth Neigwl
- Low lying pastures, poorly drained in places
- 304 m Mynydd Rhiw
- Scattered farms
- NT
- Treheli
- Camping/Caravanning
- Access via narrow track
- National Trust
- Porth Neigwl
- Llanengan
- 0 1km

North Norfolk, Holkham
- ♠ Conifers
- P Car Park
- Holkham Meals
- Nature Reserve
- West Sands
- Dunes
- Nature Reserve
- Camping & Caravanning Site
- Pedestrian Access
- Burnham Overy Staithe
- Holkham
- Wells-next-the-Sea
- 0 1km

West Dorset, Golden Cap
- Morecombelake
- Charmouth
- Black Ven
- NT
- Coastal landslips
- SW Coast Path
- 191 m Golden Cap
- Chideock
- Rich fossil collecting zones
- NT
- → Vehicle Access
- NT National Trust
- △ Camping/Caravanning Sites
- 0 1km

Fig. 6.5.3a–d

GLOSSARY

Acid deposition: deposition resulting from the emission of exhaust gases, principally SO^2 and NOx (nitrogen oxides). It takes the form of **dry** deposition (gases and particles) and **wet** deposition (rain, snow, mist and fog).

Agribusiness: heavily capitalised farming which has profit as its sole or principal motive.

Assisted Area: parts of Britain where a range of measures are regarded as necessary to encourage the growth of industry.

Biotechnology: the application of high technology to industrial processes that involve biological mechanisms and products.

Blanket Bog: bog which covers all features of low-lying relief, composed mainly of peat, and filling all hollows to considerable depth.

Blowout: hollow in sand dunes, usually formed by wind erosion.

Biological Oxygen Demand (BOD): measure of the amount of oxygen required by micro-organisms to break down pollutants.

Brown Belt: land surrounding an urban area that includes a high proportion of derelict or under-used land.

Brown earths: brown soils that develop under a cover of deciduous forest.

Brownfield sites (as distinct from greenfield sites): derelict land that could be reclaimed and used for urban or industrial development.

Bustitution: the substitution of bus services for rail services on the closure of the latter, mainly in rural areas.

CAP (Common Agricultural Policy): financial and economic policies that control the production and marketing of farm goods within the European Community.

Conference industry: the provision of the full range of conference facilities (accommodation, secretarial and communications services) in large urban centres (particularly in seaside resorts).

Convenience goods: goods which are purchased by consumers at fairly short, regular intervals e.g. groceries.

Counterurbanisation: the process of population decentralisation, whereby major urban concentrations lose population to smaller towns and rural areas.

dB (decibel): a measure of the loudness of noise.

De-industrialisation: the progressive weakening of the contribution of manufacturing industry to the national economy.

Desalination: the process whereby the dissolved solids in sea water are partially or completely removed in order to render it suitable for domestic, agricultural or industrial use.

ECU (European Currency Unit): currency unit used within the European Community.

Energy mix: the contribution of various fuels to a country's energy supply.

Enterprise Zone: vacant, and often derelict land within or near urban areas in which industrial development is encouraged through a series of financial and planning measures.

EOR (enhanced oil recovery): the use of sophisticated new techniques to extract further petroleum from oilfields that are in decline.

ESA (Environmentally Sensitive Area): areas where the natural environment is sensitive to change, particularly to change in farming. Farming techniques are encouraged that will safeguard the environment.

FGD (Flue Gas Desulphurisation): the removal of gases harmful to the atmosphere from emissions from power stations and other industrial plant.

Gabion armouring: a series of interlocking metal cages that are filled with rock fragments. They are used as a method of protection against erosion and mass movement.

General Improvement Area: urban area designated within which improvements to both housing and the urban environment may be carried out.

Gley soils: soils within which waterlogging is frequent, with a consequent lack of oxygen, and the reduction of iron compounds from the ferric to the ferrous state.

Green Belt: open, semi-rural land surrounding large urban areas in which there are severe restraints on building development.

Headland: area of unploughed land at the edge of the units of arable cultivation.

Higher order goods: goods whose purchase is irregular, and usually infrequent, where consumers require a considerable degree of choice, e.g. furniture, electrical goods and cars.

HLCA (Hill Livestock Compensatory Allowance): Financial subsidy given to farmers that operate in physically difficult environments.

Housing Action Area: area-based policy in British cities designed to stimulate improvements to housing and the environment.

Insectivorous plants: plants that consume insects as a principal part of their food intake e.g. sundew.

Intertidal flats: areas of deposition (particularly in estuaries) that are exposed at low tide.

Isostatic adjustment: crustal adjustment to accommodate either removal of material by erosion (vertical upward movement), or deposition of material where there is considerable accumulation of sediments e.g. in a delta (vertical downward movement).

Landfill site: location used for the disposal of household and industrial waste, which is often reclaimed at a later stage.

Lead-in time: a measure of the amount of time that a system takes to respond to stimulation and fluctuation within that system e.g. the time taken for an industrial system to respond to financial measures designed to stimulate it.

MAFF: Ministry of Agriculture, Fisheries and Food.

Management agreement: an agreement where farmers are compensated financially for not reclaiming land e.g. on Exmoor.

Maritime Heritage: the promotion of a town or city's maritime links.

Mass transit system: public transport system used for carrying considerable numbers of people within a large urban area.

NRA: National Rivers Authority.

One-stop shopping: opportunity to shop for a wide range of goods in one location e.g. in hypermarkets.

Podsols: soils that develop in a cool, seasonally humid, climate, often under a cover of coniferous forest cover.

Retail park: a development of retail premises, usually warehouse style buildings, on a greenfield site or reclaimed land.

Retail warehouse: premises devoted to the sale of a range of goods, often heavily discounted. Usually found in urban fringe areas or on reclaimed inner city sites.

Return time: the average length of time separating events, (e.g. floods), of similar magnitude.

Rurality: a measure of the extent which an area displays a range of rural characteristics e.g. reduced range of services, difficulty of access to services.

Shopping hierarchy: the existence of a number of different levels of shopping (characterised by a number of outlets, range and choice) in different towns.

Tax incentive: financial incentives, including reduced tax levels, or tax 'holidays', offered to firms locating in the various Development Areas or in Enterprise Zones.

Telecommuting: style of working, where staff work at home from computer terminals and fax machines linked to main offices some distance away.

Tertiarisation: the move in employment away from the manufacturing sector to the service sector.

Tidal barrage: a barrage constructed across an estuary, often for the purpose of harnessing tidal power to produce electricity.

Trans-national company (sometimes referred to as multinational company): a company that operates in several countries, whilst directing affairs from its headquarters in one of those countries.

Urban Development Corporation: a body set up by the Government, to secure the regeneration of designated land within urban areas.

Urban Programme: an organisation through which funding is made available for a range of projects within urban areas in Britain.

REFERENCES

GENERAL

Champion, A.G. and Townsend, A.R., *Contemporary Britain: A Geographical Perspective*, Edward Arnold, 1990

Cooke, P., *Localities: The Changing Urban Face of Britain*, Unwin Hyman, 1989

Healey, M.J. and Ilbery, B.W., *Location and Change; Perspectives on Economic Geography*, Oxford, 1990.

Johnston, R.J. and Gardiner, Vince(ed), *The Changing Geography of the United Kingdom*, Routledge, 1991

Price, D.G. and Blair, A.M. *The Changing Geography of the Service Sector*, Belhaven, 1989

Smith D.M., *North and South: Britain's Growing Divide*, Penguin, 1989

CHAPTER ONE: URBAN BRITAIN

Birmingham City Council, *Heartlands: A Strategy for the Regeneration of East Birmingham*, 1990

Champion, A.G. (ed), *Counterurbanisation*, Edward Arnold, 1989

Champion, A.G. and Green, A. *The Spread of Prosperity and the North-South Divide*, Booming Towns, 1990

Donnison, D. and Middleton, A., *Regenerating the Inner City: Glasgow's Experience*, Routledge, Kegan, Paul, 1987

Hall, J.M., *Metropolis Now: London and its Region* Cambridge, 1990

Leeds City Council, *South Leeds Local Plan*, 1984

Leeds City Housing Department, *Beverleys and Lindens Proposed Housing Action Area*, 1983

Leeds Development Corporation, *Fact Pack*, 1990

Liverpool City Council, *Liverpool City Centre Strategy Review*, 1987

Liverpool City Council, *Health Inequalities in Liverpool*

Merseytravel, *Annual Report and Accounts*, 1989–90

Portsmouth City Planning Department, *Portsmouth city Centre: Towards 2000*, 1988

Robson, B., *Those Inner Cities*, Oxford, 1988

Royal County of Berkshire, *Structure Plan 1990–2006*, 1991

Sainsbury, J., *Annual Report and Accounts*, 1992

South Somerset District Council, *Yeovil Area Town Plan*, 1990

CHAPTER TWO: INDUSTRIAL BRITAIN

Barrow Borough Council, *Fact File*, Barrow, 1990

Barrow Borough Council, *Furness Enterprise*, Barrow, 1990

Barrow Borough Council, *VSEL Business Profile*, Burrow, 1990

Bournemouth Borough Council Information Pack, Bournemouth, 1991

British Coal Opencast, *Opencast Coal-mining in Great Britain*, 1991

Cambridge City Council, *Employment Development Strategy: Hi-tech and Conventional Manufacturing Industry*, 1986

County of Cleveland, *British Steel builds the New Teeside*, 1984

County of Cleveland, *Economic and Social Importance of the British Steel Corporation to Cleveland*, 1983

Digest of Environmental Protection and Water Statistics, HMSO, 1990

Fife Regional Council, *Fife: A Profile*, 1991

Glenrothes Development Corporation, *Annual Report*, 1991

Hudson, R. and Sadler, D., 'State Policies and the Changing Geography of the Coal Industry in the United Kingdom' Institute of British Geographers 15(4), 1990

Motherwell District Council, *Strategy for Steel Motherwell*, 1991

Park, C.C., *Acid Rain, Rhetoric and Reality*, Routledge, 1987

Rhondda Borough Council, *Rhondda into the Future*, 1991

Rose, C. *The Dirty Man of Europe: The Great British Pollution Scandal*, Simon and Schuster, 1991

Swindon Economic Development: It's Time to Make The Connection, Thamesdown Borough Council, 1991

Teeside Development Corporation, *Teeside Opportunities*, Teeside, 1990

Telford Development Corporation, *Facts and Figures*, Telford, 1990

CHAPTER THREE: RURAL BRITAIN

Countryside Commission, *New Agricultural Landscapes*, 1974

Countryside Commission, *Agricultural Landscapes: A Second Look*, 1984

Countryside Commission, *Cannock Chase 1979–1984: A Country Park on Trial*, 1985

Countryside Commission for Scotland, *Day Trips to Scotland's Countryside: 1987–1989*, 1990

Countryside Commission for Scotland, *Conservation on Farms*, 1990

Cornwall County Council, *Investment in Cornwall*, 1991

Development Board for Rural Wales, *Mid-Wales: the new Wales*, 1991

Devon County Council, *Devon County Structure Plan: Third Alteration 1989–2001*, 1991

Devon County Council, *Community Facilities in Rural Areas*, 1987

Dorset County Council, *Dorset Heathland Strategy*, 1990

Ilbery, B.W., 'Adoption of the Farm set-aside Scheme in Britain', *Geography* **76** (1990) Part 1

National Rives Authority, *The Influence of Agriculture on the Quality of Natural Waters in England and Wales*, 1992

Peterken, C.F. and Allison, H, *Woods, Trees and Hedges: A Review of Changes in the British Countryside*, English Nature, 1989

Pye-Smith, C. and North, R., *Working the land*, Temple Smith, 1984

Tubbs, C., *The Decline and Present Status of the English Lowland Heaths and their Vertebrates*, English Nature, 1985

CHAPTER FOUR: UPLAND BRITAIN

Berry, G., *A Tale of Two Lakes*, Friends of the Lake District, 1982

Cairngorm Chairlift Company, *Lurcher's Gully; The Case for Ski Development*, 1990

Countryside Commission, *A Study of the Hartsop Valley*, 1976

Countryside Commission, *Upland Land Use in England and Wales*, 1978

Countryside Commission for Scotland, *Skiing in the Cairngorms: A Policy Paper*, 1989

English Nature, *Birds, Bogs and Forestry: the Peatlands of Caithness and Sutherland*, 1987

English Nature, *The Flow Country: the Peatlands of Caithness and Sutherland*, 1988

Forestry Commission *Policy and Procedure Paper No. 3: The Forestry Commission and Landscape Design*, Undated

MacEwen, M. and Sinclair, G., New Life for the Hills, Council for National Parks, 1983

Sinclair, G. (ed), *The Uplands Landscape Study*, Environment Information Services, 1983

Snowdonia National Park, *Snowdon Management Scheme, Management Plan*, 1981

Tompkins, S., *Forestry in Crisis*, Christopher Helm, 1989

CHAPTER FIVE: COASTAL BRITAIN

Bournemouth Borough Council, *Hengistbury Head Management Plan*, 1989

Bray, D.J., Carter, D. J. and Hooke, J.M., *Coastal Sediment Transport Study 2: Brighton to Portsmouth*, Portsmouth University, 1991

Dorset County Council, *The Fleet and Chesil Beach*, 1981

English Nature, *Sand Dunes and their Management*, 1985

English Nature, *Dungeness: A Vegetation Survey of a Shingle Beach*, 1990

English Nature, *Nature Conservation and Estuaries in Great Britain*, 1991

Humber Estuary Committee, *The Water Quality of the Humber Estuary*, 1987

National Rivers Authority, *Bathing Water Quality in England and Wales*, 1991

Portsmouth City Council, *Portsmouth Harbour Plan*, 1988

CHAPTER SIX: PROTECTED BRITAIN

Brecon Beacons National Park, *National Park Plan, First Review*, 1987

Brecon Beacons National Park, *National Park Plan, Second Review*, 1992

Burke, G., *Townscapes*, Penguin, 1976

Countryside Commission, *The New Forest Landscape*, 1986

Countryside Commission, *Protected Landscapes: The United Kingdom Experience*, 1987

Countryside Commission, *The Blackdown Hills Landscape*, 1989

Countryside Commission, *The North Pennines Landscape*, 1991

Countryside Commission, *Heritage Coasts, Policies and Priorities*, 1991

Countryside Commission, *Fit for the Future: Report of the National Parks' Review Panel*, 1991

Dorset County Council, *The Conservation of Chalk Downland in Dorset*, 1973

Insall, D.W., *Conservation in Chester*, Chester City Council, 1986

Ironbridge Gorge Museum Trust, *Teachers' Handbook*, Undated

Riley, R.C., *The Houses and Inhabitants of Thomas Ellis Owens' Southsea*, Portsmouth City Council, 1980

INDEX

Aberdeen 34, 35, 90
Acid Deposition 101, 102
Acid Rain 101, 102, 159
Afforestation 175, 177, 181, 277
Agribusiness 106, 107, 113
Allenheads 139, 141–142
Alternative source of energy 92–93
AONB (Area of Outstanding Natural Beauty) 49, 284–288
Area Centres 42
Assisted Areas 64, 65, 68, 71

Barrow-in-Furness 61–63
Bath 266–268
Bathgate 57
Bathing Water Quality 252–253
Beach Replenishment 226, 227
Beeston, Leeds 22, 24, 25
Blackbird Leys Estate, Oxford 36–37
Blackdown Hills (AONB) 286–287
Blackpool 258–259
Black-throated Diver 180
Blaenavon, South Wales 262
Blowout 246
Blue Flag Award 253, 254
Blyth 4
Bournemouth 18, 45, 215, 222–223
 – service industries 82
Botallack, Cornwall 289, 290
Bradford 9, 71
Braunton Burrows 244–247
Breckland 119–120
Brecon Beacons National Park 276–280
Brent Cross 2, 15
British Coal Enterprise 64
British Shipbuilders Enterprise 64
British Steel 56, 57
British Steel Enterprise 64
Brownfield Site 19, 45
Business Park 10, 139
Bustitution 146

Cairngorms 207, 209
Caledonian Forest 174
Cambridge,
 – Phenomenon 73
 – Science Park 73, 74
Cannock Chase 156–157
Capel Hermon 184

Carbon Dioxide emissions 98, 99, 100
Carnon, River 93
CCGP (combined cycle gas-fired power stations) 90
Channel Tunnel 71
Chapeltown, Leeds 22, 24, 25
Chase Manhattan Bank 46, 83, 84
Chesil Beach 233–234
Chester 268–271
Chlorofluorocarbons 100
City Action Teams 21
City centres 9–14
Cliff-face processes 219
Cliff-foot processes 219
Climatic Change 98, 232
Coal 68, 85–88
 – Plan for Coal 86
 – Open-cast 86, 88
Coalbrookdale 272, 273
Coalfield North, Leicestershire 88
Coastal Erosion 217–225
Coastal Management 211, 214–228
Common Agricultural Policy (CAP) 112, 114
Community Forests 134
Commuters 38, 107
Consett 52–53
Cornwall 93, 94, 149–151
Cost-benefit analysis 224, 226
Council for the Preservation of Rural England (CPRE) 158
Counterurbanisation 50–51
Country Parks 156–157
Countryside Commission 134, 135
 – for Scotland 155, 209, 283
Crail, Fife 78
Crawley 16
Crop yield and production 112

Darby, Abraham 272, 273, 274
Darlton Quarry, Derbyshire 187
Dartmoor 202
Dash for gas 88
Dawley 273
Day trips to the countryside, Scotland 155–156
De-industrialisation 52, 55–63, 64, 273
Deprivation, Inner city 24, 25, 29, 34

Derelict land 103, 105
Development Agency, Derwentside Industrial 53
Development Areas 64, 150
Development Board for Rural Wales 152, 153
Dispersants (oil) 257
Down Ampney 139–140
Dubh lochans 179, 180
Duddon Estuary 235, 241
Dungeness 91, 250–251
Dunlin 180, 242
Dutch Elm Disease 134

East Anglia 113
East Kent 71
Ebbw Vale 104–105
Ecosystem
 – peatland 179, 181
 – coniferous 181
Edinburgh 34, 35
Electricity
 – generated from coal 87, 88
 – generated from natural gas 90
Employment change 58, 62, 66, 78, 80, 81
Energy supplies 85–93
Ennerdale 195–198
Enterprise Agencies 66
Enterprise Boards 66
Enterprise Zones 2, 22, 60, 65, 66, 67
Environmental impact assessment 191
Environmentally Sensitive Areas (ESAs) 109, 117–120
Estates, council 34, 36–37
Estuaries 235–242
Estuarine reclamation 236, 239
European Community 102, 114, 115, 162
European Regional Development Fund 64
Exmoor, farming 168–171

Fareham, Hants 262
Farming, constraints on, Upland Britain 163–165
Farming, Contemporary Issues 111

INDEX

Farm Diversification Scheme   117
Farm size and structure   111, 112
Farm Woodland Scheme   109, 115, 117
Flow Country   176, 178–183
Flue-gas desulphurisation (FGD)   102
Fontmell Down, Dorset   263
Forestry Commission   69, 134, 175, 176, 178
Forestry, upland   173–183
Fuel mix   85

Garden Festivals   104
Gateshead   104
General Improvement Area (GIA)   24
Glasgow   34, 35, 47, 104
Glenrothes   76–77
Glensanda   192–193
Glenthorne Estate, Exmoor   171
Global Warming   98, 229, 232, 234
Golden Cap, Dorset   289, 290
Golden Plover   180
Golf Courses   158
Gower, South Wales   263
Great Western Railway   75
Green Belt   45, 46, 47, 48, 49
Greenfield site   47
Greenhouse Effect   98, 108
Greenhouse gases   99
Greenshank   180
Gridlock   38
Groynes   215, 218, 219, 220

Hartsop valley, Cumbria   167–168
Hayling Island   226–227
Heartlands Birmingham   29–32
Heathland   127, 135
 – Dorset   46, 135, 136–137
 – Reserves   137
Hengistbury Head, Dorset   215–224
Heritage Coasts   288
High technology   52, 72–78
Highlands and Islands Enterprise   148
Holderness   211, 212, 213
Holkham Bay, Norfolk   289, 290
Housing Action Area (HAA)   23
Housing Renewal   23
Howard, Ebenezer   45
Humber Estuary,   211, 212, 216, 254–255
Hunterston   60
Hydrocarbon resources   88–89

Improvement Grant   24

Industrial Revolution   23
Industry
 – chemical   66, 67
 – mid Wales   152–153
 – shipbuilding   4, 61–63
 – steel   55–60
Inner city   21–35
 – Conservative Government initiatives   22–23
 – Labour Government initiatives   21
Intertidal flats   239, 240
Intervention price (CAP)   113
Intervention system (CAP)   112
Irish Sea, natural gas production   90

Ironbridge   272–274

Kinmel Bay   229, 231, 232
Kirkstall Valley, Leeds   26, 27
Knoydart, Scotland   160–161

Lake District   195, 202
Lakeland   168, 184
Lakeside Centre, Thurrock   2, 15
Land quality (MAFF)   162
Landscape consultant   131
Lee Moor, Dartmoor   184
Leeds   22
 – Development Corporation   26
 – Urban Renewal Areas   24
 – Urban Renewal Programme   23
Less Favoured Areas (LFAs)   115, 162, 276, 277
Limestone, Carboniferous   185–187, 189
Lincolnshire   108
Linwood   57
Liverpool,
 – city centre   10–12
 – Garden Festival   104
 – Inner city and health   32–34
Llanwern, steel plant   57
London
 – changing pattern of service industries   80–81
 – red routes   39, 40, 41
 – road assessment studies   39
 – road plans   39
 – traffic problems   39
 – trunk routes   39
Longshore drift   218
Lurcher's Gully   208, 209

Madford, River   122
Management agreement (Exmoor)   170

Mardale Green (Lake District)   195
Maritime Heritage   236, 238
Marram grass   243, 244, 245
Mass transit system   10, 43
Mawddach Estuary   235
Meadowell Estate, North Shields   3–4, 36
Meadowhall Centre, Sheffield   2–3
Mersey Tunnel   43
Merseyrail   42
Merseyside Development Corporation   104
Merseytravel   42–43
Methane   98, 100
Moorlands Telecottage   54
Mossend   60
Motherwell   58, 59, 60
Mudeford sandspit, Dorset   216, 220

National Forest   127, 134
National parks   202, 275, 276
 – Scotland   283
National Power   85
National River Authority   200, 201, 231
National Scenic Areas, Scotland   275, 283
Nechells, Birmingham   29–30
Neighboorhood centre   20
New Forest   134, 280–282
Nitrate pollution   122–125
Nitrogen oxides   98, 102
North Sea
 – natural gas production   90
 – petroleum production   88, 89
North Shields   3
Northern Pennines (AONB)   284–286
Nuclear Power   91
 – decommissioning costs for power stations   91
 – power stations Magnox, AGR, PWR   91, 250

Oxford
 – Blackbird Leys estate   36–37
 – Park and Ride Scheme   41
Oxleas Wood   39, 132

Park and ride scheme, Oxford   41–42
Peak District   185, 186
Peak District National Park   185, 187
Pedestrianisation   10, 11, 14, 23
Petroleum   88
 – production   89

Pensarn, North Wales   230, 232
Pollution   94–105
— coastal   252
— farm   121
— perception   94
— river   96
Porchester enquiry, Exmoor   170
Porchester, Maps 1 and 2   170
Porth Neigwl (LLewyn, North Wales)   289, 290
Porthleven, Cornwall   224–226
Portsmouth
— city centre   12–13
— Harbour   236–238
Prickwillow   127
Psammoseres   243
Purbeck   261

Quality of Life   5–8, 159
Quality of Life, critera   6
Quedam Centre, Yeovil   14

Rainfall annual   198
Ravenscraig, Scotland   56–60
Red routes, London   40, 41
Regional Enterprise Grants   151
Regional Policy (Government)   64, 73
Regional Selective Assistance   151
Re-industrialisation   72
Renewable energy resources   92
Rother, River   97
Rows, The, Chester   269–270
Run-off, mean annual   198
Rural Development Commission (RDC)   147, 148
Rural services   142
Rural Railway Closures   145–146
Rurality   107–108

Sainsbury, J   16–18
Salt-marsh   239, 240, 244, 248–251
Sand couch grass   243
Sand dunes   243 et seq
Scottish Development agency   65
Sea buckthorn   267
Sea lyme grass   243–247
Selected Local Centres   142
Service employment change   80
Service Industries   79–84
Set-aside   109, 114, 158
Severn, River   241
Sewage   253
Sheffield   3
Shetland, Braer disaster   256–257
Shingle beach   234–235, 250
Shipbuilding
— Barrow-in-Furness   61–63
— Tyne   4
Shopping Malls   2, 9, 15
Silicon Glen   76
Silkin Test   186, 187
Skiing, Cairngorms   207–209
Slapton Wood   124
Snowdon   203–206
— footpaths   203–205
— summit   205–206
Snowdonia   185
Somerset Levels   118
South Molton, Devon   50, 51
Southsea   265
Spurn Head   211–213
Steel manufacture   56–60
Stoke-on-Trent   104
Stony Middleton   189, 191
Stourhead, Wiltshire   263
Strathclyde   47
Strathkelvin   47
Sulphur dioxide emissions   101, 102
Sundew   165

Target price (CAP)   113
Task forces (urban)   21
Teeside   66–68
— Development Corporation   66, 67, 68
Telecommuter   107
Telestuga   54
Teleworking   54
Telford   52
Tertiarisation   52
Thames, River   124
Thatcham, Berkshire   49
Threshold Price (CAP)   113
Tidal barrage   241
Topley Pike quarry, Derbyshire   187
Torrance, Strathclyde   48
Towyn, North Wales   229–232
Trans-national companies   72, 74
Tyneham, Dorset   259–261

Unemployment   3, 55, 59, 62, 66
Unwin, Raymond   45
Urban Development Corporations   23, 66
Urban Programme   21
Urban Renewal Areas (Leeds)   24

Village appraisal   141
Village Shops   143
VSEL (Vosper Shipbuilding and Engineering Limited) 61, 62, 63

Wash, The   238–242
Wasteland   103
Wastwater   195–198
Water Authorities   95
Water Plan (England and Wales)   200
Water Quality   94
Water Regions   95
Welsh Development Agency   65, 69
West Penwith   118
Westbourne Grove   16
Wilderness qualities   160
Wind Power   92, 93
Windrush, River   124
Wistman's Wood, Dartmoor   174
Woodhouse, Leeds   22, 24, 25
Woodland, ancient   131
Workforce (agricultural)   111, 112
World Heritage Site   183, 209, 268, 273

Yeovil   14